Confidence Intervals for Proportions and Related Measures of Effect Size

Chapman & Hall/CRC Biostatistics Series

Chapman & Hall/CRC Biostatistics Series

Adaptive Design Methods in Clinical Trials, Second Edition
Shein-Chung Chow and Mark Chang

Adaptive Design Theory and Implementation Using SAS and R
Mark Chang

Advanced Bayesian Methods for Medical Test Accuracy
Lyle D. Broemeling

Advances in Clinical Trial Biostatistics
Nancy L. Geller

Basic Statistics and Pharmaceutical Statistical Applications, Second Edition
James E. De Muth

Bayesian Adaptive Methods for Clinical Trials
Scott M. Berry, Bradley P. Carlin, J. Jack Lee, and Peter Muller

Bayesian Analysis Made Simple: An Excel GUI for WinBUGS
Phil Woodward

Bayesian Methods for Measures of Agreement
Lyle D. Broemeling

Bayesian Missing Data Problems: EM, Data Augmentation and Noniterative Computation
Ming T. Tan, Guo-Liang Tian, and Kai Wang Ng

Bayesian Modeling in Bioinformatics
Dipak K. Dey, Samiran Ghosh, and Bani K. Mallick

Biostatistics: A Computing Approach
Stewart J. Anderson

Causal Analysis in Biomedicine and Epidemiology: Based on Minimal Sufficient Causation
Mikol Aickin

Clinical Trial Data Analysis using R
Ding-Geng (Din) Chen and Karl E. Peace

Clinical Trial Methodology
Karl E. Peace and Ding-Geng (Din) Chen

Computational Methods in Biomedical Research
Ravindra Khattree and Dayanand N. Naik

Computational Pharmacokinetics
Anders Källén

Confidence Intervals for Proportions and Related Measures of Effect Size
Robert G. Newcombe

Controversial Statistical Issues in Clinical Trials
Shein-Chung Chow

Data and Safety Monitoring Committees in Clinical Trials
Jay Herson

Design and Analysis of Animal Studies in Pharmaceutical Development
Shein-Chung Chow and Jen-pei Liu

Design and Analysis of Bioavailability and Bioequivalence Studies, Third Edition
Shein-Chung Chow and Jen-pei Liu

Design and Analysis of Bridging Studies
Chin-Fu Hsiao, Jen-pei Liu, and Shein-Chung Chow

Design and Analysis of Clinical Trials with Time-to-Event Endpoints
Karl E. Peace

Design and Analysis of Non-Inferiority Trials
Mark D. Rothmann, Brian L. Wiens, and Ivan S. F. Chan

Difference Equations with Public Health Applications
Lemuel A. Moyé and Asha Seth Kapadia

DNA Methylation Microarrays: Experimental Design and Statistical Analysis
Sun-Chong Wang and Arturas Petronis

DNA Microarrays and Related Genomics Techniques: Design, Analysis, and Interpretation of Experiments
David B. Allison, Grier P. Page, T. Mark Beasley, and Jode W. Edwards

Dose Finding by the Continual Reassessment Method
Ying Kuen Cheung

Elementary Bayesian Biostatistics
Lemuel A. Moyé

Frailty Models in Survival Analysis
Andreas Wienke

Generalized Linear Models: A Bayesian Perspective
Dipak K. Dey, Sujit K. Ghosh, and Bani K. Mallick

Handbook of Regression and Modeling: Applications for the Clinical and Pharmaceutical Industries
Daryl S. Paulson

Measures of Interobserver Agreement and Reliability, Second Edition
Mohamed M. Shoukri

Medical Biostatistics, Third Edition
A. Indrayan

Meta-Analysis in Medicine and Health Policy
Dalene Stangl and Donald A. Berry

Monte Carlo Simulation for the Pharmaceutical Industry: Concepts, Algorithms, and Case Studies
Mark Chang

Multiple Testing Problems in Pharmaceutical Statistics
Alex Dmitrienko, Ajit C. Tamhane, and Frank Bretz

Randomized Clinical Trials of Nonpharmacological Treatments
Isabelle Boutron, Philippe Ravaud, and David Moher

Sample Size Calculations in Clinical Research, Second Edition
Shein-Chung Chow, Jun Shao and Hansheng Wang

Statistical Design and Analysis of Stability Studies
Shein-Chung Chow

Statistical Evaluation of Diagnostic Performance: Topics in ROC Analysis
Kelly H. Zou, Aiyi Liu, Andriy Bandos, Lucila Ohno-Machado, and Howard Rockette

Statistical Methods for Clinical Trials
Mark X. Norleans

Statistics in Drug Research: Methodologies and Recent Developments
Shein-Chung Chow and Jun Shao

Statistics in the Pharmaceutical Industry, Third Edition
Ralph Buncher and Jia-Yeong Tsay

Translational Medicine: Strategies and Statistical Methods
Dennis Cosmatos and Shein-Chung Chow

Chapman & Hall/CRC Biostatistics Series

Confidence Intervals for Proportions and Related Measures of Effect Size

Robert G. Newcombe

CRC Press is an imprint of the
Taylor & Francis Group, an **informa** business
A CHAPMAN & HALL BOOK

CRC Press
Taylor & Francis Group
6000 Broken Sound Parkway NW, Suite 300
Boca Raton, FL 33487-2742

First issued in paperback 2020

Version Date: 20120629

ISBN 13: 978-0-367-57670-7 (pbk)
ISBN 13: 978-1-4398-1278-5 (hbk)

Library of Congress Cataloging-in-Publication Data

Newcombe, Robert G.
Confidence intervals for proportions and related measures of effect size / Robert G. Newcombe.
p. ; cm. -- (Chapman & Hall/CRC biostatistics series)
Includes bibliographical references and index.
ISBN 978-1-4398-1278-5 (hardback : alk. paper)
I. Title. II. Series: Chapman & Hall/CRC biostatistics series (Unnumbered)
[DNLM: 1. Biometry--methods. 2. Data Interpretation, Statistical. WA 950]

610.72'7--dc23 2012026176

Visit the Taylor & Francis Web site at
http://www.taylorandfrancis.com

and the CRC Press Web site at
http://www.crcpress.com

Contents

Preface

During my four decades working as a statistician in a medical school, there have been major changes in the way that statistical methodology has been applied to health-related research. Both strengthened research governance frameworks and the requirements that the CONSORT statement and related guidelines impose on those seeking to publish their findings have, rightly, reinforced the need for clear thinking about the statistical issues in any study. Both the level of complexity of methodological research by statisticians and the computing power available to both "ordinary users" and users of more specialised software have increased beyond recognition. While the quality of treatment of statistical issues in published articles has probably improved over the years, serious flaws are still apparent in much published work. My greatest concern is that despite the computing revolution, the gap between methodological research performed by statisticians and the statistical methods used by researchers in applied fields has widened dramatically. While I would be the first to repudiate bibliometrics as an assessment of research output at individual or institutional level, nevertheless the fact that leading statistical journals, even those with an applied ambit, struggle to reach an impact factor of 2 indicates how little their content has generally influenced wider research practice.

During the 1980s, one particular concern was the persistent tendency for *p*-values to be the predominant medium for reporting the impact of sampling uncertainty on research findings. Leading medical statisticians, notably Doug Altman, persuaded the editors of top journals such as the *British Medical Journal* that confidence intervals give a much more interpretable expression of imprecision. While the recommended shift away from hypothesis tests towards confidence intervals has been achieved to some degree, nevertheless confidence intervals have continued to be underused. Often confidence intervals for proportions are calculated using the simplest methods, which are known to perform poorly. And the recommendation that confidence intervals are preferred begs the question, "Confidence intervals for what measures?" For continuous variables, calculating confidence intervals for means and their differences is usually straightforward. But for proportions and the measures used to compare them, this is far more problematic: three main comparative measures for comparing proportions are widely used, plus several alternatives and derivatives, with little clear guidance as to which should be used. And the most commonly used non-parametric methods to compare two series of results have generally been used in "black box" mode, merely to obtain a *p*-value; here, some kind of interpretable effect size measure would be greatly preferable, but was not available at all (for paired comparisons) or was not widely known to researchers (for unpaired ones).

The story behind this book is as follows. Around 1988, while the confidence interval issue was attracting considerable attention, a physician colleague, Huw Alban Davies, sought my advice on presenting the results of a very small comparative study with a binary outcome, which showed a very large difference between two independent groups. The proportions of positive individuals in the two groups were something like 11 of 12 and 2 of 11, and he wanted to calculate a confidence interval for the difference between these proportions. At that time I was unaware of Miettinen and Nurminen's seminal 1985 paper, and I conceded it would be difficult to calculate a good interval and I had no idea how to do so. But the issue stuck in my mind, and I started to examine how intervals for proportions and their differences were treated, first in widely used textbooks, then in research articles. I came to realise that for a difference between proportions based on paired samples, none of the widely recommended methods were sound. During summer 1990, I managed to construct a better interval, using a tail-based profile likelihood method. But I had to figure out how to perform an evaluation to demonstrate that this interval works well. This shifted my attention back to the simpler case of the binomial proportion, so I needed to review this area first, and develop criteria for optimality, which then also carried across to the unpaired and paired differences. The new methods for the paired difference, and the evaluation that showed they were greatly superior to existing methods, led to my first presentation on these issues, at the International Biometric Conference in Hamilton, New Zealand, in December 1992.

Two years later, I was about to present similar work on the unpaired difference case at the following International Biometric Conference in a different Hamilton, Ontario in August 1994. A few days before flying to Canada, the possibility occurred to me of constructing a much more computationally friendly interval by combining Wilson intervals for the two proportions by a simple "squaring and adding" process, very much as standard errors of independent means combine. I quickly re-ran the evaluation programs, and was amazed to find that this simple procedure worked just as well as the much more complicated ones I had developed. The work on proportions and their differences became three papers, which were eventually published by *Statistics in Medicine* in 1998. Subsequently I worked on several related issues, including sample-size-free generalisations of the Mann–Whitney and Wilcoxon statistics. The generalised Mann–Whitney measure had been in my mind from the late 1970s when I used it in my Ph.D. thesis as a universally applicable effect size measure to characterise and compare the predictive ability of several risk scores during pregnancy. A simple confidence interval for this measure was published by Hanley and McNeil in 1982, but as it is a kind of proportion, it would be expected that a symmetrical Wald interval would not work very well. Eventually, I developed several improved methods, and once again, one of the simpler options turned out to work very well. The generalised Wilcoxon measure is closely related—after slightly redefining the Wilcoxon test—and a good confidence interval may be developed in an analogous way.

This book is aimed at a readership comprising three overlapping groups, statistical researchers and those who review their work, applied statisticians, and researchers in the health and social sciences generally. Many readers will have used statistical methods such as confidence intervals extensively, but one of the aims is to give a deeper understanding of what happens when these methods are applied in situations far removed from the familiar Gaussian case. I hope that this book will help such researchers to report their findings in the most informative possible way. It is primarily a distillation of all my own work in this area, rather than a comprehensive review of the field, which would have meant a much lengthier book and much greater work in compiling it. In writing this book, I have been faced by a dilemma which I am sure has faced many applied statisticians when trying to write a book to summarise their research area. At one extreme, there are (or at least, used to be) "cookbooks" which are blatantly addressed at the end user, containing very little theoretical justification. At the other extreme, there are methodological expositions at a largely algebraic level, a total switch-off to all but trained statisticians. In this book I have sought to steer a course intermediate between these two extremes. The reader is assumed to be familiar with the general principles of quantitative research, in particular when applied to health-related issues. I have sought to enable such readers to understand what confidence intervals are and how they are interpreted, the link between hypothesis tests and confidence intervals, and why it is usually preferable to use confidence intervals in reporting research findings. I show how the simplest intervals for proportions perform very poorly, and describe much more reliable methods. Subsequent chapters deal with various extensions involving a wide range of more complex quantities related to proportions, in a structured, progressive manner. In each instance I have aimed to give a clear explanation of the rationale for each method. For the benefit of less statistically experienced readers, definitions of some of the terms used may be found in Appendix 1.

Illustrative numerical examples are included, with interpretation. These are predominantly health-related, many drawn from my personal professional experience, but the methods described can be applied to research in a wide range of disciplines. Little computational skill is required—highly user-friendly Excel spreadsheets are provided for most of the methods I recommend, to enable readers to apply these methods to their own empirical data very simply. In a book of this kind, it is obviously impossible to dispense with algebra altogether; while the reader is assumed to be familiar with algebraic notation, I have aimed to minimise the burden of formulae presented. Indeed, I have sought to counter the tyranny of suffix notation, which has unfortunately often lowered clarity of exposition in this field of development. Suffix notation is indisputably greatly beneficial for the matrix algebra underlying multivariate methods, but that is not the issue here. Thus for 2 × 2 tables, which underlie many of the methods considered in this book, often greater clarity is achieved by representing the four cell frequencies by a, b, c and d rather than n_{ij}, for i and j 0 and 1 or 1 and 2.

I hope that future researchers in this field will find that the presentations in these chapters set a clear, firm foundation for further development and improvement. A major emphasis is on some of the principles of the methodology of performing evaluations of coverage and related properties of confidence interval methods, the description of which has hitherto been scattered.

In writing this book I am well aware that many very important issues are barely touched on. I have dealt with modelling only by a very brief introduction to logistic regression. Overdispersion—variation larger than that predicted by a binomial model—is a crucial issue affecting many data sets involving binary and Poisson variables, but the modelling methods required to cope with this are beyond the scope of this book. The systematic review process and its statistical endpoint, meta-analysis, constitute an extremely important area of research activity, about which a huge amount has been written, to add to which is beyond the ambit of this book.

I wish to warmly commend the books listed below, to any readers unfamiliar with them.

Three general books dealing with binary data and non-parametrics:

Agresti, A. 2002. *Categorical Data Analysis,* 2nd edition. Wiley, Hoboken, NJ.

Fleiss, J. L., Levin, B. and Paik, M. C. 2003. *Statistical Methods for Rates and Proportions,* 3rd Edition. Wiley, Hoboken, NJ.

Hollander, M. and Wolfe, D. A. 1999. *Nonparametric Statistical Methods,* 2nd Edition. Wiley, New York.

Regarding the quantification of tests in medicine:

Kraemer, H. C. 1992. *Evaluating Medical Tests. Objective and Quantitative Guidelines.* Sage, Newbury Park, CA.

For explanations of the preferability of effect size measures and confidence intervals over p-values:

Altman, D. G., Machin, D. and Bryant, T. N. et al. (eds.). 2000. *Statistics with Confidence,* 2nd Edition. BMJ Books, London.

Cumming, G. 2011. *Understanding the New Statistics. Effect Sizes, Confidence Intervals, and Meta-Analysis.* Routledge, New York.

A useful general introduction to effect size measures:

Grissom, R. J. and Kim, J. J. 2011. *Effect Sizes for Research. Univariate and Multivariate Applications,* 2nd Edition. Routledge, New York.

On the same issue, several of the points that arise in relation to choice of effect size measure for the 2 × 2 table and related contexts have been considered in depth by the meta-analysis community, and apply equally to the reporting of a single study. A wealth of practical advice is found in the *Cochrane Handbook for Systematic Reviews of Interventions,* in particular Section 9.4.4.4, available from the Help menu of the freely downloadable Cochrane Collaboration systematic review software, RevMan (http://ims.cochrane.org/revman).

Additional Material is available from the CRC Web site: http://www.crcpress.com/product/isbn/9781439812785.

Acknowledgments

I am grateful to the following publishers for permission to reproduce material from their journals and publications

American Statistical Association	*American Statistician*
BMJ Group	*British Medical Journal* *Evidence Based Medicine* *Statistics with Confidence*
CONSORT group	CONSORT guidelines
Elsevier	*Addictive Behaviors*
Sage	*Statistical Methods in Medical Research*
Taylor & Francis	*Communications in Statistics—Theory & Methods*
Wiley	*Statistics in Medicine*

I wish to acknowledge numerous helpful comments from collaborators, which have made a great contribution to this book and to the published articles on which it is based, including the following: Markus Abt, Alan Agresti, Adelchi Azzalini, George Barnard, Conway Berners-Lee, Nicole Blackman, Bruce Brown, Mike Campbell, Iain Chalmers, Ted Coles, Daniel Farewell, Paolo Fina, Amit Goyal, Paul Hewson, Kerry Hood, Louise Linsell, Ying Lu, Olli Miettinen, Yongyi Min, Barry Nix, Markku Nurminen, Nicola Payne, Tim Peters, Ruth Pickering, Jenö Reiczigel, David Rindskopf, Eric Okell, Paul Seed, Jon Shepherd, Ray Watson and GuangYong Zou.

I am grateful to James Osborne and colleagues for help in using the Advanced Research Computing @ Cardiff (ARCCA) facilities for recent evaluations.

I am also grateful to several anonymous reviewers of these articles, and the anonymous reviewer of this book, whose comments have added much value.

Finally, my wife, Lynda Pritchard Newcombe, has shown great patience and understanding in the face of the pressures we have both faced in preparing our books for publication.

Author

Robert Newcombe studied mathematics at Trinity College, Cambridge, and received his Ph.D. in Medical Statistics from University of Wales in 1980. His work at the Cardiff Medical School has involved collaboration in all areas of medical and dental research and teaching in medical statistics and epidemiology. He has served as external examiner for King's College, London, the Faculty of Public Health and the Royal College of Radiologists. He is a member of the editorial board of *Statistical Methods in Medical Research,* has refereed for many statistical and clinical journals, and reviewed project applications for the NHS Health Technology Assessment Scheme and other funding bodies. He serves on the Cardiff & Vale Research Review Service and Wales Ambulance Service Trust Research & Development panels.

Acronyms

Acronym	Defined	Section Where Defined
ADR	adjusted dose ratio	14.3
AE	adverse event	11.5.3
AED	automatic external defibrillator	1.2.3
ALMANAC	Axillary Lymphatic Mapping Against Nodal Axillary Clearance trial	7.7
AM	arithmetic mean	2.4.1
ANCOVA	analysis of covariance	14.2.4
ANOVA	analysis of variance	2.2
ARR	absolute risk reduction	7.1
AUROC	area under ROC curve	12.4
A&E	accident and emergency unit	10.8.1
BCIS	Box–Cox index of symmetry	4.4.3
BMI	body mass index	2.4.3
CAM(s)	complementary and alternative medicine(s)	1.1
CC	continuity correction	3.4.1
CI	confidence interval	1.2.1
CL	confidence limit	1.2.1
CNV(s)	copy number variation(s)	14.6
COMICE	Comparative Effectiveness of Magnetic Resonance Imaging in Breast Cancer trial	
CONSORT	Consolidated Standards of Reporting Trials	10.1
CP	coverage probability	3.2
CQ	constrained quartic (model)	16.4.5
DART	Diet and Reinfarction Trial	9.2.1
DCP	dental care practitioner	2.1
df	degrees of freedom	1.5.2
DNCP	distal non-coverage probability	4.4.2
dp	decimal places	5.2.1
ECOG	Eastern Co-operative Oncology Group	7.2
ET	equal tails	3.4.6
FOBT	faecal occult blood test	12.5.3
FP	fold point	16.7
GM	geometric mean	2.4.1
H_0	null hypothesis	1.2.2
H_1	alternative hypothesis	1.2.2
HDL	high-density lipoprotein (cholesterol)	1.2.1
HIV	human immunodeficiency virus	3.3

Acronym	Defined	Section Where Defined
HM	harmonic mean	2.4.3
HPD	highest posterior density	3.4.6
HWE	Hardy–Weinberg equilibrium	6.3.1
ICI(s)	inferential confidence interval(s)	7.8
IRBR	incremental risk-benefit ratio	14.14
LAR	Levin's attributable risk	6.3.2
LNCP	left non-coverage probability	4.2
MCMC	Markov chain Monte Carlo	3.4.6
MLE	maximum likelihood estimate	1.6
MNCP	mesial non-coverage probability	4.4.2
MOVER	method of variance estimates recovery	(6.3.4) 7.3.2
MTHFR	methylenetetrahydrofolate reductase	1.1
MUE	median unbiased estimate	4.4.4, 11.4
NCP	non-coverage probability	4.4.2
NHSBSP	National Health Service breast screening programme	9.6
NLR	negative likelihood ratio	12.5.1
NNH	number needed to harm	7.6.1
NNT	number needed to treat	(6.3.3) 7.6
NNTB	number needed to treat (benefit)	7.6.1
NNTH	number needed to treat (harm)	7.6.1
NPV	negative predictive value	12.3
NRPB	National Radiological Protection Board	9.3.1
OOPEC	Office for Official Publications of the European Communities	9.3.1
OR	odds ratio	7.1
PCR	polymerase chain reaction	14.6
PDSI	patient decision support intervention	11.1
PIN	population impact number	6.3.3
PLR	positive likelihood ratio	12.5.1
PPV	positive predictive value	12.3
PRD	population risk difference	6.3.4
PSA	prostate specific antigen	12.1
PSP	parameter space point	4.2
PSZ	parameter space zone	5.1.1
Q-Q	quantile-quantile (plot)	2.4.2
RD	risk difference	6.3.3
RDUR	ratio of dose units required	14.3
RNCP	right non-coverage probability	4.2
ROC	receiver operating characteristic (curve)	(1.4.1) 12.4
RR	relative risk or risk ratio	(6.3.2) 7.1
RRR	relative risk reduction	10.1
SBP	systolic blood pressure	1.2.1
SD	standard deviation	1.2.2

Acronym	Defined	Section Where Defined
SE	standard error	1.2.1
SIR	standardised incidence ratio	6.1.3
SMR	standardised mortality ratio	6.1.3
SNB	sentinel node biopsy	7.7
SU	shifted uniform (model)	16.4.4
TAE	trial and error	13.3
WTCB	wind(ing) the clock back	14.12
ZSRs	zero-shifted ranks	16.1
ZWI	zero-width interval	3.3

1

Hypothesis Tests and Confidence Intervals

1.1 Sample and Population

Statistical analyses start with a sample of observations, and summarise the information contained in the sample. Nevertheless, the purpose of performing statistical analyses is not primarily to tell us something about that particular sample, but about a wider population from which the sample has been taken.

The enzyme methylenetetrahydrofolate reductase (MTHFR) is implicated in the aetiology of neural tube defects, and may relate to impaired vascular function and thus also affect cardiovascular risk. Circulating MTHFR levels are largely determined by the *C677T* gene, which is expressed in a recessive manner. Suppose we wish to estimate what proportion of people has the high-risk TT genotype. (In fact, for Caucasians, this proportion is approximately 10%.) McDowell et al. (1998) recruited a suitable sample of consenting people, and determined what proportion of them had the TT genotype. However, the primary concern is not with this particular sample of individuals. This proportion is reported as an estimate of the proportion that has this genotype in a wider population.

Similarly, Crawford et al. (2006) studied use of complementary and alternative medicines (CAMs) amongst 500 children attending a teaching hospital. Forty-one percent of respondents reported use of one or more CAM during the preceding year. Once again, the main concern is to determine the prevalence of CAM use amongst a wider population, not just this particular group of children.

The concept of a population is fundamental to statistical inference. However, what is the population to which the findings generalise? In the genetic study, the investigators sought to estimate the TT genotype prevalence amongst people who belong to the country in which the study took place; ethnicity may also be specified. In the CAM study, the relevant population is harder to define, as we expect use to vary more widely between populations and to change over time. Moreover, medication use differs between sick and healthy children. In a clinical trial comparing two treatment regimes, the intended application is usually very wide, potentially to future patients with a similar clinical presentation worldwide.

How should a sample be drawn from a population? The key issue is that a sample should be representative of the relevant population. This means that there are no consistent differences between the sample and the population—any differences are purely due to the chance element in selecting the sample, and are not such as to result in a consistent tendency to bias the results in a particular direction. The method of sampling that ensures representativity is random sampling. The term "random" has a specific meaning—it is not synonymous with "arbitrary" or "any." It means that all members of the eligible population are listed, then a sample of the required size is drawn without any selectivity: each member of the population has the same chance of being included in the sample. Professionals are too intelligent and too highly trained to choose at random; truly random selection can only be made by delegating the task to unintelligent mechanisms, such as physical randomisation devices including dice or lottery ball machines, or preferably computer algorithms. All such resources require meticulous design and evaluation of their achievement of randomness. Randomness is the ideal, and the derivation of statistical methods is based explicitly on this assumption, but in the real world sampling that is literally random is relatively rarely attained. At the most fundamental level, on ethical grounds, most kinds of research on humans require some element of informed consent, so wherever individuals can opt out, straightaway the assumption of randomness is violated.

With this in mind, how representative a sample are we likely to achieve? In the genetic study, this is relatively straightforward, provided the sample is not selected by clinical variables. In such studies, usually the basic demographic factors, sex and age, do not play an especially great role. The *C677T* gene is on chromosome 1, therefore we do not particularly expect the prevalence to vary by sex. Ageing does not affect genotype, so age would matter only if those with the TT genotype tend to live less long. This is why in the large-scale UK Biobank (2006) study, for example, little attention is paid to the issue of random sampling; individuals of eligible age are identified via their general practices and invited to participate.

In the CAM study, this issue is more problematic. The children studied were attending hospital, 100 as in-patients and 400 as out-patients, and these subgroups could differ importantly both in CAM use and in other characteristics. Moreover, parents would be less inclined to participate if they are giving their child some treatment which they believe clinicians would disapprove of. Our figure of 41% could be a gross underestimate of CAM use amongst children attending hospital.

Indeed, as will be seen in Section 1.7, in some situations no wider population exists as such, yet statistical methods such as confidence intervals still give a helpful appraisal of the degree of uncertainty resulting from finite sample size.

Confidence intervals are designed to convey only the effects of sampling variation on the precision of the estimated statistics and cannot control for non-sampling errors such as biases in design, conduct or analysis. Whilst there are nearly always issues regarding representativeness of samples, this issue is not particularly the focus of attention in this book. The main methods used to quantify uncertainty due to limited study size are based on the assumption that the samples of individuals studied are representative of the relevant populations.

1.2 Hypothesis Testing and Confidence Intervals: The Fundamentals

1.2.1 A Confidence Interval for a Single Group

The MTHFR and CAM studies are single-group studies. Each centred on a binary or dichotomous variable: Did the subject have the high-risk genotype or not? Were CAMs used in the preceding year or not? The natural summary statistic to summarise data for a binary variable is a proportion: 12% of the sample had the TT genotype; CAM use was reported for 41% of the children studied. Proportions, and quantities related to proportions, are the main focus of attention in this book. Strictly speaking, a proportion is a number between 0 and 1, such as 0.12 or 0.41. In practice, such a proportion would most often be presented as a percentage as above. We often interpret a proportion as a probability: our estimated probability that an untested person in our population has the TT genotype is 0.12 or 12%.

Studies may also measure quantitative or continuous variables such as systolic blood pressure. A variable of this kind is usually summarised by calculating either the mean or the median.

The prevalence of CAM use, 206 (41%) of 500, was presented with a 95% confidence interval from 37% to 46%. A confidence interval (CI) is a range of values, derived from the sample, which is intended to have a pre-specified probability—95% here—of including the population proportion. Whilst confidence intervals are often calculated by computer software, in some situations the calculation is simple enough to be practicable using a calculator. Here, the simplest interval for the proportion, known as the Wald interval, is calculated as follows. The sample proportion, 0.412, has a standard error (SE) estimated by $\sqrt{\dfrac{0.412 \times (1 - 0.412)}{500}} = 0.022$. This may be used to calculate the 95% confidence interval, which runs from $0.412 - 1.960 \times 0.022 = 0.369$ to $0.412 + 1.960 \times 0.022 = 0.455$. Both the standard error and the confidence interval express the degree of uncertainty or imprecision attaching to the observed proportion, purely because it is based on a sample of finite size. The two calculated bounds, 37% and 46%, are often referred to as confidence limits (CLs), but this is with the understanding that they are intended to be interpreted together as a pair. It is generally preferable to consider the interval as a range of plausible values for the parameter of interest rather than the two bounds themselves. The 95% criterion is arbitrary, but is so widely used that it has come to be regarded almost as a default. It is preferable to separate the two values by the word "to" or to write (37%, 46%). The common practice of using a dash is confusing, particularly when reporting confidence limits for differences, which are often negative in sign.

The interpretation of a confidence interval is as follows. Confidence intervals convey information about magnitude and precision of effect simultaneously, keeping these two aspects of measurement closely linked (Rothman and Greenland 1998). The objective is to identify the prevalence, π, in a wider population. (Usually, unknown quantities to be estimated from the data—known as parameters—are denoted by Greek letters, and ordinary Roman letters are used for observed quantities.) The 95% CI is designed to have a 95% chance of including π. The parameter π is assumed fixed, whereas the lower and upper limits are random because they are derived from the sample. We do not know the value of π; indeed the only way to get any information on it is by studying samples of suitable subjects. However, we know that there is a probability of 95% (approximately) that our interval will include the true value, whatever that value is—this is what the interval is designed to do. So we have 95% confidence that π lies between 37% and 46%. (We are not quite saying that there is a 95% chance that π lies between 37% and 46%. To make such an assertion, we would also need to take into account any information from other sources concerning π; see Section 1.6.)

A study involving 500 subjects is quite a large one and is certainly a substantial amount of work. With this in mind, the width of the interval may seem disappointing—after so much effort, the investigators were only able to narrow the prevalence down to being between 37% and 46%.

As an example of a relatively narrow confidence interval, Richardson et al. (2011) report results for a community-based health promotion programme, the Healthy Hearts programme, which was directed at adults enrolled in several general practices in Caerphilly borough, a relatively deprived former coal mining community in the South Wales valleys area. Between 2004 and 2007, the study recruited 2017 people who identified themselves as having at least one positive risk factor for heart disease (such as smoking or waist circumference greater than the recommended gender-specific maximum). Age at entry ranged approximately uniformly across the targeted 45- to 64-year-old range. Many variables related to heart disease risk were measured. The intervention consisted primarily of giving participants their 10-year Framingham-based risk scores (D'Agostino et al. 2008), followed by advice on healthy living, but they were also encouraged to contact their general practitioner who could prescribe medication such as antihypertensive drugs or statins.

Participants were encouraged to return for a follow-up visit approximately 1 year later. Seven hundred and thirty-eight of the 2017 individuals were followed up on average 1.49 years from the initial visit (range 0.84 to 2.16 years). Some of the variables studied showed clear improvements compared to the baseline level, notably smoking, fruit and vegetable intake, blood pressure, high density lipoprotein (HDL) cholesterol and Framingham score, whilst others such as exercise, body weight and waist circumference deteriorated. Several variables showed clear positive skewness and were log-transformed in all analyses, as described in Section 2.4.

Amongst the 738 participants who were followed up, the mean systolic blood pressure (SBP) at baseline was 139.1 mm Hg, with 95% interval 137.8 to 140.4 mm Hg. This interval quantifies the degree of imprecision of the sample mean, which originates in both the variation in SBP between different people and the finite sample size. Since this is a large sample, the confidence interval is narrow, indicating a high degree of precision is attained. The confidence interval must not be confused with the reference range which corresponds to the middle 95% of the observations. The reference range here is from 104 to 174 mm Hg, and is very wide. Note that the confidence limits, like the sample mean, are in the same units as the original observations.

Figure 1.1 and the first panel of Figure 1.2 illustrate how confidence interval width is related to sample size. The first panel of Figure 1.1 shows the systolic blood pressures of these 738 individuals, distinguishing men (group 1) and women (group 2). The mean SBPs amongst the 346 men and 392 women studied were 142.8 and 135.8 mm Hg, respectively. Here, we estimate the difference in mean SBP between men and women as 7.0 units, with 95% interval from 4.5 to 9.5 units.

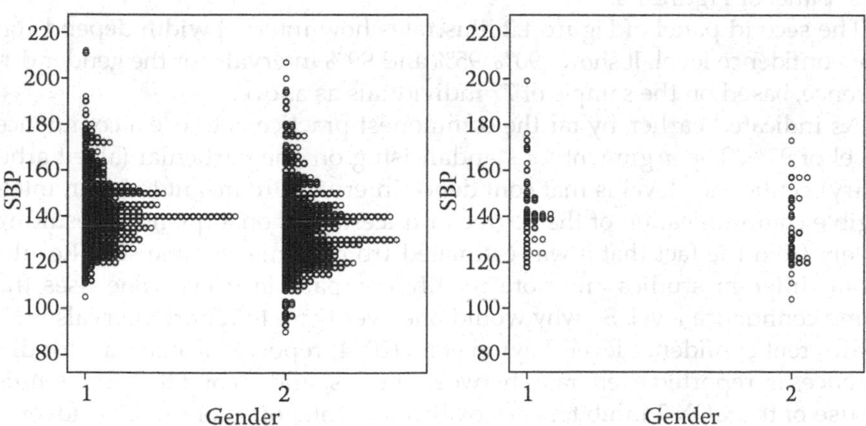

FIGURE 1.1
Baseline systolic blood pressures (mm Hg) in the Healthy Hearts study. The first panel shows results for 346 men and 392 women. The second panel shows results for a smaller sample of 45 men and 32 women. (Data from Richardson, G. et al. 2011. *British Journal of Cardiology* 18: 180–184.)

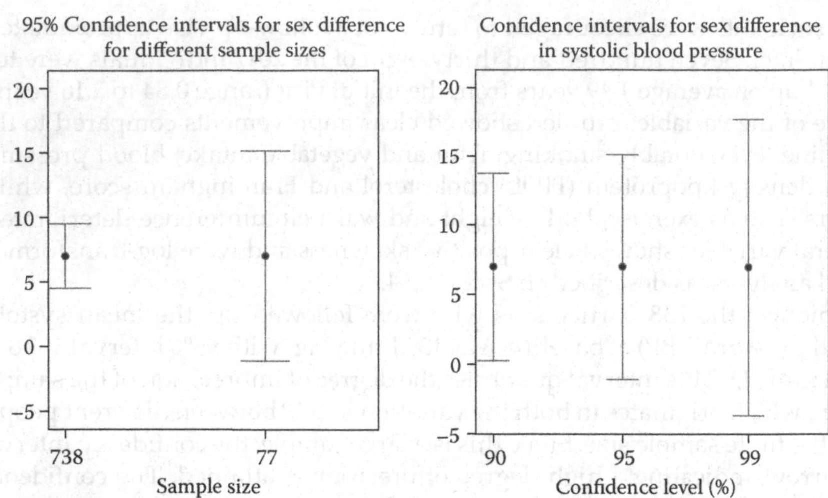

FIGURE 1.2
Influence of sample size and confidence level on confidence interval width. The first panel shows how reducing sample size widens the interval. The second panel shows how 99% intervals are wider than 90% intervals calculated from the same data.

The second panel of Figure 1.1 shows similar results for a sample approximately one-tenth as large as that in the first panel. It comprises 45 men and 32 women, with means identical to those above. Again, the difference in mean SBP between men and women is estimated as 7.0 units. The 95% interval is now much wider, from –1.1 to 15.1 units. The two intervals are shown in the first panel of Figure 1.2.

The second panel of Figure 1.2 illustrates how interval width depends on the confidence level. It shows 90%, 95% and 99% intervals for the gender difference, based on the sample of 77 individuals as above.

As indicated earlier, by far the commonest practice is to use a confidence level of 95%. The argument for standardising on one particular (albeit arbitrary) confidence level is that confidence intervals are meant to be an intelligible communication of the degree of uncertainty on a quantity, resulting solely from the fact that it was estimated from a finite sample size. Results from different studies are more readily comparable if everyone uses the same confidence level. So why would one ever want to report intervals with a different confidence level? Layton et al. (2004) report 99% intervals for differences in reported event rates between the first and second to sixth months of use of the COX-2 inhibitor celecoxib, for 14 categories of putative adverse reactions. The multiple comparison issue impacts on this situation, as described in Section 1.3. Using 99% intervals instead of the more usual 95% intervals resembles the use of Bonferroni correction of *p*-values to adjust for multiple comparisons (Dewey 2001). One could argue for intervals with confidence level 100 × (1 – 0.05/14), that is, 99.64% here. However, whilst *p*-values relate to coincidence, confidence intervals relate to measurement. From this standpoint, it is preferable to standardise on using a confidence level in this context that is more conservative than 95%, but which is not specifically linked to the number of outcomes that are relevant in the study. Otherwise, the investigators are unduly penalised for examining a range of outcomes.

1.2.2 Comparing Two Groups Using Hypothesis Tests

Most research studies compare two or more groups. Hypothesis tests, which yield *p*-values expressing the degree of statistical significance, are commonly used to appraise the evidence for a consistent difference between groups. For many years, *p*-values were the mainstay of statistical analyses in published studies in medicine and other fields of application.

Crawford et al. reported that CAM use was statistically highly significantly greater when the mother had post-secondary education than when she had not ($p < 0.001$). This may be regarded as a difference in CAM use according to the mother's educational status, or equivalently an association between CAM use and the mother's educational status. They found a similar, although less highly significant relationship to the father's educational level ($p = 0.03$). There was little evidence for a difference in use between boys and girls ($p = 0.25$).

Like confidence intervals, hypothesis tests are usually calculated by computer software, but many simpler tests are practicable using a calculator. These *p*-values were calculated using the χ^2 test, a widely used simple test which is appropriate for these comparisons involving binary variables. A variety of other hypothesis tests exist, including the *t*-test which is used to compare means of quantitative variables. Conventionally, any *p*-value below 0.05 is regarded as indicating statistical significance. What exactly should be inferred from the above *p*-values?

The *p*-value is a special kind of probability. It represents the chance that we would get a difference between the two groups being compared (e.g., boys and girls) which is at least as large in magnitude as the difference actually found, if in fact the null hypothesis (referred to as H_0) is true. In this example, the relevant null hypothesis states that there is no difference in CAM use between boys and girls (or, equivalently, no association between CAM use and gender) in the population from which the study samples are drawn. If the *p*-value is high, this indicates it is quite plausible that we would get a difference as large as the one observed, purely by chance, so we would not claim to have substantial evidence of a difference. If the *p*-value is low, such a difference would be a rare occurrence—"quite a coincidence"—if H_0 is true. We would then reject H_0 in favour of the alternative hypothesis H_1 that there is a real difference, or a real association, in the underlying population. Traditionally, H_0 is rejected whenever $p < 0.05$; we say that the α level or type I error rate of the test is 5%. This 5% criterion is entirely arbitrary, just like the use of 95% for confidence intervals; indeed as we shall see, there is a direct duality between two-sided hypothesis testing at a 5% α level and two-sided

95% confidence intervals. The lower the *p*-value, the more evidence we have against H_0. So we regard the association of CAM use with maternal post-secondary education as highly significant. The association with paternal education is also statistically significant, but the evidence for a sex difference is not significant here.

All this is completely standard practice. As we will consider in Section 1.3, the problem comes when these *p*-values get over-interpreted. "Statistically significant" is commonly regarded as synonymous with "real," and "not significant" as implying "no difference," with disregard of all other information.

The *p*-values are derived from statistical models of the data. Statistical methods, including hypothesis tests, correspond to models set up to describe the sources of variation in the data. The full model corresponding to a simple *t*-test comparing the means of two independent samples may be written as

$$x_{ij} = \begin{cases} \mu + \delta + \varepsilon_{ij} & \textit{if } i = 1 \\ \mu + \varepsilon_{ij} & \textit{if } i = 2 \end{cases} \tag{1.1}$$

Here, x_{ij} denotes the *j*th observation in group *i*, for $j = 1, 2, \ldots, n_i$, where n_1 and n_2 denote the sizes of the two samples. The population means corresponding to groups 1 and 2 are $\mu + \delta$ and μ, so this model assumes that the means of the two populations from which the samples are drawn differ by a quantity δ. This model corresponds to the alternative hypothesis H_1. The residuals ε_{ij} ($i = 1, 2$; $j = 1, 2, \ldots, n_i$) are assumed to be independently normally distributed with mean 0 and standard deviation (SD) σ. The three parameters μ, δ and σ are fitted by a general procedure known as maximum likelihood, which in this simple case reduces to least squares: these parameters are chosen to minimise the residual sum of squares, $\sum_{i,j} \varepsilon_{ij}^2$.

A reduced model is also fitted, with δ constrained to be zero. This model corresponds to the null hypothesis H_0, and necessarily results in a higher residual sum of squares. The difference between the residual sums of squares for the two models leads to a test statistic, *F*, that is equivalent to the *t*-statistic used in the familiar two-sample hypothesis test, with the same *p*-value.

1.2.3 Comparing Two Groups Using Confidence Intervals

An alternative approach is to calculate some measure of effect size, with a confidence interval, summarising the comparison between two groups. The simplest measures available are differences of means and of proportions. In Section 1.4, we will consider measures of effect size in greater depth and list the main ones available.

Fletcher et al. (2008) audited automatically recorded information on manual chest compressions delivered by ambulance staff to patients with cardiac arrest, preparatory to use of automatic external defibrillators (AEDs). It had been observed that chest compressions were frequently delivered much more rapidly than the recommended target rate of 100 compressions per minute. To this end, over a period of several months musician's metronomes were introduced to help control the compression rate. A metronome was attached to each AED with instructions to switch on at the start of any resuscitation attempt. Results were compared between 2005, in which metronomes were little used, and 2006, in which they were commonly used. For the maximum compression rate delivered per minute, in 2005 the mean was 121.1 in 164 attempted resuscitations, with standard deviation 28.0. In 2006 the mean and SD reduced to 107.9 and 15.3, based on 123 cases.

The mean reduction in maximum compression rate was 13.2 units (compressions per minute), with 95% interval 8.1 to 18.3 units. A two-sample *t*-test gave $t = 5.1$, $p < 10^{-6}$. A confirmatory non-parametric Mann–Whitney test was also performed, on account of the large difference in SD between the two series—possibly the 2005 figures reflect a mixture of compliant and non-compliant operators—nevertheless of some concern for the validity of a *t*-test. This test gave a *p*-value of 0.0001. Following introduction of the metronomes, from 2005 to 2006, the mean value for the maximum compression rate reduced by 13.2 units. The associated confidence limits, 8.1 and 18.3, express the uncertainty of this figure coming from the fact that these were only moderately large series of cases. The impact of metronome use in a wider population could have been as little as 8.1 units, or conversely as large as 18.3 units. On the null hypothesis H_0, this difference would be zero, and the analysis clearly implies that H_0 does not fit well with the observed data. Informally, the observed reduction is unlikely to have arisen by chance in choosing the two samples.

Similarly, in the Crawford et al. study above, the prevalence of CAM use was 113/259 (44%) in boys and 93/241 (39%) in girls. The difference between these proportions is +5%, with a 95% CI from −4% to +14%. Informally, the interpretation of this is as follows. The best estimate of the sex difference is that boys (or rather, boys attending hospital) are 5% more likely to have used CAMs than their female peers. Within the limits of chance variation in drawing the sample, the true difference in the underlying population could possibly be as high as +14%. Conversely, it could be 4% in the opposite direction—a 4% lower prevalence in boys than in girls, which happened to reverse in this sample. These are understood to be absolute differences.

Best practice is to show plus signs explicitly whenever a difference measure, or either of its confidence limits, turns out to be positive, along with a clear convention regarding the interpretation of the sign of the difference. It is usual practice to report follow-up minus baseline differences in a longitudinal study, and active minus control differences in a study comparing different treatment regimes. Thus, in the resuscitation example above, normally the estimated change in compression rate and its confidence limits would be reported with minus signs, unless as here the explanation of the results makes it clear that reductions are being reported.

An alternative is to report a ratio of proportions, 1.13, with 95% interval from 0.92 to 1.40. The interpretation is that the rate of CAM use in boys is apparently 1.13 times that in girls. In other words, boys appear to have a 13% higher rate of CAM use than girls, in relative terms. Within the limits of sampling variability, CAM use in boys could be anything up to 40% higher than in girls, or conversely, up to 8% lower. Here, as indeed generally, it is crucially important to clearly distinguish between absolute and relative differences when presenting results.

In these examples, when the two-sided *p*-value is below 0.05, the 95% confidence interval excludes the null hypothesis value of the chosen effect size measure, and vice versa. The null hypothesis value is 0 for a difference, 1 for a ratio, simply because when two quantities are equal, their difference is 0, but their ratio is 1. In this sense, interval estimation and hypothesis testing are complementary. The correspondence is exact when we are comparing means, but is not quite exact for most methods to compare proportions.

1.3 Why Confidence Intervals Are Generally More Informative Than *p*-Values

Much has been written on this issue, including in the very helpful books by Altman et al. (2000) and Cumming (2011b). An excellent summary is given by Gardner and Altman (2000):

- Over-emphasis on hypothesis testing—and the use of *p*-values to dichotomise results as significant or non-significant—has detracted from more useful approaches to interpreting study results such as estimation and confidence intervals.
- In medical studies, investigators should usually be interested in determining the size of the difference of a measured outcome between groups, rather than a simple indication of whether it is statistically significant.
- Confidence intervals present a range of values, on the basis of the sample data, in which the population value for such a difference is likely to lie.
- Confidence intervals, if appropriate to the type of study, should be used for major findings in both the main text of a paper and its abstract.

It is now the established policy of many leading research journals in health and other fields of application to prefer point and interval estimates of effect size to *p*-values as an expression of the uncertainty resulting from limited sample sizes. For example, the 2010 edition of the *Publication Manual of the American Psychological Association* states that interpretation of results should, wherever possible, be based on estimation (Cumming 2011a). The *p*-value is best viewed as a stand-alone measure of the strength of statistical evidence against the null hypothesis in favour of the alternative hypothesis. It is a probabilistic abstraction, which is commonly misinterpreted, in particular when it is dichotomised at 0.05 or some other conventional α level, with "significant" interpreted as "real" and "not significant" as "null."

Hypothesis testing is based on a null hypothesis, H_0 (which generally means a zero difference or absence of association) and an alternative hypothesis, H_1, which is usually equivalent to "H_0 is false." The data are collected and analysed, and the appropriate test performed. If $p > 0.05$, we accept H_0. Conversely, if $p < 0.05$, we reject H_0 as an explanation of the observed data in favour of the alternative H_1. In doing so, sometimes we reject H_0 when it is true—a type I error—or conversely, accept H_1 when it is false—a type II error.

The α level of the test is the same as the type I error rate. It is the probability of rejecting H_0 when it is true. Often, in hypothesis testing, we think of the α level as fixed at an arbitrary, conventional value, most commonly 0.05. Consequently, we would reject H_0 wrongly in 5% of tests when it is in fact true—a type I error rate of 5%.

The type II error rate β is the probability of failing to reject H_0 when it is false. This is less straightforward. We can make an arbitrary choice for α, and declare that we are working at an α level of 0.05. However, we cannot do this for both α and β together. There is a trade-off between α and β. For a given choice of α, β depends on the sample size and the actual size of the difference. Best practice is to plan the sample size to be sufficiently large to ensure that the power of the study, $1 - \beta$, is at a chosen level, usually 80% or higher (Section 1.9.2).

A type I error may be regarded as a false positive finding. This means that the data suggest a link between two variables, but in fact this is purely coincidental, a quirk of this particular sample of data—if we were to repeat the study, we would not expect a similar finding. In the same way, a type II error is a false negative result—the study fails to demonstrate a difference that really exists. This is often because the sample size used was not sufficiently large to detect an effect of only moderate size.

The consequences are as follows. Many studies in the health literature are too small to be informative. Either no thought was given to ensuring an adequate sample size, or this has not been attained due to constraints regarding study duration, participant recruitment or resources. Such studies are apt to miss detecting important effects. In the context of a small study, the correct inference from a non-significant *p*-value is not that the difference is zero, but that it should be regarded as being identified quite imprecisely. This would be better demonstrated by showing the estimated difference, accompanied by its confidence interval, although researchers may be unwilling to concede how wide the interval is.

Another consequence of the piecemeal nature of much published health-related research is that often the findings of different studies appear to conflict. In 1979 Archie Cochrane, already regarded as a leading proponent of the randomised trial as the key research design for unbiased evaluation of treatment efficacy, made a seminal observation in recognition of this phenomenon: "It is surely a great criticism of our profession [i.e., medicine] that we have not organised a critical summary, by specialty or subspecialty, adapted periodically, of all relevant randomised controlled trials." This prompted the development of systematic review as a cornerstone of evidence-based health care. The systematic review process involves clear principles for searching the literature for relevant material and filtering by eligibility criteria for relevance and methodological quality. The statistical core of the process is known as meta-analysis, a field of statistical practice that has developed greatly over the intervening decades. Whilst some rudimentary techniques are available for combining p-values from independent studies, a much more satisfactory approach is universally applied based on estimates of effect size from the studies to be combined, together with estimates of their imprecision (Cumming 2011a, b; Grissom and Kim 2011).

In the context of systematic review and meta-analysis, the common interpretation of "not significant" as "the study was too small" has led to publication bias (Newcombe 1987). This phenomenon, described as "the pernicious influence of significance testing" by Cumming (2011b), is endemic to the publication system and seriously compromises both the interpretability of the whole corpus of published research and the feasibility of unbiased meta-analyses.

Furthermore, often several hypothesis tests are carried out on the data from a study. There may be several outcome variables, more than two groups to compare, and several scheduled follow-ups. In this situation, often researchers perform a large number of hypothesis tests, and report only the ones that reach statistical significance. Many of these may well be false positive, coincidental findings, which would be better regarded with great caution. It is this multiple comparison issue, particularly, that makes p-values difficult to interpret. True, there are studies such as the Wellcome seven-disease study (Wellcome Trust Case Control Consortium 2007) that are powered to detect moderately strong gene-disease associations even when a deliberately grossly conservative criterion for statistical significance is set. However, such studies are a rarity, not least on account of the level of funding this necessitates.

A variety of techniques have been developed to compensate for the multiple comparison issue, essentially by adjusting p-values upwards. This may be satisfactory for comparisons whose p-values are extreme, such as 0.001, indeed this is the main reason why such p-values are regarded as strong evidence. However, more often, unfortunately, doing so impacts very adversely on power, and can lead to paradoxical inferences. Should we really be less impressed by finding $p = 0.01$ for a particular outcome just because in the interests of efficiency we chose to study six other outcomes as well? There is really no satisfactory answer here. These issues are deeply rooted in the practice of using a measure of coincidence on H_0, the p-value, as if it were a measure of evidence for H_1, which of course it is not.

Cumming (2011a, b) explains in depth several reasons why estimation of an effect size, with a confidence interval, should be regarded as more informative than a p-value. Verifying that research findings can be replicated is a key principle of science, yet the p-value from a study gives almost no information on the p-value that is likely to emerge if the study is replicated. Cumming summarises research evidence on statistical cognition which reveals severe and widespread misconceptions of null hypothesis significance testing, leading to poor decision making—thus, the preference for effect size measures is evidence-based. For example, dichotomous thinking, dichotomisation of inferences at an arbitrary level such as $\alpha = 0.05$, results in a cliff effect, an unreasonable, sharp drop in the degree of confidence that an effect exists when the p-value passes the α threshold. Whilst an extremely low p-value can offer some reassurance in the face of multiple comparisons as seen in Section 1.5.1, nevertheless the occurrence of a low p-value of around 0.001, such as is often reported, is likened to winning a prize in a lottery—even if we consider a very wide range of initial p-values, those values account for only about 12% of the variance of p obtained on replication.

In contrast, Cumming outlines six helpful ways to think about and interpret confidence intervals. Not all values within the interval are equally plausible—considering a 95% interval for a mean, a value at either limit is about one-seventh as likely as the interval midpoint. The development naturally assumes that the actual coverage probability of a confidence interval is the same as its nominal level. This is true for the continuous case, which is the main focus of Cumming's exposition. However, when applying these principles to the confidence intervals developed in this book for proportions and related quantities, an additional factor should be borne in mind, namely that nominal and actual coverage probabilities may be substantially different, for reasons explained in Chapter 3.

1.4 Measures of Effect Size

It is all very well for statisticians, journal editors and others to recommend use of confidence intervals as preferable to hypothesis tests as the main expression of imprecision due to limited sample sizes. Unfortunately, hitherto it has been more difficult for researchers to adopt this strategy, for several reasons. One of these relates to the fundamental but often neglected and poorly understood concept of measures of effect size, explained in detail below.

Whilst hypothesis tests are widely available in software, addressing most questions the user would want to pose, sometimes no corresponding effect size measure and associated confidence interval have been readily available to users. This point is illustrated in Section 1.4.1.

Conversely, for the task of comparing two proportions, several different measures of effect size are available, the primary ones being the difference of proportions, relative risk and odds ratio. In this situation generally there is little guidance and limited awareness which of these measures is most informative to report.

The term "measure of effect size" should be understood in a generic sense, after Grissom and Kim (2011). Confusingly, in the psychological research literature, often (as in Algina et al. 2006) the term "effect size" is understood by default to mean the standardised difference, sometimes referred to as Cohen's d, which is the mean difference divided by the within-groups SD. Similarly, in health-related research, the odds ratio has almost come to be regarded as a gold standard measure for the 2 × 2 table, although the term "effect size" tends not to be used here. However, a range of effect size measures are needed for different purposes, and imposing a single measure in all contexts is highly inadvisable. There is an extensive literature on effect size measures in the psychology domain (e.g., Kirk (1996); Olejnik and Algina (2000)), and less so in the health domain (e.g., Kraemer (2006)).

What exactly is meant by a measure of effect size? Statistical calculations produce a diversity of numerical values, including point estimates such as means and proportions; hypothesis test statistics such as t test and χ^2 test; and p-values corresponding to these. None of these quantities serves as a measure of effect size in the sense described here.

1.4.1 An Example to Illustrate the Need for Effect Size Measures and Their Associated Confidence Intervals

Fearnley and Williams (2001) studied 19 men charged with assault and referred to a forensic psychiatry service. They were rated for impulsivity using the Monroe Dyscontrol Scale (Monroe 1970, 1978) which can range from 0 to 54. Six of them had a history of head injury leading to loss of consciousness for over 48 hours. The data, presented in Table 1.1, are typical of

TABLE 1.1

Monroe Dyscontrol Scale Scores for 19 Forensic Psychiatry Referrals, by History of Head Injury

History of Head Injury	Monroe Dyscontrol Scale Scores
Yes (n = 6)	4, 36, 38, 39, 44, 45
No (n = 13)	5, 8, 12, 14, 16, 18, 19, 23, 24, 28, 29, 35, 39

Source: Data from Fearnley, D. and Williams, T. 2001. *Medicine, Science and the Law* 41: 58–62.

much work in the behavioural sciences, with small sample sizes and poor approximation to Gaussian distributional form, especially for the head injury cases.

Statistical software such as SPSS gives a Mann–Whitney *U* statistic of 62.5 for the data, with a two-tailed *p*-value of 0.039. This indicates some evidence for a difference in scores between subjects with and without head injury, but how is this difference best quantified? As discussed earlier, the *p*-value relates to the amount of evidence we have that a difference exists, not the size or consistency of the difference itself.

Consider first possible absolute measures of effect size. Clearly, the data are not well summarised by means, especially for the head injury group. So the mean difference is of little use here. As explained in Section 15.2.1, a median difference may be reported with a confidence interval, but this summary of the results has some rather disquieting features. Moreover, apart from researchers actively using the Monroe scale, few people would have a clear idea of how many points on the scale would constitute a clinically important difference.

In this situation, a relative measure of effect size is available, *U*/mn, which generalises the Mann–Whitney *U* statistic. In Chapter 15, the rationale and derivation of this measure are explained and several confidence intervals developed and evaluated. *U*/mn is equivalent to the area under the receiver operating characteristic (ROC) curve. As the *U*/mn measure does not relate exclusively to the Monroe scale, familiarity with that instrument is not a prerequisite to interpreting the calculations. For these data, *U*/mn = 0.801. The *U*/mn measure takes the value 0 or 1 in the event of no overlap between two samples, and is 0.5 when two samples are identical. A value of 0.801 indicates quite a strong association between head injury and Monroe score.

This informative measure of effect size may be obtained directly in SPSS using the ROC facility. SPSS also calculates a 95% confidence interval. For these data, this interval is 0.506 to 1.097, which unfortunately extends beyond 1, so the calculated upper limit is meaningless here. SPSS produces a warning message, but does not suggest anything more appropriate. The chosen method described in Section 15.3, method 5 of Newcombe (2006b), yields a 95% interval from 0.515 to 0.933. Here, and indeed for all possible data, this interval is boundary-respecting, and has appropriate coverage properties. These confidence limits indicate that the relationship between head injury and impulsivity in a larger population of similar men might be very strong (corresponding to 0.933), or very weak indeed (0.515), but is fairly unlikely to be an inverse relationship, with a *U*/mn value below 0.5. This appears to be the most intelligible summary of the results—in terms of the propensity for head injured individuals to have higher scores at any rate. Neither *U*/mn nor any other of the analyses considered here can bring out that there may also be a difference in spread, that occasionally head injury is accompanied by a low impulsivity score in this population.

1.4.2 How Effect Size Measures Add Value

Effect size measures when reported with their confidence intervals not only imply the corresponding *p*-values. They are also interpretable numerical measures in their own right, for which the confidence intervals show the associated degree of sampling uncertainty. This is how they add value. Point and interval estimates may be interpreted directly on a meaningful scale—in some instances, an absolute scale defined in relation to the data, in others a relative one which can be used for any variables—without specifically needing to know the associated sample sizes in order to interpret their meaning. All the information on how the limited sample size impacts on interpretation is contained in the confidence interval. This contrasts with most hypothesis-testing statistics and their associated *p*-values. In the metronome example, the *t*-statistic and its associated *p*-value indicate how much weight of evidence there is that the compression rate has reduced over time. Neither conveys any directly interpretable information on how much it has reduced.

Absolute effect size measures relating to quantitative variables, such as differences between means, have units, and are thus directly interpretable whenever the units are likely to be familiar to readers. In contrast, measures such as U/mn and the standardised difference are relative, or scaled (Barnhart et al. 2007) measures of effect size. All relative effect size measures are unit-free, and thus can meaningfully be interpreted in quite different contexts. All effect size measures that relate purely to proportions are unit-free. However, some of them are absolute, some are relative, and it is always crucial to interpret them accordingly; for example, a doubling of risk of a side effect is much more serious if the baseline risk is 10% than if it is 0.1%.

For comparing two proportions, the appropriate absolute effect size measure is the difference between them. The most widely used relative measures

TABLE 1.2

Some Effect Size Measures

Measure	Variable Types	Absolute or Relative	For Unpaired or Paired Data
Difference of means	Continuous	Absolute	Either
Difference of proportions	Binary	Absolute	Either
Ratio of proportions (relative risk)	Binary	Relative	Either
Rate ratio	Poisson	Relative	Unpaired
Odds ratio	Binary	Relative	Either
Linear interaction effect for proportions	Binary	Absolute	Either
Generalised Mann–Whitney measure U/mn	Continuous or ordinal	Relative	Unpaired
Generalised Wilcoxon measure $\Delta T/\max$	Continuous or ordinal	Relative	Paired

are ratio measures, the relative risk and the odds ratio. Table 1.2 catalogues the main effect size measures, existing and novel ones, considered in this book.

Such measures of effect size represent the fulfilment of the agenda to supplement if not supersede *p*-values with something more directly interpretable.

1.5 When Are Point and Interval Estimates Less Helpful?

1.5.1 An Example Where a *p*-Value Really Is More Informative

Occasionally, the issue really is one of deciding between two hypotheses, and it would not be particularly helpful to estimate a parameter. Roberts et al. (1988) reported uroporphyrinogen activity amongst 91 individuals in 10 kindreds affected by familial porphyria cutanea tarda. Ostensibly, families 1 to 6 showed the classic pattern in which the distribution of enzyme activity is bimodal, with low activity in about half the family members, including those already affected. Families 7 to 10 showed an apparently quite different pattern, in which no members had low levels of activity, irrespective of disease phenotype. It seemed highly plausible that the relationship between disease and enzyme was different in these kindreds, but how should this be quantified?

Several different models were fitted to the data. Each model tested the null hypothesis H_0 that the association between disease and enzyme was the same in all 10 families against the alternative H_1 that a different relationship applied in the two groups of families, 1 to 6 and 7 to 10. Each model fitted yielded a *p*-value of the order of 10^{-6} indicating that H_0 should be rejected in favour of H_1. However, these analyses took no account of the fact that this particular split of the 10 families into two groups was on *a posteriori* grounds. There are $2^{10-1} = 512$ possible ways of segregating a set of 10 families into two groups. Clearly we do not want to consider the split in which all 10 families are in the same group, so this reduces to 511. We can allow for the multiplicity of possible ways of segregating the 10 families into two groups here, by calculating a modified *p*-value $p^* = 1 - (1 - p)^{511}$. The effect of doing so is approximately the same as grossing up the *p*-value by a factor of 511. When we do so, it remains very low, at around 0.001 or below. So our data present robust evidence that the four aberrant families really do present a different relationship between enzyme and inheritance of disease.

This is an example of a situation in which a *p*-value is unequivocally the most appropriate quantity on which to base inferences. An interaction effect may be calculated, with a confidence interval, to compare the difference in mean log enzyme level between affected and unaffected family members between the two groups of families. However, this does not directly address the simpler issue of whether the disease should be regarded as homogeneous or heterogeneous. The p-value, adjusted in a simple way for multiplicity of comparisons, most closely addresses this issue, but this is the exception rather than the rule. Another possible example is the dysphagia assessment example in Section 12.1. As well, see Section 9.3 for a generalisation of single-tail-based p-values to characterise conformity to prescribed norms. Even for the drug side effect example in Section 11.5.3, the most informative analysis demonstrates the existence of a plateau less than 100% by estimating it, rather than by calculating the p-value comparing the fit of plateau and non-plateau models. For nearly all the data sets referred to in this book, it is much more informative to calculate an appropriate measure of effect size with a confidence interval.

1.5.2 An Example Where neither *p*-Values nor Effect Size Measures Are Particularly Informative

A major obstacle to the efficacy of screening programmes is non-attendance. In a telephone interview study Bell et al. (1999) studied reasons for non-attendance in 684 women in Wales who failed to attend for mammographic screening following an initial letter and a reminder letter. The respondents' stated reasons for non-attendance were initially classified into 25 categories, but these were coarsened into four groups for analysis, representing inconvenience, low motivation, mix-up (whether on the part of the screening authority or the woman, which could not always be identified), and fear (including fear of cancer and its consequences, surgery and embarrassment). Half of the non-attendances are attributable to inconvenience, with the remainder approximately equally split between the other three categories.

The women also clustered naturally into three groups relating to screening history. At the time, women in the UK were offered screening at 3-yearly intervals between the ages of 50 and 65. Consequently about one-fifth of the non-attenders were women invited for the first time. Women who had been invited to a preceding round of screening were split into those who had accepted mammographic screening previously and those who had not.

Table 1.3 shows the remarkably rich pattern of association between reasons for non-attendance and screening history. A χ^2 test gives 59.8, with 6 degrees of freedom (df) and $p < 0.001$, indicating highly significant differences in the pattern of reason for non-attendance between the three history groups. Whilst inconvenience is the commonest reason overall and for the first-round invitees and women who had ever attended before, for the never attenders low motivation is more important. The patterns in the first-round invitees and ever attenders are similar in that inconvenience predominates; but the ever attenders reported reasons in the fear cluster (especially pain) more often than the first-round invitees did, whereas low motivation was less important. Fear was more important for the ever and never attenders than for the first-round invitees. Low motivation was most important for the never attenders,

TABLE 1.3

Reasons for Women Not Attending for Breast Cancer Screening, by Screening History

	First-Time Non-Attenders	Ever Attended Before	Never Attended Before	Total
Inconvenience	71 (54%)	230 (56%)	43 (31%)	344 (50%)
Motivation	26 (20%)	45 (11%)	53 (38%)	124 (18%)
Mix-up	24 (18%)	73 (18%)	22 (16%)	119 (17%)
Fear	11 (8%)	64 (16%)	22 (16%)	97 (14%)
Total	132	412	140	684

Source: Data from Bell, T. S., Branston, L. K. and Newcombe, R. G. 1999. Survey of non-attenders for breast screening to investigate barriers to mammography. Unpublished report. Breast Test Wales, Cardiff.

least for those who had ever attended. Reasons in the confusion category accounted for a constant one-sixth of women in all history groups. Very little of the pattern would be encapsulated by any effect size measure for a table with more than two rows and columns. In such a situation, there is no adequate substitute for showing the contingency table in its entirety, with suitably chosen percentages. The p-value is needed, solely to confirm the obvious inference that the pattern is not well explained by chance.

1.6 Frequentist, Bayesian and Likelihood Intervals

Most interval estimation methods presented in this book are derived from the frequentist paradigm. When we calculate a 95% confidence interval for a proportion, this is intended to have the property that, on 95% of occasions when we calculate such an interval, it will succeed in including the true population proportion. The population proportion is regarded as fixed, although unknown, and the objective is to locate it to within a narrow margin of error. Frequentist methods are not designed to take account of any pre-existing information regarding the parameter; they merely seek to quantify what information the sample yields in relation to the parameter. This information may subsequently be combined with other information in various ways, but the primary purpose of the analysis is to summarise what this particular sample tells us about the parameter.

Alternative types of interval estimates, Bayesian and likelihood intervals, may also be calculated. The likelihood function is an important construct used in statistical modelling. It is the probability of obtaining the set of data that were actually observed, viewed as a function not of the data values but of the parameters. It is maximised when we fit a model such as Equation 1.1, leading to maximum likelihood estimates (MLEs) of the parameters. This leads to a natural alternative way to construct confidence intervals. A likelihood interval comprises parameter values on either side of the MLE, sufficiently close that the natural logarithm of the likelihood remains close to the maximum value, in fact, within 1.92 (which is $1.96^2/2$) for a 95% interval. Likelihood intervals for proportions are slightly anticonservative—the mean coverage tends to be a little below the nominal value. Nevertheless, when viewed within their own terms of reference (i.e., in relation to the likelihood function on which inferences are generally based), they incontrovertibly achieve exactly what they claim to. For this reason, it is preferable to regard likelihood intervals as attempting to achieve something slightly different from the frequentist intervals which form the majority of the methods presented in this book.

The Bayesian inference paradigm assumes that any parameter value such as a population proportion is not a fixed, absolute quantity, but itself has a frequency distribution, representing our degree of knowledge or belief about where the parameter lies. Bayesian intervals are often referred to as credible intervals, to emphasise that the interpretation is different to that of frequentist intervals. The Bayesian inference process starts with a prior distribution for the parameter of interest, and combines this with the information contained in the likelihood to produce a posterior distribution. A 95% credible interval is then formed by calculating the appropriate centiles of the posterior distribution. Sometimes the prior distribution is chosen to summarise prior information concerning the parameter. Most often, this is not done, and a diffuse prior distribution is chosen, which purports to be uninformative (i.e., does not intentionally incorporate prior information). In either case, often a prior may be chosen, known as a conjugate prior, which combines mathematically in a particularly neat way with the likelihood. For inferences about proportions, beta distributions are conjugate. Either a $B(1, 1)$ or $B(1/2, 1/2)$ prior is chosen, or a prior with different parameter values may be used to express prior knowledge. In fact interval estimates for proportions rooted in the Bayesian paradigm, using the above uninformative priors, have very favourable properties when viewed as frequentist intervals. These intervals will be examined closely in Chapter 3, alongside several frequentist methods.

1.7 Just What Is Meant by the Population?

Whilst the theory on which statistical methodology is based assumes that the study sample is drawn from some wider population, in some situations it may not be possible to arrive at a clear definition of such a population. However, this observation does not negate the use of confidence intervals, which in practice has extended to become a familiar means to display the uncertainty attaching to a parameter estimate resulting from limited sample size.

1.7.1 An Outbreak Investigation

After a private dinner party, 11 of 34 guests developed gastrointestinal illness caused by *Salmonella enteritidis*. One food item under suspicion was a chocolate mousse containing raw egg. Epidemiological investigations showed that 10 of 16 guests who ate mousse became ill, but only 1 of 18 who did not eat mousse became ill. An odds ratio of 28.3 was reported for the association between exposure and outcome, with 95% interval 3.5 to 206 using the method recommended in Section 11.3.2, indicating a strong association between consumption and illness.

Just what does a confidence interval for a measure of association mean in this situation? An outbreak of foodborne illness such as this is evidently a one-off occurrence in its own right. There is no implication whatever from these results that in the next food poisoning outbreak investigated by the same team, the infective agent or its mode of transmission would be the same, nor that when attending such an event, it is advisable to avoid eating the mousse. Unlike in the usual research paradigm in which the aim is to yield information about other similar people, here the aim is the quite different one of identifying the causal agent for this particular outbreak.

In a situation such as this the confidence interval is merely a familiar, hence readily interpretable, instrument to express the degree of uncertainty attaching to the summary measure—in this instance, the odds ratio—purely as a consequence of being based on a limited sample size. There is no question of repeating the study, or of increasing the number of individuals sampled in relation to this particular outbreak—only 34 people attended the event.

1.7.2 Sampling from a Finite Population

The derivation of most confidence interval methods presupposes that the sample is drawn from an infinitely large population, which of course is never the case. Methods closely paralleling those normally used for means, proportions and other quantities have been developed that allow for the effect of finite population size. However, such methods are rarely indicated. For example, researchers attempted to discover what proportion of a set of 150 farms were infected with a particular disease. It had been intended to sample all 150 farms in the population but only 108 could be sampled before the foot and mouth disease outbreak prevented completion of the study (Gordon 2002). Should a finite population interval be used here?

We need to consider carefully what would be the appropriate analysis had the researchers sampled 149 of 150, or indeed all 150 farms, in which situation the finite population interval would be of zero width. In practice finite population interval methods are seldom used. The reason is that the "population" is usually something intangible, in the background—a much wider, loosely defined universe of subjects or units of which those in the sample are in some sense representative. Even if we can enumerate a very finite population from which we have sampled, it is not usually the behaviour of that population that we are really concerned to quantify. The interval here is to be taken as a conventional expression of uncertainty due purely to being based on a limited number of units—the number of units in the population is irrelevant to the interpretation. It is appropriate to use finite population formulae, if the defined finite population is the ultimate goal of interest, but usually it is not.

1.8 The Unit of Data

In health and social science applications, generally the unit of data is most appropriately the individual. This is for two reasons. In most study designs (but not all), it is reasonable to assume that the responses of different subjects are statistically independent. This is a key assumption underlying the all simplest statistical analyses. This assumption is usually implicit rather than stated explicitly.

Furthermore, variation between individuals is often the dominant source of variation. Genetic, environmental and behavioural factors all contribute importantly to this. This applies even when a study involves a group of laboratory animals, which are carefully inbred to be as genetically similar as possible to one another, then reared in the same environment and supplied with the same source of food. In such situations it is readily observed that all animals are not equal; for example, some become dominant in their access to the food supply or in fights, so they are not interchangeable with one another. Obviously the potential for variation between unrelated free-living humans is far greater still.

Moreover, the objective of a statistical analysis is to generalise from the study sample to other similar individuals. Most studies involve drawing a sample of individuals. A measure of variation between individuals in the sample, such as the standard deviation, then forms the basis for both hypothesis testing and confidence interval calculations. (For binary data, no standard deviation as such is apparent, but many statistical methods for proportions amount to assuming the standard deviation is $\sqrt{p(1-p)}$). The uncertainty that these methods quantify relates to the fact that future individuals, even though they belong to a similar population, will differ from those sampled, to a degree determined by the variation between individuals.

In many studies, multiple observations per subject are collected. Many types of organs, such as eyes or kidneys, are paired. Statistical analyses that assume independence are grossly misleading. A simulation study (Newcombe and Duff 1987) investigated the consequences of disregarding such non-independence when analysing intra-ocular pressure data.

Differences reaching statistical significance at a nominal 5% α level emerged in 20%, not 5%, of randomly generated samples. Correspondingly, intervals with a nominal 95% coverage would in reality have coverage probability of only 80%. Simple analyses here would be anticonservative. Clearly, the impact of this issue is much greater when analysing data on plaque accumulated on tooth surfaces. The simplest workaround in practice is to characterise the response of each individual by a summary score at subject level, such as the average plaque index across all scoreable surfaces, or the higher of the two intra-ocular pressures. A more complicated analysis that can provide additional insight is multi-level modelling, which specifically examines variation at two or more levels.

Conversely, some types of intervention studies require that centres, not individuals, should be randomised to treatments. For example, Frenkel et al. (2001) evaluated an intervention to promote delivery of effective oral hygiene by care staff in residential homes. This involved training the care staff, who would then apply the skills they had learned to the residents. An individual resident would receive oral care from one of the care staff who was on shift at the time. So there was no question of randomising individual residents; allocation had to be at the level of the home, not the individual. In other studies, hospitals, primary care centres, schools and so forth may be randomised. Simple analyses which assume responses in different individuals to be independent are misleadingly anticonservative. Such cluster randomised studies involve special statistical methods to take account of the clustering. Once again, multi-level modelling can provide additional information.

Comprehensive description of the statistical methods that are indicated in these situations is beyond the scope of this book. Nevertheless, Chapter 9 describes a novel approach to estimating proportions based on paired organs. This may also be applied to the logically equivalent task of estimating gene frequency.

1.9 Sample Size Planning

Finally, the importance of planning the sample size to be used in a study cannot be over-emphasised. A clear rationale for sample size, in terms of either power or anticipated precision, is now a key requirement of guidelines for reporting studies such as the CONSORT statement (Schulz et al. 2010), and also of research governance requirements in some countries. Nevertheless many studies that are too small to be informative are still performed. It is often argued that processes of systematic review and meta-analysis now mean that small studies can make an important contribution. However, it is much better for a study to be definitively powered, rather than the fragmentary use of resources so often seen. Conversely, occasionally studies are unnecessarily large, which is also wasteful of financial and staff resources, and differences that are too small to matter may be labelled as statistically significant. Corresponding to the development in Sections 1.2.1 to 1.2.3, there are three basic approaches to developing a rationale for sample size.

1.9.1 Sample Size Assessment Based on a Confidence Interval for a Single Group

We first consider a study involving a single group of participants, as in Section 1.2.1. Here, we may simply seek to ensure that the parameter of interest—in the simplest case, a mean or a proportion—is estimated to within a prescribed precision.

To estimate a mean to within $\pm h$ with confidence $1 - \alpha$, we must ensure that the expected half-width h is approximately $z_{1-\alpha/2} \times \sigma/\sqrt{n}$ where $z_{1-\alpha/2}$ denotes the $1 - \alpha/2$ quantile of the standard normal distribution. For the usual 95% confidence level, $z_{1-\alpha/2} = 1.960$. This is often rounded to 2. Thus n is chosen to be $z^2_{1-\alpha/2}\sigma^2/h^2$. (Using $z^2 = 4$ instead of 3.841 here has the beneficial effect of allowing for an attrition of 4%, although this may be rather low, it is better to assess the likely degree of attrition for the specific study context.) When performing this calculation, we need to make some assumption as to the likely value of σ; this is often based on existing relevant data. If the resulting n is small, we should use a slightly larger n satisfying $n = t_{n-1,\, 1-\alpha/2} \times \sigma^2/h^2$ where $t_{n-1,\, 1-\alpha/2}$ denotes the corresponding centile of the t-distribution with n-1 degrees of freedom (see Section 2.1). The usual relationship between sample size and interval width applies: to a close approximation, halving the interval width requires a fourfold increase in sample size.

To estimate a proportion to within a specified interval half-width is more problematic, because as we shall see, the most appropriate intervals for proportions tend to be asymmetrical. If we know we will be in a situation in which the simplest asymptotic method, the Wald interval (Section 3.2) performs acceptably, it is adequate to substitute $p(1 - p)$ for σ^2, where p represents an assumed value for the proportion. We then choose n to be $4p(1 - p)/h^2$. Sometimes, investigators with no information whatever relating to the magnitude of p play safe by substituting $p = 1/2$ in this formula, which then reduces to $1/h^2$. However, this procedure cannot be recommended; it is much more cost-effective to start with a small study to help decide whether p is likely to be of the order of 50%, 10% or 1%, as this information will have a large direct effect on the sample size to be chosen, and indeed also on what would be a sensible value for h.

Calculating n as $4p(1 - p)/h^2$ relates specifically to the Wald interval. As we will see in Section 3.2, the Wald interval is known to be inappropriate when either the number of positive observations, $r = np$, or the number of negative observations, $n - r = n(1 - p)$, is small. This calculation may give us a very preliminary ballpark figure for the sample size needed. However, when investigators apply for funding to carry out a study, which is essentially linearly related to the sample size, reviewers should require a sample size calculated by a more robust method. As a more general alternative approach, we may use the Excel spreadsheet, which implements the chosen Wilson interval for a proportion (see Section 3.6). Using a trial-and-error approach, substitute possible values for n and r, and adjust until the projected interval corresponds to the degree of precision desired. This procedure is illustrated in Section 3.7.

1.9.2 Sample Size Assessment Based on Power for a Comparison of Two Groups

Two approaches are widely used to determine sample size for a comparison between two groups. The most common rationale for sample size is based on the power to detect a difference of specified size between two groups. (Sometimes, it is more relevant to think in terms of an association between a risk factor and an outcome.) A less frequently used approach is based on confidence interval width (see Section 1.9.3).

Recall that the power is defined as $1 - \beta$, the complement of the type II error rate. It is the probability that we will succeed in detecting a difference between the two groups, with a p-value lower than the chosen α level, on the assumption of the alternative hypothesis H_1, which states that such a difference really exists. We then calculate the sample size that yields the required power. For a meaningful study, power is usually set at 80% or higher. A study with a power of 50% is inadequate and just as likely to miss the difference of interest as to detect it—a hit-and-miss approach. In general, the sample size required for 80% power is approximately twice that for a study of similar design with 50% power.

Usually, equal sizes are chosen for the two study groups. This is generally, but not invariably, the best design. When the comparison involves a quantitative variable, the quantities that need to be specified are the target difference or least significant difference (LSD) δ between the groups, and some estimate of the within-groups standard deviation σ, as well as the α and β values assumed. Assuming that α corresponds to two-sided testing, the sample size for each group is

$$n = 2(z_{1-\alpha/2} + z_{1-\beta})^2/(\delta/\sigma)^2 \tag{1.2}$$

where z_k denotes the 100k centile of the standard normal distribution. The dependence on δ and σ is via the standardised difference, which is the relevant measure of effect size here—obviously only a scale-free relative measure is suitable for use in a power calculation. The chosen target difference should be large enough to be of practical importance. It should not be so large as to be beyond the bounds of credibility when the study is being planned. As in Section 1.9.1, this formula requires modification for small n to reflect use of the t-distribution.

For comparisons involving a binary outcome, once again the target difference needs to be specified. This should be done by specifying the projected values π_1 and π_2 for the proportions that will be determined for the two groups, because these also contribute to the degree of within-groups variation that the calculation assumes. Then the sample size per group is

$$n = (z_{1-\alpha/2} + z_{1-\beta})^2 \, (\pi_1(1 - \pi_1) + (\pi_2(1 - \pi_2))/\delta^2 \qquad (1.3)$$

The above formulae relate to studies aiming to demonstrate superiority of one regime over another. Different z-values are relevant to studies aiming to show either non-inferiority or equivalence.

Several types of resources are available to help the process of power assessment. The basic method is direct calculation, which involves substituting appropriate figures into formulae such as the above. Calculation may be bypassed by using published tables (e.g., Machin et al. 2009), or dedicated software such as nQuery (Statistical Solutions 2009). A particularly instructive approach to sample size assessment involves a nomogram (Altman 1982). This is a simple diagrammatic aid to sample size assessment, and was designed specifically for the comparison of two means. The diagram comprises three straight lines, which are graduated with scales corresponding to the standardised difference, the sample size and the resulting power. The user then superimposes a diagonal line which passes through the points representing any two of these quantities, and reads off the corresponding value for the third one. The nomogram is suitable for interactive, trial-and-error use, enabling the user to appraise the consequences of different choices of sample size and target difference.

1.9.3 Sample Size Assessment Based on Projected Confidence Interval Width for a Comparison of Two Groups

Alternatively, much as for the single sample case (Section 1.9.1), we can determine what sample size will lead to a confidence interval we regard as sufficiently narrow to be informative. This is done by performing a calculation similar to the intended eventual analysis for the primary endpoint, starting from hypothesised summary statistics for the data, examining what confidence width would ensue, then adjusting the sample size to achieve the desired interval width. The main difference between this and the single sample case is that instead of estimating a parameter such as a mean or a proportion within a chosen degree of precision, here we seek to estimate an appropriate effect size measure, such as a difference of proportions or an odds ratio, to the desired precision. Thus, the sample size required per group to estimate a difference of proportions within $\pm h$ with confidence $1 - \alpha$ is

$$n = z_{1-\alpha/2}^2(\pi_1(1-\pi_1)+(\pi_2(1-\pi_2))/h^2 \qquad (1.4)$$

Almost all sections of the Altman et al. book *Statistics with Confidence* (2000) strongly imply a preference for confidence intervals rather than *p*-values. In contrast, the chapter on sample size determination (Daly 2000) suggests that for comparative studies, traditional power calculations are preferable to basing the sample size on the projected confidence interval width. Similar arguments are developed by Greenland (1988) and Kupper and Hoffner (1989). Daly rightly emphasises that the target difference δ for a power calculation and the desired interval half-width h are not interchangeable. This is because the projected confidence interval calculation does not require the user to specify a figure for the power $1 - \beta$ greater than 0.5. In fact, comparing Equation 1.4 with Equation 1.3, if $h = \delta$, $z_{1-\beta} = 0$, hence the power to detect a difference of size h is simply 50%. To be equivalent to a power calculation with 80% power, the half-width h should be $\frac{1.960 + 0.842}{1.960} \times \delta$ or simply $h = \delta/0.7$.

However, in some contexts, no relevant power calculation can be performed. Flahault et al. (2005) developed a stringent approach for sample size planning for sensitivity and specificity (see Chapter 12). Sample sizes are tabulated corresponding to expected sensitivities from 0.60 to 0.99 with acceptable lower 95% confidence limits from 0.50 to 0.98, with 5% probability of the estimated lower confidence limit being lower than the acceptable level. This procedure is analogous to power calculations and satisfies Daly's criticism, although it is arguably over-conservative, because like the Clopper–Pearson interval introduced in Section 3.4.4, it is based on "exact" aggregations of tail areas. If a similar method were applied to a continuous outcome variable, the criticism of over-conservatism would not apply. Another such context in which no equivalent power calculation is available is estimation of the intraclass correlation. Zou (2012) developed methods to ensure high probabilities that (a) interval width is less than a specified quantity or (b) a specified lower limit is achieved. Similar approaches could be used for comparative studies.

1.9.4 Prior and Posterior Power

The power calculation should be performed at the planning stage. Sometimes a power calculation is performed after collecting the data, substituting the actual difference observed for the difference aimed at. This tends to occur when the investigators omitted to consider power at the planning stage, and are subsequently requested by journal referees to produce a power calculation. Such a posterior power calculation is seriously misleading, because in reality it amounts to no more than a rescaling of the *p*-value calculated from the data. Thus an *a posteriori* claim that the power of a study to detect a difference at a two-sided 5% α level was 80% implies that the observed difference divided by its standard error was $z_{1-\alpha/2} + z_{1-\beta} = 1.960 + 0.842 = 2.802$, which corresponds to a two-sided *p*-value of approximately 0.005. In one sense, it is appropriate to consider power after the study is complete and the data analysed. It is good practice to informally compare the findings with what was assumed at the planning stage—in relation to all of the factors involved, such as the target difference, inter-subject variation, and the recruitment rate and resulting number of subjects successfully recruited within the study period—and to assess the degree to which the intended power was attained as a consequence. However, the power calculation performed at the planning stage should be presented unaltered.

To some degree, similar considerations apply if the rationale for sample size is expressed in terms of anticipated confidence interval width. Unlike power, confidence intervals are designed primarily for summarising the data after they have been collected. So it is appropriate to compare the actual precision attained, as evidenced by the calculated confidence width, with what was anticipated at the planning stage. Nevertheless, the latter should not be recalculated, but should stand as the basis of a citable *a priori* justification for sample size.

2

Means and Their Differences

2.1 Confidence Interval for a Mean

In this chapter the calculation of confidence intervals is demonstrated for the mean of a continuous variable—which is not contentious—and for a difference between the means of two independent variables—which is slightly more contentious. Whilst means summarising continuous distributions are not the main concern in this book, this is the fundamental case, and many of the issues relating to proportions parallel the development for means.

Absi et al. (2011) reported results for knowledge of radiological protection issues in a series of dentists and dental care practitioners (DCPs) who attended a one-day refresher course in March 2008. Identical 16-item best-of-five multiple choice questionnaires were administered before and after the course. Matched pre- and post-course results were available for a total of 272 participants, comprising 235 dentists and 37 DCPs. Because the instrument does not incorporate negative marking, scores can potentially range from 0 to 16. Summary statistics were published for scores scaled to run from 0 to 100, but for illustration they are shown in their original integer form.

Tables 2.1 and 2.2 give frequency distributions and summary statistics for scores out of 16 before and after training, and for the change in score, based on all 272 participants. Figure 2.1 is a histogram showing the distribution of the post-training scores.

The distribution of the post-training course score is not of Gaussian form. It is bounded, discrete and negatively skewed. On account of the skewness, it would be erroneous to construct centiles from the mean and SD assuming a Gaussian distribution. A 95% reference interval calculated in the usual way as mean ± 1.96 SD is patently inappropriate because the upper reference limit is above 16. However, this degree of skewness does not invalidate construction of a confidence interval. Furthermore, although technically the scale used here is discrete, in practice when we draw a sample from a scale with so many points, behaviour is very much like when the scale is a continuous one, for purposes of inferences concerning means. (This contrasts with the severe impact of discreteness when we draw inferences based on a single value from the binomial distribution summarising a binary response variable across a

TABLE 2.1

Frequency Distributions of Radiological Protection Knowledge Scores of 272 Dental Professionals before and after a Refresher Course, and Change in Score Following Instruction

	Frequency For		
Value	**Pre-Course**	**Post-Course**	**Change in Score**
–2	–	–	1
–1	–	–	3
0	0	0	7
1	0	0	13
2	1	0	27
3	4	0	31
4	16	0	40
5	29	1	59
6	35	5	27
7	46	2	30
8	49	12	18
9	34	11	10
10	33	16	5
11	15	38	0
12	3	44	0
13	6	48	1
14	1	50	0
15	0	39	0
16	0	6	0

Source: Data from Absi, E. G. et al. 2011. *European Journal of Dental Education* 15: 188–192.

TABLE 2.2

Summary Statistics for Knowledge Scores Pre- and Post-Training, and Mean Increase in Score following Training, Including 95% Confidence Intervals

	Median	**Mean**	**Standard Deviation**	**Standard Error of Mean**	**95% Confidence Interval**
Pre-training	8	7.63	2.22	0.135	7.37 to 7.90
Post-training	13	12.33	2.22	0.135	12.07 to 12.60
Increase following training	5	4.70	2.42	0.146	4.41 to 4.99

Source: Data from Absi, E. G. et al. 2011. *European Journal of Dental Education* 15: 188–192.
Note: Based on 272 dental professionals participating in a radiological protection course.

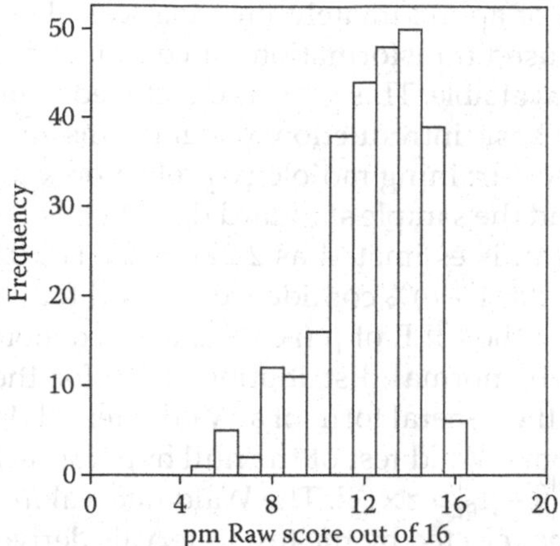

FIGURE 2.1
Histogram of radiological protection knowledge scores of 272 dental professionals following a refresher course. (Data from Absi, E. G. et al. 2011. *European Journal of Dental Education* 15: 188–192.)

sample of n individuals.) The distribution we are considering here is typical of the many real-world distributions for which inferences for means based on the Gaussian model are considered satisfactory. Such inferences, based on moderate to large sample sizes, are regarded as robust on account of a result fundamental to statistical inference, the central limit theorem.

This crucial result states that for a distribution with mean μ and variance σ^2, the sampling distribution of the mean approaches a normal distribution with mean μ and variance σ^2/n as n, the sample size, increases. This applies regardless of the shape of the original (parent) distribution. How rapidly it does so depends on the shape of the parent distribution, however. For most variables that would usually be summarised by a mean, construction of a confidence interval based on the central limit theorem is reasonable for moderate sample sizes. Where such intervals most obviously go seriously wrong is for proportions, and for many quantities derived from proportions.

The central limit theorem has repercussions other than for the sampling distribution of the mean. Many physiological parameters are determined by a great number of other factors, some identifiable, others not. A very simple model for this process is when a variable is the resultant of many independently, identically distributed influences which combine by addition. The central limit theorem implies that the resulting variable has a Gaussian distribution. Similarly, when a variable is generated by many factors which combine by multiplication, a log-Gaussian distribution ensues. In this situation, it is more appropriate to construct confidence intervals for the mean, and other analyses, on a log-transformed scale. For ratio measures such as risk ratios and odds ratios (Chapters 10 and 11), confidence intervals tend to be symmetrical, exactly or approximately, on a log scale. Log transformation is the most commonly used transformation for continuous variables, although alternatives are also available. This issue is developed in detail in Section 2.4. See Appendix 2 for a basic introduction to logarithms and exponentiation.

Returning to our post-training radiology protection knowledge scores, the mean is $\bar{y} = 12.335$ and the sample standard deviation is $s = 2.221$, so the standard error of the mean is estimated as $2.221/\sqrt{272}$ i.e., 0.135. This suggests that an approximate 100(1 – α)% confidence interval for μ may be formed as $\hat{\mu} \pm z \times \text{SE}$ where $\hat{\mu}$ is the MLE of μ (i.e., $\bar{y}$), and z denotes the 100(1 – α/2) centile of the standard normal distribution, 1.960 for the usual two-sided 95% interval. This is the general form of a Wald interval (Wald 1943) because it may be derived from a Wald test of the null hypothesis H_0: $\mu = \mu_0$. The test rejects H_0 whenever $|\hat{\mu} - \mu_0| > z \times \text{SE}$. The Wald interval includes all values of μ_0 which would not be rejected by the test. Intervals derived in an analogous way are often used for proportions (Section 3.2) and for differences between proportions, based on unpaired samples (Section 7.2) or paired samples (Section 8.2).

If we knew that the actual population standard deviation was $\sigma = 2.221$, we would report a confidence interval for the mean of the form $\bar{y} \pm z \times \sigma / \sqrt{n}$. For the usual two-sided 95% interval, $z = 1.96$, so this z-interval would run from 12.335 – 1.96 × 0.135 to 12.335 + 1.96 × 0.135; that is, from 12.071 to 12.599. Because interval width is proportional to $1/\sqrt{n}$, a fourfold increase in sample size is required to halve the width.

However, in virtually all practical applications, it is unrealistic to suppose that the standard deviation is known whilst the mean is unknown. Occasionally, there would be a specific prior expectation regarding the standard deviation. For example, standard deviation scores or z-scores are often used to express how measurements deviate from age- and sex-specific normative values. In a representative sample of some cohesively defined population, the SD is expected to be approximately 1. Similarly, the SD of a series of IQ scores might be of the order of 15, as the instrument was originally calibrated to have this degree of variation. Nevertheless, even in these situations, it is greatly preferable to use the sample data to estimate σ as well as μ. The same principle applies whenever we are estimating a difference of two means on either unpaired or paired data, as in Sections 2.2 and 2.3.

Consequently, in practice the confidence limits for the sample mean are calculated not as $\bar{y} \pm z \times \text{SE}$ but as $\bar{y} \pm t \times \text{SE}$. Here, t denotes the appropriate centile of the (central) t-distribution, a distribution closely related but not identical to the Gaussian. The multiplier t used for a 100(1 – α)% interval for the unknown variance case is larger than the corresponding z used for the known variance case, to allow for the uncertainty of σ as well as μ. The t-distribution involves an additional parameter, the number of degrees of freedom ν. In this simplest case $\nu = n - 1$. For the usual 95% interval, to a close approximation the relevant multiplier is $t_{\nu, 0.975} = 1.96 + 2.4/\nu$. This relationship implies that the actual number of degrees of freedom has little influence unless it is small. One consequence is that it remains true, to a close approximation, that a fourfold increase in sample size is required to halve interval width.

So, for our post-training scores, with 272 df, the exact value of the relevant centile of the *t*-distribution is 1.96876. The above approximation gives 1.96886, which is obviously very close here. The 95% limits are then 12.335 ± 1.96876 × 0.135, so the correct *t*-interval runs from 12.070 to 12.600. In this example, this is almost indistinguishable from the *z*-interval from 12.071 to 12.599 calculated above. The procedure to calculate the *t*-interval from a sample of raw data is implemented in all statistical software. Table 2.2 also includes confidence intervals for the mean scores pre- and post-training, and for the mean increase in score following training.

If the distributional assumptions were satisfied, then the coverage probability achieved by this interval would be exactly $1 - \alpha$, regardless of the value of the population mean μ. Whilst these assumptions are never exactly satisfied—and even if they were there would be no way to demonstrate it—in practice many data sets are sufficiently well-behaved for us to be happy that the resulting interval achieves what it claims. Subsequent chapters of the book deal with confidence intervals for proportions and related quantities. The situation there is in stark contrast to the simplicity of the Gaussian case. This applies most especially to the simplest case of a single binomial proportion (Chapter 3). Here, coverage of what is nominally a 95% interval can be very far short of 95%. Moreover, it can fluctuate widely as the parameter value, the population proportion π, varies over its range of validity from 0 to 1.

The 16 radiological protection knowledge questions cover eight areas of knowledge, with two questions on each. We could obtain separate analyses of mean scores for the resulting eight subscales. The Gaussian approximation does not apply so well to these. In particular, discreteness becomes much more important. Confidence intervals that are appropriate for means of variables which can only take the values 0, 1 or 2 are more closely related to interval methods for proportions, and will be described in Chapter 9.

2.2 Confidence Interval for the Difference between Means of Independent Samples

Suppose we want to estimate the difference between the means of independently drawn samples from two populations with means and SDs μ_i and σ_i ($i = 1, 2$). We draw a sample of size n_i from population i, resulting in a sample mean $\bar{y}_i$ and standard deviation s_i which is calculated in the usual way.

Assuming $\sigma_1 = \sigma_2 = \sigma$, the variance of $\bar{y}_1 - \bar{y}_2$ is then $\sigma^2\left(\frac{1}{n_1}+\frac{1}{n_2}\right)$.

If it turned out that $s_1 = s_2 = s$, the natural estimate of the variance of $\bar{y}_1 - \bar{y}_2$ would be $s^2\left(\frac{1}{n_1}+\frac{1}{n_2}\right)$. The 100(1 – α)% confidence limits for $\mu_1 - \mu_2$ would be

$$\bar{y}_1 - \bar{y}_2 \pm t \times s\sqrt{\frac{1}{n_1}+\frac{1}{n_2}} \tag{2.1}$$

where t denotes the appropriate quantile of the t-distribution with $n_1 + n_2 - 2$ degrees of freedom.

In practice, of course, it is very unlikely that s_1 and s_2 would be identical, especially when $n_1 \neq n_2$. Identical SDs are unusual even in the paired case, with a common sample size: in Table 2.2, the SDs pre- and post-training are actually 2.2214 and 2.2208. If s_1 and s_2 are similar, we can calculate a pooled variance estimate

$$\tilde{s}^2 = \frac{(n_1 - 1)s_1^2 + (n_2 - 1)s_2^2}{n_1 + n_2 - 2} = \frac{\sum_{i=1}^{2}\sum_{j=1}^{n_i}\left(y_{ij} - \bar{y}_i\right)^2}{n_1 + n_2 - 2} \tag{2.2}$$

and use this estimate in Equation 2.1 to calculate the confidence limits by the pooled variance method. Use of the t-distribution with $n_1 + n_2 - 2$ df is correct here.

The assumptions underlying this calculation are that the two samples are independent and random and come from approximately normal distributions with equal variances. Often, the assumption of equal variances is questionable. The applicability of the t-distribution depends on this assumption. An alternative unpooled approach does not make this assumption, but relies on an approximation by a t-distribution with a reduced number of degrees of freedom. The Satterthwaite interval, a method in the Behrens-Welch family (Armitage, Berry and Matthews 2002), is calculated directly from the standard errors for the two sample means separately as

$$\bar{y}_1 - \bar{y}_2 \pm t \times \sqrt{\frac{s_1^2}{n_1}+\frac{s_2^2}{n_2}} \tag{2.3}$$

The appropriate number of degrees of freedom here is

$$v^* = \frac{\left[\frac{s_1^2}{n_1}+\frac{s_2^2}{n_2}\right]^2}{\frac{\left(s_1^2/n_1\right)^2}{n_1 - 1}+\frac{\left(s_2^2/n_2\right)^2}{n_2 - 1}} \tag{2.4}$$

where $\max(n_1, n_2) \leq \nu^* \leq n_1 + n_2 - 2$. As noted earlier, for moderate sample sizes the actual number of degrees of freedom has only a small effect on the *t*-distribution, consequently in most situations the reduction in df does not result in an appreciable widening of the interval. Usually, what matters more is that the two approaches can impute substantially different standard errors to $\bar{y}_1 - \bar{y}_2$. Which standard error is the larger depends on whether the sample variance is larger for the larger or the smaller sample. One advantage of designing a study so that $n_1 = n_2$ is the resulting robustness of inferences to inequality of the variances in the two populations, a principle enunciated by George Box (Brown 1982).

Both approaches are widely available in statistical software. For example, the SPSS independent-samples *t*-test algorithm produces a pivot table containing *t*-tests and confidence intervals calculated by both methods. The output includes an ancillary test comparing the two sample variances. The test used here is the Levene test, which is designed to be reasonably robust to non-normality in the two samples. However, this should be regarded as an informal screening device only. It may be argued that when the comparison of the means is the endpoint of the analysis, the unpooled variance version should be chosen, by default. When the analysis goes on to use other methods from the analysis of variance (ANOVA) family which are based on pooling, to incorporate adjustment for additional potentially confounding factors, it is more helpful for the simple unadjusted two-sample comparisons to use the pooled variance approach, as then any differences between unadjusted and adjusted analyses will be purely the result of the adjustment, and not of changing from unpooled variance to pooled variance analyses. If heterogeneity of variance is a major concern, alternatives to ANOVA-based methods derived from unpooled sample variances are available in some situations.

Table 2.3 shows the output from SPSS comparing pre- and post-course knowledge scores, and changes in score following instruction, between the 37 dental care professionals and the 235 dentists. SPSS produces both the confidence intervals (95% by default, although this can be altered) and the corresponding *p*-values. There is the usual duality that the confidence interval for the difference excludes the null hypothesis value of zero exactly when the calculated *p*-value is below 0.05. It is questionable how much it matters that the Levene test reaches statistical significance for the first comparison but not for the other two. In this example, the reduction in degrees of freedom is relatively severe, since one sample is much larger than the other; nevertheless this does not make a major impact on the calculated confidence limits.

The estimated difference here may be regarded as an absolute measure of effect size. Point and interval estimates have the same units as the original data. Alternatively, the standardised difference may be reported, as a relative or scaled effect size measure.

The basic requirement for the unpaired analysis described in this section is that the two estimates must be independent. Altman and Bland (2003)

TABLE 2.3

Output from SPSS Comparing Pre- and Post-Course Knowledge Scores, and Changes in Score Following Instruction, between 37 Dental Care Professionals and 235 Dentists Who Attended a Radiological Protection Course

	Professional Qualification	*N*	Mean	SD	Std. Error Mean
Group Statistics					
am Raw score out of 16	DCP	37	7.14	1.782	.293
	Dentist	235	7.71	2.276	.148
pm Raw score out of 16	DCP	37	11.51	2.364	.389
	Dentist	235	12.46	2.174	.142
Change in raw score out of 16	DCP	37	4.38	2.385	.392
	Dentist	235	4.75	2.421	.158

		Levene's Test for Equality of Variances		*t*-Test for Equality of Means					95% Confidence Interval of the Difference	
		F	Sig.	*t*	df	Sig. (Two-Tailed)	Mean Difference	Std. Error Difference	Lower	Upper
Independent Samples Test										
am Raw score out of 16	Equal variances assumed	4.470	.035	−1.468	270	.143	−.576	.392	−1.347	.196
	Equal variances not assumed			−1.752	56.298	.085	−.576	.328	−1.233	.082
pm Raw score out of 16	Equal variances assumed	.009	.924	−2.441	270	.015	−.950	.389	−1.717	−.184
	Equal variances not assumed			−2.297	46.101	.026	−.950	.414	−1.783	−.117
Change in raw score out of 16	Equal variances assumed	.349	.555	−.877	270	.381	−.375	.427	−1.216	.467
	Equal variances not assumed			−.887	48.439	.380	−.375	.423	−1.224	.475

Source: Data from Absi, E. G. et al. 2011. *European Journal of Dental Education* 15: 188–192.
Note: SD = standard deviation.

point out that methods that assume independent samples should not be used to compare a subset with the whole group, or two estimates from the same individuals. In the former situation, we can simply compare the subgroup of interest with the remaining members of the group, using an unpaired analysis. When the two variables to be measured come from the same individuals, the paired analysis described in the next section is required.

2.3 Confidence Interval for the Difference between Two Means Based on Individually Paired Samples

Figure 2.2 is a scatter plot showing the relationship between the pre- and post-training radiological protection knowledge scores in our sample of 272 course attenders. For plotting purposes, raw integer scores have been randomly disaggregated by amounts uniformly distributed on (−0.5, +0.5) (or (−0.5, 0) for scores of 16).

This relationship is typical of such data. The great majority of the points lie above a diagonal line of equality, yet a few trainees' scores remained unaltered or even decreased, as in Table 2.1. There is less scope for improvement in candidates with relatively high pre-course scores than in those who start with poor knowledge. From the standpoint of obtaining a confidence interval for the difference between pre- and post-training mean scores, the most important feature is that the two scores show a substantial positive correlation, here +0.41. This is characteristic of individually paired data. It is

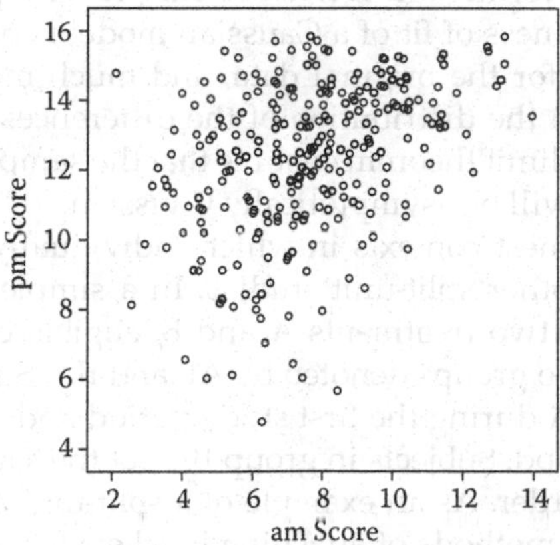

FIGURE 2.2
Relationship between the pre- and post-training knowledge scores in 272 dental professionals attending a radiological protection course. (Data from Absi, E. G. et al. 2011. *European Journal of Dental Education* 15: 188–192.)

incorrect to apply the independent-samples analysis in the previous section here. The correct paired-samples analysis imputes a smaller standard error to the difference, compared to the unpaired analysis. This results in a narrower interval, and a greater power to detect any specified size of difference as statistically significant at a chosen α level.

For the radiological protection knowledge score, the mean gain was 4.70 points. This has a 95% interval 4.41 to 4.99, as in the final row of Table 2.2. The standard error of the mean difference $\bar{\Delta}$ is $SE(\bar{\Delta}) = s_\Delta/\sqrt{n}$ where s_Δ denotes the standard deviation of the within-subjects changes. Alternatively, $SE(\bar{\Delta})$ may be calculated from the standard deviations from the two series and the correlation ρ using the formula

$$SE(\bar{\Delta}) = \sqrt{\{(s_1^2 + s_2^2 - 2\rho s_1 s_2)/n\}} \tag{2.5}$$

where n denotes the sample size (i.e., the number of pairs of observations). The required confidence limits are then given by $\bar{\Delta} \pm t \times SE(\bar{\Delta})$, where t denotes the relevant quantile of the t-distribution with $n - 1$ degrees of freedom. The paired analyses, including the corresponding t-ratio and p-value, are readily obtained using the SPSS paired t-test routine or similar facilities in other statistical software. The issue of choosing pooled or unpooled variance estimates does not arise in the paired situation.

A widely-held but erroneous view is that the above paired comparison method makes the assumption that the two series—in the above example, the pre- and post-training scores—have approximately Gaussian distributions. In fact, it is only necessary for the distribution of the differences to be approximately Gaussian. Often, the latter condition will be satisfied even though the (stronger) former one is not. For our pre- and post-course knowledge scores, the goodness of fit of a Gaussian model is noticeably better for the differences than for the original data, and much more extreme examples exist. Even when the distribution of the differences is some way from Gaussian, the central limit theorem ensures that the sampling distribution of the mean difference will be asymptotically Gaussian.

One of the commonest contexts in which individually paired data arise is in crossover and other split-unit studies. In a simple two-period crossover trial comparing two treatments A and B, eligible consenting subjects are randomised to two groups denoted by AB and BA. Subjects in group AB are given treatment A during the first study period and treatment B during the second study period. Subjects in group BA get the same two treatments, but in the opposite order. As an example of a split-unit design, Kinch et al. (1988) compared two methods of attaching brackets for orthodontic devices to teeth, using etch times of 15 or 60 seconds. In some subjects, a 15-second etch time was used for surfaces of teeth in the upper right and lower left quadrants of the mouth, with a 60-second etch time used for tooth surfaces in the other two quadrants. In other subjects, the opposite configuration was used. Both types of design have the potentially great advantage of using each subject as his or her own control, resulting in a much greater similarity between the two "series" receiving the two treatments than would be the case in a study in which participants are randomised to two parallel groups. However, this may be illusory if what happens in period 1 can affect the findings in period 2, or if the treatment administered at one anatomical site affects what is observed at the other site. In the crossover design, it is important to ensure an adequately long washout period to avoid importantly large carryover effects from the preceding treatment (or treatments, in a multi-period design). In a split-unit design comparing, say, two different substances instilled into gingival pockets, it is important to justify the assumption that the irrigant cannot leach out into the oral cavity and hence affect the response at another site.

However, the correct statistical analysis of the results from a crossover or split-unit study is not the simple paired *t*-test and the related confidence interval for the mean. Occasionally, such studies have very unequal numbers in the two groups. For example, in a crossover study evaluating a low-erosion fruit beverage (Hughes et al. 1999), 70% of subjects received the treatment sequence AB, 30% BA. In a paired *t* analysis the estimated difference in response between the two treatments would include a contribution from any difference between the two study periods. The recommended analysis (Hills and Armitage 1979) involves calculating the difference in responses between periods 1 and 2, for each subject, comparing the mean period 1 minus period 2 difference between groups AB and BA using an independent-samples analysis. If we use halved differences between responses in periods 1 and 2, comparing the means of these estimates the difference in effect between treatments A and B, with an appropriate confidence interval. Here, the paired analysis is correctly performed by using an algorithm originally designed for unpaired comparisons, applied to within-subjects differences.

The interplay between unpaired and paired analyses is slightly more complex than is sometimes imagined (see also Section 16.2).

2.4 Scale Transformation

The most important algebraic transformation of scale is the log transformation. It is applicable to variables that take positive values and exhibit positive skewness, and generally has the effect of reducing such skewness.

2.4.1 How Log Transformation Works

Table 2.4 shows analyses for testosterone levels determined by mass spectrometry in two series comprising 66 South Asian and 66 Caucasian men

TABLE 2.4

Testosterone Levels (nmol/l) in 66 South Asian and 66 Caucasian Men

	Arithmetic Mean	SD	95% Confidence Interval	Median	95% Reference Interval
Testosterone (nmol/l)					
South Asians	16.95	4.59	15.82 to 18.08	16.9	7.95 to 25.95
Caucasians	19.14	5.50	17.79 to 20.50	18.5	8.36 to 29.93
Natural log-transformed testosterone (no units)					
South Asians	2.7916	0.2871	2.7210 to 2.8621	2.8273	2.2289 to 3.3543
Caucasians	2.9131	0.2804	2.8441 to 2.9820	2.9178	2.3636 to 3.4625
	Geometric Mean		**95% Confidence Interval**	**Median**	**95% Reference Interval**
Log-transformed analyses back-transformed to original scale (nmol/l)					
South Asians	16.31		15.20 to 17.50	16.9	9.29 to 28.62
Caucasians	18.41		17.19 to 19.73	18.5	10.63 to 31.90

Source: Data from Biswas, M. et al. 2010. *Clinical Epidemiology* 73: 457–462.
Note: Analyses on untransformed and log-transformed scales.

(Biswas et al. 2010). The two groups were selected to have closely similar age distributions but were not individually matched. The first block shows summary statistics for each group on the original untransformed scale. These include the ordinary arithmetic mean (AM), the standard deviation, a 95% confidence interval calculated as in Section 2.1, the sample median, and a 95% reference interval calculated as AM ± 1.960 × SD.

The second block of the data contains corresponding analyses on a log-transformed scale. In this example, logs to the base e (natural logs) are used, although we could equally well use logs to the base 10 or some other figure. As explained in the Appendix 2, statistical analyses involving log transformation for a continuous variable are unaffected by choice of base provided the same base is used consistently throughout. When the analyses are carried out using SPSS or other statistical software, a new column of data is produced, containing log testosterone levels for each individual, then the appropriate summary statistics are produced from this new variable. The 95% confidence limits for the mean log testosterone in South Asians are calculated from this new column in the usual way as $2.7916 \pm 1.9972 \times 0.2871/\sqrt{66}$. Here, 1.9972 is the relevant *t*-value with 65 degrees of freedom.

However, these figures are not interpretable as they stand; we need to transform back to the original scale by exponentiation. The third block contains the back-transformed analyses. The mean on the log scale for the South Asians, 2.7916, transforms back to a geometric mean (GM) $e^{2.7916}$ which is 16.31. The mean on the log scale for the Caucasian group transforms back to $e^{2.9131} = 18.41$. Irrespective of distributional form, the geometric mean is always smaller than the arithmetic mean—or equal only if the variable in question takes the same value for every subject in the sample.

When the median is calculated on the log-transformed scale and transformed back to the original scale, this leads back to the original median, of course. There can be a slight distortion when the sample size is even and the sample median is calculated as midway between the middle two values in the sample.

The lower and upper limits for South Asians also transform back by exponentiation, to $e^{2.7210} = 15.20$ and $e^{2.8621} = 17.50$. The interval midpoint is 16.35, slightly higher than the geometric mean. So the confidence interval for the geometric mean is not symmetrical about the geometric mean, but extends further to the right than the left. Confidence intervals for geometric means are symmetrical on a ratio scale. The two limits may be written $16.31 \div 1.0731$ and 16.31×1.0731. The half-width factor 1.0731 is $e^{1.9972 \times 0.2871/\sqrt{66}}$.

The confidence interval for the geometric mean is shifted to the left, relative to the interval for the arithmetic mean. This is the usual pattern: on any reasonable sample size the confidence interval is quite narrow, and the predominant effect is the shift to the left due to the general inequality GM ≤ AM.

Interval location (Section 4.4) is an important, hitherto largely neglected issue in relation to confidence intervals for proportions and other quantities. This issue has an impact here also, in a rather different way. If the distribution really is lognormal, a $100(1 - \alpha)\%$ interval calculated following log transformation will have left and right non-coverage rates of $\alpha/2$. If we fail to use log transformation, the resulting interval is too far to the right, so the left non-coverage rate will be higher than it should be, and conversely the right non-coverage rate will be too low.

Testosterone, like most physiological measurements, has units, in this case nanomoles per litre (nmol/l). All summary statistics for testosterone levels are expressed in these units, and this applies to the back-transformed analyses in the third block of Table 2.4 as well as the untransformed analyses. The analyses on the log-transformed scale, in the second block, do not have units as such. If we express testosterone levels in micromoles per litre (mmol/l), all figures reported in the first and third blocks of the table are lowered by a factor of 1000. On the log-transformed scale, the means, confidence limits and medians are reduced by ln(1000) which is 6.9078. However, the standard deviations are unaltered.

The primary purpose of the study was to compare testosterone levels between the two groups of men. Table 2.5 presents the comparisons. The untransformed analysis proceeds in the usual manner for data without individual-level pairing, as described in Section 2.2. The South Asians' testosterone levels were on average 2.20 units lower than the Caucasians'. A 95% interval for this difference is 3.94 to 0.45 units reduction, with $t = -2.49$ and $p = 0.014$.

TABLE 2.5

Comparisons of Testosterone Levels (nmol/l) between 66 South Asian and 66 Caucasian Men

	Difference between South Asian and Caucasian Men			
	Mean Difference	**95% Confidence Interval**	***t*-Ratio**	***p*-Value**
Untransformed (nmol/l)	−2.20	−3.94 to −0.45	−2.49	0.014
Natural log-transformed	−0.1215	−0.2192 to −0.0238	−2.46	0.015
	Geometric Mean Ratio	**95% Confidence Interval**	***t*-Ratio**	***p*-Value**
Log-transformed analyses back-transformed to scale of ratios	0.886	0.803 to 0.976	−2.46	0.015

Source: Data from Biswas, M. et al. 2010. *Clinical Epidemiology* 73: 457–462.
Note: Analyses on untransformed and log-transformed scales.

In this instance the comparison on the log scale resulted in an almost identical *t*-test ($t = -2.46$, $p = 0.015$). The estimated difference on a log scale is −0.1215, with 95% limits −0.2192 and −0.0238. When we transform these back by exponentiation, they are interpretable as ratios. We calculate the geometric mean ratio as $e^{-0.1215} = 0.886$. This is identical to the ratio of the geometric means in the two samples, 16.95 ÷ 19.14. When we perform such a division, the numerator and denominator both have the original units, so their ratio is absolute, unit-free. The lower and upper limits transform back to $e^{-0.2192} = 0.803$ and $e^{-0.0238} = 0.976$. These figures are likewise unit-free. So the 95% interval for the geometric mean ratio is from 0.803 to 0.976. These results would normally be reported as a reduction of 100 × (1 – 0.886) = 11.4% in South Asians relative to Caucasians, with a 95% interval from 2.4% to 19.7%. These ratio figures are unaltered by expressing testosterone levels in other units such as micromoles per litre.

The results of a paired comparison following log transformation are transformed back to ratios in exactly the same way, as we will see when we consider Table 2.9.

These analyses are readily implemented using standard statistical software—up to a point. Thus SPSS includes facilities to log-transform variables, using either base e or 10, and to produce analyses on the log-transformed scale. However, it does not have an option to transform the results back. This is most conveniently done by judiciously copying output values into spreadsheet software and performing the necessary manipulations there. Unfortunately, the need to do so greatly increases the scope for error. There is a real danger that analyses performed on a log-transformed scale will be transferred to the relevant tables without any attempt to back-transform or any realisation that the results are absurd.

2.4.2 When Is Log Transformation Indicated?

Often the clearest indication of the relative goodness of fit of Gaussian models on the original and log-transformed scales is obtained by producing quantile-quantile (Q-Q) plots. The Q-Q plot usually provides the most satisfactory holistic assessment of how well the sample data is fitted by a specified statistical model. The histogram is a simpler construct. Whilst sometimes this gives perfectly clear information, as in Figure 2.1, in general histograms have two deficiencies: the eye is drawn to the fine structure of irregularity of column heights, and the apparent pattern, including this irregularity, can alter substantially if either the bin width is changed or the set of bin boundaries is offset by some fraction of the bin width. Also, use of goodness-of-fit hypothesis tests to help decide whether to transform is questionable, for reasons analogous to the argument against ancillary tests to choose between pooled and unpooled variance formulations in outlined in Section 2.2.

Figure 2.3 shows Q-Q plots for the testosterone data for the two ethnic groups combined. In this and subsequent diagrams, left-hand panels relate to the goodness-of-fit of a Gaussian model, and right-hand panels to a log-Gaussian model. If a random sample is taken from the specified distribution, the points on the Q-Q plot will then be randomly scattered close to the diagonal line of identity, very much as in the second panel. If there is systematic departure from the diagonal line, as in the first panel, this indicates a poorer fit.

Similarly, Figure 2.4 shows Q-Q plots for fasting blood glucose levels from the Healthy Hearts study (Richardson et al. 2011). Blood glucose level was the most severely positively skew variable measured in this study. For these

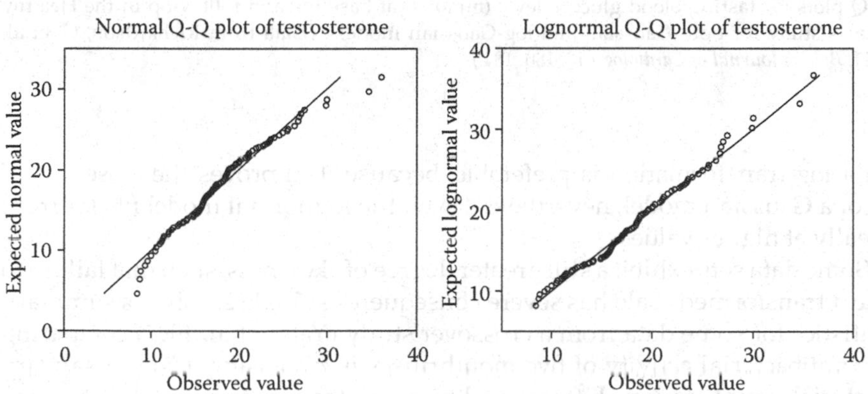

FIGURE 2.3
Q-Q plots for testosterone levels (nmol/l) in a series comprising 66 South Asian and 66 Caucasian men, for both Gaussian and log-Gaussian models. (Data from Biswas, M. et al. 2010. *Clinical Epidemiology* 73: 457–462.)

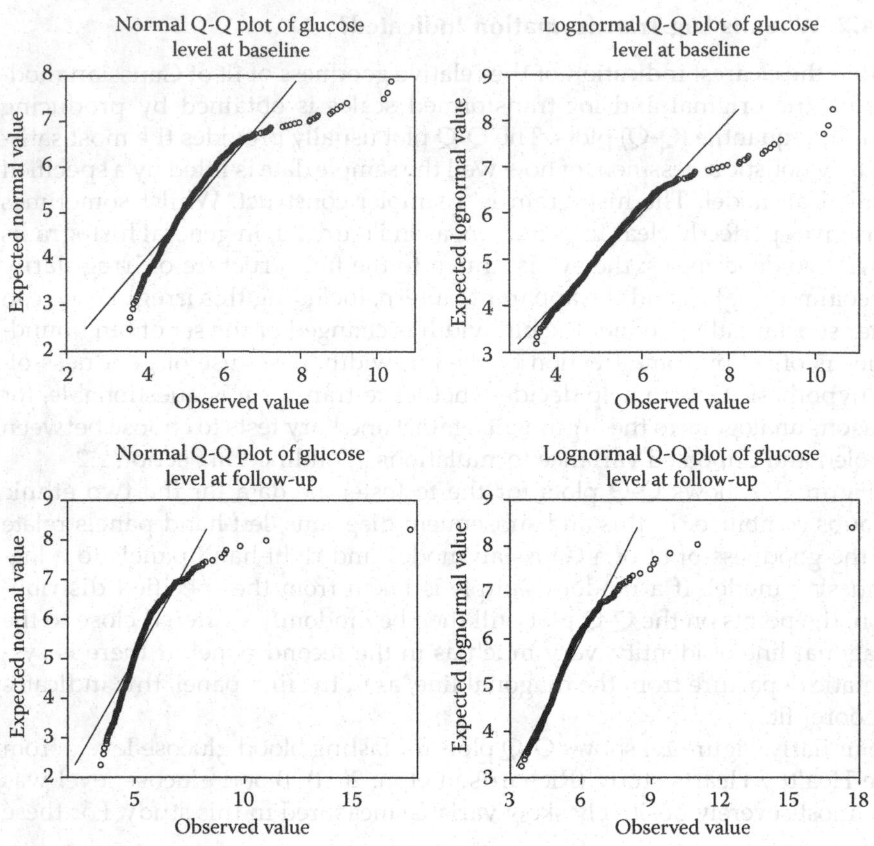

FIGURE 2.4
Q-Q plots for fasting blood glucose level (mmol/l) at baseline and follow-up in the Healthy Hearts study for both Gaussian and log-Gaussian models. (Data from Richardson, G. et al. 2011. *British Journal of Cardiology* 18: 180–184.)

data log transformation is preferable, because it improves the closeness of fit of a Gaussian model; nevertheless even the lognormal model fits far from ideally at higher values.

Some data sets exhibit a still greater degree of skewness, such that failure to use a transformed scale has severe consequences. Table 2.6 shows summary statistics for some data from a crossover study (Yates et al. 1997) comparing the antibacterial activity of five mouthrinses *in vivo* using a 7-hour salivary bacterial count design. Fifteen healthy volunteers were allocated to various sequences of the five products according to a novel Latin square design incorporating balance for first-order carryover (Newcombe 1992b). On each of five study days (separated by 6-day washout periods), saliva samples were

TABLE 2.6

Summary Statistics for Salivary Bacterial Count on Untransformed and Log-Transformed Scales from a Study Evaluating Antibacterial Mouthrinses

Scale	Mean	SD	Median	Minimum	Maximum
Untransformed	463	693	227	8	2574
Log-transformed (base 10)	2.207	0.742	2.357	0.903	3.411

Source: Data from Yates, R. et al. 1997. *Journal of Clinical Periodontology* 24: 603–609.
Note: SD = standard deviation.

taken from each participant immediately before and at several time points after a single use of the test rinse prescribed for that day, and bacterial counts determined. The bacterial count is expressed as the number of colony forming units, in millions, per millilitre of saliva. Usually the distribution of such bacterial counts is very highly positively skew—bacterial colonies tend to grow exponentially until they approach a ceiling size.

The results shown in Table 2.6 relate to samples taken immediately before rinsing with a solution containing acidified sodium chlorite with malic acid, buffered. Even more than for the radiological protection knowledge post-training scores in Table 2.2, examination of the mean and SD is sufficient to imply gross skewness. A 95% "reference range" calculated on the assumption that the distribution is Gaussian would be from −895 to 1821. The lower bound calculated in this way is clearly inadmissible, and not surprisingly the upper bound is also inappropriate. When the salivary bacterial count data are log-transformed (to base 10 in this example), the mean on the transformed scale is 2.207. This transforms back to a geometric mean of $10^{2.207}$ (i.e., 161). (The median on the log scale, 2.357, transforms back to the median on the original scale, $10^{2.357} = 227$, of course.) The 95% reference range based on the log-transformed model is delimited by $e^{2.207 \pm 1.960 \times 0.742}$ i.e., by 5.7 and 4590. These limits give a much more satisfactory summary of the data: the fitted 2.5 and 97.5 centiles are expected to encompass a rather wider range than the original series comprising only 15 results.

2.4.3 Log Transformation for Interrelated Variables

Not only can scale transformation improve conformity to Gaussian distributional form; two other frequently given reasons to transform a scale of measurement are to linearise regression relationships and to improve homogeneity of variance. It is commonly found that when a scale transformation is chosen with one of these issues in mind, it is beneficial in all three respects.

There is a fourth, less widely recognised argument favouring the use of a judiciously chosen scale transformation. It can preserve interrelationships between logically interrelated variables. Usually, as in the examples below, log transformation is the one with the desired property.

The R – R′ interval in seconds is commonly recorded in cardiological studies. The heart rate is often measured in epidemiology, with the units beats/minute. These two measures are, of course, logically interrelated simply by

$$\text{Heart rate} = 60/\text{R} - \text{R}'\ \text{interval}. \qquad (2.6)$$

The two measures are clearly equivalent. So we could express the same set of results in either heart rate or R – R′ interval units. However, the ordinary AM of a group of heart rates is not equivalent to the arithmetic mean but to the harmonic mean (HM) of their R – R′ intervals. If instead we analyse heart rate and R – R′ interval on log-transformed scales, their means (and indeed also confidence intervals) on these scales are totally equivalent, as are the geometric means transformed back to the original scale. The familiar inequality applies: HM ≤ GM ≤ AM, with equality only when all figures being averaged are identical.

Another logically interrelated set of variables comprises height, weight and body mass index (BMI). The BMI is usually defined as the body weight in kilograms divided by the square of the height in metres, and is widely used to express body weight relative to height. In the Healthy Hearts study, for most variables recorded at baseline and follow-up, the obvious approach was to analyse changes at follow-up, expressed either in absolute terms or in relative terms from an analysis on a log-transformed scale. The situation is different for body weight, because height is an obvious established determinant of weight; BMI being a familiar, widely used measure, it is appropriate to report results for BMI as well as weight here. It is advantageous to analyse the three interrelated variables, height, weight and BMI using log transformation, for two reasons.

One is the familiar observation that, in Westernised contemporary populations at any rate, the distribution of body weight, or of weight relative to height, is substantially positively skewed. Figure 2.5 shows Q-Q plots for baseline height, weight and BMI restricted to 738 subjects with follow-up data. Log transformation clearly brings weight and BMI closer to Gaussian distributional form, whilst the distribution of height is fitted well irrespective of whether log transformation is used.

The other reason is compatibility between analyses for related variables. Table 2.7 shows summary statistics for these variables at baseline and fol low-up, on the original and log-transformed scales. BMI was calculated at follow-up using the height at baseline throughout. For these analyses, we use the transformation $x \rightarrow 100\ \ln(x)$, for reasons which will become clear when we examine coefficients of variation. The coefficient of variation is often used to quantify the degree of variation in relative terms. It is defined as the SD divided by the mean, and is usually expressed as a percentage.

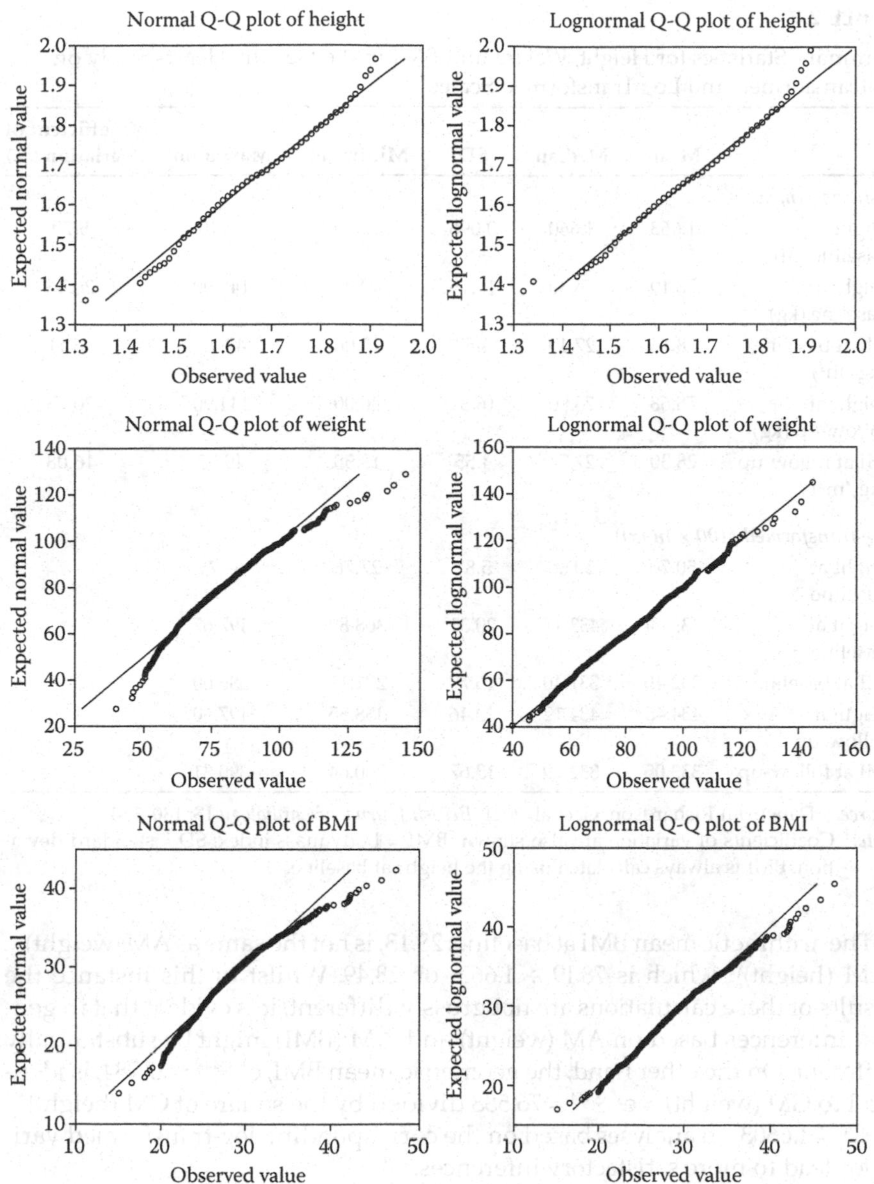

FIGURE 2.5
Q-Q plots for baseline height (m), weight (kg) and BMI (kg/m^2) in the Healthy Hearts study for both Gaussian and log-Gaussian models. (Data from Richardson, G. et al. 2011. *British Journal of Cardiology* 18: 180–184.)

TABLE 2.7

Summary Statistics for Height, Weight and BMI in the Healthy Hearts Study on Untransformed and Log-Transformed Scales

	Mean	Median	SD	Minimum	Maximum	Coefficient of Variation (%)
Untransformed						
Height at baseline (m)	1.663	1.660	0.096	1.320	1.910	5.79
Weight at baseline (kg)	78.19	76.30	16.15	40.00	145.00	20.66
BMI at baseline (kg/m^2)	28.13	27.49	4.53	15.06	47.47	16.11
Weight at follow-up (kg)	78.68	76.80	16.31	36.00	144.90	20.73
BMI at follow-up (kg/m^2)	28.30	27.74	4.55	13.55	49.55	16.08
Log-transformed (100 × ln (x))						
Height at baseline	50.70	50.68	5.81	27.76	64.71	
Weight at baseline	433.84	433.47	20.31	368.89	497.67	
BMI at baseline	332.45	331.40	15.70	271.17	386.00	
Weight at follow-up	434.45	434.12	20.46	358.35	497.60	
BMI at follow-up	333.05	332.29	15.67	260.64	390.31	

Source: Data from Richardson, G. et al. 2011. *British Journal of Cardiology* 18: 180–184.
Note: Coefficients of variation are also shown. BMI = body mass index; SD = standard deviation. BMI is always calculated using the height at baseline.

The arithmetic mean BMI at baseline, 28.13, is not the same as AM (weight) ÷ (AM (height))^2 which is 78.19 ÷ 1.663^2 or 28.49. Whilst in this instance the results of these calculations are not grossly different, it is evident that in general inferences based on AM (weight) and AM (BMI) might be substantially different. On the other hand, the geometric mean BMI, $e^{3.32446}$ = 27.784, is identical to GM (weight) = $e^{4.33844}$ = 76.588 divided by the square of GM (height) = $e^{0.5070}$ = 1.6603, so analyses based on the corresponding log-transformed variables lead to more satisfactory inferences.

How much difference does log transformation make? Table 2.8 relates to the baseline measurements in the 738 subjects who were followed up. Arithmetic and geometric means are computed for height, weight and BMI, with 95% intervals. For height, it makes a relatively small difference whether we log transform or not—whilst the results of the two analyses are not interchangeable, there is no obvious preference. For weight and BMI, choice of scale makes more difference, the geometric mean lies outside the confidence interval for the arithmetic mean and vice versa.

TABLE 2.8

Arithmetic and Geometric Means for Baseline Height, Weight and BMI in the Healthy Hearts Study, with 95% Confidence Intervals

	Arithmetic Mean	95% Confidence Interval
Untransformed		
Height (m)	1.663	1.656 to 1.670
Weight (kg)	78.19	77.02 to 79.36
BMI (kg/m^2)	28.13	27.80 to 28.46
Log-transformed (100 × ln (x))		
Height	50.70	50.28 to 51.12
Weight	433.84	432.38 to 435.31
BMI	332.45	331.31 to 335.58
	Geometric Mean	**95% Confidence Interval**
Log-transformed analyses back-transformed to original scale		
Height (m)	1.660	1.653 to 1.667
Weight (kg)	76.59	75.47 to 77.72
BMI (kg/m^2)	27.78	27.47 to 28.10

Note: BMI = body mass index.

As an example of the analysis of within-subject differences on a log-transformed scale, Table 2.9 examines changes from baseline to follow-up in these 738 individuals. Mean changes and their confidence intervals and so forth are calculated just as in Section 2.3, both for untransformed and log-transformed variables. There is an increase in weight and BMI at follow-up (similar to what is expected as a result of ageing by 1.5 years), with a p-value around 0.001, for weight and BMI alike, irrespective of whether log transformation is used. In the untransformed analyses, the t- and p-values for the weight and BMI analyses are not identical. Following log transformation, analyses for weight and BMI are necessarily identical. The geometric mean

TABLE 2.9

Analyses of Changes at Follow-Up in Body Weight and BMI in the Healthy Hearts Study, on Untransformed and Log-Transformed Scales

	Change at Follow-Up			
Measure	**Mean Difference**	**95% Confidence Interval**	***t*-Ratio**	***p*-Value**
Weight (kg)	0.495	0.223 to 0.767	3.572	0.0004
BMI (kg/m^2)	0.169	0.070 to 0.269	3.339	0.0009
100 × ln (weight)	0.606	0.250 to 0.962	3.341	0.0009
100 × ln (BMI)	0.606	0.250 to 0.962	3.341	0.0009

Note: BMI = body mass index.

weight at follow-up was greater than that at baseline by a factor of $e^{0.006061}$ i.e., 1.0061. The 95% limits are $e^{0.002499} = 1.0025$ and $e^{0.009623} = 1.0097$. In other words, body weight (irrespective of whether adjusted for baseline height) typically increased by 0.61% at follow-up, with 95% limits (0.25%, 0.97%).

In longitudinal studies with incomplete follow-up, systematic differences between subjects with and without follow-up data are of concern. Comparison of baseline characteristics between those with and without follow-up data indicated some statistically significant, nevertheless relatively small differences between these groups. Table 2.10 gives analyses comparing baseline height, weight and BMI between subjects with and without follow-up data, on both untransformed and log-transformed scales. The untransformed scale analysis gives a mean difference with a confidence interval; the log transformed analyses leads to a geometric mean ratio with a confidence interval. For height, the two analyses give very similar but not identical t-ratios and p-values: those followed up tended to be about 0.3 cm or 0.18% taller than those who were not, with a p-value of around 0.5 in both analyses. Differences between untransformed and log-transformed analyses were rather greater for BMI, however. The similarity between untransformed and log transformed analyses for height stems from the fact that the distribution

TABLE 2.10

Comparisons of Height, Weight and BMI on Untransformed and Log-Transformed Scales between 738 Subjects with and 1279 Subjects without Follow-Up Data in the Healthy Hearts Study

	Comparison at Baseline between Groups with and without Follow-Up Data			
	Mean Difference	**95% Confidence Interval**	***t*-Ratio**	***p*-Value**
Untransformed				
Height (m)	+0.003	−0.006 to +0.012	0.67	0.503
Weight (kg)	−0.792	−2.266 to +0.682	−1.05	0.292
BMI (kg/m²)	−0.456	−0.889 to −0.023	−2.06	0.039
Log-transformed (100 × ln(x))				
Height	+0.177	−0.347 to +0.702	0.66	0.508
Weight	−0.986	−2.830 to +0.858	−1.05	0.295
BMI	−1.340	−2.814 to +0.134	−1.78	0.075
	Geometric Mean Ratio	**95% Confidence Interval**	***t*-Ratio**	***p*-Value**
Log-transformed analyses back-transformed to scale of ratios				
Height (m)	1.0018	0.9965 to 1.0070	0.66	0.508
Weight (kg)	0.9902	0.9721 to 1.0086	−1.05	0.295
BMI (kg/m²)	0.9867	0.9723 to 1.0013	−1.78	0.075

Note: BMI = body mass index.

is far removed from zero. This is evidenced by the very low coefficient of variation. For variables with higher coefficients of variation, comparative analyses on untransformed and log-transformed scales may be similar (as is the case for weight here) or somewhat different (which applies to the results for BMI).

For adult human height, the degree of variation is fairly slight, relative to the typical value, so the coefficient of variation is low. The coefficient of variation is closely related to the standard deviation on the log-transformed scale—both measure the degree of variation in proportion to the typical value. The final column of Table 2.7 shows the coefficient of variation for height, weight and BMI at baseline and follow-up. These figures are clearly very similar to the standard deviations shown for the corresponding log-transformed variables, 100 × ln (height) and so forth. The smaller these two indices are, the closer they are to each other, as is apparent from the height results.

2.4.4 Other Scale Transformations

Log transformation is by far the most commonly used scale transformation for continuously distributed data, for two reasons. One is the role of the central limit theorem, which implies that many variables will be well approximated by either a Gaussian or log-Gaussian model. Empirically, many variables are better fitted by a log-transformed model, which often improves homogeneity of variance and linearity of regression relationships in tandem with distributional form. The other reason is that this is the only transformation, amongst those applicable to a continuously distributed variable on $(0, \infty)$, for which comparative results on the transformed scale transform back in an interpretable way.

Some positively skew variables include zeros or small positive values. In this situation logarithmic transformation cannot be applied as it stands: a modified transformation may be used, either $\log(x + c)$ or $\log(\max(x, c))$ for some $c > 0$.

Sometimes a case can be made for another kind of scale transformation. Obviously, analyses such as those described in Sections 2.1 to 2.3 are only altered trivially by any linear transformation of the form $x \to a + bx$. Only non-linear transformations are capable of normalising distributional form. The most important class of non-linear transformations comprises power transformations of the form x^λ for some value of λ. This family may be extended to include the log transformation by defining

$$f(x) = \begin{cases} (x^\lambda - 1)/\lambda & (\lambda \neq 0) \\ \ln(x) & (\lambda = 0) \end{cases} \tag{2.7}$$

(Box and Cox 1964)—a relationship that also plays a key role in characterising interval location in Section 4.4.3. Powers above +1 (or below −1) tend to destabilise the data and are seldom worth considering. Clearly, $\lambda = 1$ is the trivial identity transformation which is of course linear. Obvious non-trivial choices for power transformations are the square root ($\lambda = 1/2$) for the surface area of a structure such as an organ or a lesion, or the cube root ($\lambda = 1/3$) for its volume (or, by the same reasoning, its weight). When such a transformation is used, the mean and confidence limits on the transformed scale transform back straightforwardly into a location measure and an asymmetrical interval on the original scale. However, differences on the transformed scale do not transform back into anything meaningful. In contrast, the point and interval estimates for a difference on the log-transformed scale transform back into a geometric mean ratio and confidence limits for it. The reciprocal transformation, with $\lambda = -1$, is sometimes encountered. As argued above, use of the log transformation may lead to more cogent inferences for a variable whenever this would be a natural reparametrisation. In an individually paired study, scale transformations are usually applied to the raw data, not the paired differences.

Confidence intervals following scale transformation then back-transformed to the original scale, and intervals for differences on a log scale back-transformed to a ratio scale, are important special cases of the substitution method outlined by Daly (Chapter 6). In Daly's 1998 article, published in an epidemiological journal, not surprisingly all the illustrative examples concern quantities related to proportions, but transformation methods for confidence intervals for means and their differences follow the same principle.

2.5 Non-Parametric Methods

In some situations, no transformation of scale is adequate to achieve approximate Gaussian distributional form. Indeed, many variables are inherently ordinal, and there may not be the implication that scales are equally spaced. Consequently, non-parametric methods were developed as robust alternatives to the unpaired and paired *t* methods.

The most important non-parametric methods are the tests for location shift for two independent samples, originating in the work of Wilcoxon (1945) and Mann and Whitney (1947), and for paired samples (Wilcoxon 1945). The history of statistical inference is, unfortunately, littered with ambiguities of terminology. Here, and also in Chapters 15 and 16, following the most common usage we refer to the unpaired test as the Mann-Whitney test even though an equivalent test was suggested earlier by Wilcoxon for the unnecessarily restricted case of equal sample sizes in the two groups. All references to Wilcoxon's test should be understood to relate to the matched-pairs signed-ranks test.

The Mann–Whitney and Wilcoxon procedures are generally regarded as primarily hypothesis tests. Confidence intervals for differences between medians and median differences are available in some statistical software, including Minitab and the software, CIA, distributed with *Statistics with Confidence* (Altman et al. 2000). Unfortunately, they are rather awkward constructs, as illustrated in the forensic psychiatry example (see Section 15.2.1), and consequently are not widely used.

The Mann–Whitney and Wilcoxon procedures are ranking methods. In the Mann–Whitney test comparing two samples of size m and n, the two samples are merged, and the data are represented by their ranks, 1, 2,…, $m + n$ based on the combined series. In the paired Wilcoxon test comparing two series each of size n, the paired differences are calculated, just as in the paired t-test. The absolute values of the differences are ranked from 1 to n, then the sums of ranks associated with positive and negative differences are accumulated.

Non-parametric methods are often conceptualised as invariant under any monotonic transformation of scale or scale-free. In fact the method used in the unpaired case, the Mann–Whitney method, is truly scale-free, whilst the Wilcoxon approach used for the paired case is not quite scale-free—it can yield slightly different results if an algebraic transformation of scale such as log transformation is used first.

Ranking may be regarded as a special kind of non-linear transformation. Unlike the algebraic scale transformations considered in Section 2.4 it is context-dependent, hence there is the potential for the non-transitive behaviour known as the Condorcet paradox, as demonstrated in Section 15.2.3.

2.6 The Effect of Dichotomising Continuous Variables

The practice of dichotomising continuous (and ordinal) variables is widespread. This is acceptable for display purposes, particularly when there is a conventionally accepted threshold value. For example, in the analysis of large perinatal data sets, it is customary to report the proportion of babies weighing less than 2.5 kg, for an entire series of births or for subgroups defined by antecedent factors. However, in comparative analyses, the practice of dichotomising at an arbitrary value, not chosen in the light of the data, results in a serious loss of statistical power.

As an extreme example, in Section 1.2.3 we considered data on the maximum rate at which ambulance staff delivered chest compressions to cardiac arrest patients (Fletcher et al. 2008). Results were compared between 2005, when metronomes were little used, and 2006, when they were commonly used. The mean reduction in maximum compression rate was 13.2 compressions per minute, with 95% interval 8.1 to 18.3 units, $t = 5.1$, $p < 10^{-6}$. A Mann–Whitney test gave $p = 0.0001$. The maximum compression rate recommended is 100 compressions per minute. The proportion of resuscitation attempts with maximum compression rate above 100 reduced from 97/164 (59%) in 2005 to 62/123 (50%) in 2006. A chi-square test based on this dichotomised measure gives $\chi^2 = 2.17$, $p = 0.14$. The particularly great loss of statistical significance here is largely because the distribution of maximum compression rate had a sharp mode at 100 in both years, the majority of values classed as satisfactory were actually at 100. This was not figure preference in recording the data, which was captured automatically by the apparatus used. The fairest interpretation is that, whilst there was not a detectable impact on the proportion of resuscitation attempts involving excessively rapid compressions as such, the introduction of metronomes was followed by a very substantial lowering of the distribution of maximum compression rate. The highest compression rate recorded reduced from 260 per minute in 2005 to 180 in 2006. It is reasonable to regard either of these figures as far more dangerous than a rate of 105 compressions per minute, although if we dichotomise the data at 100 we effectively disregard this.

Sometimes a dichotomy point is selected after inspecting the data. This practice can also have a serious adverse impact on the inferences drawn from the data, by exaggerating the importance of any difference or association in the data.

3

Confidence Intervals for a Simple Binomial Proportion

3.1 Introduction

In this chapter we will consider the simple proportion. This is the basic building block from which most of the measures developed subsequently in this book are derived. We aim to determine the frequency of a binary characteristic in some population. We take a sample of size n. This most often means n individuals—it can also refer to some other kind of unit, but for purposes of exposition, with health-related applications particularly in mind, we will assume that the unit of sampling is the individual person. The standard assumptions are made, namely that sampling is representative, and all units sampled are statistically independent of one another. We find that r out of the n individuals are positive for the characteristic of interest. We report that the proportion is $p = r/n$. This is called the empirical estimate, and is also the maximum likelihood estimate based on the binomial distribution, defined in Section 3.4.4. It serves as the obvious point estimate of the theoretical proportion or population proportion π. The empirical estimate is an unbiased estimate of π in the usual statistical sense that π is the average value expected for p. The larger the sample size n, the better p is as an estimate of π, in the sense of approximating closely to π.

For example, in Chapter 1 we considered the frequency of use of complementary or alternative medicines (CAMs) among $n = 500$ children. The number of children who had used CAMs was $r = 206$. We calculate $p = 206/500 = 0.412$, which we may round to 41%.

Throughout this book, references to proportions should be understood to be of the form described above. This usage must be distinguished from measures such as the proportion (or equivalently percentage) of body weight identified as adipose tissue (fat). Such continuous variables that are expressed in percentage terms are analysed using the methods in Chapter 2.

A proportion may be regarded as the mean of a binary variable. If we give each individual in the sample a score of 1 if they are positive for the characteristic of interest, and 0 if negative, then the total of the n scores is simply r, and their mean is p. So, in our example, $r = 206$ children had used CAMs and $n - r = 294$ had not, so the mean is $(206 \times 1 + 294 \times 0)/500 = 0.412$, which is identical to the proportion.

Accordingly, the reader might imagine that the statistical methodology used to analyse means, described in Chapter 2, could be used to analyse proportions also. As we shall see, this is far from the case. There are two major differences between analysing means of continuous variables and analysing proportions, relating to the issues of boundedness and discreteness.

When we analyse a continuous variable assuming Gaussian distributional form, we are assuming implicitly that the scale on which observations lie is unbounded, extending to infinity in both directions. While this assumption is almost invariably unrealistic for real data, often the true constraints on the support space are remote enough not to distort coverage properties. On the other hand, the support scale for a proportion is bounded: by definition a proportion has to lie between 0 and 1, yet the simplest confidence interval suffers from the deficiency that calculated lower or upper limits can fall outside this range.

Furthermore, in the continuous case, the true population mean μ, the sampled observations $y_1, y_2,\ldots, y_n$ and the sample mean $\bar{y}$ can all take any value on a continuous scale. Likewise, when we are considering proportions, the population proportion π can take any value between 0 and 1. In contrast, the random variable R representing the number of individuals who turn out to be positive can only take a discrete set of possible values $r = 0, 1, 2,\ldots, n - 1$ or n. Consequently, the sample proportion p can only take one of a discrete set of values, $0, 1/n, 2/n,\ldots, (n - 1)/n$ or 1.

As a consequence of discreteness and boundedness, the simplest calculation methods for proportions generally do not perform adequately, and better methods are required. Furthermore, similar issues affect interval estimation for other quantities related to proportions, to a greater or lesser degree. See Newcombe (2008c) for a general discussion of some of the issues in relation to proportions and their differences. The methods recommended throughout this book are designed to obviate the deficiencies of the simplest methods.

3.2 The Wald Interval

The simplest confidence interval for a proportion in scientific use is constructed in much the same way as the confidence interval for a mean, based on the standard error. The formula usually used to calculate the simple asymptotic interval is $p \pm z \times \mathrm{SE}(p)$. Here, p denotes the empirical estimate of the proportion, r/n, with standard error $\mathrm{SE}(p) = \sqrt{(pq/n)}$, q denotes $1 - p$ and z denotes the appropriate centile of the standard Gaussian distribution, 1.960 for the usual two-sided 95% interval.

Although its use is known to go back at least to Laplace (1812), this interval is commonly referred to as the Wald interval (Wald 1943), as in Section 2.1. The Wald test rejects H_0 whenever $|\hat{\pi} - \pi| > z \times \text{SE}(\hat{\pi})$. Here, $\hat{\pi}$ is the maximum likelihood estimate of π, namely $p = r/n$, and $\text{SE}(\hat{\pi})$ is the standard error evaluated at $\hat{\pi}$; namely, $\sqrt{(pq/n)}$ as above. The Wald interval consists of all values of π which would not be rejected by the test.

The maximum likelihood estimator $\hat{\pi}$ tends to a normal distribution with mean π and standard deviation $\sqrt{(\pi(1 - \pi)/n)}$. This is an asymptotic property, derived from the central limit theorem. In most contexts, this would mean that the desired properties are expected to apply as n becomes very large, but the approximation involved may not be adequate for moderate values of n. Here, the issue is more complex, since the approximation can also be poor when π is close to 0 or 1. Accordingly, the Wald interval is an acceptable approximation when the sample size is large and the proportion is well away from 0 or 1. The simplicity and widespread availability of Wald intervals generally in statistical software have resulted in widespread use. But unfortunately, in practice the Wald interval for the binomial proportion generally has very poor properties.

The coverage of the Wald interval is very deficient. The coverage probability (CP) is the probability that the interval encloses the true population value, π. When this is evaluated numerically or by simulation, it is found that the Wald interval is anticonservative, with a very unsatisfactory achieved coverage, often much lower than the nominal coverage level the interval claims to give. In a detailed evaluation (Newcombe 1998a), in which the nominal coverage level $1 - \alpha$ was set at 0.95, the attained coverage was only 0.881 on average, whilst for some values of π, it is close to zero.

The Wald interval is not boundary-respecting, in two respects. Upper and lower limits calculated as $U = p + z \times \text{SE}$ and $L = p - z \times \text{SE}$ can extend beyond 0 or 1. This boundary violation is eliminated by defining the interval as comprising all values of π_0 within the range of validity $0 \leq \pi_0 \leq 1$ which would not be rejected by the Wald test. But this does not obviate the other problem, degeneracy when $r = 0$ or $r = n$.

Location is also very unsatisfactory. The non-coverage probabilities in the two tails are very different. The location of the interval is too distal—too far out from the centre of symmetry of the scale, 0.5. The non-coverage of the interval is predominantly mesial. The deficiencies of the Wald interval in regard to coverage and location are considered in greater detail in the evaluation in Section 5.2.

Numerous publications attest to the poor performance of the Wald interval. It is often held that its shortcomings arise because it is an asymptotic method, and the normal approximation to the binomial distribution breaks down when small numbers are involved. But other asymptotic interval methods are available, such as the Wilson (1927) score interval and also beta intervals as described below, which perform very well. It seems preferable to regard all of the shortcomings of the Wald interval as arising from the fact that it is based on the simplest Gaussian approximation, leading to a simple symmetrical form. As we will see in Section 3.3, something different is called for, particularly when r or $n - r$ is small.

In Chapter 2 we developed a Wald interval for the mean of a variable that has a continuous distribution, based on assumed Gaussian distributional form. An interval of the form $\bar{y} \pm z \times \text{SE}$ with $\text{SE} = \sigma/\sqrt{n}$ would be appropriate if the population standard deviation σ is given. But in practice this is virtually never the case, so we use a slightly wider interval $\bar{y} \pm t \times \text{SE}$, where $\text{SE} = s/\sqrt{n}$ is based on the sample standard deviation, s. The standard error of a proportion is estimated as $\text{SE}(p) = \sqrt{(p(1-p)/n)}$, as above. Given the sample size n, this is completely determined by the value of the sample proportion p, there is no additional dispersion parameter to estimate. This is why the Wald formula involves the simple multiplier z, and not t. The confidence intervals described in the rest of this book are calculated using either the z-value or the desired coverage $1 - \alpha$ itself.

The Wald interval has the usual property that interval width is proportional to $1/\sqrt{n}$, implying that, as in Section 2.1, a fourfold increase in sample size is required to halve interval width.

For the CAM use example, $p = 206/500 = 0.412$, so $\text{SE}(p) = \sqrt{(0.412 \times 0.588/500)} = 0.022$. Then the Wald 95% interval is calculated as $0.412 \pm 1.960 \times 0.022$ i.e., 0.369 to 0.455. A 99% interval would be rather wider, $0.412 \pm 2.576 \times 0.022$ or 0.355 to 0.469. In this instance, with a fairly large denominator and a proportion somewhere in the middle of the range from 0 to 1, the Wald interval is quite satisfactory, and the alternative methods described below yield very similar intervals. It is when numbers are small that the deficiencies of the Wald method become evident.

3.3 Boundary Anomalies

Tuyl (2007, p. 17) argues that "the most important property of a method is that it produces sensible intervals for any possible data outcome," quoting Jaynes (1976, p. 178): "The merits of any statistical method are determined by the results it gives when applied to specific problems."

Turnbull et al. (1992) tested saliva samples from 385 people within 3 months of release from prison and reported that 19 (4.9%) were positive for human immunodeficiency virus (HIV) antibodies. They reported confidence intervals for the proportions who were antibody-positive in the whole series and in several subgroups. Table 3.1 shows these intervals calculated by two methods: Wald intervals as reported by Turnbull et al. and Clopper–Pearson (1934) exact intervals as subsequently suggested by Newcombe (1992a).

TABLE 3.1

Prevalence of HIV Antibodies in Selected Groups of Ex-Prisoners

		HIV Antibody-Positive		
			95% Confidence Interval (%)	
Group	**Number Tested**	**Number (%)**	**Wald**	**Clopper–Pearson**
Injectors:	148	15 (10.1)	5.3 to 15.0	5.8 to 16.2
Men	103	8 (7.8)	2.6 to 12.9	3.4 to 14.7
Women	45	7 (15.6)	5.0 to 26.1	6.5 to 29.5
Non-injecting women	29	1 (3.4)	−3.1 to 10.1	0.1 to 17.8
Homosexual/ bisexual men	20	0	0.0 to 0.0	0.0 to 16.8
Others	188	3 (1.6)	−0.2 to 3.4	0.3 to 4.6

Source: Data from Turnbull, P. J., Stimson, G. V. and Dolan, K. A. 1992. Prevalence of HIV infection among ex-prisoners in England. *British Medical Journal* 304: 90–91.

Comparing the two sets of intervals, we note that the Wald method often produces boundary anomalies when numbers are small. It produces a zero width interval when $r = 0$, and can yield a lower limit below 0 for small non-zero values of r. Furthermore, in both cases the calculated Wald lower and upper limits are lower than the corresponding Clopper–Pearson limits—a difference in interval location, an issue we return to in Section 4.4.

It is clear from the formula for SE(p) than when $p = 0$ (and also when $p = 1$), the standard error imputed to the proportion is zero. Consequently the Wald method produces a degenerate or zero width interval (ZWI). This occurs irrespective of whether we choose $z = 1.960$ for the usual 95% interval, or some higher value such as 3.291 for a 99.9% interval. The context here makes it particularly obvious why it would be absurd to treat this as a meaningful confidence interval for the prevalence in this group. Homosexual and bisexual males were the classical risk group for HIV infection, and there is no reason to expect prisoners to be exempt. If HIV infection has ever been reported in the population that we are considering, then a zero upper limit is unacceptable—we know that the true proportion cannot be zero. The fact that we failed to find any positives in a small series does not imply that there never would be any.

Conversely, overshoot or boundary violation occurs when the calculated lower limit is below 0. For a 95% interval, this occurs whenever $r = 1$ or 2, and also when $r = 3$ except when $n < 14$. For an example, see the analysis for non-injecting women in Table 3.1. In this situation, it is often recommended to truncate the interval to lie within the interval [0, 1] – the limits would then be reported as $\max(0, L) = 0$ and $\min(1, U) = 0.101$. Truncation cannot affect coverage properties. Nevertheless this practice is unsatisfactory, a poor workaround when perfectly practicable methods are available that produce sensible intervals. We know that π cannot be zero here, so 0 is uninterpretable as a lower limit. Closely adjoining values are also highly implausible. An anonymous reviewer of a paper suggested using an interval of the form $0 < \pi \le U$ in this context, but this too is unsatisfactory: the subtle distinction from $0 \le \pi \le U$ would be lost on a non-mathematician user, and it is reasonable to want to quote tangible values as lower and upper limits. What is required here is a lower limit a little above zero, determined by an evidence-based algorithm. Notwithstanding the argument in Section 1.2.1 that it is preferable to consider the interval as a range of plausible values rather than the two bounds themselves; nevertheless, it is reasonable to report a non-zero lower limit in this situation.

Similar anomalous behaviour in relation to the upper boundary of the support space, 1, occurs for proportions at or close to 1, when r is equal to or close to n. It is desirable that the sensitivity and specificity of a screening or diagnostic test (Chapter 12) should be very high, and Wald intervals are equally inappropriate for high proportions such as these. All the confidence interval methods considered in this chapter are equivariant (Blyth and Still 1983), meaning that limits for $n - r$ out of n are the complements of those for r out of n.

Furthermore, when we consider derived quantities in later chapters, additional, more subtle types of boundary violation and other anomalous behaviours sometimes appear, such as latent overshoot or non-monotonicity.

3.4 Alternative Intervals

In recognition of the deficiencies of the Wald interval, a wide variety of alternative intervals for proportions have been proposed. A bibliography by Khurshid and Ageel (2010) lists numerous publications on confidence intervals for both the binomial proportion and the closely related Poisson parameter, and quantities derived from them. Most methods described in this chapter derive from frequentist theory, and were designed primarily to give improved coverage. Most also obviate or greatly reduce problems of degeneracy and boundary violation. When a boundary-respecting method is used such as Clopper–Pearson (1934) in Table 3.1, the lower limit is a small positive value when r is small, and when $r = 0$, the calculated upper limit is positive. Bayes intervals also have appropriate performance and are boundary-respecting.

3.4.1 The Continuity-Corrected Wald Interval

A continuity correction (CC) is often recommended to improve the coverage of the Wald interval. The interval is simply widened at each end by 1/(2n), becoming $p \pm \{z\sqrt{(pq/n)} + 1/(2n)\}$. Doing so naturally increases the coverage probability, although not always to the right degree. For example, for two of the problematic proportions in Table 3.1, 0 out of 20 and 1 out of 29, the continuity-corrected Wald upper limits are 0.025 and 0.118, respectively, which fall well short of those calculated by the score method (0.161 and 0.172) or the exact method (0.168 and 0.178).

Equally naturally, improved coverage is attained at the cost of increased average interval width, by exactly $1/n$. However, the lengthening produces additional instances of boundary violation. Indeed, when $r = 0$, instead of a degenerate interval we get an interval from $-1/(2n)$ to $1/(2n)$, which overshoots the boundary at 0. If, as suggested above, we regard truncation of overshoot as unsatisfactory, we must regard simple additive-scale symmetry about the empirical estimate $p = r/n$ as an undesirable property, which lies at the root of the problems attaching to the Wald and similar methods. Most of the methods introduced below forego such symmetry, and achieve the boundary-respecting property as a consequence.

3.4.2 The Delta Logit Interval

The **logit** transformation of scale is widely used in statistical analyses relating to proportions, notably **logistic regression**, a modelling method described briefly in Chapter 11. The logit of a proportion π is defined as $\ln(\pi/(1-\pi))$. The standard error of logit(p) may be derived from the standard error of p by using a widely applicable general approach known as the delta method. Let $f(\pi)$ denote a function of π with suitable regularity properties. Then a delta interval for $f(\pi)$ is based on an estimate for the standard error of $f(p)$ derived from SE(p) and $\left|\frac{df}{d\pi}\right|$ evaluated at $\pi = \hat{\pi}$. As in Section 3.2, by the central limit theorem the maximum likelihood estimator $\hat{\pi} = p = r/n$ has an asymptotic normal distribution with mean π and standard deviation $\sqrt{(\pi(1-\pi)/n)}$. Let $f(\pi) = \text{logit}(\pi) = \ln(\pi/(1-\pi))$. Then $\frac{df}{d\pi} = \frac{1}{\pi} + \frac{1}{1-\pi} = \frac{1}{\pi(1-\pi)}$, which when evaluated at $\pi = \hat{\pi}$ is $\frac{1}{pq}$.

The standard error of logit(p) is then given by

$$\left|\frac{df}{d\pi}\right| \times \text{SE}(p) = \frac{1}{pq} \times \sqrt{\frac{pq}{n}} = \frac{1}{\sqrt{npq}} \tag{3.1}$$

The delta limits for logit(π) are then $\ln(p/q) \pm z\,(npq)^{-1/2}$. These limits may be transformed back to the proportion scale, leading to the delta logit limits for π

$$1/\{1 + \exp\{-\ln(p/q) \pm z\,(npq)^{-1/2}\} \tag{3.2}$$

These limits are not identical to the Wald limits for π.

Intervals derived in this way are often used for ratio measures comparing two independent samples, namely ratios of proportions (Section 10.4.1), ratios of rates (Section 10.9) and odds ratios (Section 11.3.1). The general properties of this approach are discussed in detail by Agresti (2002, p. 74–75) and Fleiss, Levin and Paik (2003, p. 36–37).

Like the Wald approach, the delta method is asymptotic, and is not guaranteed to perform satisfactorily when n is small or π is close to 0 or 1. The above formulae obviously fail to cope with the cases $p = 0$ or 1, and none of several modifications suggested for this situation are satisfactory (Newcombe 2001a). Inability to produce a meaningful interval in boundary cases must be regarded as a serious flaw. While the delta method logit interval cannot be recommended in its own right, nevertheless it has interesting links to several other intervals (Section 10.9).

3.4.3 The Wilson Score Interval

The simplest acceptable alternative to the Wald interval is the score interval (Wilson 1927). The score limits are also asymptotic and are closely related to the Wald formula. But they incorporate a subtle modification which results in greatly improved performance even when n is small or p or q is at or near zero. The standard error imputed to p is $\sqrt{(\pi(1-\pi)/n)}$ instead of $\sqrt{(p(1-p)/n)}$. We obtain lower and upper limits by solving for π the equation

$$p = \pi \pm z \times \sqrt{(\pi(1-\pi)/n)} \tag{3.3}$$

A confidence interval for any parameter θ delimited by the two roots of an equation $\tilde{\theta} = \theta \pm z \times \text{SE}$, where $\tilde{\theta}$ denotes an empirical estimate of θ and SE denotes its standard error, may be termed an inversion interval. Inversion is a fruitful method to derive well-behaved intervals, and is also used in Sections 6.2, 15.2, 15.3 and 16.4.

For the binomial proportion, we solve the quadratic equation

$$\pi^2 (1 + z^2/n) - \pi (2p + z^2/n) + p^2 = 0 \tag{3.4}$$

with two solutions U and L given by

$$\{2r + z^2 \pm z\sqrt{(z^2 + 4npq)}\}/\{2(n + z^2)\} \tag{3.5}$$

where as usual q denotes $1 - p$. These are the upper and lower Wilson limits. Such closed form methods are easily programmed using any spreadsheet or programming language.

The Wilson interval cannot produce boundary anomalies: when $r > 0$, the calculated lower limit is always positive, whilst when $r = 0$, the lower limit is 0 and the upper limit reduces to $z^2/(n + z^2) > 0$. It is not symmetrical on an additive scale, except of course when $p = 0.5$, the centre of symmetry of the scale, in which situation all the intervals described here (apart from intervals based on the Poisson distribution) are symmetrical because they are equivariant. Instead, the Wilson interval is symmetric on a logit scale (Newcombe 1998a), a property which is all the more remarkable in that it remained unnoticed until 1994. This is an appealing alternative form of symmetry, considering that Agresti and Min (2001) identify the logit as the natural parameter for the binomial distribution.

The logit scale symmetry property is trivial when $p = 0$ (leading to $L = 0$) or $p = 1$ (whence $U = 1$). Away from these boundary cases, it means that

$$\{U/(1-U)\}\{L/(1-L)\} = \{p/(1-p)\}^2. \tag{3.6}$$

This result is easily demonstrated either numerically or algebraically as follows. The product of the roots of the quadratic equation 3.4 defining L and U is $LU = p^2/(1 + z^2/n)$. Similarly, by the equivariance property, lower and upper limits for $1 - \pi$ are $1 - U$ and $1 - L$ such that $(1 - L)(1 - U) = q^2/(1 + z^2/n)$. Since $1 + z^2/n$ is strictly positive, Equation 3.6 follows directly by division.

The Wilson limits are simple algebraic functions of n, r and z, and are thus flexible for use in deriving confidence intervals for related quantities. In particular, in Chapter 7 we introduce the square-and-add method which can be used to derive a confidence interval for $\pi_1 - \pi_2$ from intervals for π_1 and π_2. This procedure can be further developed to construct intervals for differences of differences of proportions (Section 14.2), and to compare sensitivity and specificity simultaneously between two diagnostic or screening tests (Section 12.5). The Wilson limits, which involve square roots, are the obvious candidates for such processes, rather than beta limits such as Clopper–Pearson or Bayes ones. Also, a conditional interval for the odds ratio for paired data (Section 11.6) may be derived from a Wilson interval. The resulting interval for the odds ratio is then symmetrical on a multiplicative scale, as is often appropriate for such ratio measures.

3.4.4 The Clopper–Pearson Exact Interval

The intervals described in the preceding sections are asymptotic intervals, involving approximations to Gaussian distributional form which work best for large numbers. The rationale for the so-called exact interval (Clopper and Pearson 1934) is quite different. It is designed to be strictly conservative: it ensures that the coverage probability will be at least the nominal level, $1 - \alpha$, for any value of the population proportion π between 0 and 1.

The term "exact" is used in a way that can be highly confusing, for reasons considered in Section 4.2.1. The distinctive characteristics of methods labelled as exact are that they are based directly on tail areas of the binomial distribution which underlies how proportions operate, and that these tail areas are accumulated in a particular way.

In the usual notation, n denotes the size of the sample, π denotes the true population proportion, and R denotes the random variable representing how many of the n individuals sampled are positive. Then the probability that R takes a particular value, j, is

$$p_j = \Pr\left[R = j\right] = \binom{n}{j} \pi^j (1-\pi)^{n-j} \tag{3.7}$$

for $j = 0, 1, \ldots, n$.

The binomial coefficient $\binom{n}{j}$ is defined in the usual way as $\frac{n!}{j!(n-j)!}$, where $n! = n \times (n-1)!$ for $n \geq 1$ and $1! = 0! = 1$.

If r of the n individuals is positive, the lower and upper limits are the solutions of

$$\sum_{j:r \leq j \leq n} p_j = \alpha/2 \tag{3.8}$$

and

$$\sum_{j:0 \leq j \leq r} p_j = \alpha/2 \tag{3.9}$$

respectively.

These binomial cumulative probabilities are closely related to beta distribution probabilities, as may be demonstrated by integration by parts. Consequently, in practice the interval endpoints are calculated using the beta distribution as

$$L = QB(r, n-r+1, \alpha/2) \text{ and } U = QB(r+1, n-r, 1-\alpha/2) \tag{3.10}$$

(Julious 2005). Here, $QB(a, b, q)$ denotes the q quantile ($0 \leq q \leq 1$) of the beta distribution with parameters a and b, such that

$$\int_0^{QB} \frac{\Gamma(a+b)}{\Gamma(a)\Gamma(b)} x^{a-1}(1-x)^{b-1}\, dx = q \tag{3.11}$$

But we must set $L = 0$ when $r = 0$, and $U = 1$ when $r = n$.

So, for example, if $r = 0$, $L = 0$, and U satisfies $p_0 = (1-\pi)^n = \alpha/2$, hence $U = 1 - (\alpha/2)^{1/n}$. Similarly, if $r = n$, $U = 1$ and $L = (\alpha/2)^{1/n}$. So the Clopper–Pearson 95% interval for 0 out of 10 is 0 to 0.308, and the 95% interval for 10 out of 10 is 0.692 to 1.

The Clopper–Pearson interval ensures strict conservatism by imposing a stronger constraint, that neither the left nor the right non-coverage probability can exceed $\alpha/2$. This results in an interval that is unnecessarily wide to achieve the desired coverage, in other words over-conservative. This issue is considered in depth in Section 4.2. One suggestion for shortening the Clopper–Pearson interval (Hanley and Lipmann-Hand 1983) is that in the boundary cases $r = 0$ and $r = n$ the α value (or the equivalent z value) could be replaced by that for a one-tailed interval which will consequently be shorter. The argument against doing so is that we normally seek to report a conventional margin of error extending in both directions from the point estimate whenever possible, with use of the same formula for all values of r, hence in effect z, not $1 - \alpha$, is the invariant. For Wilson and other intervals that involve calculation using z, exactly the same z is used in boundary and non-boundary cases; such intervals are not over-conservative, and shortening in boundary cases would impair their performance.

Indeed, the suggestion to shorten the Clopper–Pearson interval in boundary cases does not go far enough: the interval may be shortened in a related way for all values of r, not merely the boundary ones, resulting in the mid-p interval below.

3.4.5 The Mid-*p* Interval

The mid-p accumulation of tail areas (Lancaster 1949; Stone 1969; Berry and Armitage 1995) was developed to help mitigate the known conservatism of statistical procedures such as the Fisher exact test and the Clopper–Pearson interval that are based on exact accumulations of tail areas. Specifically, the mid-p lower and upper limits L and U are such that

$$\text{(i) if } L \le \pi \le p, \quad kp_r + \sum_{j:r<j\le n} p_j \ge \alpha/2 \tag{3.12}$$

$$\text{(ii) if } p \le \pi \le U, \quad \sum_{j:0\le j<r} p_j + kp_r \ge \alpha/2 \tag{3.13}$$

Here, substituting $k = 0.5$ defines the mid-p limits. If we put $k = 1$, we revert to the Clopper–Pearson limits.

The mid-p interval is contained within the Clopper–Pearson interval, and has coverage and width properties fairly similar to the Wilson score interval. If $r = 0$, $L = 0$ and $U = 1 - \alpha^{1/n}$. Similarly, if $r = n$, $U = 1$ and $L = \alpha^{1/n}$. These are identical to the limits suggested by Hanley and Lipmann-Hand. However, in non-boundary cases the mid-p limits are computationally complex, requiring iterative solution—unlike the Clopper–Pearson ones, no simplification using the inverse beta function is applicable here.

3.4.6 Bayesian or Beta Intervals

As an alternative to frequentist methods such as those considered above, Bayesian credible intervals may be used. For a comprehensive review of their application to proportions and related quantities, see Agresti and Hitchcock (2005). Bayesian intervals may be chosen because of a conceptual preference for the Bayesian inferential paradigm, or in order to incorporate prior information explicitly. However, when viewed according to frequentist criteria, Bayesian intervals have favourable properties: they are boundary-respecting, and when an uninformative prior is used, align mean coverage closely or in some circumstances exactly with $1 - \alpha$ (Carlin and Louis 1996; Agresti and Min 2005a). Bayesian methods are also used to obtain credible intervals for parameters in much more complex situations. Often Markov chain Monte Carlo (MCMC) methods provide a readily available solution when no computationally feasible frequentist solution is available.

Lindley (1965) and others have observed that for Bayesian intervals, there are infinitely many ways to "spend" the nominal α non-coverage. The choice affects interval location predominantly, although it also has some effect on coverage. Two obvious criteria are highest posterior density (HPD) and equal tails (ET). Agresti and Min (2005a) noted the drawback that HPD intervals are not invariant under monotone scale transformation. This makes them less suitable for quantities such as the single proportion, for which favourable alternative transformed scales are available; namely, logit and angular as in Section 3.4.8. Moreover, it is the ET, not the HPD interval that corresponds to the desideratum that a frequentist interval should have right and left non-coverage each $\alpha/2$, and usually MCMC leads to ET intervals. (The ET criterion is achieved exactly with respect to the posterior distribution, but this does not necessarily imply that the attained non-coverage probabilities in a frequentist sense will be equal.)

It is well known that prior distributions chosen from the beta family have the advantage of being conjugate for inferences concerning a single proportion. Accordingly, a Bayesian interval for a proportion is usually calculated using a beta prior with parameters a and b, denoted by $B(a, b)$. Most often a symmetrical, uninformative, diffuse prior is used with $a = b$, either uniform $B(1,1) \equiv U[0,1]$, or $B(0.5, 0.5)$, after Jeffreys (1946). The posterior distribution is then $B(r + a, n - r + b)$. This leads to a point estimate (either the mean or the median) which is shrunk mesially towards the scale midpoint 0.5. Agresti and Hitchcock (2005) identify the uniform prior with Laplace (1774), leading to a posterior mean $(r + 1)/(n + 2)$ as perhaps the first example of a shrinkage estimator.

Apropos these priors, at first sight, the Jeffreys prior, with density $\to \infty$ as $\pi \to 0$ or 1, looks as if it would push a proportion further away from 0.5, but this is illusory—it simply does not pull mesially as strongly as $U[0,1]$—the fallacy is to imagine that multiplying the likelihood by the $U[0,1]$ density, which is constant at 1, would leave a proportion at the empirical estimate.

In the notation of Section 3.4.4, the tail-based limits may be written

$$L = QB(r + a, n - r + b, \alpha/2) \text{ and } U = QB(r + a, n - r + b, 1 - \alpha/2). \quad (3.14)$$

In fact the Clopper–Pearson interval contains both the $B(1,1)$ interval (Walters 1985) and also the $B(0.5, 0.5)$ interval. For both uniform and Jeffreys priors, when $r = 0$ the HPD lower limit is zero but the ET lower limit is positive.

Of course, an informative prior may be chosen, to incorporate prior information regarding π explicitly. This would most often be asymmetrical, in which case it could pull point and interval estimates in either a mesial or a distal direction. This contrasts with use of an uninformative prior which is generally symmetric about the midpoint and thus pulls mesially. Informative priors usually also take the beta form, $B(a, b)$, with a and b generally unequal and not small. Use of a restricted, non-conjugate prior such as $U[0, 0.1]$ is quite practicable using MCMC, but this has a disadvantage similar to the conventional argument that disfavours one-sided hypothesis tests (Ruxton and Neuhäuser 2010): supposing that on the contrary the data strongly suggested $\pi \gg 0.1$, what then? With this as well as computational convenience in mind, we will restrict attention to beta priors.

In view of the favourable performance of Bayesian intervals in a frequentist sense, they should be regarded as confidence interval methods in their own right, with theoretical justification rooted in the Bayesian paradigm but pragmatic validity from a frequentist standpoint. Bayes intervals with a conjugate beta prior and hence a beta posterior may thus be termed beta intervals, a term that also encompasses the Clopper–Pearson interval. These intervals are readily calculated using software for the incomplete beta function, which is included in statistical packages and also spreadsheet software including MS Excel. As such, they should now be regarded as computationally of closed form, although less transparent than Wald methods.

3.4.7 Modified Wald Intervals

The empirical estimate p is an unbiased estimate of the population proportion π, in the usual statistical sense that the expectation of p given n, $E[p|n]$ is π. However, when $r = 0$, many users of statistical methods are uneasy with the idea that $p = 0$ is an unbiased estimate. The range of possible values for π is the interval from 0 to 1. Generally, this means the open interval $0 < \pi < 1$, not the closed interval $0 \le \pi \le 1$, as usually it is already known that the event sometimes occurs and sometimes does not. As the true value of π cannot then be negative or zero, but must be greater than zero, the notion that $p = 0$ should be regarded as an unbiased estimate of π seems highly counter-intuitive.

Largely with this issue in mind, alternative estimators known as shrinkage estimators are available. The commonest form of shrinkage estimator is $p_\psi = (r + \psi)/(n + 2\psi)$ for some chosen pseudo-frequency $\psi > 0$. Essentially,

ψ observations are added to the number of "successes" and also to the number of "failures". The resulting estimate p_ψ is intermediate between the empirical estimate $p = r/n$ and 0.5, which is the midpoint and centre of symmetry of the support scale from 0 to 1. The degree of shrinkage in a mesial direction towards 0.5 is great when n is small and minor for large n. Bayesian analyses of proportions also lead naturally to shrinkage estimators, with $\psi = 1$ and 0.5 aproximating to the most widely used uninformative conjugate priors, the uniform prior $B(1,1)$ and the Jeffreys prior $B(0.5, 0.5)$ as described above.

The midpoint of the score interval on the ordinary additive scale is a shrinkage estimator with $\psi = z^2/2$, which is 1.92 for the default 95% interval. With both this and Bayesian intervals in mind, Agresti and Coull (1998) proposed a modified Wald interval of the form

$$p_\psi \pm z \sqrt{\{p_\psi(1 - p_\psi)/(n + 2\psi)\}} \tag{3.15}$$

where ψ is set to 2, irrespective of the nominal coverage level. This is also a great improvement over the Wald method, and is computationally and conceptually very simple. It reduces but does not eliminate the boundary violation problem: zero width intervals cannot occur, but overshoot occurs with zero or small values of r or $n - r$. A variety of alternatives can be formulated, with different choices for ψ (for example, $\psi = 1$ is appropriate when comparing two proportions as in Chapter 7), or using some modification to n other than $n + 2\psi$ as the denominator of the variance.

3.4.8 Other Intervals

In Section 3.1 we identified the empirical estimate $p = r/n$ with the maximum likelihood estimate $\hat{\pi}$. The latter is defined as the value of π that maximises the likelihood function Λ, which for a binomial variate is proportional to $\pi^r(1 - \pi)^{n-r}$. A very general approach to interval estimation involves identifying a range of values over which $2 \ln \Lambda$ is not less than z^2 short of the maximum value attained when $\pi = \hat{\pi}$. When this approach is applied to estimation of the simple binomial proportion, the resulting likelihood interval is fully boundary-respecting, with very favourable location properties. Unfortunately it is slightly anticonservative on average. We will see in Chapters 7 and 8 that this also applies to related intervals for unpaired and paired differences of proportions based on a pure profile likelihood approach to eliminate nuisance parameters.

One of the difficulties besetting interval estimation for quantities such as proportions is that the variance is not constant across the range of permissible parameter values. In some interval estimation contexts, use of a variance stabilising transformation leads to a computationally simple asymptotic interval, which performs much better than imputing a standard error to the parameter on its original scale. Variance stabilising transformation based intervals are usually closed form. The angular or arcsin (inverse sine) transformation is recognised to be variance stabilising for proportions. Several arcsin intervals have been suggested, the simplest being

$$\left(\sin^2\left\{\max\left(-\frac{\pi}{2}, \sin^{-1}\left(\sqrt{p}\right)-\frac{z}{2\sqrt{n}}\right)\right\}, \sin^2\left\{\min\left(\frac{\pi}{2}, \sin^{-1}\left(\sqrt{p}\right)+\frac{z}{2\sqrt{n}}\right)\right\}\right) \tag{3.16}$$

When defined with care as above, this interval is properly boundary respecting; a continuity correction may be incorporated if greater conservatism is desired.

The mid-p interval represents one approach to shortening the Clopper–Pearson interval, by aiming to align mean coverage with $1 - \alpha$. Several other intervals have been developed which are labelled shortened intervals (Blaker 2000; Reiczigel 2003), which are designed to maintain minimum coverage at $1 - \alpha$ but substantially shrink interval length. The Clopper–Pearson is the shortest interval that guarantees both right and left non-coverage will be at most $\alpha/2$ for all values of π. But a considerable reduction in interval width can be achieved if we impose the less strong condition that the overall non-coverage should be at most α. However, shortened intervals are much more complex, both computationally and conceptually. They also have the disadvantage that what is optimised is the interval, not the lower and upper limits separately. Consequently they are unsuitable when interest centres on one of the limits rather than the other, and have erratic and sometimes anomalous location properties, as demonstrated in Table 6 of Newcombe (2011b).

When r is small or zero, an interval for $p = r/n$ is often constructed by assuming that the numerator has a Poisson distribution. In this situation, the Poisson approximation is greatly preferable to the Gaussian approximation inherent to the Wald interval, and enables mental calculation of approximate limits in some cases. Nevertheless this procedure cannot be recommended, for reasons explained in Section 6.1.1.

3.5 Algebraic Definitions for Several Confidence Intervals for the Binomial Proportion

In this section we bring together the formulae for two-sided $100(1 - \alpha)\%$ confidence intervals for the simple binomial proportion $p = r/n$ calculated by 13 methods. These intervals are shown for several exemplar datasets in Table 3.2, with the boundary violations of Wald and related intervals truncated. They are the ones used in an evaluation (Newcombe 2007c) which was designed primarily with interval location in mind. The methodology and

TABLE 3.2

Confidence Limits for Several Illustrative Proportions Calculated by 13 Methods as Defined in Section 3.5

Method		1	2	3	4	5	6	7	8	9	10	11	12	13
r	*n*	Wald	Wald CC	Wilson	Wilson CC	Clopper–Pearson	Mid-*p*	Wald +1	Wald +2	Wald + 3	Uniform TB	Uniform HPD	Jeffreys TB	Jeffreys HPD
0	10	0	<0[a]	0	0	0	0	<0[a]	<0[a]	<0[a]	0.0023	0	0.00005	0
		0[a]	0.0500	0.2775	0.3445	0.3085	0.2589	0.2397	0.3262	0.3788	0.2849	0.2384	0.2172	0.1708
1	10	<0[a]	<0[a]	0.0179	0.0052	0.0025	0.0050	<0[a]	<0[a]	0.0378	0.0228	0.0063	0.0110	0.0004
		0.2859	0.3359	0.4042	0.4588	0.4450	0.4035	0.3775	0.4292	0.4622	0.4128	0.3675	0.3813	0.3308
5	10	0.1901	0.1401	0.2366	0.2014	0.1871	0.2120	0.2171	0.2381	0.2550	0.2338	0.2338	0.2235	0.2235
		0.8099	0.8599	0.7634	0.7986	0.8129	0.7880	0.7829	0.7619	0.7450	0.7662	0.7662	0.7765	0.7765
0	250	0	<0[a]	0	0	0	0	<0[a]	<0[a]	<0[a]	0.0001	0	0.000002	0
		0	0.0020	0.0151	0.0189	0.0146	0.0119	0.0117	0.0187	0.0249	0.0146	0.0119	0.0100	0.0076
1	250	<0[a]	<0[a]	0.0007	0.0002	0.0001	0.0002	<0[a]	<0[a]	0.0004	0.0010	0.0002	0.0004	0.000007
		0.0118	0.0138	0.0223	0.0256	0.0221	0.0196	0.0189	0.0251	0.0308	0.0220	0.0188	0.0185	0.0155
2	250	<0[a]	<0[a]	0.0022	0.0014	0.0010	0.0013	<0[a]	0.0004	0.0026	0.0025	0.0012	0.0017	0.0006
		0.0190	0.0210	0.0287	0.0317	0.0286	0.0262	0.0253	0.0311	0.0365	0.0285	0.0253	0.0254	0.0222
3	250	<0[a]	<0[a]	0.0041	0.0031	0.0025	0.0031	0.0004	0.0026	0.0049	0.0044	0.0029	0.0034	0.0020
		0.0255	0.0275	0.0347	0.0376	0.0347	0.0323	0.0313	0.0368	0.0420	0.0345	0.0314	0.0317	0.0285
4	250	0.0004	<0[a]	0.0062	0.0051	0.0044	0.0051	0.0026	0.0049	0.0074	0.0065	0.0049	0.0054	0.0039
		0.0316	0.0336	0.0404	0.0432	0.0405	0.0381	0.0371	0.0423	0.0473	0.0403	0.0372	0.0376	0.0345
125	250	0.4380	0.4360	0.43849	0.4365	0.4363	0.4382	0.4383	0.43851	0.4388	0.4384	0.4384	0.4383	0.4383
		0.5620	0.5640	0.56151	0.5635	0.5637	0.5618	0.5617	0.56149	0.5612	0.5616	0.5616	0.5617	0.5617

Source: Newcombe, R. G. 2011b. *Communications in Statistics—Theory & Methods* 40: 1743–1767. With permission.

[a] Indicates boundary aberrations.

results of this evaluation are shown in full in Section 5.2; the results for interval location are also reported in Newcombe (2011b).

1. *Wald interval, no continuity correction.* Upper and lower limits are calculated as $U = p + z\sqrt{(pq/n)}$ and $L = p - z\sqrt{(pq/n)}$, and are then truncated to lie within [0, 1]. Here $q = 1 - p$ and z denotes the appropriate centile of the standard normal distribution: $z = 1.9600$ for the usual 95% interval.
2. *Wald interval with continuity correction.* Upper and lower limits are calculated as $p \pm \{z\sqrt{(pq/n)} + 1/(2n)\}$. Incorporating the continuity correction avoids degenerate intervals, but increases the propensity to overshoot the boundaries at 0 and 1. Truncation is applied in the same way when this occurs.
3. *The Wilson score interval* (Wilson 1927) is delimited by the two solutions of the quadratic

$$p = \pi \pm z\sqrt{\{\pi(1-\pi)/n\}};\ \text{namely, } \{2r + z^2 \pm z\sqrt{(z^2 + 4rq)}\}/\{2(n + z^2)\}.$$

   This method is boundary-respecting: the two calculated limits always lie in the interval [0,1], and degenerate intervals cannot occur.
4. A *continuity-corrected score interval* (Fleiss, Levin and Paik 2003) comprises all θ such that $|p - \theta| - 1/(2n) \le z\sqrt{\{\theta(1-\theta)/n\}}$. The lower and upper limits are often expressed in closed form:

$$L = [2r + z^2 - 1 - z\sqrt{\{z^2 - 2 - 1/n + 4p(n(1-p) + 1)\}}]/\{2(n + z^2)\} \quad (3.17)$$

$$U = [2r + z^2 + 1 + z\sqrt{\{z^2 + 2 - 1/n + 4p(n(1-p) - 1)\}}]/\{2(n + z^2)\} \quad (3.18)$$

   However, when $p = 0$, Equation 3.17 can produce a lower limit greater than zero; conversely, when $p = 1$, Equation 3.18 can produce a upper limit below 1. Consequently, if $p = 0$, L must be taken as 0; if $p = 1$, then $U = 1$. The resulting interval is boundary-respecting, but does not guarantee strictly conservative coverage properties.
5. The *Clopper–Pearson exact interval* (Clopper and Pearson 1934) is defined in terms of "exact" accumulations of tail areas as in Equations 3.8 and 3.9. The limits may be obtained as in Equation 3.10, or iteratively, or by use of the F distribution:

$$L = \left[1 + \frac{n - r + 1}{rF_{2r,2(n-r+1),1-\alpha/2}}\right]^{-1},\ U = \left[1 + \frac{n - r}{(r+1)F_{2(r+1),2(n-r),\alpha/2}}\right]^{-1} \quad (3.19)$$

6. *The mid-p interval* (Lancaster 1949; Berry and Armitage 1995) is defined in terms of accumulations of tail areas as in Section 3.4.5 with $k = 0.5$.

7–9. *Pseudo-frequency methods* calculate a mesially shrunk point estimate $p_\psi = (r + \psi)/(n + 2\psi)$, for some chosen pseudo-frequency $\psi > 0$. This is then substituted in the Wald formula, giving the limits $p_\psi \pm z \sqrt{\{p_\psi (1 - p_\psi)/(n + 2\psi)\}}$. Agresti and Coull (1998) recommended using $\psi = 2$, although in Brown et al. (2001), ψ is set at $z^2/2$. Here, we examine the effect of choosing $\psi = 1$, 2 and 3 as methods 7, 8 and 9, respectively.

10–13. Four *Bayesian intervals* defined in Section 3.4.6 are included. HPD as well as tail-based intervals are evaluated, even though they are not invariant on scale transformation, specifically because in this evaluation the emphasis is on interval location, and the HPD methods achieve optimal shortness explicitly by shifting location.

Method 10: uniform prior, tail-based

Method 11: uniform prior, HPD

Method 12: Jeffreys prior, tail-based

Method 13: Jeffreys prior, HPD

3.6 Implementation of Wilson Score Interval in MS Excel

The Wilson score interval is readily programmed using spreadsheet software, as in the first block of the accompanying spreadsheet CIPROPORTION .xls, which is displayed in Figure 3.1. The advantages of such an implementation are discussed in detail in Section 4.5. The greatest advantage is that it is straightforward to construct such a resource in a highly user-friendly form. Unfortunately, as far as software is concerned, user-friendliness is usually in the eyes of the developer, not the user. It is anticipated that readers will find this spreadsheet, and similar ones developed for related tasks, genuinely easy to use.

When the user opens the file CIPROPORTION.xls, calculations are displayed for an exemplar dataset, with input values $r = 81$ and $n = 263$. (Although these are large numbers, the method is equally suitable for any valid (n, r) pair, including when $r = 0$ or n.) These two cells, C12 and E12, are displayed in bold, and are unlocked. So is the cell, E7, which indicates the nominal confidence level, shown as a percentage (95 here). All numerical cells pertaining to this calculation, apart from these three, are locked (some which the user doesn't need to access (here, in column J) are also hidden) and are derived from the input cells using formulae. These formulae propagate changes, in the usual Excel manner.

Spreadsheet CIPROPORTION. Confidence intervals for proportions and differences.					
This spreadsheet performs CI calculations for proportions and their differences using good methods.					
To perform these calculations, replace values in **bold** as appropriate.					
Two-sided confidence level required			**95**	%.	
Single proportion. (Wilson EB. J Am Stat Assoc 1927, 22, 209–212).					
Observed proportion	**81**	out of	**263**	i.e.	0.307985
95 % confidence interval		0.255289	to	0.36621	

FIGURE 3.1
The first block of the spreadsheet CIPROPORTION.xls, which calculates a Wilson score interval for a simple binomial proportion.

So, to use this calculation resource, the user then simply replaces the input values for the exemplar calculation with the corresponding values for his or her own data. The results of the calculation for the user's data then appear in place of the corresponding output for the exemplar data.

This spreadsheet calculates the interval for one dataset summarised by n and r. For a simple closed-form method such as this, it is also straightforward to set up software to perform the calculations simultaneously for a large number of (n, r) pairs. This can also easily be done in Excel, or using an equivalent SPSS macro. A rather more tedious approach is to use CIPROPORTION.xls for each (n, r) pair in turn, then copy the calculated lower and upper limits to some other location using Paste Values.

3.7 Sample Size for Estimating a Proportion

Many publications describing methods to calculate the sample size needed to estimate a proportion within a desired degree of precision begin with a specified interval width $\Delta = U - L$ or a half-width $h = \Delta/2$; see for example Krishnamoorthy and Peng (2007). Such approaches presuppose the Wald interval and disregard the fact that better intervals are not symmetrical about the empirical estimate. It is preferable to examine intervals for the hypothesised proportion with a range of sample sizes, and choose n to yield an acceptable degree of precision in both directions from the anticipated value.

Thus, suppose that an investigator seeks to estimate a proportion that is believed to be of the order of 0.1. The investigator anticipates that it is realistic to expect to obtain a sample of size 50 within a reasonable period of study. For a proportion of 5 out of 50, the Wald 95% interval is from 0.0168 to 0.1832. This is symmetrical about the empirical estimate, 0.1, as it always will be provided truncation at 0 or 1 is not required. The corresponding interval width is obviously 0.1832 – 0.0168 = 0.1664.

Consider what would happen if one of the better methods were used. The Agresti–Coull interval is likewise symmetrical about the point estimate. But others are not. For example, the corresponding Clopper–Pearson interval is 0.0333 to 0.2181. We could calculate the interval width here as 0.2181 – 0.0333 = 0.1848. But this does not mean a margin of error of 0.0924 on either side of the empirical estimate. In this example, overall the Clopper–Pearson interval is wider than the Wald interval. It extends further from the empirical estimate in a mesial direction (towards 0.5), but it extends less far from the empirical estimate in a distal direction (towards 0). Most good interval methods, including Clopper–Pearson, mid-*p*, Wilson, Bayes and arcsin algorithms produce intervals that are generally not symmetrical about the empirical estimate. When such methods are used, specifying a desired interval width Δ does not mean that the interval will estimate the proportion with a margin of error of $\Delta/2$ on each side. This will not matter too much when we seek to estimate a proportion that is somewhere in the middle of the range to a high degree of precision, so that the sample size required is large—any reasonable method would give a nearly symmetrical interval. But there will be an important degree of asymmetry when we seek to estimate a low proportion such as disease prevalence, or a high one such as test sensitivity or specificity, and the sample size available is not abundant.

Interval asymmetry is inherent to these better methods. A more realistic approach is to use software that calculates the interval corresponding to any specified values of n and r, then examine the intervals for the specified p with a range of values of n. The sample size may be chosen such that the lower limit, the upper limit or both correspond to what is considered an informative level of precision. For a proportion that is substantially below 0.5, the interval will extend less far to the left than to the right, in absolute terms—but in relative terms, it is being estimated more precisely at the upper limit. So it is appropriate that different margins of precision should be considered on the two sides of the point estimate.

The argument set out above applies particularly to a binomial proportion estimated from a single sample. Different issues with regard to symmetry apply to different measures. It is inappropriate to estimate a ratio measure such as a relative risk or an odds ratio by specifying $\mathbf{E}(U-L)$, so why should it necessarily be appropriate to do so for a binomial proportion? In contrast, for the parallel problem of determining a sample size to estimate a difference of two proportions to a desired degree of precision, the intervals calculated by improved methods are much closer to symmetry, and it makes little difference whether we specify the whole interval width $\Delta = U - L$ or equal margins of error $\Delta/2$ on either side of the observed difference.

For example, in 2000 a researcher sought to determine the sample size required to estimate the specificity with which chest pain patients transported to hospital by emergency ambulance are selected not to undergo thrombolysis. In the eventual sample size calculation she anticipated a specificity of 0.98, and calculated that a 95% interval estimating this to within a margin of error of 0.02 would require 188 patients, using the simple Wald formula. At an earlier stage, she actually said that she sought to estimate this specificity of 0.98 to within 0.05, although on reflection admitted that it would be illogical to do so! A more reasonable interpretation of such a margin of error would be to aim for a lower limit at 0.93. With a sample of size 100, the calculated 95% Wilson limits for a proportion of 0.98 are (0.9300, 0.9945), so one could argue that this is the sample size needed to ensure a specificity at least 93%. Using the less anomalous-looking specification of a margin of error of 0.02, a sample size of 188 leads to Wald limits (0.9600, 1.0000), but the Wald interval performs very poorly in such situations, and should definitely not be used here. The Wilson limits are (0.9484, 0.9924) here, so the above sample size of 188 is inadequate. The appropriate sample size is actually nearly twice as great, 369, leading to anticipated 95% Wilson limits (0.9600, 0.9901). This calculation is easily done using the spreadsheet CIPROPORTION.xls, by replacing the contents of the numerator cell (C12) by the formula =0.98*E12, then varying the denominator E12 until the calculated limits both satisfy the specified definition of margin of error.

4

Criteria for Optimality

4.1 How Can We Say Which Methods Are Good Ones?

In Chapter 3 we introduced several confidence interval methods for binomial proportions. In subsequent chapters we will examine various intervals for differences of proportions and other derived quantities. Consequently we need to consider what criteria are relevant for choosing between alternative methods.

Two main types of criteria for optimality can be identified, which may be classed informally as structural and performance criteria, although the demarcation between the two categories is not rigid. There is general (mainly implied) agreement with regard to the desirability of certain obvious structural criteria, and of two basic performance criteria: the achieved coverage should be close to the nominal level, $1 - \alpha$, and intervals should be sufficiently wide to achieve this, but no wider. Two additional closely interrelated downstream issues, computational ease and transparency, are considered in Section 4.5.

4.1.1 Structural Criteria for Optimality

The structural criteria generally assumed are explained by Blyth and Still (1983) and Santner and Duffy (1989). Blyth and Still postulated several desirable properties for a confidence region for a scalar parameter, in particular, for a binomial proportion $p = r/n$. While the desirability of these properties is obvious, they are satisfied by most, but not all confidence interval methods for proportions or related quantities.

The confidence region for any scalar parameter should be an interval. This means that for any two values contained in the confidence region, all intermediate values should also be included. This property is implicit in the terminology "confidence interval." While this is an obvious criterion, it is notoriously violated by the confidence region for the number needed to treat in the non-significant case, as described in Section 7.6. In two dimensions, there is the obvious analogous desirable property that a confidence region should be convex. Yet when a shortened confidence region is obtained for two parameters simultaneously (Reiczigel et al. 2008), the resulting region can actually include "holes."

Another obvious property is that the interval should be equivariant as defined in Section 3.3; that is, invariant under relabelling of successes and failures. This is violated by the practice of deriving intervals for $p = r/n$ from Poisson-based intervals for the numerator (Section 6.1.1).

Also, the calculated lower and upper limits should be monotonic functions of both r and n. This property is equally obviously desirable, in the interest of consistency. But it may be violated unwittingly, especially when a hybrid of two different interval methods is used. For example, software might use one method if $n \le 40$ or an alternative if $n > 40$. This suggestion masquerades as use of two different methods depending on the value of n. But we then fail to appraise what really happens. This algorithm is better regarded as a hybrid of two different methods. The criterion for choosing between the two methods can depend on n, on r, on p, or on these parameters jointly. Whatever form the criterion takes, for such a hybrid interval it is unlikely that the resulting transition would always be smooth. In this instance, when the criterion for choice is framed in terms of n, it is advisable to check that the transition is monotonic for several possible values of r, at least for the usual $1 - \alpha = 0.95$.

In more complex contexts, there are further possible ways in which some kind of non-monotonic behaviour can occur. For example, in Chapter 8, an easily computed interval for the paired difference of proportions, method 10 of Newcombe (1998c), may be derived from the Wilson interval for the simple proportion. Unlike other methods for this situation, it distinguishes between the two concordant cells in the paired 2×2 table, and can display a curious non-monotonic behaviour as their split varies, holding their sum constant. This should not be regarded as an absolute contra-indication to the method, but rather, as one criterion to be balanced against the others.

Santner and Duffy (1989) listed similar desirable properties. Their criteria also include symmetry about 0.5 when n is even and $r = n/2$, although this is simply a corollary of equivariance. Appropriately, they postulated monotonicity in α as well as in n and in r. (Strict monotonicity is not implied—everything is defined in terms of $\le$ signs and the terms non-increasing and non-decreasing—but presumably the intention is that equality is permitted only in boundary cases; namely, $r = 0 \Rightarrow L = 0$ for all n and α, etc.)

4.1.2 Performance Criteria for Optimality

Choice among confidence interval methods depends crucially on choice of performance criteria. The following criteria are suggested.

The mean coverage (Section 4.2) should be slightly conservative, slightly above $1 - \alpha$. (As explained there, the minimum coverage across all parameter space points is not so crucial.)

Expected interval width (Section 4.3) should be examined as well as coverage—adequacy of coverage is not to be at the expense of excessive width. Widths of intervals by two or more methods should be compared holding the sample size and parameter(s) of interest fixed.

With regard to interval location (Section 4.4), normally we want mean mesial and distal non-coverage probabilities to be similar as far as this can be achieved. In this section we introduce two indices. MNCP/NCP expresses mesial non-coverage as a proportion of total non-coverage. As a rule of thumb, values between 0.4 and 0.6 are desirable. The Box–Cox index of symmetry expresses the symmetry of location of a calculated interval about the point estimate.

We wish to avoid anomalous behaviour, in particular, at or near extreme values. This issue is discussed in detail in Section 3.3 for the simple proportion. But this does not imply that we totally reject methods such as pseudo-frequency intervals on these grounds. When we consider intervals for more complex derived quantities, other kinds of anomalous behaviour can arise, which are discussed in subsequent chapters.

4.2 Coverage

For any confidence interval method for the single proportion, the coverage probability depends on both π and n. The dependence of coverage on π has a well-known characteristic shape which is very far from being monotonic, let alone smooth. It is illustrated in Figure 4.1 which relates to the Wilson score interval with $n = 10$ and $1 - \alpha = 0.95$. Similar graphs are to be found in Santner and Duffy (1989), Vollset (1993) and Brown et al. (2001), among others. Brown

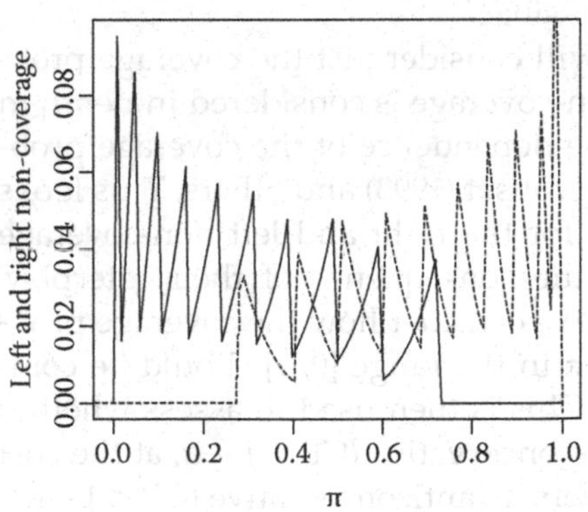

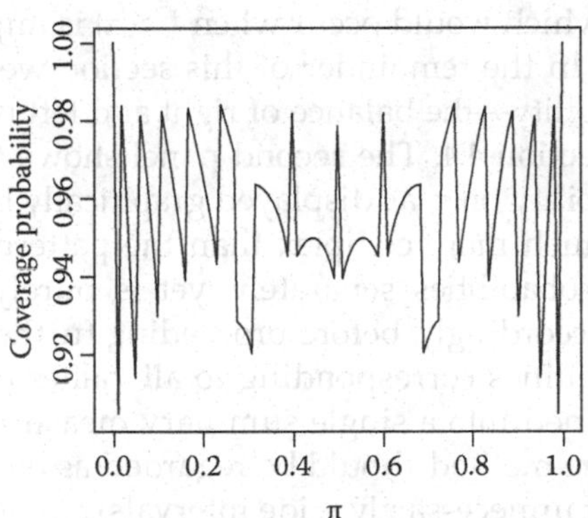

FIGURE 4.1
Coverage properties of Wilson score interval for $n = 10$ and $1 - \alpha = 0.95$. The first panel shows left (solid curve) and right (broken curve) non-coverage probabilities as functions of the theoretical proportion π. The second panel shows the resulting coverage probability as a function of π. (From Newcombe, R. G. and Nurminen, M. M. 2011a. *Communications in Statistics—Theory & Methods* 40: 1271–1282. With permission.)

et al. noted that the dependence on n for fixed π is also erratic, an observation which prompted them to describe some (n, π) combinations as "lucky" and others as "unlucky." While they made this comment particularly in relation to the Wald interval, all confidence interval methods for $p = r/n$ display a qualitatively similar irregularity of coverage. These terms might equally be used in relation to the Clopper–Pearson exact interval: a parameter space point for which coverage is only slightly higher than the nominal level might be regarded as lucky, and one at which coverage is much higher as unlucky in that the interval is redundantly wide.

This issue of irregular coverage behaviour pervades all evaluations of confidence interval methods for proportions and related quantities. Consequently it is important that any evaluation study to assess coverage properties of confidence interval methods should involve a large number of parameter space points (PSPs). In Chapter 5 we describe in detail how this may be done. In some simpler situations such as for the simple proportion $p = r/n$, exact calculations are feasible. Monte Carlo methods are also acceptable in these contexts, and are practically essential in more complex ones.

In the first panel of Figure 4.1, the solid curve represents the left non-coverage probability (LNCP), and the broken curve the right non-coverage probability (RNCP), as functions of π. These quantities are defined in the obvious way, as probabilities of non-coverage at the left and right hand ends of the interval. The Wilson method, like all methods considered here, has the equivariance property that the limits for $(n - r)/n$ are complements of those for r/n; consequently the two curves are mirror images. Each curve consists of $n + 1$ segments, separated by n discontinuity points. For the solid curve, the final segment for $\pi \geq 0.7225$ is a straight line at $y = 0$. This is because the interval for 10/10 is (0.7225, 1), so whenever $\pi \geq 0.7225$, left non-coverage (which would occur when $L > \pi$) is impossible.

In the remainder of this section we will consider just the coverage probability—the balance of right and left non-coverage is considered in depth in Section 4.4. The second panel shows the dependence of the coverage probability on π, as displayed graphically by Vollset (1993) and others. This looks much more complex than the patterns for the right and left non-coverage probabilities separately, yet is purely the consequence of their interplay. Accordingly, before proceeding further we consider how the coverage probabilities corresponding to all values of π in the range [0, 1] should be combined into a single summary measure. This is then used to assess whether the method should be regarded as over-conservative (CP $> 1 - \alpha$, at the cost of unnecessarily wide intervals) or conversely anticonservative (CP $< 1 - \alpha$).

4.2.1 Just What Is the Nominal Coverage, $1 - \alpha$, Meant to Represent?

Some authors such as Anderson and Burstein (1967), Pratt (1968), Fujino (1980), Blyth (1986), Angus (1987) and Vollset (1993) have taken the view that the Clopper–Pearson method should be regarded as the gold standard interval for a proportion, implying that the only rationale for other intervals is greater computational convenience. While this is a defensible position, there are highly plausible arguments to the contrary. Firstly, we note that the Clopper–Pearson method itself should now be regarded as computationally convenient, as a beta interval (Section 3.4.6), so the argument that the exact interval should be paramount would imply no other interval need be considered. And any term used in antithesis to "exact" risks being construed as pejorative (Newcombe 1998a). However, the term "exact" risks being misinterpreted, as it could be construed in several different senses:

(i) Strictly conservative.
(ii) Being based on a precise enumeration of an assumed underlying binomial distribution, not on an asymptotic Gaussian approximation.
(iii) Use of a multiplier 1 for the probability of the outcome observed, in contrast to the mid-*p* approach (Section 3.4.5) where the probability of the observed outcome contributes to the tail probability with coefficient 0.5.
(iv) Attaining a coverage probability of precisely the nominal $1 - \alpha$ for all π (and n) constituting the parameter space.

The Clopper–Pearson interval fulfils the first three criteria. No method can achieve exactness in sense (iv), on account of the discontinuous behaviour of the coverage probability as π moves past the lower or upper limit corresponding to any possible $r = 0, 1, \ldots, n$ as in Figure 4.1. Yet this is the meaning which consumers of information presented in the form of confidence intervals are likely to infer.

The crucial issue is whether the nominal coverage probability $1 - \alpha$ is meant to be a minimum or an average—bearing in mind this discontinuous behaviour. Some confidence interval methods seek to align the minimum coverage probability with $1 - \alpha$, others the mean coverage probability. For example, the Clopper–Pearson interval specifically aims to align the minimum coverage with $1 - \alpha$, and is the simplest interval that achieves this; the related interval involving a mid-*p* definition of tail areas aligns mean coverage approximately with $1 - \alpha$. The strict conservatism of the Clopper–Pearson interval is achieved at the cost of a greater width than the corresponding mid-*p* interval. Similarly, when a continuity correction is applied to a Wald or Wilson interval, the increased coverage is obtained at the cost of an obvious increase in width. This trade-off between conservatism and width is practically universal—so we must consider very carefully whether it is really the minimum coverage that we want to align with $1 - \alpha$.

Two main arguments clinch the case for aligning mean coverage with $1 - \alpha$. One argument relates to current statistical practice, in particular the widespread use of Bayesian models for more complex problems. Another, more persuasive argument is set out in Section 4.2.2 below.

In the related context of interval estimation for the Poisson parameter (Section 6.2), Cohen and Yang (1994) argue, quoting from Cox and Hinkley (1974, p. 216), that " 'if we were using confidence limits as a general guide, … then it would be sensible to arrange' that the coverage probability for nominally 95% confidence intervals averages around 0.95 rather than having 0.95 as a lower bound." In a similar vein, the seminal paper by Brown, Cai and DasGupta (2001) was followed by contributions from several discussants, culminating in a response by the authors. Several very pertinent issues were aired, including coverage and also transparency (see Section 4.5). Apropos coverage, Casella and Corcoran & Mehta argued against aligning the mean coverage probability with the nominal $1 - \alpha$. Brown et al. disagreed, saying, "It seems more consistent with contemporary statistical practice that a γ% CI should cover the true value *approximately* γ% of the time." Note that the references cited at the beginning of this section relate to work published about 20 to 45 years ago. Modern statistical practice has moved very much in the direction of regarding $1 - \alpha$ as representing average coverage: often, complex Bayesian models are fitted using MCMC software such as WinBugs, in which case implicitly the coverage achieved is an average. Acceptance of the Bayes interval in this situation when it is the mean, not minimum coverage that aligns with $1 - \alpha$, provides a further argument for the acceptability of a mean coverage criterion in simpler situations.

4.2.2 Why the First Dip in Coverage May Be Disregarded

With the above argument in mind, and following much current statistical thinking, rather than regarding the Clopper–Pearson exact interval as a gold standard, it is more appropriate to produce narrower, less conservative intervals that align the mean coverage (defined in some way) with $1 - \alpha$. Doing so presupposes that the existence of short segments of the support space in which the coverage probability is substantially lower than the nominal value need not be viewed with disquiet. This stance may be justified by the following argument (Newcombe and Nurminen 2011a).

When the Wilson score interval is used, there are several dips in coverage to well below the nominal $1 - \alpha$ level (Figure 4.1). The deepest dips are the first and last ones, which are mirror images. The first relates to the lower limit L_1 corresponding to $r = 1$. As π increases from 0 towards L_1, right non-coverage cannot occur, but the left non-coverage probability increases towards a supremum of $1 - (1 - L_1)^n$. For the exact interval, this quantity is identically $\alpha/2$, but for several other methods, it is greater than α.

While this phenomenon has been levelled as a criticism against the Wilson interval, it also applies to the most commonly used Bayes intervals, especially when a uniform prior is used. Table 4.1 shows L_1 and the corresponding minimum coverage probability $(1 - L_1)^n$ for five confidence interval methods, for small and large values of n. The minimum coverage probability does not depend heavily on the sample size n here.

TABLE 4.1

Lower Confidence Limit L_1 Corresponding to the Binomial Proportions 0.1 and 0.004 for Selected Methods, and Corresponding Minimum Coverage When π Is Just below L_1

	$n = 10$		$n = 250$	
Method	L_1	**Minimum CP**	L_1	**Minimum CP**
Uniform *B*(1,1)	0.0228	0.794	0.00097	0.785
Wilson score	0.0179	0.835	0.00071	0.838
Jeffreys *B*(0.5, 0.5)	0.0110	0.895	0.00043	0.898
Mid-*p*	0.0050	0.951	0.00020	0.951
Clopper–Pearson	0.0025	0.975	0.00010	0.975

Source: Newcombe, R. G. and Nurminen, M. M. 2011a. *Communications in Statistics—Theory & Methods* 40: 1271–1282. With permission.
Note: CP = coverage probability.

For such values of n, the Wilson 95% interval has a minimum coverage in the range 0.835–0.838 which occurs when π is just below L_1. (Its minimum coverage is a little lower still for $n < 10$, for instance 0.832 for $n = 5$.) For example, with $n = 10$ as in Figure 4.1, $L_1 = 0.0179$. If the prior belief is that π is distributed uniformly between 0 and some value slightly above L_1, say 0.02, the mean of the coverage probabilities for this part of the parameter space would be only 0.923. But a sample size of only 10 would not give adequate information on such a low proportion. To warrant sampling just 10 individuals would presuppose that the true proportion could be as high as 0.2 (at least), and averaging coverage from 0 to 0.2 gives a mean coverage of 0.954. The left-hand ends of the corresponding graphs for larger values of n take a similar form, as in Vollset (1993), and a similar argument applies: L_1 is $0.0177/n$ plus a term of the order of $1/n^2$, and the chosen sample size might be judged reasonable if the expected number of positives was at least 2, but not if it was only 0.2.

Figure 4.2 plots the moving average of coverage probabilities of the Wilson score interval for $n = 10$ and $1 - \alpha = 0.95$ over the range $(\pi - 0.1, \pi + 0.1)$ against π. This is shown superimposed on a graph of the attained coverage at π, for π in the range 0.1 to 0.9, as in Figure 4.1. Thus the moving average value plotted for $\pi = 0.1$ is 0.954, as above. Comparing the two curves it is clear that the effect of examining such moving averages is to virtually eliminate the dips below $1 - \alpha$. Naturally, calculating moving averages has the effect of greatly flattening out the variation, so that the anticipated mean coverage now ranges only from 0.949 to 0.962. There is a dip to around 0.949 for π or $1 - \pi$ slightly below 0.4 only. For approximately 95% of the range of π from 0.1 to 0.9, the anticipated mean coverage based on a moving average over the range $(\pi - 0.1, \pi + 0.1)$ is at least at the nominal level of $1 - \alpha = 0.95$. A smoothed curve such as the moving average provides a more realistic assessment of the

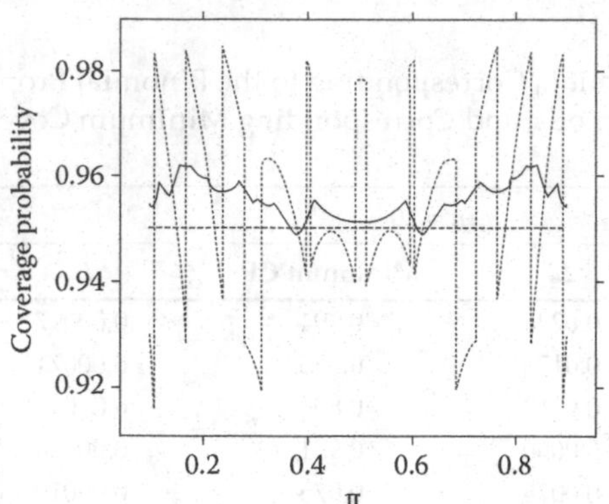

FIGURE 4.2
Coverage probability of Wilson score interval for $n = 10$ and $1 - \alpha = 0.95$ as a function of π. The broken curve shows the coverage probability calculated directly. The solid curve shows the moving average over the range $(\pi - 0.1, \pi + 0.1)$ superimposed upon this, for $0.1 < \pi < 0.9$. (From Newcombe, R. G. and Nurminen, M. M. 2011a. *Communications in Statistics—Theory & Methods* 40: 1271–1282. With permission.)

coverage achieved in practice by the score interval, and this pattern corresponds realistically to what we should want a confidence interval algorithm to achieve.

Thus when we use a moving average approach, the irregular behaviour of the coverage probability is largely smoothed out, coverage is now generally a little above the nominal $1 - \alpha$ level for the Wilson interval. Similar diagrams can be produced analogously for other interval methods, and may be regarded as another, complementary way to determine which methods should be regarded as having appropriate coverage characteristics. Little is gained by producing such plots for the Wald interval, for which the coverage probability is generally considerably below $1 - \alpha$, nor for the Clopper–Pearson interval, for which the coverage probability is invariably at least $1 - \alpha$.

Figures 4.3 to 4.6 present coverage characteristics for the case $n = 10$ for four intervals, the continuity-corrected Wald interval, tail-based Bayes intervals with uniform $B(1,1)$ and Jeffreys $B(0.5, 0.5)$ priors, and the Agresti–Coull interval using a pseudo-frequency $\psi = 2$. In each instance, the first panel shows the coverage probability plotted as a function of π, for the full range (0, 1). In the second panel, the plot goes from $\pi = 0.1$ to 0.9. The broken curve shows the unsmoothed coverage probability. The superimposed solid curve shows the moving average coverage over $(\pi - 0.1, \pi + 0.1)$ plotted as a function of π, which is naturally very much smoother.

It is evident from Figure 4.3 that examining moving average coverage does not obviate the deficient coverage of the continuity-corrected Wald interval.

The moving average plots suggests satisfactory coverage for the tail-based Bayes interval if a uniform prior is used (Figure 4.4), but not if a Jeffreys prior is

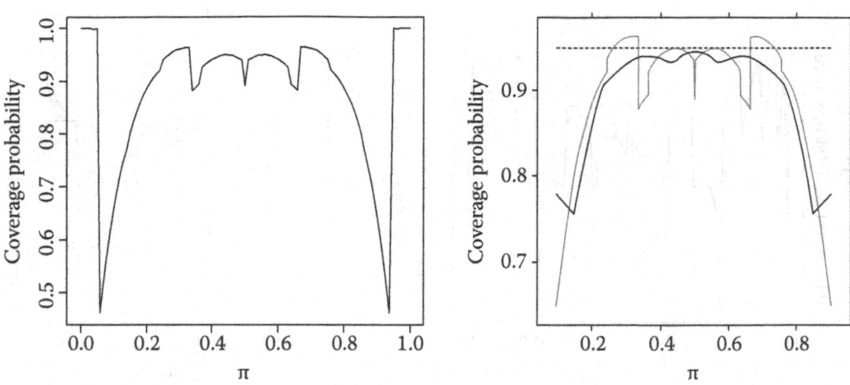

FIGURE 4.3
Coverage probability of the continuity-corrected Wald interval as a function of π, for $n = 10$. In the second panel the solid curve shows the moving average coverage over $(\pi - 0.1, \pi + 0.1)$ plotted as a function of π.

chosen (Figure 4.5). This finding may be linked to the average coverages of these two methods in the relevant parameter space zones in Table 5.1. In Tables 5.3 and 4.4 we will see that the Jeffreys interval has appropriate location whereas the interval with uniform prior is too mesially located. This finding is counterbalanced by the demonstration here that from the standpoint of weighted mean coverage, the uniform prior is preferable, for small n at any rate.

Finally, the moving average plot for the Wald + 2 method (Figure 4.6) indicates that from this standpoint this interval should certainly be regarded as having adequate coverage—indeed, slightly higher than necessary—counterbalanced of course by its susceptibility to boundary violation.

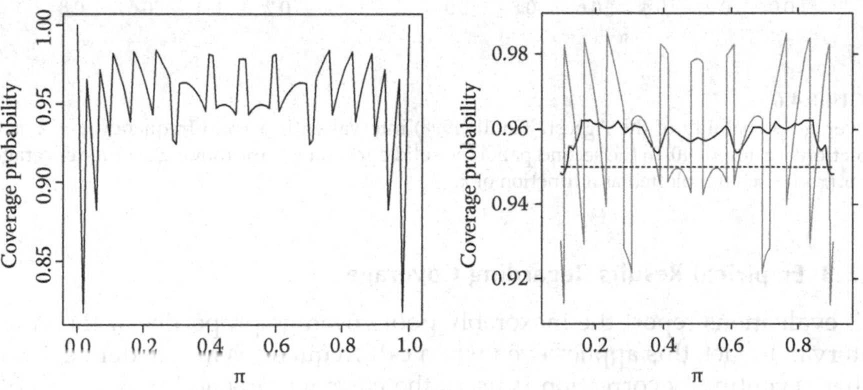

FIGURE 4.4
Coverage probability of the tail-based Bayes interval with uniform $B(1, 1)$ prior as a function of π, for $n = 10$. In the second panel the solid curve shows the moving average coverage over $(\pi - 0.1, \pi + 0.1)$ plotted as a function of π.

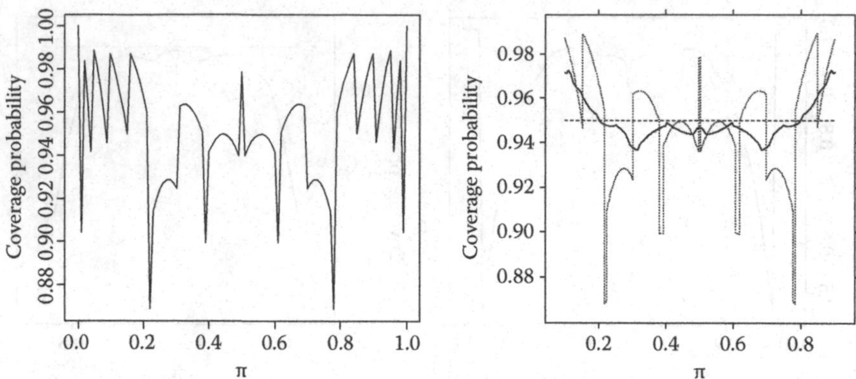

FIGURE 4.5
Coverage probability of the tail-based Bayes interval with Jeffreys $B(1/2, 1/2)$ prior as a function of π, for $n = 10$. In the second panel the solid curve shows the moving average coverage over $(\pi - 0.1, \pi + 0.1)$ plotted as a function of π.

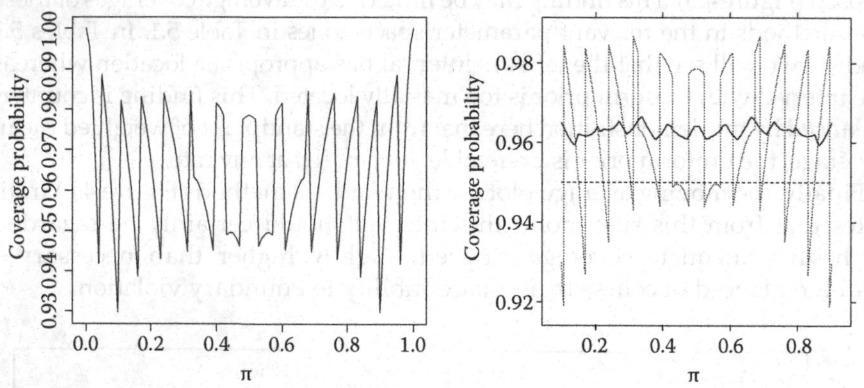

FIGURE 4.6
Coverage probability of the Agresti–Coull (1998) interval with pseudo-frequency $\psi = 2$ as a function of π, for $n = 10$. In the second panel the solid curve shows the moving average coverage over $(\pi - 0.1, \pi + 0.1)$ plotted as a function of π.

4.2.3 Empirical Results Regarding Coverage

All evaluations report the inexorably poor coverage properties of the Wald interval. In fact, this applies even when cell frequencies are moderate. Even when a continuity correction is used, the coverage probability is unacceptably low for many values of π, and the mean coverage falls well short of the nominal value.

The Clopper–Pearson interval always has a coverage probability of at least $1 - \alpha$. Because coverage varies erratically with π, this can only be achieved with an average coverage much higher than the nominal value—in Newcombe (1998a), the mean coverage was 0.971 for an interval with nominal 95% coverage. The Clopper–Pearson interval may thus be regarded as overconservative; in other words, simply unnecessarily wide.

In contrast, the Wilson interval had a mean coverage of 0.952 in Newcombe (1998a). The minimum coverage was 0.832, as explained in Section 4.2.2.

The full results for the Newcombe (2007c) evaluation follow in Section 5.2. In brief, for large n and central π, all 13 methods perform similarly, as expected, with little to choose between them: the mean coverage is close to the nominal $1 - \alpha$, interval location is satisfactory, interval widths are similar and boundary aberrations do not occur. For smaller n, and especially smaller π, differences between methods are much greater. Overall, the mean coverage was about right for Wilson, mid-p, Wald + 2 and Bayes intervals, with the Jeffreys interval achieving slightly more consistent mean coverage than the interval with uniform prior. The mean coverage of the Wilson interval ranged from 0.940 to 0.962 in nine contrasting parameter space zones.

Wilson, mid-p and Bayes intervals with uniform prior are suitable if the aim is to align the mean coverage probability with $1 - \alpha$. The Clopper–Pearson interval meets the aim of aligning minimum coverage with $1 - \alpha$.

4.3 Expected Width

In any comparative evaluation of different confidence interval methods for the binomial proportion, interval width (sometimes referred to as length) is crucially related to the sample size n, and (almost invariably) to the observed proportion p also. Consequently, interval widths are only meaningfully compared when these parameters are held constant. A further prerequisite is truncation of any boundary overshoot.

As noted in Section 4.2, the coverage probability is a very erratic function of π for given n, as in Figure 4.1, and is also a very erratic function of n for fixed π. In marked contrast, expected width is generally a smooth function of both the population proportion and the sample size. Figure 4.7 shows the dependence of the interval width on the proportion, for Wilson intervals for $n = 10$.

The upper, broken curve shows the relationship of the interval width to the observed proportion p. For the Wilson interval this is $z\{z^2 + 4np(1-p)\}^{1/2}/(n + z^2)$. The only values of p that are possible here are 0, 0.1, 0.2,..., 1, nevertheless as in Vollset (1993) it is convenient to plot interval width as a function of a continuously varying parameter p. The lower, solid curve shows mean interval width as a function of the theoretical proportion π, as in Böhning (1994), Agresti and Coull (1998) and Brown et al. (2001). Whatever method is chosen, the dependence of mean width on π is a polynomial of order n. The

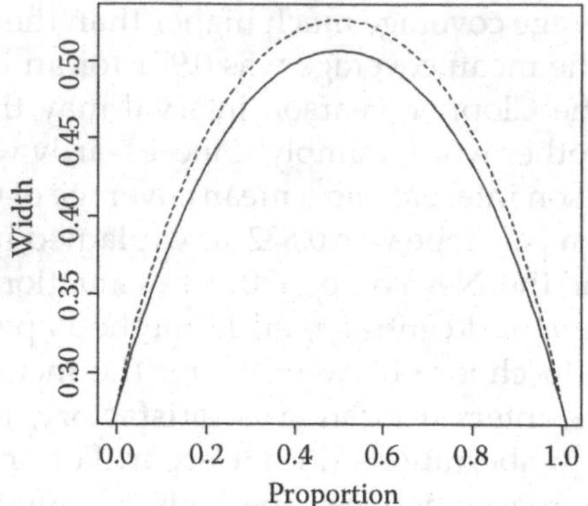

FIGURE 4.7
Plot of interval width by proportion for Wilson interval with $n = 10$. The upper, broken curve represents actual interval widths corresponding to $r = 0, 1,\ldots, n$. The lower, solid curve plots average interval width by the theoretical proportion π.

plot of actual width by p generally resembles that for the Wilson interval in Figure 4.7, for any method defined by an algebraic function. For equivariant methods, both curves have a characteristic shape, symmetrical about a peak at $p = 0.5$, corresponding closely to the simple dependence of the Wald standard error $\surd\{p(1-p)/n\}$ on p.

It is striking, and characteristic, that the broken curve always lies above the solid one. As the proportion approaches 0 or 1, both curves tend to the interval width $W_0 = U_0 - L_0$ corresponding to $r = 0$ or equivalently n. W_0 takes a positive value for any boundary-respecting interval method, but is zero for the Wald interval (truncated to lie within [0, 1] if necessary). Conversely, at $\pi = 0.5$, the mean interval width is not that for $r = n/2$ but is a weighted average of this and widths of the narrower intervals corresponding to values of r on either side of $n/2$. Both plots are informative, but numerical values for interval width are not interchangeable.

Vollset (1993) showed how, for small sample sizes only, the degree of dependence of width on the proportion can be substantially different for different methods: width was most heavily dependent on p for arcsine, likelihood and Wald intervals, but was almost constant for the delta logit interval. As an extreme example, sometimes proportions derived from simple sample surveys such as election exit polls are reported by the media accompanied by a margin of error calculated simply as $1/\surd n$. This formula is given by Goertzel (2010) with the wise advice, "This is the easy one, you should try to learn to use it in your head". The rationale for this extraordinarily simple procedure is that the Wald interval half-width $z\surd(pq/n)$ is maximised when $p = 0.5$, and is then $z/(2\surd n)$ which is just below $1/\surd n$ for the usual 95% nominal confidence level with $z = 1.96$. These approximations would not be wildly wrong for elections in the United States and United Kingdom in which two main parties with approximately equal support dominate, although there would be severe distortion and indeed the potential for boundary anomalies for fringe parties. The resulting interval $p \pm 1/\sqrt{n}$ is always wider than the 95% Wald interval, much wider for small p, resulting in less anticonservative coverage and no ZWIs, at the expense of greater width. This interval is hardly worthy of scientific study, although as it is amenable to mental calculation it is sometimes useful to give an instant appraisal of the degree of imprecision attaching to a result, or of the adequacy of a proposed sample size when limited prior information on π is available.

The dependence of mean interval width on n is also regular. Generally, interval width is approximately proportional to $1/\sqrt{n}$, in line with the usual dependence of standard errors of the form $\sigma/\sqrt{n}$ or $\sqrt{\{p(1-p)/n\}}$ on n. Although when attention is restricted to intervals for the boundary case $r = 0$, the upper limit produced by a boundary-respecting method tends to approximate to the form c/n. As noted in Section 3.4.3, when $r = 0$, the Wilson upper limit reduces to $z^2/(n + z^2)$. Following Rümke (1974), Hanley and Lippman-Hand (1983) observed that the 95% one-sided upper Clopper–Pearson limit $1 - \alpha^{1/n} \sim c/n$ with $c = -\ln(\alpha) = 3.00$, leading to the "rule of three" upper limit $3/n$. If we define the 95% interval for an extreme case in the same way as in non-extreme ones, with intended tail area $\alpha/2 = 0.025$ or $z = 1.96$, c takes the value $-\ln(\alpha/2) = 3.69$ for the Clopper–Pearson interval, or $z^2 = 3.84$ for the Wilson interval as in Section 3.4.3, an approximate rule of four. At first sight, this form of dependence on n seems very strange—the interval width is proportional to $1/n$, apparently, not to $1/\sqrt{n}$. But note that as n increases, the outcome $r = 0$ becomes less likely to occur, whatever the value of π.

An important consequence of this regularity of dependence is that in any evaluation comparing the performance of different methods, there is no particular advantage in examining interval width for a large number of parameter space points. It may sometimes be most convenient to do so as a by-product of an evaluation of coverage. But it is sufficient to calculate expected width for a small number of combinations of n and π spanning a reasonable range of values; for example $n = 10$, 50 and 250 with $\pi = 0.05$, 0.25 and 0.45. Both approaches are equally valid ways to examine expected interval width. With either approach, following Zou and Donner (2004), we may display interval widths explicitly for one chosen method, then express widths of other intervals as multiples of the width for the chosen method, as in Table 5.2. Furthermore, when π is assumed to have a uniform prior distribution, for any given n expected interval width may be derived from the tabulated intervals for $r = 0, 1,..., n$ simply by averaging their widths.

When considering confidence intervals for means and their differences as in Chapter 2, the natural measure of interval width is simply the numerical difference between the upper and lower limits, $\Delta = U - L$. However, for many other quantities, Δ is not the only possible expression of interval width, and indeed is sometimes not the most appropriate one. For example, an odds ratio ω (Chapter 11) may be estimated as 2.5, with 95% interval $2.5 \div 2 = 1.25$ to $2.5 \times 2 = 5.0$. This is how confidence intervals for odds ratios work—the sampling distribution of log ω, not ω, is approximately symmetrical, so the simplest method involves the standard error of log ω, not the standard error of ω itself. If we interchange the status of the "exposed" and "unexposed" groups, ω becomes 0.4, with 95% interval 0.2 to 0.8. Which of these intervals is wider? They convey identical information, so both should be regarded as equal in width. On the log scale, of course, they are. Exactly the same issue applies to other ratio measures including relative risks and rate ratios (Chapter 10) and the ratio of the geometric means of two samples assuming lognormal distributions as in Section 2.4.

Furthermore, the ordinary empirical estimate $\tilde{\omega}$ can range from 0 to ∞. When $\tilde{\omega} = 0$, the corresponding lower limit should be 0; when $\tilde{\omega} = \infty$, an upper "limit" of ∞ should be reported. Consequently, unless pseudo-frequency or Bayesian methods are used to produce point and interval estimates with shrinkage towards 1, the expected interval width on a log scale may well be infinite. It may be more sensible to evaluate expected width on an alternative transformed scale, as $\frac{U}{1+U} - \frac{L}{1+L}$ which is both bounded and invariant on interchanging the status of the two groups. Likewise for relative risks and rate ratios.

Similarly, confidence interval widths for proportions depend on p, but in a different way for linear and logit scales. For example, Wilson 95% intervals for 50/100 and 5/100 are (0.4038, 0.5962) and (0.0215, 0.1118). On the ordinary additive scale, the former is wider. However, logit transforming these limits gives −0.3895 to 0.3895 and −3.816 to −2.073, so on a logit scale, the latter is wider. Close to the boundary at zero, the logit transformation resembles the log transformation. What this means is that in absolute terms less uncertainty is imputed when a proportion is near the boundary than when it is in the middle of the scale, but in relative terms the uncertainty is greater. Only a severely misfitted interval such as $p \pm \sqrt{1/n}$ has a width independent of p, and that is on an absolute scale, which is not the only possible criterion, and not the best one.

The problems with the Wald interval arise not because it is too narrow, but because interval width depends on p in an inappropriate manner and location is inappropriate. Sometimes other interval methods attain increased coverage at the cost of increased width. A similar trade-off between coverage and width applies to intervals for other measures.

4.4 Interval Location

The issue of location has received relatively little attention in the published literature on interval estimation for proportions and so forth. In the frequentist paradigm, confidence intervals comprise parameter values that are not rejected by a two-tailed hypothesis test, where the critical region is intended to have equal probabilities in the two tails. Similarly, Bayesian intervals are delimited by $\alpha/2$ and $1 - \alpha/2$ quantiles of the posterior distribution. But neither of these properties guarantees that the left and right non-coverage probabilities of the resulting interval will be balanced. Accordingly, the left and right non-coverage probabilities achieved by the interval may be reported, as in Figure 4.1. However, for many parameters it is preferable to characterise mesial and distal non-coverage, as explained below. In this section two indices characterising interval location are described. One relates to simulation studies of coverage. The other expresses how symmetrically the calculated lower and upper limits are placed in relation to the observed proportion. Both characterisations presuppose that the interval method under consideration is equivariant as defined in Section 3.3 (i.e., that the limits for $n - r$ out of n are the complements of those for r out of n). For methods that are not equivariant—notably, when a Poisson approximation is used as in Section 6.1.1—only left and right non-coverage can be calculated.

4.4.1 General Principles

The Wald interval is not only seriously anticonservative and boundary-violating, it also produces intervals that are too far out from the midpoint of the scale, 0.5. We say that Wald intervals are too distally located, so that mesial non-coverage predominates. Conversely, Wilson intervals are too mesially located, meaning that distal non-coverage predominates. Mesial and distal non-coverage appropriately generalise left and right non-coverage in the context of a support space with an obvious symmetry.

It is necessary first to define the support space for the parameter of interest. In this section the parameter of interest is denoted generically by θ. At this stage we are primarily considering the case of a single proportion, so that $\theta \equiv \pi$, but we will also consider comparative measures. For many parameters, the support space is a bounded, closed interval, with a centre of symmetry μ at the midpoint. Thus for the simple proportion, the support is [0, 1], with centre of symmetry 0.5. For differences of proportions (Chapters 7 and 8), and also the generalised Wilcoxon measure $\Delta T/\max$ (Chapter 16), the support is [–1, 1], with centre of symmetry 0. In Chapter 15, the U/mn measure derived from the Mann-Whitney U-statistic ranges from 0 to 1 with centre of symmetry 0.5; an equivalent measure, Somers' D ranges from –1 to 1 with centre of symmetry 0.

For other parameters, the support space can extend to infinity. For the difference of means of two Gaussian distributions, theoretically the support space is the open interval $(-\infty, +\infty)$, with centre of symmetry 0. For ratio scale parameters such as a ratio of proportions or an odds ratio (Chapters 10 and 11), the support space is $(0, +\infty)$. Here, the value corresponding to the usual null hypothesis, 1, cannot be regarded as a centre of symmetry. In this situation, usually the log of the ratio has more satisfactory properties, with support space (−∞, +∞) and centre of symmetry 0. Often a natural scale transformation of the support space preserves the centre of symmetry. Thus for the single proportion, the support space is [0, 1], with centre of symmetry $\mu_{linear} = 0.5$. Following logit transformation, the support space becomes the closed interval [−∞, +∞], with centre of symmetry $\mu_{logit} = 0$ directly corresponding to $\mu_{linear} = 0.5$.

In any evaluation of coverage properties in these situations, if a symmetrical distribution on the entire support space is chosen as a pseudo-prior for θ, then it is gratuitous that the resulting left and right non-coverage probabilities will be equal (to within the limits of chance variation). Comparison of these quantities would be totally uninformative with regard to the appropriateness of interval location. It is preferable to restrict the evaluation to one-half of the support space, as in Newcombe (1998a, b, c, 2006b, 2007c). Right and left non-coverage are then identified as mesial and distal (or vice versa), a much more informative classification.

Table 4.2 and Figure 4.8 show how the direction of non-coverage is defined. Suppose first that $\theta \neq \mu$. The first two rows of the table are relevant in an evaluation of coverage for a single proportion such as Newcombe (1998a), in which a pseudo-prior for θ limited to the lower half of the parameter space, $\theta < \mu$, is the obvious choice. The final two rows are relevant in an evaluation of coverage for contrast measures such as differences between proportions (Chapters 7 and 8) or the U/mn and $\Delta T/\max$ measures (Chapters 15 and 16), in which a pseudo-prior for the relevant parameter θ limited to the upper half of the parameter space corresponding to effects in a positive direction, $\theta > \mu$, is the natural choice.

Mesial and distal directions are defined in relation to the true value of θ. Non-coverage is distal whenever the interval is located too mesially to include θ, and mesial whenever the interval is located too distally to include θ. This applies irrespective of which of the calculated limits is closer to μ. It is straightforward to apply the definition in an evaluation in which θ is limited to one half of the support space, as in the examples above—we simply count occurrences

TABLE 4.2

Circumstances in Which Mesial and Distal Non-Coverage Are Imputed

True Parameter Value θ	Relevant Confidence Limit	Interpretation	Type of Non-Coverage
$\theta \le \mu$	$U < \theta$	CI too far out from μ	Mesial ≡ right
$\theta < \mu$	$L > \theta$	CI too close to μ	Distal ≡ left
$\theta > \mu$	$U < \theta$	CI too close to μ	Distal ≡ right
$\theta \ge \mu$	$L > \theta$	CI too far out from μ	Mesial ≡ left

Source: Newcombe, R. G. 2011b. *Communications in Statistics—Theory & Methods* 40: 1743–1767. With permission.

Note: CI = confidence interval. θ denotes the true value of the parameter. μ is used here to denote the centre of symmetry of the support space for θ. For a bounded support space, this is the midpoint.

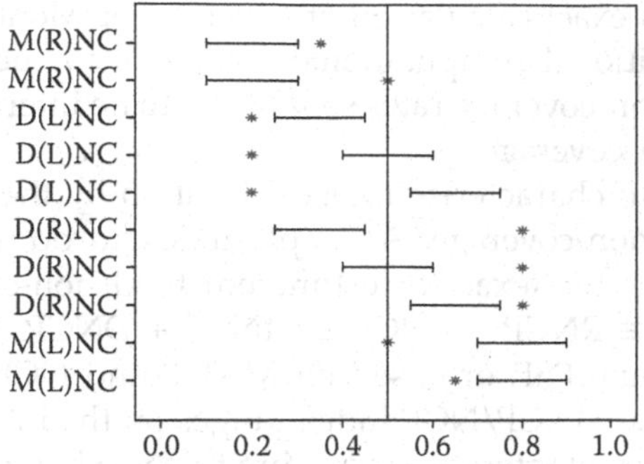

FIGURE 4.8
Mesial and distal non-coverage illustrated for the simple proportion. The centre of symmetry of the support space is at $\mu = 0.5$. For each case, the asterisk denotes the true parameter value π and the error bar represents the interval (L, U). MNC = mesial non-coverage; DNC = distal non-coverage; RNC = right non-coverage; LNC = left non-coverage.

of right and left non-coverage, defined as above as non-coverage at the right and left ends of the interval, and then identify these as mesial and distal non-coverage respectively (assuming $\theta < \mu$, and vice versa in the case $\theta > \mu$).

An equivalent definition (Newcombe 2001c, 2003a) is that non-coverage is distal if (and only if) the interval is on the same side of the parameter θ as the midpoint μ is. When $\theta = \mu$, the calculated interval cannot possibly be too mesially located, but can be too distal, in which case non-coverage is mesial. Accordingly, the cases in which $\theta = \mu$ merge into lines 1 and 4 of Table 4.2.

4.4.2 Interval Location in Simulation Studies of Coverage

Returning to our definitions of mesial and distal non-coverage, it is assumed by default that we want the frequencies of these occurrences to be equal (approximately or on average), resulting in an interval termed "central" by Clopper and Pearson (1934). Most intervals for π have been developed with the implicit intention of having this symmetry property. Indeed, the obvious symmetry of the simple Wald formula, untruncated, belies its grossly asymmetric coverage properties. As already explained, it is trivially true that, with a $U[0,1]$ pseudo-prior, the left and right non-coverage probabilities will be equal. But that is unhelpful. We would like the same to apply with a $U[0, 0.5]$ pseudo-prior, which would imply equality of the mesial and distal non-coverage probabilities (MNCP and DNCP). But it is no more possible to achieve this exactly than to align coverage exactly with $1 - \alpha$, and for the same reason. It is therefore important to evaluate mesial and distal coverage properties, as distinguished in Figure 4.8, as well as the overall coverage probability. In Chapter 5 we consider in detail how evaluation studies are performed to assess coverage properties of confidence interval methods, using a number of parameter space points (PSPs). Whether such an evaluation involves exact calculation of coverage or Monte Carlo methods, no substantial additional computational complexity is needed to accumulate left and right non-coverage rates separately, to be interpreted in terms of mesial or distal non-coverage.

It is then useful to characterise interval location by the balance between mesial and distal non-coverage. A simple index to do so is MNCP/NCP where NCP denotes the (exact or estimated) total non-coverage probability, NCP ≡ 1 − CP ≡ RNCP + LNCP ≡ MNCP + DNCP. This ratio may be calculated for a single PSP, or based on MNCP and NCP averaged over a number of PSPs. The MNCP/NCP index ranges on the interval [0,1], and 0.5 corresponds to a central interval. Assuming this is what we want, we might classify an MNCP/NCP ratio as satisfactory if it is between 0.4 and 0.6; the interval is too mesially located if the ratio is below 0.4, and too distal if it is above 0.6. Although it should be borne in mind that for a small sample size, or a parameter value towards the extremity of the range, it is often not possible to achieve balanced mesial and distal non-coverage. Table 4.3, derived from Newcombe (1998a), shows mean mesial and distal non-coverage probabilities, and the resulting MNCP/NCP ratio, for five methods, all based on the same sample of parameter space points. These results show the clear gradation from grossly predominantly mesial non-coverage (Wald) to predominantly distal (Wilson).

In Section 3.2 we identified the simple symmetry of the Wald interval as the root of both its coverage and boundary violation problems: it is problematic specifically because it is additively symmetric about the empirical estimate. Pseudo-frequency methods avoid these deficiencies by being additively symmetrical not about the empirical estimate but about a mesially shrunk point estimate. The Wilson interval overcorrects, by producing symmetry on a logit scale. Some beta intervals, including Clopper–Pearson as in Table 4.3 and also intervals with a Jeffreys $B(0.5, 0.5)$ prior, are about right,

TABLE 4.3

Balance of Mesial (Left) and Distal (Right) Non-Coverage Probabilities for 95% Intervals for the Binomial Proportion by Five Methods

Interval Method	Mean MNCP	Mean DNCP	MNCP/ NCP	Interpretation
Wald	0.1014	0.0172	0.855	Much too distal
Likelihood	0.0285	0.0238	0.545	Satisfactory—slightly too distal
Mid-*p*	0.0196	0.0233	0.458	Satisfactory—slightly too mesial
Clopper–Pearson	0.0127	0.0163	0.438	Satisfactory—slightly too mesial
Score	0.0162	0.0317	0.338	Substantially too mesial

Source: Newcombe, R. G. 2011b. *Communications in Statistics—Theory & Methods* 40: 1743–1767. With permission.

Note: MNCP = mesial non-coverage probability; DNCP = distal non-coverage probability; NCP = non-coverage probability.

as are mid-p intervals. The issue of boundary anomalies is linked to that of location, and looms largest for the single proportion: it also has repercussions for differences of proportions and so forth, but in those situations overt boundary anomalies are relatively infrequent.

4.4.3 Box–Cox Symmetry Index for Interval Location

The index MNCP/NCP is a property of a confidence interval method in the context of an evaluation of coverage properties based on one or more parameter space points, and examines mesial vis-à-vis distal coverage of the true parameter value. It cannot be applied to the confidence interval for a particular proportion $p = r/n$. A complementary index λ defined in Newcombe (2011b) may then be used, the Box–Cox index of symmetry (BCIS), defined as the power that must be used in a Box–Cox transformation in order to make the interval symmetrical about p.

The starting point is as follows:

> The Wald interval is symmetrical on the natural untransformed linear scale and is too distal.
>
> The Wilson interval is symmetrical on a logit scale and is too mesial.

The BCIS takes the linear scale symmetry of the Wald interval as one (fairly) extreme case, represented by $\lambda = 1$, and the logit scale symmetry of the Wilson interval as a (fairly) extreme case in the opposite direction, represented by $\lambda = 0$. Generally, intervals with good location properties are expected to be intermediate, with λ rather above 0. The two values defining the scale, 0 and 1, are not absolute limits—indeed, for a Bayes interval with uniform prior, λ takes a small negative value, indicating greater mesial shift than even the Wilson interval. An analogy may be drawn with the Celsius temperature scale: values of λ outside [0, 1], and temperatures outside the range 0°C to 100°C can occur; in each instance the two defining values are simply convenient, easily interpreted, fairly extreme values.

Specifically, for x ranging on (0, 1), consider the Box–Cox transformation family (1964) introduced in Section 2.4.4:

$$f(x) = \begin{cases} (x^{\lambda} - 1)/\lambda & (\lambda \neq 0) \\ \ln(x) & (\lambda = 0) \end{cases} \tag{4.1}$$

Then define

$$F(\lambda) = 2\{f(p) - f(1-p)\} - f(L) - f(U) + f(1-L) + f(1-U) \tag{4.2}$$

This is a decreasing function of λ, and the value of λ that makes $F(\lambda) = 0$ is readily determined by interval bisection. Table 4.4 shows λ calculated

TABLE 4.4

Box–Cox Symmetry Index λ for 95% Intervals for the Binomial Proportion Calculated by Several Methods

n	p	$B(1, 1)$ Prior	Wilson Score	$B(0.5, 0.5)$ Prior	Clopper–Pearson	arcsin	Wald
20	0.05	−0.109	0	0.168	0.259	0.448	1
20	0.25	−0.033	0	0.172	0.222	0.396	1
20	0.45	−0.026	0	0.168	0.210	0.386	1
100	0.05	−0.035	0	0.202	0.250	0.434	1
100	0.25	−0.015	0	0.180	0.203	0.393	1
100	0.45	−0.013	0	0.173	0.193	0.383	1
500	0.05	−0.017	0	0.210	0.232	0.432	1
500	0.25	−0.011	0	0.182	0.192	0.392	1
500	0.45	−0.010	0	0.175	0.183	0.382	1

Source: Newcombe, R. G. 2011b. *Communications in Statistics—Theory & Methods* 40: 1743–1767. With permission.

for nine representative proportions with six confidence interval methods, including Wald and Wilson for which $\lambda \equiv 1$ and 0, respectively.

For the other four interval methods, λ does depend on n and p, but only to a slight degree. For each method, the nine examples give quite similar values of λ. The simple arcsin interval is designed to be symmetrical on a square root scale, approximately although not exactly, so we would expect λ to be around 0.5. In fact it is always rather lower than this, around 0.4. Other methods are more mesial, with Clopper–Pearson and Jeffreys around 0.2. Remarkably, the Bayes interval with uniform prior has $\lambda < 0$, shrinking towards 0 for large n. This indicates greater mesial shift than even the Wilson interval—in line with its known location properties, although the differences in MNCP/NCP between these two in the 2007 evaluation were mostly very small.

When only point estimates and corresponding calculated limits for a few cases are available, calculating the BCIS bypasses the need to wonder whether examples such as this are consistent with the assertion that the Wilson interval is more mesially located than the Wald interval. While of course the location properties of these two interval methods are by now very clear, the same argument would apply to less well understood methods.

An index of this form is designed to be used when $0 < L < p < U < 1$. It cannot be meaningfully calculated for boundary cases, for any interval method. Note also that additive and logit scale symmetry are not necessarily in antithesis: for the case $p = 0.5$, any equivariant method will yield an interval that is symmetric on both scales, and $F(\lambda)$ is then identically 0. For this reason, the BCIS is inapplicable when $p = 0.5$, and it becomes difficult to estimate numerically when p approaches 0.5. Nevertheless it is adequate to evaluate the BCIS for, say, $p = 0.05$, 0.25 and 0.45, for chosen small, medium and large values of n, as in Table 4.4.

There is a remarkably close relationship between MNCP/NCP and the BCIS (Newcombe 2011b). Indeed, the BCIS, when it is applicable, has two major advantages over MNCP/NCP. Firstly, it has greater stability. In the best-behaved situations, with large n, the BCIS is an excellent proxy for MNCP/NCP. For other parts of the parameter space, the BCIS is nearly constant but MNCP/NCP varies greatly with n and π. Furthermore, it is straightforward to calculate the BCIS from the point estimate and sufficiently precisely calculated lower and upper limits, thus it is readily determined without the need to re-run an extensive evaluation of coverage properties. The choice of parameter space points for an evaluation of coverage properties is just as arbitrary as the choice of exemplar proportions for which to calculate BCIS. This suggests regarding the BCIS, rather than MNCP/NCP ratio, as the most appropriate measure of location.

The BCIS was developed specifically for the simple binomial proportion. It is also applicable to the U/mn measure developed in Chapter 15, as this is a special kind of proportion, which can range from 0 to 1 with a neutral point at 0.5; some methods that have been proposed have additive scale symmetry, others logit scale symmetry. There is no obvious applicability to measures such as differences or ratios of proportions.

It is strongly recommended that future studies evaluating coverage properties of intervals for proportions and related quantities should also assess interval location.

4.4.4 Implications for Choice of Point Estimate

Much has been written elsewhere on the desiderata for point estimates in general, and this is not the place to go into this issue in great detail.

It is good practice to report a point estimate that corresponds in some sense to the chosen confidence interval, and not just the calculated lower and upper limits. If this is not done, there is a real possibility that the reader will assume that all intervals are symmetrical and that the point estimate is therefore $(L + U)/2$.

While the issues of coverage and width are inapplicable to point estimates, the issue of location applies to point estimates just as it does to intervals. Indeed, sometimes the difference between several possible point estimates is well appraised from the perspective of location, in particular for Bayes intervals. Also, in an evaluation context, suppose that a particular interval method is found to produce a great degree of mesial shift. It is then helpful to appraise to what degree this is due to the mesial shift of the corresponding point estimate.

When estimating a proportion in the frequentist paradigm, the empirical estimate, which is identical to the MLE, is the obvious choice. But, as already described, mesially shrunk estimates have often been proposed—partly with extreme cases in mind—either via pseudo-frequencies, or by taking the arithmetic midpoint of the Wilson interval, or by an appeal to the properties of Bayes intervals.

For Bayes intervals, there are three obvious choices of point estimate: the posterior mean, median and mode. With a conjugate $B(a, b)$ prior, the posterior mode is normally $\eta = (r + a - 1)/(n + a + b - 2)$. When an uninformative, uniform prior $B(1, 1)$ is used, this reduces to the empirical estimate or MLE, $p = r/n$. When a Jeffreys $B(0.5, 0.5)$ prior is used, if $r = 0$ the posterior is a monotonically decreasing function, so the mode is at zero; similarly when $r = n$ the mode is at 1. The posterior mean $(r + a)/(n + a + b)$ incorporates a deliberate shrinkage towards the scale midpoint 0.5, and is thus more mesially located than the mode. For most distributions, including the beta posterior distribution here, the median is intermediate between the mode and the mean.

An alternative perspective is to consider the limiting case of a (central) 0% confidence interval, calculated by any of the methods described here. For many of these, the 0% interval is a single point, either identical to the empirical estimate $p = r/n$ (Wald, also Wilson as noted above), or mesially displaced, as in pseudo-frequency and Bayes methods. Thus 0% tail-based and HPD intervals are degenerate, at the posterior median and mode, respectively. However, the 0% Wald and score intervals with continuity correction, and also Clopper–Pearson 0% intervals are not degenerate. In these cases the upper limit for r/n is identical to the lower limit for $(r + 1)/n$. We can conceptualise this in two ways, as a hybrid of two different beta-Bayes intervals with different parameters, and also by considering the definition in terms of tails. Another tail-based interval, mid-p is degenerate when $1 - \alpha$ is set to 0, but does not reduce to any recognisable point estimate. It is shifted mesially, most markedly for $r = 1$ (or $n - 1$), relative to the empirical estimate, but is equal to it when $r = 0$ or n.

When constructing ratio measures such as the ratio of two proportions (Chapter 10) or the odds ratio (Chapter 11), the use of mesially shrunk point estimates is particularly desirable as it obviates problems arising from division by zero. This may be achieved by adding pseudo-frequencies to cells, or by a Bayesian approach. Parzen et al. (2002), Lin et al. (2009) and Carter et al. (2010) describe an alternative mesially shrunk point estimate, the median unbiased estimate (MUE), and corresponding deterministic bootstrap intervals for the odds ratio and the difference and ratio of proportions suitable for small samples, based on MUEs for the two proportions. The MUE and the corresponding bootstrap approach are described in Section 11.4.

4.5 Computational Ease and Transparency

Wherever possible, it is desirable that methods are easily implemented using widely available software. Gilks (2004) argued in a very different context for accessibility: "... unless methodological research results in a web-based tool or, at the very least, downloadable code that can be easily run by the user, it is effectively useless." In Section 3.6 we saw how a simple spreadsheet implementation in MS Excel is very convenient for calculating the Wilson score interval.

Why choose Excel to develop such resources? There are many reasons why Excel is an ideal platform for resources to help ordinary researchers post-process results from statistical packages to produce more appropriate representations of their results. To most computer users, Excel is already available and familiar. There is no additional cost, no need to install or become familiar with new software. Above all, it is straightforward to construct such a resource in a genuinely highly user-friendly form, as described in Section 3.6.

Obviously closed-form methods are easiest to program. But in practice methods involving a single depth of iteration, such as the preferred interval for the *U*/mn measure that generalises the Mann–Whitney statistic (Chapter 15) can be implemented in closed form by using a fixed number of interval bisections—typically 40 iterations yield precision around 10^{-12} which is more than adequate. The same applies to the relative risk and odds ratio after transformation to a bounded scale, as in Chapters 10 and 11, and the generalised Wilcoxon measure in Chapter 16.

These resources are designed to trap inappropriate input data. For CIPROPORTION.xls, when inappropriate starting values are entered, calculations result in cells displaying codes such as #DIV/0!, #NUM! and so forth, which is clear enough to give the message! However, doing so does not crash the software—when the inadmissible values are overwritten with acceptable ones, no harm is done, the correct answer appears as usual. For some other calculations, such as relative risks or odds ratios, a division by zero can legitimately occur, leading to an infinite point estimate. A good method, such as the one implemented in the spreadsheet ODDSRATIOANDRR.xls, will then display a finite lower limit in the usual way, and #DIV/0! for the point estimate and upper limit, correctly implying that these should be reported as ∞. This contrasts markedly with the RISK option for the SPSS CROSSTABS procedure, which does not even display a point estimate for a relative risk or odds ratio that is infinite or even zero.

Most users will be happy to use such resources in black-box mode. The more mathematically inclined can unprotect the sheet and disclose concealed workings by changing the numerical format for all cells to General.

All of the above is in stark contrast to the principle that spreadsheet software is not recommended for primary statistical analysis starting from raw data, one record per case. For this purpose, use of dedicated statistical software is strongly recommended. The minimum requirements are that the software should constrain the format of the dataset, so that normally analyses are performed on a rectangular data array in which rows represent cases and columns variables, preferably in numerical rather than text format. It is also important that missing values are denoted either by blank cells or by dedicated values that are declared as missing by the user, and are then excluded from statistical analyses in a way that would be appropriate on the assumption that missing status gives no information on what the missing values are likely to be.

While Excel is ideal for implementing closed-form methods and for preliminary development work, evaluation of coverage and so forth requires higher precision. All the author's evaluations reported in this book used double precision Fortran programming, based on uniform random variates generated by algorithm AS183 (Wichmann and Hill 1985) then transformed as necessary. In Chapters 15 and 16, algorithms for normal and beta distributions (Griffiths and Hill 1985; Vlachos 2005) are also used.

Another important, often disregarded issue is transparency. This is not quite the same issue as avoiding unnecessary computational complexity. It could be argued that computational complexity as such does not really matter, if the algorithm is made available to ordinary users in widely used software. But transparency does matter. In Section 4.2.1 we considered some of the points raised with regard to coverage in the discussion following the Brown et al. (2001) paper. It is also noteworthy that in response to Corcoran and Mehta's assertion that: "Simplicity and ease of computation have no role to play in statistical practice," Brown et al. argue that: "… conceptual simplicity, parsimony and consistency with general theory remain important secondary conditions to choose among procedures with acceptable coverage and precision". In this respect, as well as in relation to coverage, the author agrees with the position of Brown et al., as evidently do discussants Agresti and Coull, who even consider the Wilson formula to be offputting to some. Considering the widely diverging stances on this issue, and bearing in mind that a major purpose of statistical methodology is to communicate a comprehensible quantitative appraisal of the precision of empirical findings, simplicity and transparency of statistical methods is desirable, whenever it can be achieved without important loss of performance. The more that end users—both researchers and those who read what they produce—can see just how statistical methods work to achieve their objectives, the more confidently they will use them.

A similar issue applies to choice of comparative measures (Deeks et al. 2008, Section 9.4.4.4).

5

Evaluation of Performance of Confidence Interval Methods

5.1 Introduction

In the previous chapter we examined various possible criteria for optimality of confidence intervals. Here, we consider how evaluation studies are conducted to yield pertinent information on coverage and other properties of confidence interval methods.

Many studies have been published which report evaluations, often comparative, of various confidence interval methods for quantities related to proportions. The most numerous evaluations relate to the binomial proportion $p = r/n$. Notable examples include Vollset (1993), Newcombe (1998a, 2007c), Agresti and Coull (1998) and Brown et al. (2001).

5.1.1 Systematic Evaluation Methodology

As observed in Chapter 2, confidence interval methods for means and their differences, in the admittedly simplistic case of known variances, achieve exactly the nominal coverage level. We have seen that this is far from the case for intervals for proportions and related quantities. Accordingly, there is a strong argument that any publication which introduces a new method should also include some evaluation of coverage and related properties.

Evaluation of coverage involves identification of a number of parameter space points (PSPs), which should be reasonably representative of those likely to be encountered in real-world applications. For the simplest case of the binomial proportion $p = r/n$, in principle all parameter values between 0 and 1 are relevant. Good methods should be equivariant, so by restricting the evaluation to $0 < \pi < 0.5$ we may also obtain an assessment of location properties. A confidence interval algorithm for the difference between two proportions p_1 and p_2 should cope with the extreme case $p_1 = 1, p_2 = 0$, nevertheless a good evaluation of coverage properties will recognise that in practice large differences are very infrequent. As explained in Section 11.1 we would rarely want to report the ratio of two proportions when both are large.

Nevertheless, for means of equally spaced trinomial variables (Section 9.2), there are several contexts of application, for which quite different regions of the parameter space are plausible. Thus here it is necessary to validate that good coverage properties extend throughout the parameter space.

Some kind of evaluation of coverage is then carried out for each PSP in turn, and the results presented numerically or graphically. The chosen PSPs can be either chosen purposively or randomly generated. Two approaches are possible. In simpler contexts it is feasible to calculate left and right non-coverage for each PSP in turn, either exactly or by generating a large number of random datasets corresponding to that PSP and counting how often each type of non-coverage occurs. When just a small number of PSPs are used, the results may then be presented for individual PSPs. But it is preferable to use a large number of PSPs, so naturally some form of summarisation is then required. Thus in the original evaluation (Newcombe 1998a) a representative set of 96000 PSPs were generated randomly. Coverage properties were determined by simulation for each PSP, then summarised across all PSPs.

In more complex settings this may not be practicable. An alternative approach as in Newcombe (2007c), which is more efficient and in some situations computationally far more tractable, is to generate a very large number of PSPs, with just one random dataset per PSP, then determine whether right or left non-coverage has occurred, and summarise across all the PSPs. What is lost is information on minimum coverage across all possible values of π for any given n. But the more complex approach in Newcombe (1998a) only gives an upper bound for this: moreover, in this simplest application, this can be determined directly for some interval methods, as described in Section 4.2.2. If it is accepted that mean coverage is more important than minimum coverage across all PSPs, the simpler approach leads to appropriate estimates of mean coverage and distal and mesial non-coverage probabilities. However, this would be unacceptable to those who take the stance that coverage should be construed as minimum coverage—from that standpoint, it would be obligatory to calculate coverage at the PSP level, then summarise across PSPs.

Whichever approach is used, the evaluation should also count instances of each possible kind of boundary anomaly for Wald and related methods.

It is also important to appraise performance in different parameter space zones (PSZs). Newcombe (1998a) presented results aggregated across all PSPs, with some results also for selected regions of the parameter space, which were not exhaustive. In the 2007 evaluation, several parameter space zones of interest were first selected, then a large representative sample of PSPs was generated for each zone. Findings for different PSZs are reported alongside one another, but are not combined numerically. The chosen PSZs should represent meaningful possible combinations of parameter values, but as in both these evaluations, need not abut.

Within the chosen parameter space or zone, the natural choice of pseudo-prior for sampling a proportion π is a uniform distribution. For more complex quantities such as ratios of proportions or odds ratios, choice of a suitable pseudo-prior is more contentious, for two reasons: these quantities are more naturally considered on a log scale, and the support scale extends to infinity.

Unless otherwise specified, all evaluations described here relate either exclusively or primarily to intervals designed to have a nominal 95% two-sided coverage level. Some studies such as Newcombe (1998a, b) report subsidiary results for other confidence levels also, typically 90% and either 99% or 99.9%, and for parameter space points involving large sample sizes and small proportions, as often encountered in epidemiology.

5.1.2 Parametrisation

In general, the definitions of confidence interval methods can involve three types of parameters.

Given parameters are typically sample sizes such as n for the simple binomial proportion π, or m and n for a comparison of two proportions π_1 and π_2.

There is a single *parameter to be estimated*—in the above examples, π or $\delta = \pi_1 - \pi_2$.

Also, for more complex cases, there can be additional parameters, which can be broadly classed as either *nuisance parameters* or *instrumental parameters*. For example, when the quantity of interest is an unpaired or paired difference of proportions, the sum (or equivalently the average) of the two proportions has an obvious role as a nuisance parameter. Sometimes a nuisance parameter may be eliminated by conditioning on a suitable sufficient statistic. There can be additional parameters of an instrumental nature, such as the correlation measure ϕ in the paired difference case (see Section 8.4). There is no firm distinction between these—we think of a nuisance parameter as something to eliminate, either by substituting some estimate or preferably by using a profile likelihood approach, whereas an instrumental parameter is an additional parameter needed both to fully specify and then to drive the evaluation of coverage and so forth for that parameter space point.

In the evaluation of various intervals for the unpaired effect size measure U/mn generalising the Mann–Whitney U statistic (Newcombe 2006b; see Chapter 15), the concept of a parameter space point is extended to include different distributional models used to generate the Monte Carlo data. There were 336 combinations of seven sample sizes for the two groups and eight values of the parameter of interest θ with six distributions used. Strictly, in this evaluation we should refer to 56 PSPs and 336 cells, although by slight abuse of terminology these 336 cells were referred to interchangeably as PSPs—in normal usage a "parameter" is understood to mean a variable on a continuous, ordinal or binary scale, rather than a nominal, classificatory variable distinguishing between qualitatively different distributional models.

5.1.3 Interval Width

For inferences relating to means, interval width depends on the sample sizes and standard deviations, but not on the means themselves. In contrast, for the simple binomial proportion, the expected interval width depends grossly on π as well as n, although unlike coverage, it varies smoothly with these parameters. Hence it is only meaningful to compare average widths of intervals produced by different methods holding these parameters constant. This is best achieved by comparing the average widths for different interval methods when applied within the same parameter space zone.

The exact mean interval width for $p = r/n$ given n and $\pi \sim U[0, 1]$ may simply be obtained, without recourse to Monte Carlo methods, as the average of the interval widths corresponding to $r = 0, 1,\ldots, n$. But this does not help to evaluate the dependence of interval width on π. For a specific parameter space point (n, π), it is practicable to evaluate expected width exactly for fairly small values of n, as in Newcombe (1998a), but this becomes unwieldy for large n. Often it is more convenient to assess mean interval width as a by-product of the Monte Carlo process used to evaluate coverage. As indicated in Section 4.3, evaluation of expected width must be based on intervals that have been truncated at the boundary of the support space whenever necessary; moreover for ratio-scale quantities such as ratios of proportions or odds ratios, interval width is only meaningful on a suitably transformed scale.

5.2 An Example of Evaluation

The evaluation presented in this section (Newcombe 2007c) was performed primarily in order to examine location properties of various confidence interval methods for the simple binomial proportion $p = r/n$. Results relating to location are published as Newcombe (2011b).

5.2.1 Methods

Thirteen confidence interval methods were evaluated, including pseudo-frequency methods with $\psi = 1$, 2 and 3, and Bayes intervals with uniform $B(1, 1)$ and Jeffreys $B(0.5, 0.5)$ uninformative conjugate priors. Both tail-based (invariant, closed-form) and highest posterior density (shortest) intervals were included. These methods are defined in detail in Section 3.5 and illustrative examples are shown in Table 3.2. As well as location, coverage, average width and boundary aberration properties were also evaluated.

Section 4.2 argues for aligning mean rather than minimum coverage with $1 - \alpha$. This implies the need to calculate the mean coverage probability by averaging or integrating over some pseudo-prior for π, and possibly over a range of values of n also. This is conveniently achieved by a Monte Carlo simulation process, involving randomly sampled parameter space points and either one or many randomly generated data points for each such PSP. For the single proportion, an obvious choice for π is the uniform distribution $U[0, 1]$. A pseudo-prior restricted to $[0, 0.5]$ is preferable, enabling right and left non-coverage to be interpreted more informatively as mesial and distal non-coverage. The pseudo-prior $U[0, 0.5]$ was used in an earlier evaluation of several intervals for $p = r/n$ (Newcombe 1998a), but in the 2007 evaluation, only three shorter ranges were used, as described below.

In recognition of the fact that different methods can perform very differently according to the values of n and π, we consider their performance in nine parameter space zones, defined as follows.

Sample size n:	Small (5 to 10)	Medium (45 to 50)	Large (245 to 250)
Proportion π:	Extreme (0 to 0.05)	Intermediate (0.2 to 0.25)	Central (0.4 to 0.45)

For example, zone small extreme (SE) comprises parameter space points with n from 5 to 10 and π ranging from 0 to 0.05, and similarly for the remaining 8 combinations of ranges of sample sizes and proportions. These zones were chosen as representative of most situations likely to arise in practice. We use ranges of values of both parameters n and π, for two reasons. One is Brown et al.'s demonstration of the existence of "lucky" and "unlucky" combinations of these parameters. The other issue is that some previous evaluations such as Ghosh (1979) used round values of n and π, resulting in an integer value for the expectation of R, $ER = n\pi$. Although the observed value, r, must be an integer, there is no reason why ER should be; due to discrete behaviour, it is theoretically possible that in this situation performance may not be typical. Consequently in this and the author's other evaluations this is obviated by using randomly generated real numbers for proportion parameters, although no evidence has emerged that this would actually matter.

It is conceded that sample sizes of the order of 5 to 10 are smaller than would be recommended, even for central proportions, and certainly for extreme ones—it is difficult to imagine that any method could perform excellently in zone SE. It is important to include such small values of n in an evaluation because they are often encountered in practice nonetheless. The above zones for π, together with their complements, serve as an effective proxy for a uniform distribution on $[0, 1]$.

Tables were produced listing intervals for each method for each of these 18 values of n with $r = 0, 1,\ldots, n$. For each zone, for each of the six constituent values of n in turn 1,600,000 values of π were generated randomly from a uniform distribution on the relevant range, resulting in 9,600,000 PSPs of the form (n, π) per zone. For each PSP, a single random variate $R = r \sim B(n, \pi)$ was generated. Each randomly generated π was compared with the prestored intervals for r out of n for the 13 methods, noting whether right or left non-coverage or boundary anomalies occur. Results were then aggregated at the PSZ level. This degree of replication estimates right and left non-coverage rates of the order of 0.025 within 0.0001 with 95% confidence for each PSZ, hence coverage figures are unlikely to be in error by more than 0.0001 and are appropriately rounded to four decimal places.

5.2.2 Results

Tables 5.1 to 5.4 summarise the performance of the 13 methods in the nine chosen zones of the parameter space. For large n and central π, all 13 methods perform similarly, as expected, with little to choose between them: the mean coverage probability is close to the nominal 0.95, interval location is satisfactory, interval widths are similar, and boundary aberrations do not occur. For smaller n, and especially smaller π, differences between methods are much greater.

Considering first coverage (Table 5.1), the Wald method performs very poorly in several zones, including LE as well as SE, ME and SI, driven largely by the high frequency of zero width intervals. Incorporating a continuity correction results in better yet erratic coverage properties, remaining poor in zones SI and ME, but excessively high in zone SE as the continuity correction ensures that low values of π are almost always covered.

The coverage of the Wilson interval is quite close to 0.95 for all nine zones, although rather low in zone SE. Incorporating a continuity correction leads to coverage well above 0.95 throughout.

As is well recognised, from the standpoint of average coverage the Clopper–Pearson interval is unnecessarily conservative, especially for small n or π. The analogous interval using mid-p tail areas substantially mitigates this, with mean coverage remaining above 0.95 throughout.

The evaluation confirms the claim (Agresti and Coull 1998) that expected coverage is about right for the Wald + 2 interval. However, coverage tends to fluctuate for Wald + 1 and Wald + 3, as described below.

With a Jeffreys prior (J), the tail-based interval has generally reasonable mean coverage properties. The corresponding highest posterior density interval is substantially anticonservative in zones SI and SC. When a uniform prior (U) is used, the tail-based interval has reasonable mean coverage except in zone SE. The uniform-prior highest posterior density interval is conservative there, but somewhat anticonservative in zone SC. The contrast between the coverages of the tail-based intervals with uniform and Jeffreys priors sheds light on why the moving average coverage plot for the former (Figure 4.4) is satisfactory whilst that for the latter (Figure 4.5) is not. Apparently it is more crucial to avoid low coverage closer to the scale midpoint.

For each of the nine zones, average interval width was calculated as a by-product of the Monte Carlo simulation. For each sample size range, the exact mean interval width for a $U[0, 1]$ prior was also calculated for each n, then averaged over the six values of n in the relevant range. In Table 5.2 the mean absolute width is displayed for method 6 (mid-p) to illustrate the dependence

TABLE 5.1

Mean Coverage of 95% Intervals for the Binomial Proportion by 13 Methods for 9 Parameter Space Zones

Zone	SE	SI	SC	ME	MI	MC	LE	LI	LC
Range of n	5–10	5–10	5–10	45–50	45–50	45–50	245–250	245–250	245–250
Range of π	0–0.05	0.2–0.25	0.4–0.45	0–0.05	0.2–0.25	0.4–0.45	0–0.05	0.2–0.25	0.4–0.45
Method									
1 Wald	0.1673	0.8203	0.8795	0.6202	0.9323	0.9406	0.8581	0.9466	0.9480
2 Wald CC	0.9999	0.8282	0.9289	0.7870	0.9521	0.9567	0.9099	0.9551	0.9552
3 Wilson	0.9404	0.9616	0.9515	0.9535	0.9511	0.9503	0.9534	0.9503	0.9499
4 Wilson CC	0.9834	0.9851	0.9847	0.9791	0.9684	0.9654	0.9722	0.9585	0.9570
5 Clopper-Pearson	0.9919	0.9886	0.9821	0.9877	0.9669	0.9645	0.9734	0.9582	0.9568
6 Mid-p	0.9838	0.9802	0.9618	0.9793	0.9526	0.9516	0.9590	0.9505	0.9502
7 Wald + 1	0.9959	0.9736	0.9358	0.9944	0.9496	0.9466	0.9665	0.9497	0.9491
8 Wald + 2	0.9730	0.9609	0.9715	0.9786	0.9550	0.9515	0.9678	0.9511	0.9501
9 Wald + 3	0.7984	0.9405	0.9819	0.9196	0.9525	0.9556	0.9408	0.9507	0.9509
Bayes intervals:									
10 Uniform TB	0.8588	0.9654	0.9627	0.9325	0.9516	0.9516	0.9471	0.9504	0.9502
11 Uniform HPD	0.9765	0.9583	0.9231	0.9804	0.9460	0.9453	0.9580	0.9491	0.9489
12 Jeffreys TB	0.9653	0.9624	0.9392	0.9716	0.9491	0.9491	0.9514	0.9498	0.9497
13 Jeffreys HPD	0.9918	0.9099	0.8985	0.9663	0.9393	0.9422	0.9437	0.9480	0.9484

Note: CC = continuity-corrected; TB = tail-based; HPD = highest posterior density.

TABLE 5.2

Mean Width of 95% Intervals for the Binomial Proportion by 13 Methods for 9 Parameter Space Zones, by Simulation; Mean Absolute Widths Are Shown for Method 6 (mid-p); for Other Methods, the Width Is Shown as a Multiple of That for Method 6

Zone	SE	SI	SC	S	ME	MI	MC	M	LE	LI	LC	L
Range of n	5–10	5–10	5–10	5–10	45–50	45–50	45–50	45–50	245–250	245–250	245–250	245–250
Range of π	0–0.05	0.2–0.25	0.4–0.45	0–1	0–0.05	0.2–0.25	0.4–0.45	0–1	0–0.05	0.2–0.25	0.4–0.45	0–1
Mean width as a multiple of that of method 6												
Method 1 Wald	0.181	0.844	1.017	0.816	0.591	1.004	1.014	0.980	0.914	1.001	1.003	0.997
2 Wald CC	0.372	0.992	1.174	0.973	0.703	1.094	1.091	1.067	1.001	1.040	1.036	1.038
3 Wilson	0.997	0.923	0.904	0.926	1.087	0.983	0.977	0.988	1.031	0.997	0.995	0.998
4 Wilson CC	1.220	1.079	1.044	1.086	1.278	1.066	1.048	1.076	1.135	1.035	1.028	1.039
5 Clopper-Pearson	1.157	1.099	1.087	1.103	1.139	1.070	1.062	1.072	1.083	1.036	1.031	1.037
6 Mid-p	1.000	1.000	1.000	1.000	1.000	1.000	1.000	1.000	1.000	1.000	1.000	1.000
7 Wald + 1	0.918	0.968	0.980	0.965	0.995	1.002	0.995	1.001	1.021	1.000	0.999	1.001
8 Wald + 2	1.135	0.968	0.917	0.973	1.249	0.999	0.977	1.011	1.105	1.000	0.995	1.004
9 Wald + 3	1.229	0.930	0.855	0.944	1.434	0.994	0.960	1.015	1.177	0.999	0.991	1.007
Bayes intervals:												
10 Uniform TB	1.026	0.937	0.915	0.941	1.070	0.985	0.981	0.989	1.022	0.997	0.996	0.998
11 Uniform HPD	0.895	0.897	0.901	0.898	0.963	0.977	0.980	0.976	0.984	0.995	0.996	0.995
12 Jeffreys TB	0.860	0.928	0.943	0.924	0.924	0.985	0.989	0.982	0.977	0.997	0.998	0.996
13 Jeffreys HPD	0.710	0.874	0.921	0.868	0.813	0.976	0.989	0.968	0.937	0.996	0.998	0.993
Mean absolute width												
for method 6	0.3706	0.5405	0.6123	0.5299	0.0977	0.2328	0.2742	0.2209	0.0378	0.1036	0.1225	0.0976

Note: Precisely calculated mean intervals width for a $U[0, 1]$ prior averaged over each range of n are also shown. Intervals are truncated in the event of overshoot. CC = continuity-corrected; TB = tail-based; HPD = highest posterior density.

on n and π for a gold standard method which achieves very good coverage properties without excessive width. Widths of other intervals are shown as multiples of that for the mid-*p* method.

As expected, relative differences in interval width are more marked for small n and/or π, nevertheless even in zone L it is clear that the continuity-corrected and exact intervals are about 4% wider than necessary. In all nine zones continuity-corrected Wilson and Clopper–Pearson intervals are wider than mid-*p*, whilst Bayes intervals are almost always somewhat narrower than mid-*p*.

Table 5.3 shows that for all 13 methods, interval location as expressed by MNCP/NCP was always very good, between 0.46 and 0.54, in zone LC, and in the satisfactory range, 0.4 to 0.6, in zone MC. However, for smaller n or π, interval location is often less satisfactory.

The Wald interval was too distal for extreme and also intermediate proportions, even for large n. Using a continuity correction does not correct this asymmetry. However, in zone SE, left and right NCPs were 0.0005 and 0.8322 without and 0.00012 and 0 with continuity correction, all driven largely by the high frequency of a zero numerator, leading to a degenerate, zero width interval for method 1 or an interval attributable purely to the continuity correction when method 2 is used.

Wilson intervals, without and with continuity correction, tend to be too mesial, particularly when π or n is small. However, in zone SC, interval location is satisfactory only when the continuity correction is not used.

Coverage was never balanced in zones SE, SI and ME. Barring these, the Clopper–Pearson, mid-*p* and Jeffreys tail-based intervals generally have satisfactory location properties. Zone LE is challenging: Wald intervals, without and with continuity correction, and the Jeffreys highest posterior density interval are much too distal, whereas Wilson intervals, without and with continuity correction, Wald+2 and 3, and the tail-based uniform-prior interval are much too mesial.

Of the pseudo-frequency methods, Wald + 1 is best for symmetry, but Wald + 2 for overall coverage. The Wald + 2 interval was developed as a simplifying approximation to the Wilson interval, and tends to be rather more mesial than it. The Wald + 3 interval is simply too mesial.

For Bayesian intervals, when a symmetrical conjugate prior is used, both the tail-based and highest posterior density approaches aim to achieve symmetrical coverage, but do so in quite different ways. These results enable us to examine which best achieves this, from a frequentist standpoint. When a uniform prior is used, the highest posterior density interval achieves better symmetry, as well as greater narrowness, at the cost of non-invariance. With a Jeffreys prior, the preference is far from clear.

Apropos boundary anomalies, degeneracy occurs only with the Wald interval, when $r = 0$ (or n), the frequency of which depends crucially on n and π. Overshoot is frequent for all Wald-based methods, including pseudo-frequency methods, when π or n is small (Table 5.4).

TABLE 5.3

Mean Mesial (Right) Non-Coverage as a Proportion of Mean Total Non-Coverage Probability for 95% Intervals for the Binomial Proportion by 13 Methods for 9 Parameter Space Zones

Zone	SE	SI	SC	ME	MI	MC	LE	LI	LC
Range of n	5–10	5–10	5–10	45–50	45–50	45–50	245–250	245–250	245–250
Range of π	0–0.05	0.2–0.25	0.4–0.45	0–0.05	0.2–0.25	0.4–0.45	0–0.05	0.2–0.25	0.4–0.45
Method									
1 Wald	0.9994	0.9092	0.5492	0.9961	0.7671	0.5691	0.9596	0.6362	0.5327
2 Wald CC	0.0000	0.9515	0.6173	0.9978	0.7889	0.5760	0.9635	0.6416	0.5342
3 Wilson	0.0000	0.0000	0.4648	0.0000	0.3902	0.4754	0.2111	0.4532	0.4904
4 Wilson CC	0.0000	0.0000	0.2813	0.0000	0.3495	0.4677	0.1670	0.4472	0.4887
5 Clopper-Pearson	0.0000	0.0000	0.3829	0.0000	0.4755	0.4953	0.3816	0.4952	0.4997
6 Mid-p	0.0000	0.0000	0.5070	0.0000	0.4955	0.5004	0.4305	0.4999	0.5010
7 Wald + 1	0.0000	0.0782	0.5391	0.0000	0.5694	0.5171	0.6315	0.5415	0.5104
8 Wald + 2	0.0000	0.0000	0.2924	0.0000	0.3341	0.4633	0.1720	0.4421	0.4878
9 Wald + 3	0.0000	0.0000	0.1333	0.0000	0.1500	0.4082	0.0233	0.3463	0.4649
Bayes intervals									
10 Uniform TB	0.0000	0.0000	0.4286	0.0000	0.3855	0.4738	0.1971	0.4508	0.4897
11 Uniform HPD	0.0000	0.0564	0.5195	0.0000	0.5282	0.5072	0.4930	0.5162	0.5047
12 Jeffreys TB	0.0000	0.2934	0.5587	0.0142	0.5016	0.5022	0.4636	0.5005	0.5010
13 Jeffreys HPD	0.0000	0.6285	0.5054	0.7173	0.6440	0.5343	0.7502	0.5651	0.5159

Source: Newcombe, R. G. 2011b. *Communications in Statistics—Theory & Methods* 40: 1743–1767. With permission.

Note: CC = continuity-corrected; TB = tail-based; HPD = highest posterior density.

TABLE 5.4

Frequency of Degeneracy and Overshoot for 95% Intervals for the Binomial Proportion by Relevant Methods for 9 Parameter Space Zones

Zone	SE	SI	SC	ME	MI	MC	LE	LI	LC
Range of n	5–10	5–10	5–10	45–50	45–50	45–50	245–250	245–250	245–250
Range of π	0–0.05	0.2–0.25	0.4–0.45	0–0.05	0.2–0.25	0.4–0.45	0–0.05	0.2–0.25	0.4–0.45
Degeneracy									
Method 1 Wald	0.8322	0.1636	0.0281	0.3783	0.00001	0.0000	0.0805	0.0000	0.0000
Overshoot									
Method 1 Wald	0.1664	0.6234	0.4567	0.5636	0.0036	0.0000	0.2412	0.0000	0.0000
2 Wald CC	0.9998	0.9239	0.7353	0.9807	0.0122	0.0000	0.4017	0.0000	0.0000
7 Wald + 1	0.9816	0.4850	0.1673	0.8501	0.0008	0.0000	0.2417	0.0000	0.0000
8 Wald + 2	0.8631	0.2016	0.0335	0.6689	0.0001	0.0000	0.1609	0.0000	0.0000
9 Wald + 3	0.4013	0.0522	0.0041	0.3783	0.00001	0.0000	0.0805	0.0000	0.0000

Note: CC = continuity corrected.

5.3 Approaches Used in Evaluations for the Binomial Proportion

5.3.1 Methodology Used in the Author's Previous Evaluation

Newcombe (1998a) evaluated coverage, location, expected width and boundary aberrations for seven frequentist methods, including Wald and Wilson without and with continuity correction, exact and mid-p intervals for $p = r/n$.

There were 96,000 representative parameter space points generated randomly; for each n from 5 to 100, there were 1000 values of π sampled from $U[0, 0.5]$. Tables were generated for each method for $n = 5, 6, \ldots, 100$ and $r = 0, 1, \ldots, n$. For each n and π, the binomial probabilities $\Pr[R = r \mid n, \pi]$ were generated for each r from 0 to n. These were used to produce precisely computed probabilities of left and right non-coverage, boundary violation and degeneracy for each method for each PSP by summation. These were then summarised across the 96000 randomly generated PSPs, and for meaningful subsets of the parameter space with $5 \le n \le 10$, $90 < n \le 100$, $0 \le \pi \le 0.05$, $0.45 \le \pi \le 0.5$, $0 \le n\pi < 5$ and $45 \le n\pi < 50$. This approach estimates left and right non-coverage of the order of 0.025 to within 0.001 with 95% confidence.

Results were also produced for 90% and 99% intervals, and for proportions with large denominators and small to moderate numerators such as often encountered in epidemiology. The findings were generally similar to those in the main evaluation.

Exact expected interval width was reported for nine PSPs ($n = 5$, 20 and 100 with $\pi = 0.5$, 0.2 and 0.05) and more simply, for these three values of n with uniform π.

Similar methodological issues arise in evaluations for intervals for various measures derived from or related to proportions (see Sections 7.5.2, 8.5.3, 9.2.5, 14.2.3, 15.5 and 16.5).

5.3.2 Methodology Used in Other Evaluations

Vollset (1993) evaluated 12 intervals for the single proportion. Implicitly starting from the standpoint that the Clopper–Pearson method was the agreed gold standard, he developed clear evidence that other less conservative methods such as the Wilson interval perform well within their own terms of reference. Results are presented mainly graphically, without recourse to Monte Carlo methods—an approach that is practicable for $p = r/n$ but less so for more richly parametrised contexts. The main expression of performance comprises plots of coverage probabilities of nominal 95% intervals for selected sample sizes against π for $n = 10$, 100 and 1000, similar to the right-hand panel of Figure 4.1.

Vollset also investigated alternative methods which aim to mitigate anti-conservatism by substituting Clopper–Pearson exact limits in boundary cases. The effect of this modification is to greatly improve the coverage of several intervals, Wald without and with continuity correction, delta logit, likelihood and arcsine, for which the pattern of coverage below $1 - \alpha$ is not predominantly the result of the "first dip" effect described in Section 4.2.2. This modification hardly affects the coverage properties of the Wilson and Bayes intervals, for which this first dip effect is characteristic: it occurs because the lower limit for $r = 1$ is too high, not because the upper limit for $r = 0$ is too low.

Interval widths for all possible r are plotted against r for each method and sample size, with mean widths for $\pi = 0.5$ and 0.1 indicated. Percentage differences from the Clopper–Pearson limits as gold standard (relative errors) are plotted against sample size for each method. The minimum distance from the boundary (i.e., $\min(r, n - r)$) to guarantee relative errors not exceeding 1%, 5% or 10% is tabulated for each method for the two values of π.

Agresti and Coull (1998) introduced pseudo-frequency intervals as described in Section 3.4.7. They evaluated the mean coverage of several methods including these for uniform, bell-shaped and skew beta pseudo-priors for π, for five sample sizes from 5 to 100. Most results shown relate to the uniform pseudo-prior, but similar results were obtained with the others. Mean coverage probabilities were calculated and plotted against the sample size. Several novel evaluative measures were introduced. For nominal 95% intervals, $\Pr[CP < 0.9]$ and $\Pr[0.93 < CP < 0.97]$, the proportions of the parameter space with coverage probabilities below 90% and between 93% and 97%, were calculated. An interesting measure $\Pr[CP_A$ closer to $1 - \alpha$ than $CP_B]$ was used to compare coverage of two methods A and B based on exactly the same PSPs. However, like several other measures it presupposes equal but opposite discrepancies from $1 - \alpha$ should be regarded as equally serious, which is debatable— such an index is only really interpretable once it has been established that overall coverage is very similar for the two methods, it cannot be interpreted as a stand-alone measure. A uniform-weighted root mean square discrepancy of the coverage probability from the nominal level $1 - \alpha$ was also calculated. Expected width was calculated for beta as well as uniform pseudo-priors, and plotted against π for given n. Similar criteria were used in subsequent studies relating to unpaired and paired differences of proportions (Agresti and Caffo 2000; Agresti and Min 2005b).

Brown et al. (2001) evaluated Wald, Wilson, Jeffreys and Agresti–Coull intervals, and considered several others including modifications to Wilson and Jeffreys intervals near the boundary. The presentation of results is mainly graphical, similar to Vollset (1993). As noted in Section 4.2, they demonstrated that the coverage probability is a highly unsmooth function of π for given n and vice versa. In contrast, Figure 6 of the article shows how the coverage probability for four selected algebraically formulated methods, averaged over a $U[0, 1]$ pseudo-prior for π, are smooth functions of n, with an unequivocal ordering of their coverage probabilities. Mean absolute errors, defined by $\int \left| CP - (1 - \alpha) \right| d\pi$ are plotted against n for these methods. A measure of this kind weights undercoverage and overcoverage equally, which is not necessarily desirable. Expected interval width is plotted as a function of π for a typical n, and average expected width over $\pi \sim U[0, 1]$ plotted against n.

These authors recommended Wilson or Jeffreys intervals for n up to 40 and Agresti–Coull intervals for n over 40. As noted in Section 4.1.1, the drawback to such hybrid intervals is the potential for unsmooth or even non-monotonic behaviour as n crosses the boundary at 40: it would be illogical to report a higher value for a calculated limit for r out of 41 than for the same r out of 40, but there is little reassurance that this cannot occur. At the least, smoothness is lost.

Böhning and Viwatwongkasem (2005) calculated coverage probabilities for selected n for Wald, Wilson and various modified Wald intervals, for $\pi = 0.0001, 0.0002, \ldots, 0.9999$, by simulation with 10000 runs. These were then used to evaluate the variance of the coverage probability across these 9999 π values as well as their mean and minimum.

Pires and Amado (2008) evaluated performance of 20 interval methods by calculating coverage and expected width for each method for 5000 values of π equally spaced in [0.0001, 0.5], for $n = 10, 11, \ldots, 1000$ and $\alpha = 0.05$ and 0.01, as functions of n and π. Coverage was plotted as a function of π for selected n. Minimum and mean coverage (presumably over these πs) are plotted as functions of n. Expected interval width (also presumably averaged over these πs) was expressed as a multiple of that for the Clopper–Pearson method for selected n.

Pires and Amado imply (p. 177) that exactly computed coverage probabilities are preferable to results based on Monte Carlo evaluations. That is what one would expect, but the reality is not so simple. Monte Carlo evaluations can explore problems of greater complexity, so sometimes it would be prohibitive to set up all possible tables required for an exact evaluation.

Zhou et al. (2008) evaluated coverage of 95% intervals calculated by four methods using 80000 PSPs representing each combination of eight values of n from 10 to 100 with 10000 values of π from 0.000099 to 0.999999. The coverage probability and expected width were calculated for each PSP, and plotted against π for selected sample sizes. Then for each n, the mean absolute error, proportion of PSPs with coverage below 0.93 and expected width averaged over π were calculated and plotted against n.

5.3.3 Some Further Methodological Points

Coverage evaluated for a large number of PSPs can be summarised by a variety of essentially equivalent types of plots. These include histograms, such as those of Mehta and Walsh (1992) who used histograms with bin width 0.00333, an unusual yet very effective choice. Also box plots may be used, or tadpole-like plots as in Santner et al. (2007). These authors presented corresponding plots for coverage and width side by side, which is sensible considering the usual inverse relationship between these. The primary criterion in this evaluation was closeness of coverage to the nominal 1–α, which was chosen to be 90% here. This was interpreted from the standpoint that strict conservatism is desirable, so that coverage was evaluated for 70000 PSPs, and the proportion of these with coverage below 0.9 was reported.

Interval width is most often summarised by the mean. It is obviously both convenient and appealing that such a measure can be calculated directly as an average of the widths of all possible intervals. Nevertheless width may equally be characterised by other measures, such as the median, minimum, maximum and quartiles as in Peskun (1993). This of course corresponds to the approach taken when box plots are used for presentation. Peskun also reported proportions of PSPs for which one interval method yields narrower intervals than an alternative method.

Bonett and Price (2011) considered the proportion of PSPs with non-coverage > 1.5α, as well as minimum and mean coverage and average width. While considering coverage properties of intervals with various nominal α levels, a ratio criterion of this kind is appropriate: we would want to regard an absolute 1% shortfall in coverage as less serious when the nominal coverage is 90% than when it is 99%.

5.4 The Need for Illustrative Examples

Many published papers that develop novel confidence interval methods go straight from the algebraic derivation to an evaluation of coverage properties. It is helpful to report calculated lower and upper limits for each method for some exemplar datasets—in the simplest case, the binomial proportion, this would mean examples with small and large denominators and with 0, 1 and large numerators. This is for two reasons. The reader can readily appraise some properties of the different methods by comparing the calculated intervals obtained by different methods. Also, when anyone seeks to program these methods, for either routine practical use or further research, this helps ensure they implement the algorithms correctly. In a journal article, the table of illustrative results for exemplar datasets should be placed as early as possible.

Furthermore, for any context more complex than the simple binomial proportion, it is helpful to present actual data arising from a real context that the intended readership can identify with. The titles of many statistical journals refer to specific fields of application such as medicine or psychology, and it is then appropriate that illustrative examples should be chosen accordingly.

6

Intervals for the Poisson Parameter and the Substitution Approach

6.1 The Poisson Distribution and Its Applications

Section 3.4.4 introduced the binomial distribution (Equation 3.7) underlying inferences concerning a simple proportion π, which is normally estimated by $p = r/n$. The Poisson distribution is a closely related model, widely used to represent data on counts of events. It represents the limiting case of the binomial distribution as the group size $n \to \infty$ while the expected number of events $ER = n\pi = \mu$ is held constant at a finite value and $\pi = \mu/n \to 0$. Many applications in epidemiology involve a large population size with a low incidence or prevalence of disease, leading to a small to moderate number of cases positive for the disease. Under these circumstances, the Poisson distribution serves as a simplified proxy for the binomial. The fact that the Poisson distribution has a single parameter, μ was a great advantage when users relied on extensive tables to obtain confidence limits. For example, *Documenta Geigy Scientific Tables* (Diem and Lentner 1970) used 19 pages to list Clopper–Pearson intervals for π, but only two pages for exact intervals for μ. Correspondingly, it is simpler to evaluate coverage properties and so forth in the Poisson case.

The probability function of the Poisson distribution is

$$\Pr\left[R = r\right] = \frac{\mu^r e^{-\mu}}{r!} \text{ for } r = 0, 1, 2 \ldots . \qquad (6.1)$$

Its parameter μ is the mean of the distribution, and can take any positive value. There is no reason why μ should be integer valued. While π cannot exceed 1, usually μ is considerably greater than 1; a μ below 1 implies that the study was much too small to produce any reasonable estimate.

In contrast to the binomial case, the number of positive cases, R can take any non-negative integer value, with no upper limit. However, $\Pr[R = r]$ declines rapidly for $r \gg \mu$, so the unboundedness of the range is of little practical importance.

Just as for the binomial parameter, a variety of confidence intervals have been suggested for the Poisson parameter. In a Poisson distribution with mean μ, the variance is also μ. Consequently the Wald limits are delimited by $r \pm z \times \sqrt{r}$. More refined methods are introduced in Section 6.2.

6.1.1 Binomial Intervals via the Poisson Parameter

Several other quantities are closely related to the Poisson parameter, and any interval chosen for μ will readily translate into intervals for these derived quantities. Sometimes a confidence interval for the binomial proportion π is constructed by assuming that the numerator r has a Poisson distribution. Lower and upper limits for μ are then divided by n to produce limits for p. This practice cannot be recommended, for two reasons. The resulting interval for p is wider than the interval calculated using the corresponding purpose-built method for p. The standard error imputed to p is $\sqrt{p/n}$ on the Poisson model and $\sqrt{pq/n}$ on the binomial model, where $q = 1 - p$. Consequently, the interval for π derived from the Poisson Wald interval is wider than the binomial Wald interval, by a factor $1/\sqrt{q}$. A similar relationship applies to other interval methods. The resulting widening will be unimportant only if the proportion being estimated is very small. Furthermore, Poisson-based intervals do not have the equivariance property enjoyed by all other intervals we have considered: interval location is unlikely to be satisfactory, and the Box–Cox index of symmetry is inapplicable. The only cogent advantage of this approach is that Poisson limits for $r = 0, 1, 2 \ldots$ provide a convenient rapid approximation to limits derived correctly from the binomial distribution.

6.1.2 Incidence Rates

Rothman and Greenland (1998) describe a comparison of breast cancer incidence between two groups of women. In a cohort of women undergoing X-ray fluoroscopy during treatment for tuberculosis, 41 cancers developed in $M = 28{,}010$ person-years at risk, a rate of 0.00146 per person-year. This is most conveniently displayed using a suitable radix, as 1.46 per 1000 person-years. In a control series who had not had X-ray fluoroscopy, there were 15 cancers in 19,017 person-years, a rate of 0.79 per 1000 person-years. The 95% Wald limits for the incidence in the exposed group are $1000 \times (41 \pm 1.96 \times \sqrt{41}) \div 28{,}010$ i.e., 1.02 to 1.91 per 1000 person-years. Similarly, the 95% Wald interval for the incidence in the unexposed group is $1000 \times (15 \pm 1.96 \times \sqrt{15}) \div 19{,}017$ i.e. 0.39 to 1.19 per 1000 person-years.

6.1.3 Indirect Standardisation

The breast cancer comparison in the previous section is subject to the criticism that no information is presented on the degree of comparability of the exposed and unexposed series. Obviously the sex distributions are identical, but the two series could well differ in other important respects, most crucially age. It is well recognised that for practically all conditions age is a major determinant of incidence and mortality rates. For comparisons of mortality and morbidity between two or more series, some adjustment for age, at least, is obligatory.

Indirect standardisation is widely used to adjust mortality comparisons for age, leading to the standardised mortality ratio (SMR) measure. Whenever data are available for both males and females, best practice is to construct separate SMRs for the two sexes. Analogously, for comparisons of disease incidence, a standardised incidence ratio (SIR) is used.

The SMR compares the number of deaths that occur in a specified population, such as a cohort exposed to an industrial hazard, to the number expected in a cohort of the same size and age distribution over the same follow-up period exposed to the age and sex specific mortality rates currently prevalent in the whole population. Specifically, SMR = 100 × Obs/Exp where Obs and Exp denote observed and expected numbers of deaths. The radix 100 is conventionally incorporated, to express risk as a percentage of that for the general population. The SMR for a cohort of interest may then be compared to the whole population (null hypothesis SMR = 100), or between two different groups.

A confidence interval for the SMR is derived from an interval for the observed number of deaths, based on a Poisson model. Rittgen and Becker (2000) presented data for several causes of mortality in a large cohort of male iron foundry workers. There were 322 deaths attributed to lung cancer and 28 to liver cancer. Expected deaths from these two malignancies, taking into account the cohort's age distribution, were 253.18 and 12.44, respectively. The lung cancer SMR is 100 × 322 ÷ 253.18 or 127, with Wald limits 100 × (322 ± 1.96 × $\sqrt{322}$) ÷ 253.18 i.e., 113 to 141. Similarly, the liver cancer SMR is 225, with Wald limits 142 to 308. These results indicate substantially increased mortality for both malignancies, although inferences relating to causation are subject to the usual limitations regarding comparability for other factors, notably smoking, and healthy worker selection effects.

6.2 Confidence Intervals for the Poisson Parameter and Related Quantities

Several methods are available to calculate a confidence interval for the Poisson parameter μ. Generally, interval methods for μ correspond to the various ones for π described above, but do not use the information on the denominator, n.

As indicated above, the Wald limits are $r \pm z \times \sqrt{r}$. This is a useful result, as it gives an approximate expression of the degree of uncertainty of any count of events which is amenable to mental arithmetic—it is generally easy to calculate $\sqrt{r}$ to the nearest integer and double it. This formula implies that as μ increases, the interval width increases in absolute terms, but decreases in relative terms—a property that applies also to other interval methods.

However, the Wald interval for the Poisson parameter obviously shares the boundary violation propensity of its binomial counterpart. When no events are observed, $r = 0$, a zero-width interval at 0 is produced, even though the data are perfectly compatible with small positive values for μ. When $r = 1$, the calculated 95% Wald limits are (–0.96, 2.96); truncation at 0 is inadequate here, for the same reason as in the binomial case. Consequently the Wald interval cannot be recommended here either, except as a useful first approximation or when r is anticipated to be very large.

Analogous to the Wilson interval for the binomial parameter π, we can calculate an inversion interval for μ with lower and upper limits satisfying $\mu \pm z \times \sqrt{\mu} = r$. The resulting limits are $r + z^2/2 \pm z\sqrt{(r + z^2/4)}$. This interval is fully boundary-respecting; for example when $r = 0$ the limits are 0 and $U_0 = z^2 = 3.84$, and the mean coverage is close to the nominal level α. The interval is symmetrical about r on a log scale. This property is the Poisson analogue of the logit scale symmetry of the Wilson interval for the binomial proportion. However, the location of the inversion interval for μ is very poor. Left and right non-coverage probabilities for the inversion interval are displayed graphically by Macdonald (1997) for μ ranging from 0 to 20 and from 0 to 100. For the inversion interval, the left non-coverage is always $\alpha/2$ at most, whereas the right non-coverage is nearly always greater than $\alpha/2$, indicating an unsatisfactory asymmetry of coverage.

The Macdonald website shows corresponding graphs for another fully boundary-respecting interval based on an exact aggregation of tail areas, akin to the Clopper–Pearson interval for π. The lower and upper limits may be obtained iteratively, or in closed form from gamma or chi-square distributions as described below. The exact limits for $r = 0$ are 0 and $U_0 = 3.69$. When $\mu < U_0$, only left non-coverage can occur; this property applies to any boundary-respecting interval for μ. But when $\mu > U_0$, left and right non-coverage of the exact interval are well-balanced.

Just as in the binomial case, the exact interval is strictly conservative; consequently the mean coverage and interval width should be considered unnecessarily large. Cohen and Yang (1994) investigated coverage properties of a mid-p interval for μ. This is defined in the usual way in terms of tail areas; it cannot be programmed in closed form. For $\mu < U_0 = 2.996$, once again only left non-coverage can occur, and the interval is conservative. For $\mu > U_0$, the mean coverage is close to $1 - \alpha$. Some results for left and right non-coverage were included, confirming that location is appropriate just as for the exact interval. Cohen and Yang also suggested modified Wald limits $\{\sqrt{(r + 0.5)} \pm z/2\}^2$ as a very simple, remarkably close approximation to the mid-p limits.

Both the exact and mid-p limits, together with several approximate intervals, are readily obtained for any r (up to 10,000 at least), from a web calculator

(Sullivan 2007). This resource was designed to calculate confidence intervals for the SMR, but an interval for μ is obtained by setting the expected number of events to 1.

Alternatively, a Bayesian interval based on a conjugate gamma prior for μ is fully boundary-respecting and programmable in closed form. The gamma distribution is usually defined in terms of a shape parameter and an inverse scale or rate parameter, denoted here by *a* and *b*. The pdf is then $\frac{b^a}{\Gamma(a)}x^{a-1}e^{-bx}$, with support $0 < x < \infty$. Generally, an uninformative prior would be used, $\Gamma(a, b)$, with small, positive values for *a* and *b*, typically 0.001. A simpler, practically indistinguishable but theoretically less satisfactory option is to set $a = b = 0$, resulting in an improper prior distribution for μ. Lower and upper limits are then obtained as the $\alpha/2$ and $1 - \alpha/2$ quantiles of the gamma distribution with parameters $a + r$ and $b + 1$:

$$L = QG(a + r, b + 1, \alpha/2) \text{ and } U = QG(a + r, b + 1, 1 - \alpha/2). \quad (6.2)$$

Here, $QG(a, b, q)$ denotes the *q*-th quantile ($0 \le q \le 1$) of the gamma distribution with parameters *a* and *b*, such that

$$\int_0^{QG} \frac{b^a}{\Gamma(a)} x^{a-1}e^{-bx}\,dx = q \quad (6.3)$$

The exact limits likewise may be derived from the gamma distribution:

$$L = QG(r, 1, \alpha/2) \text{ and } U = QG(r + 1, 1, 1 - \alpha/2) \quad (6.4)$$

or equivalently by halving the $\alpha/2$ and $1 - \alpha/2$ quantiles of the chi-square distributions with $2r$ and $2(r + 1)$ degrees of freedom, respectively. The lower limits for the exact and improper Bayes intervals are identical, but the upper exact limit for $R = r$ is the upper improper Bayes limit for $R = r + 1$.

Table 6.1 shows 95% limits for μ calculated by seven methods for the cases $r = 15$ and 41 (as in the X-ray fluoroscopy data) and $r = 322$ (lung cancers in iron foundry workers). For example, with $r = 15$, the lower and upper exact limits may be derived from the respective quantiles of the chi-square distribution as 16.791/2 and 49.480/2, respectively. The results are arranged in increasing order of the lower limit. There is a great consistency of pattern here. Interval width is generally similar for six of the methods, whilst the exact interval is wider—about one "unit" wider than the mid-*p* interval. The most obvious differences between the other six methods relate to interval location. The Wald limits are located furthest to the left, and the inversion limits are the furthest to the right. This is the Poisson counterpart of the observation that for most intervals for the binomial proportion, the Box–Cox index of symmetry introduced in Section 4.4.3 lies between 1 (corresponding to the Wald interval) and 0 (corresponding to the Wilson interval which

TABLE 6.1

95% Confidence Limits for the Poisson Parameter μ Calculated by Seven Methods for Three Exemplar Numerators

	r = 15 (Rothman and Greenland 1998)		r = 41 (Rothman and Greenland 1998)		r = 322 (Rittgen and Becker 2000)	
	Lower Limit	Upper Limit	Lower Limit	Upper Limit	Lower Limit	Upper Limit
Wald	7.41	22.59	28.45	53.55	286.8	357.2
Exact	8.395	24.74	29.42	55.62	287.8	359.2
Improper Bayes	8.395	23.49	29.42	54.46	287.8	358.1
Proper Bayes	8.405	23.51	29.45	54.52	288.1	358.5
Mid-p	8.72	24.19	29.81	55.08	288.2	358.6
Modified Wald	8.74	24.18	29.83	55.09	288.3	358.7
Inversion	9.09	24.75	30.22	55.62	288.7	359.1

Note: Proper Bayes interval used a Γ(0.001, 0.001 prior).

is obtained by "inverting" the Wald interval). The modified Wald limits are very close to the mid-p ones.

As noted above, for the cohort of women exposed to X-ray fluoroscopy and their controls, the 95% Wald intervals for the incidences are 1.02 to 1.91 and 0.39 to 1.19 per 1000 person-years. The corresponding mid-p intervals are 1.03 to 1.97 and 0.46 to 1.27. Similarly, for the iron foundry workers, the 95% Wald intervals for the SMRs for lung and liver cancer are 113 to 141 and 142 to 308. The corresponding mid-p intervals are 114 to 142 and 153 to 321. The usual principle applies: the smaller the numerator, the greater the divergence between different methods.

6.3 Widening the Applicability of Confidence Interval Methods: The Substitution Approach

In the preceding section, a confidence interval for the Poisson parameter was used to obtain intervals for three related quantities: a binomial proportion, a rate and a standardised mortality ratio. Each calculation is a special case of a method commonly used to derive a confidence interval for one quantity from an interval for a simpler one. This approach was termed the substitution method by Daly (1998) and is sometimes referred to as the plug-in approach. This conceptually and computationally simple method obtains a confidence interval for a function $f(p)$ of a quantity p for which a confidence interval is available. If L and U denote the calculated limits for p, then $f(L)$ and $f(U)$ delimit the corresponding interval for $f(p)$.

The method may be used whenever f is a monotonic function of p. This means that as p increases, either $f(p)$ always increases, or $f(p)$ always decreases. When the function f is monotonic increasing, then $f(L)$ and $f(U)$ are the lower and upper limits for $f(p)$. Conversely, if f is monotonic decreasing, then $f(U)$ and $f(L)$ serve as lower and upper limits for $f(p)$.

The applicability of this approach is not restricted to quantities related to counts and proportions, but these provide the most obvious area of application, including all the examples reported by Daly and the ones outlined below. On the natural assumption that the boundaries of the range of validity for $f(p)$ correspond directly to those for p, a boundary-respecting interval method for p results in a boundary-respecting interval for $f(p)$—clearly an important advantage for quantities related to counts and proportions.

6.3.1 Gene Frequency

The purest illustrative example from Daly's paper comes from population genetics. On assumption of Hardy–Weinberg equilibrium (HWE), which is usual for human populations (but see Section 9.2.2), the frequency ζ of a rare recessive gene may be estimated simply as the square root of the birth incidence of homozygosity. Reilly et al. (1993) reported an incidence of $r = 17$ cases of Wilson's disease in a series of $n = 1{,}240{,}091$ births in Ireland, a rate of $p = 0.0000137$ or 13.7 cases per million births. The 95% score interval for this simple proportion is 8.6 to 22.0 per million. Taking square roots, the estimated gene frequency $\tilde{\zeta}$ is 3.7 per thousand, with 95% interval 2.9 to 4.7 per thousand.

Daly reported a slightly different interval, 2.8 to 4.7 per thousand, based on an interval for p derived from an exact Poisson interval for the numerator. For such a low proportion, Poisson intervals are barely distinguishable from their binomial counterparts. Daly noted that the usual interval is $\tilde{\zeta} \pm z\sqrt{\left(1-\tilde{\zeta}^2\right)/4n}$. While this formula gives a satisfactory interval from 2.8 to 4.6 per 1000 here, it is not always boundary-respecting. For example, for large n, when $r = 0$, the calculated 95% limits for ζ are approximately $\pm 1/\sqrt{n}$, whilst for $r = 1$, the 99% limits are approximately $\left(-0.3\tilde{\zeta}, 2.3\tilde{\zeta}\right)$. In such cases a boundary-respecting substitution interval is particularly advantageous. When $r = 0$, the 95% score interval for ζ is from 0 to approximately $1.96/\sqrt{n}$, whilst when $r = 1$, the 99% score limits for ζ are approximately $0.34/\sqrt{n}$ and $2.92/\sqrt{n}$. Unlike the standard-error-based intervals, these intervals are evidently meaningful to interpret.

Daly also considered extending the gene frequency estimation take into account a population inbreeding coefficient, and stated that the substitution method then enables confidence limits for the gene frequency to be calculated easily. But the sampling imprecision of the inbreeding coefficient should also be considered, whether this came from the same data set as the prevalence or some other source.

6.3.2 Levin's Attributable Risk

Many measures of effect size derived from a 2 × 2 contingency table constitute potential applications of the substitution method. One considered by Daly is the Levin attributable risk (LAR) measure. Table 6.2 presents illustrative data on infant mortality by birthweight. The LAR is defined as $(p_T - p_2)/p_2$ where p_T denotes the mortality in the entire series and p_1 and p_2 denote mortality in the groups with low and normal birthweights. For this example, the LAR is calculated as 0.563, suggesting that (if it were possible) aligning the birthweights of the 5215 low birthweight cases with their normal birthweight counterparts would reduce mortality by 56.3% (in relative terms).

Daly calculated confidence intervals for the LAR using two existing methods, then proposed a simpler alternative based on substitution. The LAR may be expressed as a function of the prevalence of low birthweight in the series, here 5215/72730 or 0.0717, and the risk ratio RR = p_1/p_2 which is 18.959:

$$\text{LAR} = \frac{\text{Prev} \times (\text{RR}-1)}{1+\text{Prev} \times (\text{RR}-1)} = \frac{1}{1+1/\{\text{Prev} \times (\text{RR}-1)\}} \tag{6.5}$$

Daly calculated a 95% interval for the relative risk as 16.807 to 21.387, using the simple asymptotic method based on the standard error of ln RR described in Section 10.4.1. These values are then substituted together with the observed prevalence of 0.0717, to give a 95% interval for the LAR as 0.531 to 0.594. These limits are very similar to those produced by existing methods.

Daly observed that of course, the limits obtained depend on which formula is used for the confidence interval for the relative risk. When a better method is used—the Miettinen–Nurminen method (1985) described in Section 10.4.2—the 95% limits for the RR alter very slightly to 16.806 to 21.384. The resulting interval for the LAR is once again 0.531 to 0.594, to three

TABLE 6.2

Illustrative Data for Levin's Attributable Risk Example; Infant Mortality among Whites in New York City, by Birthweight, 1974

	Observed Proportions		
Outcome at 1 year	**Birthweight ≤ 2500 g**	**Birthweight > 2500 g**	**Total**
Dead	$p_1 = 0.1185$	$p_2 = 0.0063$	$p_T = 0.0143$
Alive	$q_1 = 0.8815$	$q_2 = 0.9937$	$q_T = 0.9857$
Total births	5215	67515	72730

Source: Based on Fleiss, J. L., Levin, B. and Paik, M. C. 2003. *Statistical Methods for Rates and Proportions*, 3rd Edition. Wiley, Hoboken, NJ, p. 128. Reprinted from Daly, L. E. 1998. *American Journal of Epidemiology* 147: 783–790.

decimal places; on such large numbers, the method chosen for the RR makes no perceptible difference to the interval for the LAR.

6.3.3 Population Risk Difference and Population Impact Number

For data similar to the LAR example, Bender and Grouven (2008) defined the population risk difference (PRD) as the numerator of the LAR, namely $p_T - p_2$. The reciprocal of this measure is defined as the population impact number (PIN) (Heller and Dobson 2000). Here, the PRD is simply the product of the prevalence of exposure and the risk difference (RD) $p_1 - p_2$. Bender and Grouven recommended that for fixed prevalence of exposure, interval estimation for the PRD and PIN can be based on a suitable interval for the RD, and recommended the square-and-add Wilson method developed in Section 7.3.1, as used in Bender's interval for the number needed to treat (NNT) (2001).

6.3.4 Assumptions and Limitations of the Substitution Approach

The examples considered above highlight some of the issues attaching to the substitution approach in general. The above examples are presented as if only a single parameter was involved. In fact, in Sections 6.3.2 and 6.3.3, additional parameters come into play. It is assumed that only the variation in parameter p_1 matters, and variation in p_2 and any other contributory quantities can be disregarded. This is satisfactory only if p_1 really is the dominant source of variation. While this is the case in some examples, it should not be taken for granted without careful investigation. If this property does not hold, the methods introduced in later chapters are applicable. In some situations, the square-and-add or MOVER approach introduced in Chapter 7 is appropriate. In more complex applications, this does not work, and the more general Propagating Imprecision (PropImp) algorithm (Chapter 13) is required.

Monotonicity is an important issue for all these approaches. Daly stated, "It is difficult to imagine a practically useful measure in medical or epidemiological applications for which this condition (i.e., monotonicity) will not hold". Setting aside the obvious deduction from this statement that the NNT should not be regarded as a useful measure—one has to agree with Daly's conclusion, for the one-parameter case—but not otherwise. Sometimes monotonicity over a meaningful sub-interval of the allowable range of values is adequate, as we will see in relation to the seizure prophylaxis example used to introduce the PropImp algorithm in Chapter 13.

7

Difference between Independent Proportions and the Square-and-Add Approach

7.1 The Ordinary 2 × 2 Table for Unpaired Data

The 2 × 2 contingency table is very widely used to display the comparison between two independently sampled proportions or the association between two binary variables. Table 7.1 shows three representations of the 2 × 2 contingency table used to compare a binary outcome variable between two independent groups 1 and 2. In this chapter, and also in Chapter 10 which deals with the ratio of independent proportions, we are assuming that the column totals, m and n, are fixed. For example, the data could come from a randomised trial in which m subjects are allocated to treatment 1 and n subjects are allocated to treatment 2. In statistical parlance, inferences are conditional on the two marginal totals, m and n. This contrasts with the situation in a cross-sectional study, in which often only the table total is fixed. In Chapter 8 we consider a very different 2 × 2 table representing paired binary data.

In Table 7.1, a out of m subjects in group 1, and b out of n subjects in group 2, are positive for the outcome of interest. Just as in Chapter 3, the simple sample proportions $p_1 = a/m$ and $p_2 = b/n$ are used as empirical estimates to characterise the proportions positive in the two groups. They can take values ranging from 0 to 1.

Two natural measures may be used to compare p_1 and p_2. The difference $t = p_1 - p_2$ can take values ranging from –1 to 1. It is often referred to as the risk difference (RD), or the absolute risk reduction (ARR), as in Newcombe (2007d), although the latter terminology, like the number needed to treat (NNT) considered in Section 7.6, really presupposes that the direction of the difference is known; it is awkward to express the results in this way if the effect of treatment turns out to be detrimental.

The ratio $p_1 \div p_2$ is known as the relative risk or risk ratio (RR), which can take positive values from 0 to ∞. A less intuitive, nevertheless very important third contrast measure is the odds ratio (OR) which is defined as $\{p_1/(1-p_1)\}/\{p_2/(1-p_2)\}$. It is most conveniently calculated for the simple 2 × 2 table as the

TABLE 7.1

Notation for a Difference between Independent Proportions

2 × 2 contingency table representation:

	Observed frequencies		Observed proportions		Theoretical proportions	
	Group 1	Group 2	Group 1	Group 2	Group 1	Group 2
Positive	a	b	p_1	p_2	π_1	π_2
Negative	c	d	$1-p_1$	$1-p_2$	$1-\pi_1$	$1-\pi_2$
Total	m	n	1	1	1	1

Parameter of interest $\theta = \pi_1 - \pi_2$
Nuisance parameter $\eta = (\pi_1 + \pi_2)/2$

cross-product ratio, ad/bc. This too takes positive values from 0 to ∞. When p_1 and p_2 are identical, RD = 0, but RR and OR are both 1. The relative risk and odds ratio are considered in detail in Chapters 10 and 11.

As before, we are not so much concerned with our two samples in their own right, as in the information they give us on the populations from which they are drawn. Almost throughout this chapter, we denote the theoretical proportions corresponding to p_1 and p_2 by π_1 and π_2. We let t denote the observed difference, $p_1 - p_2$, and θ denote its theoretical value, $\pi_1 - \pi_2$. The exception is in Section 7.7 which relates to Bayesian intervals, for which this distinction is not made.

We assume that neither π_1 nor π_2 is identically 0 or 1, otherwise there is no question of drawing inferences about them from our data. As usual, p_1 and p_2 are unbiased estimates of π_1 and π_2. Also, the observed risk difference, $t = p_1 - p_2$ is an unbiased estimate of $\theta = \pi_1 - \pi_2$.

Traditionally, hypothesis testing was the mainstay of statistical inferences, including for the comparison of independent proportions. By default, the null and alternative hypotheses considered are H_0: $\pi_1 = \pi_2$ and H_1: $\pi_1 \neq \pi_2$. The test used when cell frequencies are sufficiently large is the well-known chi-square (χ^2) test. Several equivalent formulae are available. The simplest procedure is to calculate

$$X^2 = \frac{(a+b+c+d)(ad-bc)^2}{(a+b)(c+d)(a+c)(b+d)} \tag{7.1}$$

and obtain a p-value by referring to the χ^2 distribution with 1 degree of freedom. An equivalent, more generally applicable formulation calculates expected frequencies based on the assumption that the overall proportion positive $(a + b)/(m + n)$ applies equally to the two groups. X^2 is then calculated as $\sum (Obs - Exp)^2 / Exp$ where Obs and Exp denote observed and expected frequencies based on the marginal totals assuming H_0, with summation across all four cells of the table. Even though only the upper tail of the χ^2 distribution is used, this leads to a two-sided test, because $X^2 = 0$ only when p_1 and p_2 are identical, and as they move apart in either direction this leads to an increase in X^2. Unlike the *t*-test statistic, χ^2 is a squared measure, which is always positive and hence does not incorporate information on the direction of the difference, only its magnitude.

Being an asymptotic test, the χ^2 test is appropriate only when cell frequencies are not small. A continuity correction is sometimes incorporated to mitigate anticonservatism. For small cell frequencies, the Fisher exact test is generally recommended whenever the lowest of the expected frequencies used in the χ^2 test is less than some value such as 5. The Fisher exact test involves summation of precisely calculated probabilities of the various possible tables, conditional on the observed column totals m and n and also the row totals $a + b$ and $c + d$. These probabilities come from the hypergeometric distribution: the probability of the observed table is $\dfrac{\binom{a+b}{a}\binom{c+d}{c}}{\binom{m+n}{m}}$.

This contrasts with the χ^2 test which may be derived conditioning on either these four marginals or else just the two column totals m and n. Also, the Fisher exact test tends to be over-conservative, and hence underpowered. An alternative test based on a mid-*p* accumulation of tail areas (Lancaster 1949) downweights the contribution from the probability of the observed table, and is therefore less conservative.

The χ^2 and Fisher exact tests are presented in detail in elementary statistics textbooks. Our concern in this book is rather with better interpreted measures of effect size such as the risk difference or relative risk, and confidence intervals calculated to express the sampling imprecision attaching to them.

7.2 The Wald Interval

The simplest confidence interval for the difference of independent proportions is the Wald interval. This is closely related to the Wald interval for the single proportion case, and shares its drawbacks, although to a lesser degree.

As usual, we let z denote the $1 - \alpha$ quantile of the standard normal distribution, so that $z = 1.960$ for the default interval with intended 2.5% right and left non-coverage. Also, let $q_i = 1 - p_i$, for $i = 1$ and 2. Then the simple Wald interval, without continuity correction, may be calculated as

$$p_1 - p_2 \pm z\sqrt{\frac{p_1 q_1}{m} + \frac{p_2 q_2}{n}} \tag{7.2}$$

or equivalently

$$\frac{a}{m} - \frac{b}{n} \pm z\sqrt{\frac{ac}{m^3} + \frac{bd}{n^3}}. \tag{7.3}$$

For example, in Crawford et al. (2006), the prevalence of use of CAMs was 113/259 (43.6%) in boys and 93/241 (38.6%) in girls. The absolute difference in CAM use between boys and girls is estimated here as +5.0%, with 95% interval from −3.6% to +13.7%. These limits are very close to those obtained using other methods.

However, in a study comparing the performance of several intervals for the RD (Newcombe 1998b), the Wald interval had a mean attained coverage of only 0.881, substantially lower than the nominal 0.95; for some parameter space points the attained coverage approaches zero.

The simplest way to mitigate such anticonservatism is to incorporate a continuity correction, just as for the single proportion. The continuity-corrected Wald interval is calculated as

$$p_1 - p_2 \pm \left\{ z\sqrt{\frac{p_1 q_1}{m} + \frac{p_2 q_2}{n}} + \frac{1}{2m} + \frac{1}{2n} \right\} \tag{7.4}$$

or equivalently

$$\frac{a}{m} - \frac{b}{n} \pm \left\{ z\sqrt{\frac{ac}{m^3} + \frac{bd}{n^3}} + \frac{1}{2m} + \frac{1}{2n} \right\}. \tag{7.5}$$

This method resulted in a satisfactory mean coverage of 0.962, but with a minimum coverage of only 0.514, it can hardly be considered satisfactory either.

This may be illustrated using data from an Eastern Co-operative Oncology Group (ECOG) clinical trial. These data appear not to have been published in the oncology literature, but are cited by Parzen et al. (2002). This was a Phase II study, a feasibility study designed to give some preliminary information on whether a novel treatment modality warrants definitive evaluation in an adequately powered Phase III trial. It was a two-arm, randomised clinical trial evaluating the safety and efficacy of two chemotherapy treatments in patients with advanced large bowel cancer. In all, $m = 14$ subjects were allocated to treatment 1 and $n = 11$ to treatment 2. For the primary outcome, tumour shrinkage by at least 50%, unfortunately no successes were observed on either treatment, so $a = b = 0$. While the main conclusion was that neither treatment warranted further evaluation, it is important to be able to report a confidence interval comparing the outcomes in the two groups in such a situation, to show how much uncertainty attaches to the point estimate for the difference.

For these data, the Wald interval without continuity correction is degenerate—Equation 7.2 gives 0 for both lower and upper limits, irrespective of the value of z or α used. Particularly when we consider the very small sample sizes used here, we would expect to report a wide interval, certainly not a degenerate one.

The upper and lower limits with continuity correction are rather better here, ±0.081, but this is still unrealistically narrow—it is unsatisfactory that the interval width derives solely from the continuity correction.

Parzen et al. (2002) also reported data for life-threatening treatment toxicity, with rates of 2/14 and 1/11 in the two groups. The Wald interval for this difference is from –0.1980 to +0.3019. There is no obvious abnormality of behaviour here, although the interval is rather narrower than the square-and-add Wilson interval described below; just as for the single proportion, the Wald interval tends to be too narrow when proportions are close to 0 or 1.

Anomalous behaviour can occur also when t is at or close to ±1. For example, keeping m and n at 14 and 11, suppose we observe $a = 14$ and $b = 0$. Here, too, we get a degenerate, zero-width interval, this time at +1, regardless of z or α, even though it is perfectly possible that $\pi_1 - \pi_2 < 1$. On the other hand, when $a = 14$ and $b = 1$, the calculated 95% limits are (0.7392, 1.0790). Also, when $a = 13$ and $b = 1$, the calculated 95% limits are (0.6207, 1.0546). Even in the latter case, in which none of the four cell frequencies a, b, c or d is zero, the resulting interval can violate the boundary at +1. Similar behaviour occurs at or near the lower boundary at –1. As in the single proportion case, such boundary overflow would usually be truncated, but this is unsatisfactory. Use of a continuity correction exacerbates both the frequency and severity of boundary violation.

Several better intervals have been formulated. Many of these, including the score intervals introduced in Section 7.4.1, are based on a re-parametrisation of the model in terms of the parameter of interest, $\theta = \pi_1 - \pi_2$ and a nuisance parameter $(\pi_1 + \pi_2)/2$ denoted here by η. The simplest method is to replace η by its maximum likelihood estimate $\hat{\eta} = \frac{p_1 + p_2}{2}$ and develop an interval for θ analogous to the Wilson interval for the binomial proportion. This procedure is flawed, when $p_1 = p_2 = 0$ or $p_1 = p_2 = 1$ it fails to produce an interval. Several superior intervals based on this principle are introduced in Section 7.4, but a simpler approach is also effective here.

7.3 The Square-and-Add or MOVER Approach

7.3.1 The Square-and-Add Wilson Interval

In the previous chapters we saw how a simple modification to the Wald interval—in which the variance imputed to the observed proportion is based on the theoretical proportion π instead of the observed proportion p—obviates boundary anomalies and results in greatly improved coverage and location properties. There are several ways to achieve a similar effect here. The simplest of these, the square-and-add approach, closely mimics the relationship between the Wald intervals for p_1, p_2 and t. The most straightforward application is to produce an interval for a difference between two separately estimated parameters, as in Newcombe (1998b). But the approach is of much wider applicability. Zou et al. (2009a) describe the application to an arbitrary linear function of proportions. Important special cases of this relate to interaction effects on an additive scale (Newcombe 2001c; Zou 2008b, see Section 14.2), and a diagram to compare sensitivity and specificity simultaneously between two screening or diagnostic tests (Newcombe 2001b, see Section 12.5).

To illustrate the approach, suppose that as in Section 7.2 the two proportions to be compared are 113/259 = 0.4363 and 93/241 = 0.3859. The straightforward empirical estimate of the difference is 0.0504. Assuming a default nominal 95% confidence level, Wald intervals for these proportions are (0.3759, 0.4967) and (0.3244, 0.4474). Variances of independent quantities combine simply by adding, consequently standard errors of independent quantities combine simply by squaring and adding—followed, of course, by taking the square root to return to the original scale:

$$\mathrm{SE}(p_1 - p_2) = \sqrt{\{\mathrm{SE}(p_1)^2 + \mathrm{SE}(p_2)^2\}} \tag{7.6}$$

Thus the simplest confidence interval for the difference is obtained as in Section 7.2 by using this relationship as (−0.0358, +0.1366).

The square-and-add relationship is very pervasive in mathematics, being the basis of Pythagoras' theorem in geometry (for a novel demonstration see Newcombe 2010b), its trigonometric counterpart $\sin^2 + \cos^2 = 1$ and the Euclidean metric in 2, 3 or higher-dimensional space, and Fermat's last theorem in number theory, as well as the statistical application to standard errors.

We note that the lower and upper limits for $p_1 - p_2$ may be expressed in terms of those for p_1 and p_2. Let (L_i, U_i) denote a $(1 - \alpha)$ interval for p_i $(i = 1, 2)$. For the present, we will assume that these are calculated by the Wald method, and disregard continuity correction and boundary violation issues. Then it is easy to see that the Wald limits for $\pi_1 - \pi_2$ may be expressed as

$$L = p_1 - p_2 - \sqrt{(p_1 - L_1)^2 + (U_2 - p_2)^2} \tag{7.7}$$

and

$$U = p_1 - p_2 + \sqrt{(U_1 - p_1)^2 + (p_2 - L_2)^2} \tag{7.8}$$

Wald intervals are symmetrical, so several similar formulae for L and U can be formed. The formulae above are chosen because U should correspond to $U_1 \geq p_1$ with $L_2 \leq p_2$, whilst L should correspond to $L_1 \leq p_1$ with $U_2 \geq p_2$.

Equations (7.7) and (7.8) suggest a simple, quite general method to obtain a confidence interval for a difference between two independent quantities. We calculate confidence limits for π_1 and π_2 using a better method, then substitute into the above formulae to obtain limits for $\pi_1 - \pi_2$. Whenever a closed-form method such as the Wilson score interval is used for π_1 and π_2, the resulting square-and-add interval for $\pi_1 - \pi_2$ is of course also of closed form.

The square-and-add Wilson method produces satisfactory intervals, regardless of whether cell frequencies are large or very small or zero. For the case when all cell frequencies are large, for example 113/259 – 93/241, the 95% interval is –0.0358 to +0.1353. This is similar to the Wald interval which we calculated above as –0.0358 to +0.1366.

In the Eastern Co-operative Oncology Group example above, applying the above square-and-add process to the difference in success rates between the two regimes, 0/14 – 0/11, gives an interval from –0.2588 to +0.2153. This contrasts markedly with the degenerate Wald interval in which both calculated limits are zero. Note that this interval extends slightly further into negative values than into positive ones. This is reasonable, because we have weaker information for π_2 than for π_1.

For life-threatening treatment toxicity, with rates of 2/14 and 1/11 in the two groups, the square-and-add Wilson interval for the difference is from –0.2524 to +0.3192. This is rather wider than the Wald interval.

Several interval estimation methods for proportions are available which are boundary-respecting and have satisfactory coverage characteristics. The natural choice for squaring and adding here is the Wilson score interval, which involves square roots, and is easily programmed. The resulting method was published as method 10 of Newcombe (1998b). A square-and-add interval based on Clopper–Pearson (1934) exact intervals was described by Fagan (1999). It is easily shown that when a boundary-respecting interval method is used for π_1 and π_2, the squaring and adding process preserves this property. Consequently the approach is particularly suited to intervals related to proportions. The square-and-add Wilson method yields a mean coverage slightly higher than the nominal $1 - \alpha$ (Newcombe 1998b), with a mean coverage of 0.960 for nominal 95% intervals.

The Wilson square-and-add interval is readily programmed using spreadsheet software such as MS Excel, as in the second block of the accompanying spreadsheet CIPROPORTION.xls.

The issue of interval location also applies to constructs such as differences between proportions. Here, distal non-coverage occurs when the interval is located too close to the neutral point at 0 to cover the true difference $\pi_1 - \pi_2$. For the difference between proportions, in contrast to the single proportion case, not only the square-and-add Wilson interval but also the Wald interval tends to be too mesially located. As noted in Section 4.4.3, only the MNCP/NCP index, not the Box–Cox index of symmetry, is applicable to difference measures.

7.3.2 A Theoretical Justification

While the plausibility of the square-and-add approach should be intuitively clear to statisticians, nevertheless a firmer basis for its use is required. There are two strands to this.

Donner and Zou (2002) and Zou and Donner (2008a) have provided a theoretical justification for this approach in terms of local variance estimates reconstructed from the (L_i, U_i), both for $i = 1$ and 2, summarised in the acronym MOVER (which stands for method of variance estimates recovery) proposed by Zou (2008b). Essentially, L_i and U_i are used to derive local variance estimates for π_i, separately for π_i ranging from L_i to p_i and from p_i to U_i. For the lower tail of the interval for π_i, the variance estimate recovered from L_i is simply $(p_i - L_i)^2/z^2$. Similarly, the variance estimate appropriate for the upper tail is $(U_i - p_i)^2/z^2$. These recovered variance estimates are then used to construct lower and upper limits for $\theta = \pi_1 - \pi_2$ in an obvious manner as follows.

As in Table 7.1, let π_i denote the parameter of interest for population i, $i = 1, 2$, with point estimate $\hat{\pi}_i = p_i$. Assuming that $\hat{\pi}_1$ and $\hat{\pi}_2$ are independently distributed, approximate two-sided $1 - \alpha$ limits (L, U) for $\theta = \pi_1 - \pi_2$ are often calculated as

$$\hat{\theta} \pm z\sqrt{\hat{v}\left(\hat{\pi}_1\right) + \hat{v}\left(\hat{\pi}_2\right)} \tag{7.9}$$

where $\hat{\theta} = \hat{\pi}_1 - \hat{\pi}_2$ and $\hat{v}\left(\hat{\pi}_i\right)$, $i = 1, 2$ are the corresponding variance estimators. This procedure performs well only if the sample sizes are large enough to ensure that the sampling distributions of $\hat{\pi}_i$ are close to normal. When this is not the case, the above interval does not reflect the asymmetry of the underlying sampling distributions. It is preferable to use separate estimates for $v\left(\hat{\pi}_i\right)$ close to the two limits L and U. Estimating $\hat{v}\left(\hat{\pi}_i\right)$ at L or U iteratively is equivalent to inverting a test statistic to obtain a confidence interval. Thus we may regard the $1 - \alpha$ limits (L, U) as the minimum and maximum values of θ that satisfy

$$\frac{\{(\hat{\pi}_1-\hat{\pi}_2)-(\pi_1-\pi_2)\}^2}{v(\hat{\pi}_1)+v(\hat{\pi}_2)}<z^2 \tag{7.10}$$

Let (L_i, U_i) denote the two-sided $1-\alpha$ interval for π_i, $i=1,2$. Among the plausible parameter values provided by these two sets of limits, L_1-U_2 is close to L and U_1-L_2 is close to U. Thus, to obtain L we estimate $\hat{v}(\hat{\pi}_1)=(\hat{\pi}_1-L_1)^2/z^2$ using $\pi_1=L_1$ and $\hat{v}(\hat{\pi}_2)=(U_2-\hat{\pi}_2)^2/z^2$ using $\pi_2=U_2$. Conversely, to obtain U we estimate $\hat{v}(\hat{\pi}_1)=(U_1-\hat{\pi}_1)^2/z^2$ using $\pi_1=U_1$ and $\hat{v}(\hat{\pi}_2)=(\hat{\pi}_2-L_2)^2/z^2$ using $\pi_2=L_2$. Then

$$L=\hat{\theta}-z\sqrt{\frac{(\hat{\pi}_1-L_1)^2}{z^2}+\frac{(U_2-\hat{\pi}_2)^2}{z^2}}=\hat{\theta}-\sqrt{(\hat{\pi}_1-L_1)^2+(U_2-\hat{\pi}_2)^2} \tag{7.11}$$

and similarly

$$U=\hat{\theta}+\sqrt{(U_1-\hat{\pi}_1)^2+(\hat{\pi}_2-L_2)^2} \tag{7.12}$$

This procedure does not require any specific underlying distributions for $\hat{\pi}_i (i=1,2)$, but only separate confidence intervals that have coverage levels close to nominal. It leads to the traditional interval for θ if the sampling distributions for the $\hat{\pi}_i$ are symmetric, but is more general because the symmetry assumption is not required.

Complementary to this, for all the applications described here, the appropriateness of coverage properties may be demonstrated by a Monte Carlo evaluation such as in Section 7.5.

The square-and-add and MOVER approaches as described here are evidently identical, yet neither is the original use of the method. Zou and Donner (2008a) pointed out that methods similar to MOVER were proposed by Howe (1974) for approximate intervals for the mean or sum of two independent random variables, and in relation to variance components by Graybill and Wang (1980) and Burdick and Graybill (1992); see also Lee et al. (2004).

7.3.3 Other Applications

The MOVER or square-and-add approach has been used in a variety of applications. Other applications relating to differences between independent proportions in addition to the ones described above relate to cases where proportions are based on clustered data (Lui 2004a; method L4 in Paul and

Zaihra 2008), or estimated from pooled samples (Biggerstaff 2008). The approach extends in an obvious manner to the sum of two independently estimated proportions or other quantities, to their average, or to any linear combination of two or more of them.

The approach may also be adapted for a paired comparison of proportions by using an estimate of the correlation, preferably incorporating a novel form of continuity correction (Newcombe 1998c, see Section 8.4). Further related methods for differences and ratios of paired proportions are developed by Tang et al. (2010a, b). Methods developed for paired differences between proportions can be adapted to produce intervals for variables that can take the values 0, 1 or 2, such as paired organ procedure rates; comparisons of such rates between two series are then easily derived by squaring and adding (Newcombe 2003a, see Chapter 9).

Further recent applications by Zou and co-workers include comparisons of correlations and intra-class correlations (Zou 2007; Ramasundarahettige et al. 2009), lognormal data (Zou et al. 2009b, c), risk assessment (Zou 2009d), an application to ratio measures in general (Zou and Donner 2010a), inverse sampling (Zou 2010c), and Bland–Altman limits of agreement with multiple observations per individual (Zou 2011).

The fact that the MOVER algorithm can be used to obtain a well-behaved interval for the difference between two quantities does not imply that a simple difference is necessarily an interpretable measure to compare them. For example, it is perfectly practicable to use MOVER to calculate an interval for the difference between two relative risks. However, it is questionable whether this is an interpretable measure: there is no obvious reason to regard a difference between relative risks of 2.5 and 2 as in any sense equivalent to a difference between 0.75 and 0.25. For reasons explained in Section 10.3, the ratio of two relative risks would be a more appealing measure. Similar arguments apply to a range of other possible effect size measures, such as the generalisation of the Mann–Whitney statistic introduced in Chapter 15.

Sometimes the word "hybrid" has been used to refer to square-and-add or MOVER intervals. It seems preferable to reserve this term for composite methods in which different formulae are used depending on the sample data, as discussed in Section 4.1.1—with the implication that coverage properties and so forth may well not vary smoothly when the data changes by just one observation.

7.3.4 The MOVER Approach Generalised to Ratio Measures

The square-and-add approach may also be applied to transformations of π_1 and π_2 to develop new intervals for the relative risk and odds ratio. The resulting methods are likewise naturally conditional on m and n. Thus we may obtain confidence intervals for π_1 and π_2, use these to derive intervals for $\ln \pi_1$ and $\ln \pi_2$, and combine to obtain an interval for $\ln RR = \ln \pi_1 - \ln \pi_2$, which is then back-transformed to produce an interval for the RR. This approach was also suggested by Zou and Donner (2008a), based on Wilson intervals for π_1 and π_2.

Similarly, intervals for logit π_1 and logit π_2 may be combined to obtain an interval for the odds ratio, using any reasonable intervals for π_1 and π_2 separately. Squaring and adding delta intervals for logit π_i (Section 3.4.2) simply leads to the familiar Woolf (1955) delta interval (Section 11.3.1).

However, the above approach is not generally the best way to apply the MOVER principle to obtain an interval for the ratio of two quantities. Zou and Donner (2010a) give a purpose-built general formula for the ratio case, which also incorporates a correlation measure to take account of non-independence. A simpler version, appropriate for the more frequent situation in which the two parameters are estimated independently, was given by Zou et al. (2010b). Suppose the quantity we wish to estimate is defined by $R = q_1/q_2$, where q_1 and q_2 are two quantities which are estimated independently. Then a MOVER interval for the quantity $q_1 - Rq_2$ is delimited by the quantities

$$L = \hat{q}_1 - R\hat{q}_2 - \sqrt{\left(\hat{q}_1 - L_1\right)^2 + \left(RU_2 - R\hat{q}_2\right)^2} \tag{7.13}$$

$$U = \hat{q}_1 - R\hat{q}_2 + \sqrt{\left(U_1 - \hat{q}_1\right)^2 + \left(R\hat{q}_2 - RL_2\right)^2} \tag{7.14}$$

We obtain a lower limit R_L for the ratio by setting $L = 0$:

$$R_L = \frac{\hat{q}_1\hat{q}_2 - \sqrt{\left(\hat{q}_1\hat{q}_2\right)^2 - L_1U_2(2\hat{q}_1 - L_1)(2\hat{q}_2 - U_2)}}{U_2\left(2\hat{q}_2 - U_2\right)} \tag{7.15}$$

Similarly, we obtain an upper limit R_U for the ratio by setting $U = 0$:

$$R_U = \frac{\hat{q}_1\hat{q}_2 + \sqrt{\left(\hat{q}_1\hat{q}_2\right)^2 - U_1L_2(2\hat{q}_1 - U_1)(2\hat{q}_2 - L_2)}}{L_2\left(2\hat{q}_2 - L_2\right)} \tag{7.16}$$

This formulation has the considerable advantage that it remains applicable in the event of zero point estimates or calculated limits, with the understanding that division of a non-zero numerator by a zero denominator results in an infinite confidence limit of appropriate sign.

While most of the applications of the MOVER algorithms described in this book concern quantities related to proportions, these algorithms, just like substitution, are equally applicable to the continuous case. Zou et al. (2010b) showed how this procedure, together with the usual linear MOVER formulation (Section 7.3.2) facilitates calculation of confidence intervals for 12 quantities that arise in the one-way random effects model. The latter include quantities of the form σ/μ where μ denotes the mean and σ denotes either the between-groups or the within-groups variance. A similar approach may be used to construct an interval for the standardised difference $(\mu_1 - \mu_2)/\sigma$ comparing two samples from Gaussian distributions, or the corresponding measure Δ/s_Δ appropriate to the paired case, as illustrated in Section 14.11, as a more convenient alternative than recourse to the non-central *t*-distribution (see also Li et al. (2010); Donner and Zou (2010); Zou and Donner (2010a); and also Dube et al. (2008), see Section 14.6).

For clarity, we term the square-and-add approach used to construct an interval for $q_1 - q_2$ MOVER-D, meaning MOVER on the difference scale. When Equations 7.11 and 7.12 are used to obtain an interval for $\ln(q_1) - \ln(q_2)$ and hence for $R = q_1/q_2$, this is called MOVER-DL (MOVER-D on the log scale). The purpose-built method, using Equations 7.15 and 7.16, is MOVER on the ratio scale (MOVER-R). Chapter 14 presents many applications of the various MOVER approaches, and also of an alternative algorithm, Propagating Imprecision (PropImp) introduced in Chapter 13 which may be used even when the parameters q_1 and q_2 contribute to the quantity of interest $F(q_1, q_2)$ in an interlocked manner.

Unfortunately, the "trick" above which turns MOVER-D into MOVER-R does not appear to extend to give a purpose-built interval for a product, $q_1 \times q_2$. In this situation, an interval analogous to MOVER-DL may be constructed via $\ln(q_1) + \ln(q_2)$. Alternatively, MOVER-R may be applied to q_1 and $1/q_2$, or to $1/q_1$ and q_2. In general, the intervals produced by these alternative approaches are similar but not identical.

7.4 Other Well-Behaved Intervals for the Difference between Independent Proportions

7.4.1 Reparametrisation and Score Intervals

The evaluation study reported by Newcombe (1998b) also included several other fully boundary-respecting intervals. Mee (1984) suggested a formulation for a score interval. A slightly more conservative version of this was developed by Miettinen and Nurminen (1985), who showed this method aligns mean coverage approximately with $1 - \alpha$, with reasonable balance between right and left non-coverage. The Miettinen–Nurminen interval has substantially better coverage than the Mee interval when judged by the proportion of parameter space points for which the attained coverage is at least the nominal value (Santner et al. 2007; Newcombe and Nurminen 2011a).

7.4.2 True Profile Likelihood Interval

In Section 3.4.8 we saw that a likelihood-based interval for the binomial proportion π may be obtained by identifying a range of values over which 2 ln Λ is not less than z^2 short of the maximum value attained when $\pi = \hat{\pi}$. This approach extends to differences between proportions and similar contexts. We express the likelihood in terms of θ and η, then determine the profile estimate of η given θ, denoted by η_θ, as the maximum likelihood estimate of η for given θ. This is a function of θ. We slot this into the likelihood function, which then becomes a function of θ only. We obtain a likelihood interval for θ from this by iteration, as in the single-parameter case.

For the unpaired difference, the log likelihood reduces (within an additive constant) to

$$\ln \Lambda = a \ln (\eta + \theta/2) + b \ln (\eta - \theta/2) + c \ln (1 - \eta - \theta/2) + d \ln (1 - \eta + \theta/2) \quad (7.17)$$

with the understanding that terms corresponding to empty cells are omitted. The constraints $0 \le \pi_i \le 1$ ($i = 1, 2$) translate into restricting evaluation of ln Λ only within the bounding rhombus

$$\frac{1}{2}|\theta| \le \eta \le 1 - \frac{1}{2}|\theta| \quad (7.18)$$

The four diagonal boundaries of this are $\eta = -\theta/2$; $\theta/2$; $1 - \theta/2$; and $1 + \theta/2$. The likelihood is zero on each of these boundaries, unless the corresponding cell entry (*a*, *b*, *c* or *d*, respectively) is zero.

It is necessary to distinguish four situations, according to the pattern of zero cells.

NZ: No zero cells; all four edges precipitous.

OZ: One zero cell (e.g., $c = 0$, $abd > 0$); the upper right edge $\eta = 1 - \theta/2$ is now available.

RZ: Two cells in same row zero (e.g., $a = b = 0$, $cd > 0$); the lower edges $\eta = \pm\theta/2$ are now available.

DZ: Two cells on same diagonal zero (e.g., $b = c = 0$, $ad > 0$); the right-hand edges $\eta = \theta/2$ and $1 - \theta/2$ are now available.

These represent all meaningful situations, since *m* and *n* must be greater than zero.

7.4.3 Tail-Based Exact and Mid-*p* Profile Intervals

Unfortunately, like the likelihood interval for the single proportion, true profile likelihood intervals in general are recognised to be anticonservative (Cox and Reid 1992). This applies to the intervals for unpaired differences of proportions (Miettinen and Nurminen 1985), and also to the paired difference case (Chapter 8). Nevertheless, in both contexts, the profile likelihood can also be used to construct tail-based intervals for both unpaired and paired differences, to which this criticism does not apply.

Thus we can use substitution of η_θ for η as the basis for calculating a tail-based profile interval for θ, analogous to the Clopper–Pearson interval introduced in Section 3.4.4. This calculation involves two nested iterative processes—one to obtain η_θ as a function of θ, and the other to align the relevant sum of tail probabilities with the desired $1 - \alpha/2$—we cannot bypass this process by using beta functions as in the analogous calculation for the Clopper–Pearson interval.

Not surprisingly, the resulting tail-based profile interval for θ is rather conservative, with a mean coverage of 0.968 in Newcombe (1998b), although with a minimum coverage of 0.942 it does not guarantee strict conservatism. An analogous, rather less conservative interval may be obtained by the mid-*p* approach, as in Section 3.4.5. This interval gave mean coverage 0.959, very similar to the square-and-add Wilson method, but with a more favourable minimum coverage of 0.913, compared to 0.867 for the square-and-add Wilson method. As in the single proportion case (Section 4.2), good methods that align mean coverage with the nominal $1 - \alpha$ have a moving average coverage probability much closer to $1 - \alpha$ than the minimum coverage probability, suggesting that we need not be over-concerned about minimum coverage probabilities of this magnitude. The relative ease of application and transparency of the square-and-add Wilson method appear to outweigh concerns about such minimum coverage rates which are arguably theoretical rather than of any practical importance.

7.4.4 Intervals Based on Mesially Shifted Estimates

For the difference between independent proportions, several approaches are available which are based on mesially shifted estimates for the 2 proportions. While the most cogent rationale for such methods relates to the need to avoid division by zero when calculating ratio measures, the relative risk and odds ratio, these methods have nevertheless been recommended for the risk difference also. Three such methods are considered here.

In Section 3.4.7 we considered modified Wald intervals for the binomial proportion, which are obtained by adding a pseudo-frequency ψ to the numbers of successes and failures. For the single proportion, Agresti and Coull (1998) recommended using an interval with $\psi = 2$. Taking this process further, Agresti and Caffo (2000) suggested that for the unpaired difference case, a satisfactory interval for t could be obtained by adding a pseudo-frequency $\psi = 1$ to each of the four cells a, b, c and d (and hence 2 to both column totals, m and n) and calculating a Wald interval. In terms of aligning mean coverage with $1 - \alpha$, these are certainly the best choices for these two cases. Just as in the single proportion case, boundary violation can sometimes occur. This is likely to be very infrequent here as it can only arise when t is at or close to ± 1.

A deterministic bootstrap interval based on median unbiased estimates is described in Section 11.4, alongside similar intervals for the odds ratio and relative risk.

Bayes intervals are considered in Section 7.7, where a modified interval incorporating a non-negativity constraint is developed.

7.5 Evaluation of Performance

Just as for the simpler case of the single binomial proportion, it is necessary to evaluate coverage and related properties of the various methods that have been proposed. The results presented in this section summarise those reported by Newcombe (1998b), and relate to 11 methods defined in Section 7.5.1, of which methods 8 to 11 were original. The results relate to coverage, width, location and boundary anomaly properties. This evaluation is extended in Chapter 13 to include several additional intervals obtained by squaring and adding and also the PropImp algorithm.

7.5.1 Definitions of 11 Methods

1. The simple asymptotic (Wald) limits without continuity correction are

$$\hat{\theta} \pm z\sqrt{ac/m^3 + bd/n^3}. \tag{7.19}$$

2. The Wald limits with continuity correction are

$$\hat{\theta} \pm \left(z\sqrt{ac/m^3 + bd/n^3} + \left(1/m + 1/n\right)/2\right). \tag{7.20}$$

3. Methods 3 and 4 in Newcombe (1998b) are two methods introduced by Beal (1987) which produce intervals of the form $\theta^* \pm w$ where

$$\theta^* = \frac{\hat{\theta} + z^2 v(1 - 2\hat{\eta})}{1 + z^2 u}, \tag{7.21}$$

$$w = \frac{z}{1+z^2u}\sqrt{u\left\{4\hat{\eta}(1-\hat{\eta})-\hat{\theta}^2\right\}+2v(1-2\hat{\eta})\hat{\theta}+4z^2u^2(1-\hat{\eta})\hat{\eta}+z^2v^2(1-2\hat{\eta})^2}, \tag{7.22}$$

and $\hat{\eta}$ is a Bayes posterior estimate of η.

The parameters u and v are defined as $(1/m \pm 1/n)/4$.

For method 3, $\hat{\eta}$ is defined as the Haldane estimator $(a/m + b/n)/2$.

4. Beal's Jeffreys–Perks interval is as method 3, but with

$$\hat{\eta} = \frac{1}{2}\left(\frac{a+0.5}{m+1}+\frac{b+0.5}{n+1}\right). \tag{7.23}$$

5. The Mee (1984) score interval comprises all θ satisfying

$$|\hat{\theta}-\theta| \le z\sqrt{\lambda\left\{\frac{(\eta_\theta+\theta/2)(1-\eta_\theta-\theta/2)}{m}+\frac{(\eta_\theta-\theta/2)(1-\eta_\theta+\theta/2)}{n}\right\}} \tag{7.24}$$

where $\lambda = 1$.

6. The Miettinen and Nurminen (1985) score interval is as method 5, with $(m + n/(m + n - 1))$.
7. The true profile likelihood interval (Miettinen and Nurminen 1985) comprises all θ satisfying

$$a\ln\frac{\eta_\theta+\theta/2}{a/m}+b\ln\frac{\eta_\theta-\theta/2}{b/n}+c\ln\frac{1-\eta_\theta-\theta/2}{c/m}+d\ln\frac{1-\eta_\theta+\theta/2}{d/n}\ge-\frac{z^2}{2} \tag{7.25}$$

omitting any terms corresponding to empty cells.

8. The profile likelihood interval based on exact tail areas is delimited by $L \le \theta \le U$ where

(i) if $L \le \theta \le \hat{\theta}, kP_x + \sum_{1\ge\xi>x} P_\xi \ge \frac{\alpha}{2}$ (7.26)

(ii) if $\hat{\theta} \le \theta \le U, kP_x + \sum_{-1\le\xi<x} P_\xi \ge \frac{\alpha}{2}$ (7.27)

with $P_\xi = \Pr[A/m - B/n = \xi \mid \theta, \eta_\theta]$, $x = a/m - b/n$, and $k = 1$.

9. Profile likelihood method based on mid-*p* tail areas: as method 8, but with $k = 0.5$.
10. Square-and-add limits based on the Wilson score method for the single proportion, without continuity correction are $L = \hat{\theta} - \delta, U = \hat{\theta} + \varepsilon$ where

$$\delta = \sqrt{(a/m - l_1)^2 + (u_2 - b/n)^2}, \tag{7.28}$$

$$\varepsilon = \sqrt{(u_1 - a/m)^2 + (b/n - l_2)^2}, \tag{7.29}$$

l_1 and u_1 are the roots of

$$|\pi_1 - a/m| = z\sqrt{\pi_1(1-\pi_1)/m}, \tag{7.30}$$

and l_2 and u_2 are the roots of

$$|\pi_2 - b/n| = z\sqrt{\pi_2(1-\pi_2)/n}. \tag{7.31}$$

11. Square-and-add limits based on the Wilson score method for the single proportion, with continuity correction are as method 10, but l_1 and u_1 are defined by

$$\left\{\pi_1 : |\pi_1 - a/m| - 1/(2m) \le z\sqrt{\pi_1(1-\pi_1)/m}\right\}. \tag{7.32}$$

Similarly l_2 and u_2 delimit the interval defined by

$$\left\{\pi_2 : |\pi_2 - b/n| - 1/(2n) \le z\sqrt{\pi_2(1-\pi_2)/n}\right\}. \tag{7.33}$$

Note that if $a = 0$, $l_1 = 0$; if $c = 0$, $u_1 = 1$ and so forth, as in Section 3.5.

Table 7.2 shows 95% intervals calculated by the 11 methods for eight selected combinations of *a*, *m*, *b* and *n*, which cover all four combinations of zero cells, NZ, OZ, RZ and DZ as defined in Section 7.4.2. These intervals illustrate how the boundary anomalies of overflow and inappropriate tethering can occur when the simpler methods are used.

TABLE 7.2

95% Limits for Selected Contrasts between Independent Binomial Proportions, Calculated by 11 Methods

	(a)	(b)	(c)	(d)	(e)	(f)	(g)	(h)
Contrast	**56/70–48/80**	**9/10–3/10**	**6/7–2/7**	**5/56–0/29**	**0/10–0/20**	**0/10–0/10**	**10/10–0/20**	**10/10–0/10**
Wald intervals								
1. Without CC	+0.0575, +0.3425	+0.2605, +0.9395	+0.1481, +0.9947	+0.0146, +0.1640	0.0000*, 0.0000*	0.0000*, 0.0000*	+1.0000*, +1.0000	+1.0000*, +1.0000
2. With CC	+0.0441, +0.3559	+0.1605, >+1.0000*	+0.0053, >+1.0000*	–0.0116, +0.1901	–0.0750, +0.0750	–0.1000, +0.1000	+0.9250, >+1.0000*	+0.9000, >+1.0000*
Beal intervals								
3. Haldane	+0.0535, +0.3351	+0.1777, +0.8289	+0.0537, +0.8430	–0.0039, +0.1463	0.0000*, +0.0839	0.0000*, 0.0000*	+0.7482, +1.0000	+0.6777, +1.0000
4. Jeffreys–Perks	+0.0531, +0.3355	+0.1760, +0.8306	+0.0524, +0.8443	–0.0165, +0.1595	–0.0965, +0.1746	–0.1672, +0.1672	+0.7431, >+1.0000*	+0.6777, +1.0000
Score intervals								
5. Mee	+0.0533, +0.3377	+0.1821, +0.8370	+0.0544, +0.8478	–0.0313, +0.1926	–0.1611, +0.2775	–0.2775, +0.2775	+0.7225, +1.0000	+0.6777, +1.0000
6. Miettinen–Nurminen	+0.0528, +0.3382	+0.1700, +0.8406	+0.0342, +0.8534	–0.0326, +0.1933	–0.1658, +0.2844	–0.2879, +0.2879	+0.7156, +1.0000	+0.6636, +1.0000

Intervals based on profile likelihood								
7. True profile	+0.0547, +0.3394	+0.2055, +0.8634	+0.0760, +0.8824	+0.0080, +0.1822	−0.0916, +0.1748	−0.1748, +0.1748	+0.8252, +1.0000	+0.8169, +1.0000
8. Exact	+0.0529, +0.3403	+0.1393, +0.8836	−0.0104, +0.9062	−0.0302, +0.1962	−0.1684, +0.3085	−0.3085, +0.3085	+0.6915, +1.0000	+0.6631, +1.0000
9. Mid-*p*	+0.0539, +0.3393	+0.1834, +0.8640	+0.0470, +0.8840	−0.0233, +0.1868	−0.1391, +0.2589	−0.2589, +0.2589	+0.7411, +1.0000	+0.7218, +1.0000
Square-and-add Wilson intervals								
10. Without CC	+0.0524, +0.3339	+0.1705, +0.8090	+0.0582, +0.8062	−0.0381, +0.1926	−0.1611, +0.2775	−0.2775, +0.2775	+0.6791, +1.0000	+0.6075, +1.0000
11. With CC	+0.0428, +0.3422	+0.1013, +0.8387	−0.0290, +0.8423	−0.0667, +0.2037	−0.2005, +0.3445	−0.3445, +0.3445	+0.6014, +1.0000	+0.5128, +1.0000

Note: * = Aberrations (limits beyond ±1 or inappropriately equal to $\hat{\theta}$). CC = continuity correction.

7.5.2 Evaluation

In the original evaluation Newcombe (1998b) assessed 11 frequentist intervals for a difference between two independent proportions. The main evaluation of coverage was based on a sample of 9200 parameter space points defined by four quantities: the two relevant sample sizes m and n, ranging from 5 to 50; a nuisance parameter $\eta = (\pi_1 + \pi_2)/2$, and the parameter of interest, $\theta = \pi_1 - \pi_2$.

There were 230 of the 2116 possible (m, n) pairs chosen purposively, comprising all 46 pairs with $m = n$ together with 92 pairs with $m \neq n$ and the corresponding reversed pairs. For each of $m = 5, 6, \ldots, 50$, two values of $n \neq m$ were chosen, avoiding duplicates and mirror-image pairs. These were selected in such a way that m and n should be uncorrelated and that distributions of $|m - n|$ and the highest common factor should be very close to those for all 2070 available unequal pairs. The rationale for considering the highest common factor is that when tail areas are defined in terms of $a/m - b/n$, the difference between mid-p and exact limits could be great when m and n are equal but relatively small when they are coprime (i.e., have no common factor greater than 1). Thus the chosen set of unequal (m, n) pairs may be regarded as representative, in all important respects, of all 2070 possible ones; these were used together with a deliberate over-representation of diagonal pairs, in view of their commonness of occurrence, to give a set of (m, n) pairs which may reasonably be regarded as typical.

For each of the 230 (m, n) pairs, 40 (η, θ) pairs were chosen, with $\theta = \lambda\{1 - |2\eta - 1|\}$ and η and $\lambda \sim U(0, 1)$, all sampling being random and independent. This resulted in a set of θ values with median 0.193, quartiles 0.070 and 0.393.

For each chosen PSP, frequencies of all possible outcomes were determined as products of binomial probabilities. The achieved probability of coverage of θ by nominal 95% intervals calculated by each method was computed exactly by summating all appropriate non-negligible terms. Minimum as well as mean coverage probabilities were also examined. Mesial and distal non-coverage rates were computed similarly, as were incidences of overt overshoot for methods 1 to 4, and latent overshoot for methods 3 and 4. Probabilities of inappropriate tethering for methods 1 and 3 were computed by examining frequencies of occurrence of cases RZ and DZ, in conjunction with whether m and n were equal. Additionally, results were reported for defined subsets of the parameter space, also averages for subsets defined by several criteria: whether m and n were equal, coprime or intermediate; the smaller of the two sample sizes, the minimum expected cell frequency, ψ and θ. Some results were also reported for nominal 90% and 99% intervals, and for very large denominators with very small proportions.

Expected interval width was calculated exactly for 95% intervals by each method, truncated to lie within [−1, +1] where necessary, for nine selected PSPs with $\pi_1 = \pi_2 = 0.5$ or 0.01 and m and n 10 or 100.

The evaluation was subsequently expanded to provide some results for intervals developed using the PropImp algorithm as described in Chapter 13 (Newcombe 2011c).

In response to criticisms of score methods (Santner et al. 2007), Newcombe and Nurminen (2011a) extended the results of this evaluation to include coverage for 95% and 90% intervals for all 9200 PSPs and for selected sample size pairs, calculating the minimum coverage probability and proportion of PSPs with coverage $>1 - \alpha$ as well as the mean coverage for 95% and 90% intervals.

7.5.3 Coverage, Location and Width

The results of the evaluation may be summarised as follows. Among the asymptotic methods, both the Miettinen–Nurminen score interval and the square-and-add Wilson interval have satisfactory coverage properties. The tail-based methods 8 and 9 align minimum and mean coverage closely with the nominal $1 - \alpha$.

Table 7.3 shows that the coverage probability of nominal 95% intervals, averaged over the 9200 parameter space points, ranged from 0.881 for method 1 to 0.979 for method 11. As well as method 1, method 3 was also anticonservative on average, and method 7 slightly so; method 8 was slightly conservative, whereas other methods had appropriate mean coverage rates.

The maximum CP of method 1 in this evaluation was only 0.966; for all other methods some parameter space points have CP = 1. The coverage probability of method 1 is arbitrarily close to 0 in extreme cases, either MNCP can approach 1 (when $\eta \sim 0$ or 1 and $\theta \sim 0$) or DNCP can (when $\eta \approx 0.5$ and $\theta \sim 1$), due to ZWIs at 0 and 1, respectively. The continuity correction of method 2, although adequate to correct the mean CP, yields an unacceptably low min CP of 0.5137 in this evaluation. DNCP approaches $\exp(-0.5) = 0.6065$ with $n \ll m$ but $n \to \infty$, $\theta = 1/2\ (1/m + 1/n) + \varepsilon$, $\pi_1 = 1 - \varepsilon$; a similar supremum applies to MNCP. The coverage of method 2 exhibited appropriate symmetry; for method 1, overall, distal non-coverage predominated in this evaluation; see Section 13.6.3.

Method 3, like method 1, can have CP arbitrarily close to 0, but only DNCP can approach 1, as $\eta \to 1$ (if $m \le n$) or $\eta \to 0$ (if $m \ge n$), due to inappropriate tethering at 0. Method 4 eliminates this deficiency as well as the low mean CP, and reduces the preponderance of distal non-coverage.

Methods 5 and 6 have generally very similar coverage properties to each other, with overall coverage similar to method 4. There was one PSP with coverage probability only 0.852; substantial distal non-coverage can also occur. These methods exhibited very good symmetry of coverage.

Coverage of method 7 was symmetrical but slightly anticonservative on average, with a minimum coverage of 0.830. Values of either DNCP or MNCP around 0.14 can occur.

Even method 8 fails to be strictly conservative, the minimum coverage in this evaluation being 0.942. Here both MNCP (0.028) and DNCP (0.030) exceeded the nominal $\alpha/2$; there are other parameter space points for which either exceeds 0.03. (This contrasts with the performance of the analogous method for the paired case (Newcombe 1998c) for which min CP was 0.955, and DNCP (but not MNCP) was always less than 0.025.) The lowest coverage

TABLE 7.3

Estimated Coverage Probabilities for 95% Intervals for Differences between Independent Binomial Proportions, Calculated by 11 Methods

	Coverage		Mesial Non-Coverage		Distal Non-Coverage		Mean MNCP ÷ Mean NCP
	Mean	Minimum	Mean	Maximum	Mean	Maximum	
Wald intervals							
1. Without CC	0.8807	0.0004	0.0417	0.7845	0.0775	0.9996	0.350
2. With CC	0.9623	0.5137	0.0183	0.4216	0.0194	0.4844	0.486
Beal intervals							
3. Haldane	0.9183	0.0035	0.0153	0.0656	0.0664	0.9965	0.187
4. Jeffreys–Perks	0.9561	0.8505	0.0140	0.0606	0.0299	0.1418	0.319
Score intervals							
5. Mee	0.9562	0.8516	0.0207	0.1484	0.0231	0.1064	0.473
6. Miettinen–Nurminen	0.9584	0.8516	0.0196	0.1484	0.0220	0.1064	0.471
Intervals based on profile likelihood							
7. True profile	0.9454	0.8299	0.0268	0.1440	0.0278	0.1384	0.491
8. Exact	0.9680	0.9424	0.0149	0.0308	0.0170	0.0317	0.467
9. Mid-*p*	0.9591	0.9131	0.0197	0.04996	0.0212	0.0470	0.482
Square-and-add Wilson intervals							
10. Without CC	0.9602	0.8673	0.0134	0.0660	0.0264	0.1327	0.337
11. With CC	0.9793	0.9339	0.0061	0.0271	0.0147	0.0661	0.293

Note: Based on 9200 parameter space points with $5 \le m \le 50$, $5 \le n \le 50$, $0 < \psi < 1$ and $0 < \theta < 1 - |2\psi - 1|$ as in Section 7.5.2. CC = continuity correction.

obtained for its "mid-p" analogue, method 9, was 0.913. Both these methods yielded symmetrical coverage.

For method 10, the lowest coverage obtained in the main evaluation was 0.867; this and other extremes arose entirely as distal non-coverage. MNCP can approach 0.1685, corresponding to the limiting non-coverage for the score method for the single proportion when $\theta = L_1 - \varepsilon$, $\pi_2 = \varepsilon$, m is small and $n \to \infty$, where L_1 is the lower score limit for $1/m$. Method 11 produced its lowest CP, 0.934, entirely from distal non-coverage; this compares unfavourably with 0.949 for the corresponding method for the single proportion. Both methods 10 and 11 yielded intervals erring towards mesial location.

Variation in expected interval width (Table 7.4) between different methods is most marked when any of the expected frequencies $m\pi_1$, $m(1 - \pi_1)$, $n\pi_2$ or

TABLE 7.4

Average Width of 95% Intervals Calculated by 11 Methods, for Selected Parameter Space Points

m	10	10	10	100	100	100	100	100	100
n	10	10	10	10	10	10	100	100	100
π_1	0.01	0.5	0.95	0.01	0.5	0.95	0.01	0.5	0.95
π_2	0.01	0.5	0.05	0.01	0.5	0.05	0.01	0.5	0.05
Wald intervals									
1. Without CC	0.125	1.148	0.504	0.193	1.138	0.567	0.551	1.019	0.940
2. With CC	0.480	1.424	0.716	0.528	1.340	0.743	0.774	1.093	1.096
Beal intervals									
3. Haldane	0.115	1.057	0.904	0.553	1.087	0.825	0.544	1.009	0.971
4. Jeffreys–Perks	0.636	1.062	0.907	0.798	1.092	0.922	0.720	1.009	0.971
Score intervals									
5. Mee	1.001	1.070	0.885	0.999	1.018	0.996	0.992	1.009	0.969
6. Miettinen–Nurminen	1.038	1.094	0.916	1.005	1.022	1.001	0.996	1.012	0.972
Intervals based on profile likelihood									
7. True profile	0.666	1.105	0.721	0.679	1.067	0.815	0.742	1.014	0.945
8. Exact	1.119	1.217	0.977	1.025	1.084	1.005	1.027	1.049	1.025
9. Mid-p	0.946	1.124	0.854	0.905	1.069	0.961	0.894	1.016	0.960
Square-and-add Wilson intervals									
10. Without CC	1.000	1.000	1.000	1.000	1.000	1.000	1.000	1.000	1.000
11. With CC	1.234	1.138	1.203	1.227	1.127	1.196	1.185	1.050	1.106
Actual average width for method 10	0.5627	0.7231	0.4773	0.3289	0.5430	0.3522	0.0895	0.2707	0.1264

Note: All widths are expressed relative to the average width for the square-and-add Wilson interval without continuity correction (CC).

$n(1 - \pi_2)$ is low. The width is then least for method 1 or 3, largely on account of the high frequency of degeneracy.

7.5.4 Boundary Anomalies

For the single proportion case, two types of boundary anomaly can occur when Wald or related methods are used, namely degenerate, zero width intervals and overshoot, meaning that one or both calculated limits lie outside the acceptable range. These anomalies can also occur when related methods for the difference of proportions are used. Furthermore, as a result of the greater complexity of the two-parameter case, additional anomalous behaviours can occur, namely latent overshoot and unilateral tethering.

An interval is said to be tethered to a point estimate for the parameter, $\tilde{\theta}$, when one or both of the calculated limits L and U coincides with $\tilde{\theta}$. Most often, the point estimate in question is the maximum likelihood estimate $\hat{\theta}$. In the extreme case, where $\hat{\theta} = +1$, it is appropriate that $U = \hat{\theta}$, likewise that $L = \hat{\theta}$ when $\hat{\theta} = -1$. Otherwise this is an infringement of the principle that a confidence interval should represent some margin of error on both sides of $\hat{\theta}$, and is counted as adverse. Bilateral point estimate tethering, $L = \hat{\theta} = U$, constitutes a degenerate zero-width interval and is always inappropriate.

For the unpaired difference of proportions, tethering can only occur in case RZ (two zeros in the same row) for methods 1 and 3, and in case DZ (two zeros on the same diagonal) for methods 1, 3 and 4, exemplified in Table 7.2, contrasts (e) to (h). The Haldane method produces appropriate, unilateral tethering in case DZ. In case RZ, it produces unilateral tethering if $m \neq n$ (e.g., for 0/10–0/20), and a ZWI at $\hat{\theta} = 0$ if $m = n$, both of which are inappropriate. Method 1 produces a totally inappropriate ZWI at 0 in case RZ, and a ZWI at +1 (or −1) in case DZ, for which unilateral tethering would be appropriate. In case DZ, the Jeffreys–Perks method reduces to the Haldane method if $m = n$, otherwise produces overt overshoot.

Overt overshoot occurs when either calculated limit is outside [−1, +1]. $U = +1$ is not to be regarded as aberrant when $\hat{\theta} = +1$, and correspondingly at −1. Methods 1 to 4 are liable to produce overt overshoot, in which case we truncate the resulting interval to be a subset of [−1, +1]; instances in which overt overshoot would otherwise occur were counted in the evaluation.

Latent overshoot can occur when Beal methods are used. Methods 3 and 4 substitute an estimate $\hat{\eta}$ for ψ which is formed without reference to θ, and very often one or two of the implied parameters $\tilde{\pi}_{1L} = \hat{\eta} + L/2$, $\tilde{\pi}_{2L} = \hat{\eta} - L/2$, $\tilde{\pi}_{1U} = \hat{\eta} + U/2$, $\tilde{\pi}_{2U} = \hat{\eta} - U/2$ lie outside [0, 1]. This anomaly is termed latent overshoot: overt overshoot always implies latent overshoot, but latent overshoot can occur in the absence of overt overshoot, when inherent bounds for θ are not violated but the bounding rhombus $|\theta| \leq 2\eta \leq 2 - |\theta|$ is; the formulae still work, and do not indicate anything peculiar has occurred. In this evaluation the frequency of occurrence of any latent overshoot (irrespective of whether involving one or two implied parameters, and of whether overt overshoot also occurs) is obtained for methods 3 and 4, using the chosen $\alpha = 0.05$. Unlike overt overshoot, latent overshoot cannot effectively be eliminated by truncation: as well as affecting coverage, such truncation can produce inappropriate point estimate tethering, or even an interval that excludes $\hat{\theta}$. It appears to be a consequence of losing the simplicity of using information concerning π_1 and π_2 separately.

In the evaluation, overt overshoot was common using methods 1 (mean probability 0.0270) and 2 (0.0594), less so for methods 3 (0.0012) and 4 (0.0052). It occurs with probability approaching 1 for method 2 as $\theta \to 1$, and also for method 4 provided $m \neq n$. For method 1, the maximum overshoot probability is lower, 0.7996, since as $\theta \to 1$, Pr [ZWI at 1] $\to$ 1. Latent overshoot occurred with probability around 0.5 for methods 3 and 4; some parameter space points produced latent overshoot with probability 1.

The Haldane method produced unilateral point estimate tethering with probability 0.0413 when $m \neq n$, and a ZWI at 0 with probability 0.0471 when $m = n$. Method 1 produced a ZWI at 0 in both these cases, and a ZWI at 1 with probability 0.0035. For both methods the probability of inappropriate tethering approaches 1 as $\eta \to 0$ or 1.

7.6 Number Needed to Treat

Proposed by Laupacis et al. in 1988, the NNT is often used in medical publications. It is defined as the reciprocal of the absolute risk reduction. With an appropriate sign convention—for example, if successful outcome of treatment is labelled as the positive outcome, and groups 1 and 2 are the intervention and control groups—this is simply $1/\theta$, which is usually estimated by $1/t$. It is often held to be a more intuitive measure for clinicians to grasp than a difference between proportions. However, for most effect size measures, a larger value means a larger effect. But the larger the effect of treatment, the smaller the NNT, and vice versa. This hardly seems intuitive.

Like any other measure of effect size, the NNT should normally be reported with a confidence interval expressing its sampling imprecision, and this is where the deficiency of this approach becomes apparent. Confidence limits for the NNT may be obtained simply as reciprocals of the limits calculated for θ. This approach is one of the simplest applications of the substitution method introduced in Chapter 6. Equivalently, applying either the MOVER-DL or MOVER-R algorithms introduced in Section 7.3 with $q_2 = t$, with confidence limits L_2 and U_2, and $q_1 = L_1 = U_1 = 1$, leads to confidence limits $1/U_2$ and $1/L_2$ for the NNT. Obviously, it is preferable for the interval for θ to be calculated by a good method, such as the square-and-add Wilson interval, as in Bender (2001). Nevertheless, for clarity of illustration, we use the symmetrical Wald interval here. (Calculating reciprocals of Wald limits presupposes that any boundary overshoot has already been truncated, but this issue does not apply to the figures considered here.) Suppose first that the success rates on the two treatments are 32 out of 64 (50%) and 24 out of 96 (25%). The estimated difference here is +0.2500, with 95% Wald interval +0.1000 to +0.4000. Inverting these figures, we calculate NNT = 1/0.2500 = 4, with 95% interval from 1/0.4000 = 2.5 to 1/0.1000 = 10. Note that the lower and upper limits for the NNT come from the upper and lower limits for θ, and that the interval for the NNT is much more asymmetrical than any two-sided interval we have encountered hitherto. But these are the only issues—provided that, as here, the difference is statistically significant….

7.6.1 To Infinity and Beyond

It is in the non-significant case, where the interval for θ includes zero, that the confidence interval for the NNT behaves anomalously (Newcombe 1999). Firstly, obviously the NNT has nothing to offer for data such as our Eastern Co-operative Oncology Group example, in which both numerators are zero. But there are serious problems even when none of the cell frequencies is small. Suppose we wish to compare the success rates on two treatments, 47 out of 94 (50%) and 30 out of 75 (40%). The estimated difference here is +0.1000. The 95% Wald interval is −0.0500 to +0.2500. Naively inverting these figures results in an NNT of +10, with 95% interval −20 to +4.

We noted in the first example that the lower and upper limits for the NNT come from the upper and lower limits for θ. This does not apply here. Instead, the calculated interval actually excludes the point estimate, +10. Conversely, it includes impossible values between −1 and +1, including zero. The explanation is that the calculated limits, −20 and +4, are correct. But the interval that these limits encompass must not be regarded as a confidence interval for the NNT. In fact, it comprises values of $1/\theta$ that are *not* consistent with the observed data. The correct confidence region is not a finite interval, but consists of two separate intervals extending to infinity, namely from $-\infty$ to −20 and from +4 to $+\infty$ (Figure 7.1). While a confidence region of this nature is perfectly comprehensible to a mathematician, one would not expect clinicians to be comfortable with it as an expression of sampling uncertainty.

In the event that p_1 and p_2 are identical, the point estimate of the NNT is then not only infinite but also of indeterminate sign. If, for example, the limits for θ are −0.1 and +0.1, the confidence region for the NNT comprises two intervals, from $-\infty$ to −10 and from +10 to $+\infty$. This will be a very rare occurrence when m and n are coprime and π_1 and π_2 mesial, but could easily occur if m and n are equal and small.

An NNT of −20 would indicate that if 20 patients are treated with the new treatment, one fewer would have a good outcome than if they all received the standard treatment. A negative number needed to treat has been called

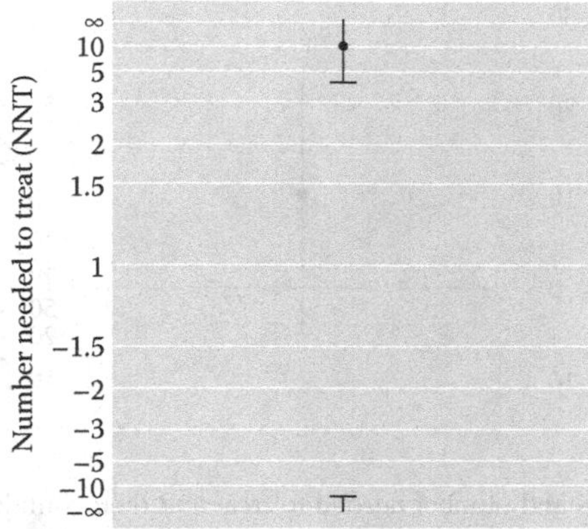

FIGURE 7.1
Confidence region for the NNT in the non-significant case. (Reproduced from Altman, D. G. 1998. *British Medical Journal* 317: 1309–1312. With permission.)

the number needed to harm (NNH) by McQuay and Moore (1997). These authors concluded that in the case of a non-significant difference it is not possible to get a useful confidence interval, and so only a point estimate is available. But it can be argued that it is even more important to report a confidence interval in the non-significant case than when there is clearly statistically significant benefit.

Altman (1998) suggested that NNT and NNH are not good abbreviations, and that the number of patients needed to be treated for one additional patient to benefit or be harmed are more appropriately denoted NNTB and NNTH, or NNT(benefit) and NNT(harm). But no such attempt to use more helpful labelling gets around the issue that the use of the NNT leads to an excessively convoluted confidence region in the frequent and important case of a non-significant difference. A much more palatable, less fraught way to present differences of proportions and their confidence limits is simply to express them as percentages, just as is routinely done for proportions themselves. It is recommended that differences of proportions, expressed in percentage terms, should be used in preference to the NNT as the primary display of results. Thus, in our second example, we would report that the additional benefit by using treatment 1 (compared to the control treatment 2) is estimated to be 5%, with 95% interval from −10% (indicating a difference of 10% disfavouring treatment 1) to +20% (a difference of 20% favouring treatment 1). Here, as in all such situations, it is crucial to make it clear that all these figures convey information about the absolute difference. Relative measures, based on ratios of proportions, are considered in Chapters 10 and 11.

Figure 7.2 illustrates an appropriate display that incorporates the NNT alongside the absolute difference. Here, the primary vertical axis is labelled to represent the absolute risk reduction, expressed in percentage terms, with

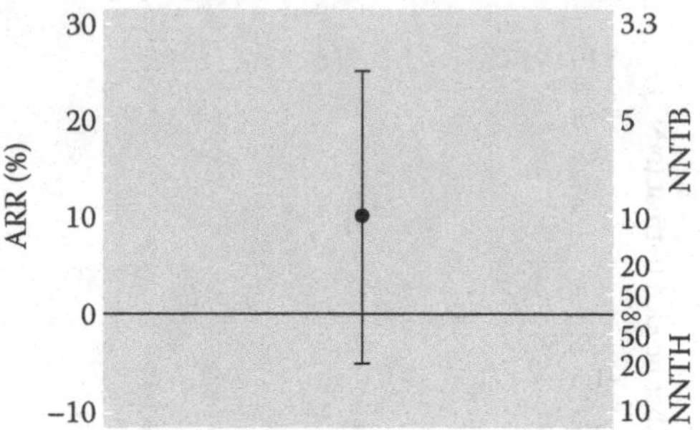

FIGURE 7.2
Relation between the ARR and number needed to treat and their confidence intervals (NNTB = number needed to treat (benefit); NNTH = number needed to treat (harm)) for the same example as in Figure 7.1. (Reproduced from Altman, D. G. 1998. *British Medical Journal* 317: 1309–1312. With permission.)

a simple linear scale. There is also a secondary vertical axis on which equivalent information is expressed in terms of the NNT measures.

Lui (2004b) specifically linked the difficulty of interpreting the NNT in the singular case explicitly to the non-monotonicity of the reciprocal function *f*. He suggested that this measure should be used only when prior knowledge clearly determines the rank ordering of response rates of different treatments, and that if the interval for the risk difference crosses 0, it should be truncated at 0. This is unsatisfactory and bias-prone. Such prior information can only be based on sampling, and hence invariably is subject to some uncertainty from sampling variation. The probability of this occurrence is simply the type II error rate β, which we can never assume to be zero. Lui refers to examples in which the control group receives an active drug A and the intervention group receives either A plus another active drug B, or a higher dose of A, yet it is unsafe to assume that B could not interfere with the action of A, or that dose-response curves are necessarily monotonic. Moreover, the NNT was designed for use in randomised trials, and Lui's requirement implies that such a trial would not be in equipoise. From this standpoint, it might be a viable option in a context such as a placebo-controlled experimental gingivitis study in healthy volunteers, but not when substantial benefits to ill patients are at stake. Ironically, Lui's criterion is more likely to be satisfied in an observational study, for which the reporting of an NNT would not normally be considered. It is satisfactory only if the null hypothesis is a non-starter i.e., in a situation in which it would be meaningless to perform the hypothesis test, and the only purpose is estimation. Indeed, what if the point estimate, or even the whole of the interval for the risk difference, were the wrong side of zero?

As an example of how NNTs should not be reported, a systematic review by Tramèr et al. (1997) (quoted by Lui 2004b) included several parallel-groups trials evaluating the early anti-emetic efficacy of ondansetron compared with intravenous droperidol. For the first trial included (Heim et al. 1994), the event rates on the two regimes were 30/50 and 34/50. The difference of proportions here is −0.0800, with a Wald 95% interval from −0.2675 to +0.1075, which was truncated as above to −0.2675 to 0. Inverting these figures, they reported an NNT of −12.5, with confidence interval −3.7 to ∞. Taken literally, this interval does not include the point estimate, −12.5, but does include inadmissible values between −1 and +1. Several other similarly incorrect intervals were reported. The confidence region for the NNT based on the Wald interval for $\theta = \pi_1 - \pi_2$ here should comprise two intervals, one from −∞ to −3.7, the other from 1/0.1075 = +9.3 to +∞. The Bender (2001) 95% confidence region for the NNT derived by inverting the Wilson square-and-add limits for θ is very similar here, comprising two intervals, one from −∞ to −3.9, the other from +9.5 to +∞.

Stang et al. (2010) concede the shortcomings of the NNT as a scientific expression of research findings, and consider some issues to be considered when the NNT is used to translate findings to consumers. They conclude that when study results are presented in terms of NNTs the basic information about which treatments are compared, the treatment and follow-up periods, and the direction of the effect should be given; over-rounding of NNTs should be avoided; and in more complicated situations of confounding or varying follow-up times, the use of more sophisticated methods is required, which may increase the potential for misinterpretation.

7.7 Bayesian Intervals

A Bayesian interval for a difference of proportions may also be calculated. We saw in Section 3.4.6 that it is straightforward to obtain a Bayesian interval for a single proportion when a conjugate beta prior $B(\alpha, \beta)$, usually either uniform $B(1,1)$ or Jeffreys $B(1/2, 1/2)$ is used. Unfortunately, this ease of calculation does not carry over to the difference between two independent proportions. Assuming that identical priors $B(\alpha_0, \beta_0)$ are used for p_1 and p_2 independently, the posterior density for $t = p_1 - p_2$ on the support [−1, 1] is given by the following expression (Newcombe, 2007a, modified from Hashemi et al. 1997):

$$f(\Delta) = \frac{1}{k}\int_L^U p_1^{\alpha_1-1}(1-p_1)^{\beta_1-1}(p_1+t)^{\alpha_2-1}(1-p_1-t)^{\beta_2-1}\,dp_1 \qquad (7.34)$$

Here, $\alpha_1 = \alpha_0 + a$, $\alpha_2 = \alpha_0 + b$, $\beta_1 = \beta_0 + c$ and $\beta_2 = \beta_0 + d$, and the constant k is $B(\alpha_1, \beta_1) \times B(\alpha_2, \beta_2)$. The limits of integration are $L = 0$, $U = 1 - t$ for $t > 0$ and $L = -t$, $U = 1$ for $t \le 0$. While the required centiles (2.5 and 97.5 by default) for an equal-tails interval are readily determined from the posterior density in the usual way, calculating the posterior density requires numerical integration here. Unless a dedicated program is available, MCMC provides a more readily accessible route to obtain the lower and upper limits.

Logically, prior beliefs may relate to p_1 and p_2 or to their difference. When independent uniform priors are used, so that $\alpha = \beta = 1$, the resulting prior for t is the triangular distribution on $[-1, 1]$. This is the simplest choice, and leads to a satisfactory interval when m and n are large. Nevertheless, it is unsatisfactory for the case where sample sizes are small, and hence the prior has a greater influence to make the assumption that p_1 and p_2 are independent. This assumption seems highly implausible here (just as in the evaluation of coverage properties of several frequentist intervals by Santner et al. 2007)—we would expect that a measure such as $t = p_1 - p_2$ (or perhaps p_1/p_2 or the corresponding odds ratio) would be fairly constant across different series even if $\eta = (p_1 + p_2)/2$ is not. So one could argue for using a prior expressing beliefs in relation to t and η rather than p_1 and p_2–especially if an informative prior is desired.

An extreme situation arises when the difference t logically cannot be negative. Here, the Bayesian approach has an important advantage. Newcombe (2007a) described the estimation of the false negative rate of intra-operative axillary sentinel node biopsy (SNB) for staging of clinically early breast cancer. The false negative rate here is expressed as a proportion of all cases with successful localisation of one or more sentinel node, not as a proportion of all node positive cases, because in the absence of a definitive axillary clearance procedure, the actual number of node positive cases cannot be identified.

The context was the Axillary Lymphatic Mapping Against Nodal Axillary Clearance (ALMANAC) study (Mansel et al. 2006; Clarke et al. 2004), a large randomised trial of SNB conducted in several centres in the United Kingdom. Patients of either sex, aged under 80 years, scheduled for either wide local excision or mastectomy for clinically node-negative, invasive breast cancer irrespective of tumour size were eligible. As explained in Section 10.7.1, two sentinel node localising agents were used, resulting in a localisation rate that was high but not 100%.

Two sources of information on the false negative rate were available. Direct information was available from a preliminary validation phase in which all patients underwent SNB followed by axillary nodal clearance or sampling. Of 803 patients with successful sentinel node localisation, 19 (2.4%) were classed as false negatives. Indirect information was also available from the randomised phase. Ninety-seven (25.4%) of 382 control patients undergoing axillary clearance had positive axillae. In the experimental group, 94/366 (25.7%) were apparently node-positive. Taking a simple difference of these proportions gives a point estimate of –0.3% for the proportion of patients who had positive axillae but were missed by SNB. This estimate is clearly inadmissible.

In this situation a Bayesian analysis yields interpretable point and interval estimates. The analysis considered the single proportion estimate from the validation phase; the difference between independent proportions from the randomised phase, both unconstrained and constrained to non-negativity; and combined information from the two parts of the study. To make inferences from the randomised phase with the restriction $t \geq 0$, the posterior density on the restricted parameter space becomes

$$f^*(t) = f(t) \Big/ \int_0^{+1} f(u)\,du \qquad (7.35)$$

Figure 7.3 shows the relevant posterior distributions, and Table 7.5 shows the corresponding point and interval estimates. Posterior means and medians were similar for the validation and randomised phases separately and combined, all between 2% and 3%, indicating similarity rather than conflict between the two data sources.

In this example, although two treatment groups were compared in a randomised trial, the outcome related to detection of sentinel node metastases, not outcome of treatment, and taking reciprocals to express results in terms of the NNT would not be a meaningful measure here in any case—

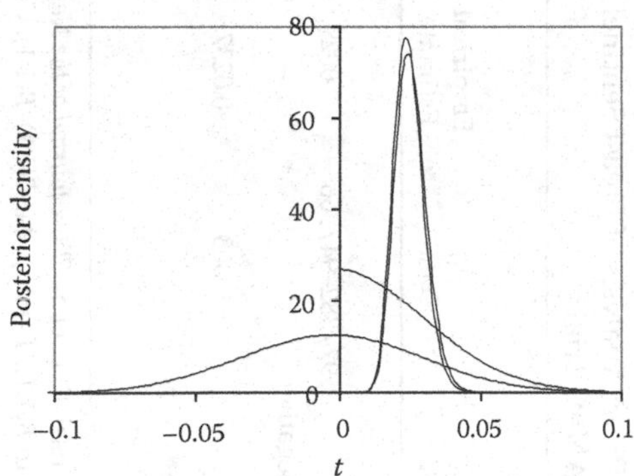

FIGURE 7.3
Posterior densities for the false negative rate defined as a proportion of all cases, based on the ALMANAC trial of sentinel node biopsy for staging breast cancer. Uniform ($B(1,1)$) priors used. Shallow bell-shaped curve with peak density 13: randomised phase, unconstrained. Half bell-shaped curve with peak density 27: randomised phase, restricted to $t \geq 0$ for admissibility. Narrow bell-shaped curve with peak density 74: validation phase, with single uniform prior. Narrow bell-shaped curve with peak density 77: randomised and validation phases combined. (From Newcombe, R. G. 2007a. *Statistics in Medicine* 26: 3429–3442.)

TABLE 7.5

Bayesian (Uniform Prior) Analysis of Breast Sentinel Node Biopsy False Negative Rate from Randomised and Validation Phases of the ALMANAC Trial

	Data	Empirical Estimate	Prior	Bayesian Posterior Estimates				
				Mean	Median	Mode	Tail-Based Interval	HPD Interval
Randomised phase								
Unconstrained	97/382–94/366	–0.0029	T	–0.0029	-0.0029	–0.0029	–0.0654 to 0.0594	–0.0653 to 0.0594
Constrained non-negative			T	0.0243	0.0204	0	0.0009 to 0.0693	0 to 0.0604
Validation phase	19/803	0.0237	U	0.0248	0.0245	0.0237	0.0153 to 0.0367	0.0146 to 0.0358
			T	0.0248	0.0244	0.0236	0.0152 to 0.0366	0.0146 to 0.0357
Combined			T	0.0240	0.0237	0.0229	0.0148 to 0.0353	0.0142 to 0.0345

Source: Data from Mansel, R. E. et al. 2006. *Journal of the National Cancer Institute* 98: 1–11.
Note: Priors: *U* uniform *B*(1,1); T triangular *B*(1,1)–*B*(1,1). Constrained to be non-negative (≡ *B*(1,2)) except in first row.

the Bayesian solution presented above is clearly preferable here. Tuyl (2007, p. 30–34) gives a similar example related to muon decay.

7.8 Interpreting Overlapping Intervals

There is a well-known relationship between the statistical significance of a difference and the confidence interval for it. Consider first the case of the means of independent samples. When the *t*-test comparing the means gives a two-sided *p*-value below α, the two-sided $1 - \alpha$ interval for the difference of means excludes the null hypothesis value, 0. When the hypothesis test does not reject the null hypothesis of equality at the specified α level, the corresponding confidence interval includes zero.

When we turn to comparisons of proportions, a similar result holds generally, although not quite invariably, because the most usually applied hypothesis tests and confidence intervals do not correspond exactly, as they do for comparisons of means. For differences of proportions, once again the null hypothesis value is 0; for ratios or odds ratios, the relevant value is 1.

The issue of what can be inferred from the overlap or non-overlap of confidence intervals for means, proportions and so forth derived from two independent groups is an elementary one, yet it has important implications for the development of confidence intervals for measures that contrast two groups. Clearly, non-overlap of the confidence intervals for two quantities is a stronger condition than the interval for the difference excluding zero. Realising this, Tryon (2001) and Tryon and Lewis (2009) suggested calculating inferential confidence intervals (ICIs) for the two quantities using a modified *z*-value of $1.9600/\sqrt{2} = 1.3859$, which corresponds to a modified $1 - \alpha$ of 0.8342. The two intervals would be expected to overlap only if the proportions differ significantly at the conventional 5% α level. Unfortunately, displaying two separate ICIs constructed in this way risks their being taken out of context and misinterpreted in a grossly anticonservative manner, with a non-coverage rate of 1 in 6 instead of 1 in 20.

Notwithstanding this arguably crucial drawback to the use of ICIs per se, there are two closely related approaches to which this disadvantage does not apply. Cumming (2009) described the visual assessment of the significance of differences from error bars, which is illustrated by Figure 7.4. Furthermore, these considerations lead naturally towards developing the Propagating Imprecision (PropImp) algorithm introduced in Chapter 13.

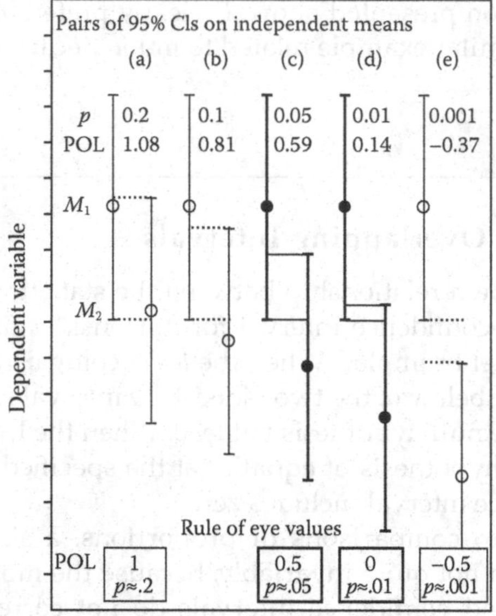

FIGURE 7.4
Illustration of a procedure for determining statistical significance approximately from the degree of overlap between confidence intervals (CI) for two independently estimated quantities, M_1 and M_2. The proportion of overlap (POL) is the degree of overlap expressed as a proportion of the length of a single arm of an interval. (Reproduced from Cumming, G. 2009. *Statistics in Medicine* 28: 205–220. With permission.)

7.9 Sample Size Planning

In Section 3.7 we considered the planning of the sample size required to estimate a single proportion within a prescribed degree of precision. It is for the single proportion that the deficiencies of the Wald method are glaring. A specification that a quantity is to be estimated to within $\pm h$ with 95% confidence does not square with the recognition that good intervals for proportions are far from symmetrical in form. Confidence intervals for differences between proportions are so much more symmetrical that it is usually reasonable to specify an interval half-width as above. However, in comparative studies, the sample size requirement should be based primarily on a conventional power calculation. This is for the reason set out by Daly (2000) and explained in Section 1.9.3: if the confidence interval based calculation is regarded as the primary justification for the sample size, there is a serious possibility that the study will have a power of only 50% to detect the size of difference that the investigators have in mind. It is preferable to perform the power calculation, then use the second block of the spreadsheet CIPROPORTION.xls to determine the resulting anticipated interval width.

8

Difference between Proportions Based on Individually Paired Data

8.1 The 2 × 2 Table for Paired Binary Data

A 2 × 2 table may also used to display paired binary data. Table 8.1 gives the notation used for the paired 2 × 2 table, including several parameters pertinent to different methods considered. The interpretation of paired and unpaired 2 × 2 tables is totally different, hence clear labelling and description of the table contents is crucial.

For example, Armitage, Berry and Matthews (2002, p. 123) compared the ability of two culture media to detect tubercle bacilli. Of 50 sputum specimens studied, 32 were classed as positive using medium A, and 22 were positive using medium B. The absolute difference between these proportions is (32 – 22)/50 = +0.2. However, to calculate a confidence interval for this difference, we also need to know how the 32 positives using medium A relate to the 22 positives using medium B. In fact only two of those specimens that were positive by B were negative by A. So here, the four cell frequencies are $a = 20$, $b = 12$, $c = 2$ and $d = 16$.

In this situation the McNemar (1947) test is used to test the null hypothesis $\theta \equiv \pi_2 - \pi_3 = 0$. In a matched comparative retrospective study in which a series of cases of a disease and a series of healthy controls are compared for history of some binary exposure, θ does not summarise the findings in a meaningful way; association between outcome and exposure is expressed by the odds ratio which is most often estimated by $\hat{\omega} = b/c$, the null hypothesis then corresponds to $\omega = 1$. In both situations, $\gamma = \pi_2 + \pi_3$ plays an important role as a nuisance parameter; when interest centres on θ, it imposes the constraint $|\theta| \leq \gamma$. Contexts in which θ rather than ω is a useful characterisation include the following.

- *Method comparison studies for a binary attribute: assessment of bias of one method relative to the other, as in the tuberculosis example above.* Similarly, Hope et al. (1996) evaluated a new faecal occult blood test together with two existing ones on the same subjects, and gave confidence intervals for differences in sensitivity and specificity, using method 10 described below. Simultaneous comparison of sensitivity and specificity is clearly desirable; a graphical method to do so is developed in Section 12.5.

- *Analysis of a binary outcome variable in crossover or split-unit designs* (e.g., Vaeth and Poulsen 1998). As usual, it is important that prior information precludes important carry-over effects. Here, if numbers of subjects in the two treatment order groups are equal, the analysis may be simplified by disregarding the order in which the

TABLE 8.1

Notation for Comparison of Proportions from Paired Data

2 × 2 table representations:

Frequencies

		Second Classification		
		+ve	−ve	Total
First	+ve	a	b	$a+b$
classification	−ve	c	d	$c+d$
	Total	$a+c$	$b+d$	n

Theoretical Proportions

		Second Classification		
		+ve	−ve	Total
First	+ve	π_1	π_2	$\pi_1+\pi_2$
classification	−ve	π_3	π_4	$\pi_3+\pi_4$
	Total	$\pi_1+\pi_3$	$\pi_2+\pi_4$	1

Four-row representation:

First Classification	Second Classification	Observed Frequencies	Observed Proportions	Theoretical Proportions
+ve	+ve	a	$p_1 = a/n$	π_1
+ve	−ve	b	$p_2 = b/n$	π_2
−ve	+ve	c	$p_3 = c/n$	π_3
−ve	−ve	d	$p_4 = d/n$	π_4
Total		n	1	1

Unconditional model:

Parameter of interest $\theta = (\pi_1 + \pi_2) - (\pi_1 + \pi_3) = \pi_2 - \pi_3$

Nuisance parameter $\gamma = \pi_2 + \pi_3$

Conditional model:

Conditional probability $\xi = \pi_2/(\pi_2 + \pi_3)$

McNemar odds ratio $\omega = \pi_2/\pi_3$

Other parameters:

Phi coefficient $$\phi = \frac{\pi_1\pi_4 - \pi_2\pi_3}{\sqrt{(\pi_1 + \pi_2)(\pi_3 + \pi_4)(\pi_1 + \pi_3)(\pi_2 + \pi_4)}}$$

Inverse tetrachoric odds-ratio $\eta = \pi_2\pi_3/\pi_1\pi_4$

Diagonal split parameters $\nu = \pi_1/(\pi_1 + \pi_4)$

ξ as above.

treatments were administered. The methods evaluated here are then applicable.

- *Progression over time, either naturally or in association with some intervention.* Herpes simplex virus sheds spontaneously, sporadically, in antibody-positive subjects. Lewis (1993) planned to study 40 antibody-positive subjects immediately before and after third molar extraction. He anticipated detection of herpes simplex virus in oral rinse in two subjects pre-operatively and 10 subjects post-operatively.

In these situations θ is estimated by $\hat{\theta} = (a+b)/n - (a+c)/n = (b-c)/n$, which is the maximum likelihood estimate as well as the obvious empirical one. Newcombe (1998c) evaluated 10 interval methods for θ, defined in Section 8.5.1. Two additional methods are included here, from Newcombe (2003a) and Tango (1998).

The important case of the odds ratio in a retrospective case-control study is developed in detail in Section 11.6. The usual estimate of the odds ratio for paired data, $\hat{\omega} = b/c$, arises from a distribution which is conditional on $b + c$. The conditional approach is used here to eliminate a nuisance parameter by conditioning on a statistic that is sufficient for it. Such a sufficient statistic is available for the odds ratio, but not for the difference or ratio of proportions. Furthermore, a confidence interval for an unpaired difference of proportions should not be conditional on the sum of the two numerators, because this approach fails to produce an interval in the case RZ (other than one emanating solely from a continuity correction). For our present objective of obtaining a confidence interval for a paired difference of proportions, the issue of conditioning is more obviously crucial. When interest centres on the odds ratio, the estimate $\hat{\omega}$ defined above is often appropriate. But there is no reason why inferences for the difference in proportions should be conditional on $b + c$, and conditional interval methods for θ that condition on $b + c$ such as methods 3 and 4 below perform very poorly; here, only unconditional intervals have sensible properties. Indeed, even the use of the conditional odds ratio estimator has been seriously queried (Liang and Zeger 1988).

In many respects, the findings for the paired difference case reflect those for the unpaired difference. Profile likelihood intervals based on exact and mid-*p* tail areas perform well, as does a score interval developed by Tango (1998). However, a modified square-and-add method based on Wilson intervals for the marginal proportions works reasonably well but is not as good as the purpose-built Tango method.

8.2 Wald and Conditional Intervals

The Wald limits for θ are

$$\frac{b-c}{n} \pm \frac{z}{n}\left\{\sqrt{b+c-\frac{(b-c)^2}{n}}\right\} \tag{8.1}$$

(Altman 1991, p. 236–241). The standard error in this expression is calculated *without* assuming H_0 (Fleiss, Levin and Paik 2003, p. 378).

Just as for the single proportion and unpaired difference, this simple asymptotic interval is not boundary-respecting; for some combinations of n, γ and θ, it usually violates the [−1, +1] bounds for θ. A continuity correction may be incorporated, as in method 2. As in other applications, doing so does not obviate inappropriate behaviour at or near boundaries of the parameter space.

Perhaps in recognition of this deficiency, conditional methods were developed. The null hypothesis of marginal homogeneity may be expressed as $\pi_1 + \pi_2 = \pi_1 + \pi_3$, which simplifies to $\pi_2 = \pi_3$. It is usually tested conditional on the observed split into $b + c$ discordant (informative) pairs and $a + d$ concordant (uninformative) ones. It is then expressed in terms of the conditional probability $\xi = \pi_2/(\pi_2 + \pi_3)$, resulting in either an asymptotic or an exact test of H_0: $\xi = 0.5$. Given that boundary-respecting intervals for a single binomial proportion have been known since 1927, it might reasonably be supposed that such an interval for ξ would then transform to a well-behaved interval for θ by substitution. However, such conditional intervals frequently incur a different problem, namely inappropriate point estimate tethering, as explained in Newcombe (1998c): the calculated interval unnecessarily fails to act as a two-sided "margin of error" about the point estimate. This behaviour can be seen for three of the exemplar datasets in Table 8.2 for the two conditional intervals.

Various pseudo-frequency intervals have also been developed. Agresti (2002, p. 411) suggested adding $k = 1$ to each cell of the 2 × 2 table before computing Equation 8.1. Bonett and Price (2011) identified cases where this adjustment can have a true coverage probability of 0. Agresti and Min (2005b) dismissed the Agresti (2002) recommendation as suboptimal and recommended using $k = 0.5$ instead, with truncation in the event of boundary violation. Bonett and Price proposed an alternative adjusted Wald interval

$$\tilde{p}_2 - \tilde{p}_3 \pm z\sqrt{\left\{\tilde{p}_2 + \tilde{p}_3 - \left(\tilde{p}_2 - \tilde{p}_3\right)^2\right\}/(n+2)} \tag{8.2}$$

where $\tilde{p}_2$ and $\tilde{p}_3$ are the Laplace estimates $(b + 1)/(n + 2)$ and $(c + 1)/(n + 2)$, and presented results suggesting that performance is better than the Agresti–Min interval and similar to the computationally intensive Tango score interval.

8.3 Intervals Based on Profile Likelihoods

Just as for the unpaired difference, several types of interval involve substituting γ_θ, the maximum likelihood estimate of the nuisance parameter γ conditional on θ. This leads to a true profile likelihood interval (method 7), and tail-based intervals based on exact and mid-*p* aggregations of tail areas (methods 5 and 6). In line with the results for the unpaired difference, these methods have coverage slightly lower than the nominal $1 - \alpha$, much higher than it, and close to $1 - \alpha$, respectively.

The form of the likelihood function is not identical in the paired and unpaired cases. For the paired case, the likelihood may be expressed using the parameters defined in Table 8.1 as

$$\Lambda \propto \nu^a(1-\nu)^d(1-\gamma)^{a+d}\left(\frac{\gamma+\theta}{2}\right)^b\left(\frac{\gamma-\theta}{2}\right)^c. \tag{8.3}$$

By the sufficiency principle, the terms in ν, the parameter determining the *a*:*d* split, may be disregarded as not contributing to inferences concerning θ. Thus here, just as in the Wald and conditional methods, the distinction between the two diagonal cells, with frequencies *a* and *d*, is regarded as uninformative, and the likelihood is essentially trinomial. Newcombe (1998c) describes the behaviour of the likelihood function in four cases defined by the pattern of zeros in the data:

1. $a + d > 0$, $b > 0$ and $c > 0$
2. $a + d > 0$, one of b and c zero
3. $a = d = 0$, $b + c > 0$
4. $a + d > 0$, $b = c = 0$

The profile estimate γ_θ takes a simpler form in the cases with zero cells, and in some instances the resulting intervals are closely related to intervals for binomial parameters. But all three methods require nested iteration to cope with the usual case 1, and thus are not amenable to closed-form implementation on spreadsheet platforms.

8.4 Score-Based Intervals

While it is technically possible to use any of the methods described in Chapter 7 to calculate an interval for θ as if $(a + b)/n$ and $(a + c)/n$ came from independent samples, it would clearly be incorrect to do so: just as for inferences relating to differences between means, purpose-built methods that take the non-independence into account are obviously called for. In Section 2.3 we saw how the standard error for a paired difference of two means may be obtained from the standard errors of the two means and the correlation. A similar adjustment may be used to develop an interval for the paired difference of proportions, using a quantity usually denoted by ϕ. This is the Pearson product-moment correlation based on binary data (Fleiss, Levin and Paik 2003, p. 98). It is not particularly good as a measure of association per se, but it is what is needed here to develop an interval for the paired case. Consider the 2×2 table representation of the data, at the top of Table 8.1. Regarding this table as an ordinary (not paired) 2×2 table, the usual chi-square test statistic for independence of row and column classifications (H_0: $\eta = 1$ as defined in Table 8.1) may be calculated using Equation 7.1. We then define $\hat{\phi} = \text{sgn}(ad - bc) \times \sqrt{X^2/n}$. When any of the four marginal frequencies $a + b$, $c + d$, $a + c$ or $b + d$ is zero, $X^2 = 0$, hence it is reasonable to substitute 0 for $\hat{\phi}$ in the event of one or more zero cells. Then $\hat{\phi}$ is used to adjust the square-and-add Wilson interval for the unpaired difference, leading to method 8 below. The resulting interval for θ remains within $[-1, +1]$ whatever value in this range is substituted for ϕ. In method 9, a conventional continuity correction is incorporated in each score interval.

In the absence of a continuity correction the mean coverage is very close to $1 - \alpha$, at the expense of serious dips of coverage, especially when ν is close to 0.5 and n is small. This suggests an alternative approach, incorporating a continuity correction in $\hat{\phi}$. After some experimentation, a continuity correction to the numerator of $\hat{\phi}$, applied only when $\hat{\phi} > 0$ was developed as method 10.

These three methods have an unusual property: the a:d split is not disregarded. The value of $\hat{\phi}$ is maximal when a or d is $[(a + d)/2]$ and reduces monotonically as $a/(a + d) \to 0$ or 1. Often, although not invariably, the lower and upper limits for θ display a similar monotonicity property, as shown in Table 8.3.

As described in Chapter 9, any interval method for the paired difference of two proportions may be reshaped into a corresponding interval for the mean of a variable that can take only the values 0, 1 or 2. In that situation, the cell frequencies a and d cannot be identified, only their sum. So, to apply method 10 to this situation, we need to make an arbitary choice for the a:d split. The obvious choice is the symmetrical one, using $h = (a + d)/2$ in place of both a and d. Because h is then used in a closed-form algebraic formula, not a tail-based one, it does not matter here if $a + d$ is odd and h is not an integer.

The same marginal homogenisation links the Scott (1955) version of the kappa statistic for the 2 × 2 table to the more familiar version without marginal homogenisation formulated by Cohen (1960). The effect of marginal homogenisation is far from trivial, in both of these contexts. Accordingly, the marginal-homogenised version of method 10 is included in the evaluation in Section 8.5 as a separate method, method 11.

Method 10 is readily programmed using spreadsheet software, as in the third block of the accompanying spreadsheet CIPROPORTION.xls. This may also be used to obtain the method 11 interval in an obvious manner.

Contemporaneous with Newcombe (1998c), Tango (1998) described a purpose-built asymptotic score interval for θ, which is defined as method 12 below. While the limits do not have obvious closed form, they can be derived by a single-depth iteration, consequently a spreadsheet implementation is available, TANGO.xls.

Subsequently, Tang et al. (2005) described exact and approximate unconditional intervals θ derived by inverting a score test. Tang et al. (2010a) evaluated several intervals for the paired difference of proportions based on MOVER relative to the Tango score interval.

The issue of comparison of two non-independent proportions also arises in the context of inferences from a 2 × 2 table with a structural zero, as in Tang and Tang (2003), Tang et al. (2004) and Shi et al. (2009), but this is a totally different inference problem.

8.5 Evaluation of Performance

Just as for the single proportion and unpaired difference cases, it is necessary to evaluate coverage, width, location and boundary anomaly properties of various methods. The results for methods 1 to 10 are derived from Newcombe (1998c). In developing and evaluating optimal methods for the mean of a variable taking the values 0, 1 or 2 (Newcombe 2003a), it was necessary to extend these findings by obtaining corresponding results for methods 11 and 12 as defined below.

8.5.1 Definitions of 12 Methods

1. The simple asymptotic (Wald) limits without continuity correction are $\hat{\theta} \pm z \times \mathrm{SE}$ where $\hat{\theta} = (b-c)/n$, $\mathrm{SE} = \left\{\sqrt{b+c-(b-c)^2/n}\right\}/n = \sqrt{((a+d)(b+c)+4bc)/n^3}$ and z is the $1 - \alpha/2$ point of the standard normal distribution.

2. The Wald limits with continuity correction (Fleiss, Levin and Paik 2003, p. 378) are

$$\hat{\theta} \pm (z \times \text{SE} + 1/n) \tag{8.4}$$

where SE is defined as for method 1 above.

3. Conditional method based on exact interval for ξ (Liddell 1983; Armitage, Berry and Matthews, 2002, p. 122): $(2L_\xi - 1)\hat{\gamma}$ to $(2U_\xi - 1)\hat{\gamma}$ where $\hat{\gamma} = (b+c)/n$ and (L_ξ, U_ξ) is a Clopper–Pearson interval for $\xi = \pi_2/(\pi_2 + \pi_3)$, defined as in Section 3.4.4.

4. An alternative conditional interval may be defined as above, but using a mid-p interval (Section 3.4.5) for ξ.

5. Unconditional profile likelihood method based on exact tail areas: interval for θ is $[L, U]$ such that

(i) $$\text{if } L \le \theta \le \hat{\theta}, kP_x + \sum_{x<\xi\le n} P_\xi \ge \alpha/2 \tag{8.5}$$

(ii) $$\text{if } \hat{\theta} \le \theta \le U, kP_x + \sum_{-n\le\xi<x} P_\xi \ge \alpha/2 \tag{8.6}$$

where $P_\xi = \Pr[B - C = \xi \mid \theta, \gamma_\theta]$, $x = f - g$, and $k = 1$; γ_θ denotes the MLE of γ given θ; B and C denote the random variables of which b and c are the realisations.

6. Unconditional profile likelihood method based on mid-p tail areas: as method 5, but with $k = 0.5$.

7. True profile likelihood method: the interval comprises all $\theta \in [-1, +1]$ satisfying

$$(a+d)\{\ln(1-\gamma_\theta) - \ln(1-\hat{\gamma})\} + b\{\ln(\gamma_\theta + \theta) - \ln(\hat{\gamma} + \hat{\theta})\} + c\{\ln(\gamma_\theta - \theta) - \ln(\hat{\gamma} - \hat{\theta})\} \ge -\frac{1}{2} z^2. \tag{8.7}$$

Terms corresponding to $a + d$, b or c zero are omitted.

8. Square-and-add method based on Wilson score interval for the single proportion, no continuity correction. Interval is $\hat{\theta} - \delta$ to $\hat{\theta} + \varepsilon$ where δ and ε are the positive values $\delta = \sqrt{dl_2^2 - 2\hat{\phi} dl_2 du_3 + du_3^2}$, $\varepsilon = \sqrt{du_2^2 - 2\hat{\phi} du_2 dl_3 + dl_3^2}$.

Here

$$dl_2 = (b + c)/n - l_2, \; du_2 = u_2 - (b + c)/n$$

where l_2 and u_2 are roots of $|\xi - (b+c)/n| = z\sqrt{\xi(1-\xi)/n}$.

Likewise

$$dl_3 = (a + d)/n - l_3,\ du_3 = u_3 - (a + d)/n$$

where l_3 and u_3 are roots of $|\xi - (a+d)/n| = z\sqrt{\xi(1-\xi)/n}$.

Also $\hat{\phi} = (ad - bc)/\sqrt{(a+b)(c+d)(a+c)(b+d)}$, but $\hat{\phi} = 0$ if the denominator is 0.

9. Square-and-add method using continuity-corrected score intervals: as above, but l_2 and u_2 delimit the interval

$$\left\{\xi : |\xi - (b+c)/n| - 1/(2n) \le z\sqrt{\xi(1-\xi)/n}\right\}. \quad (8.8)$$

However, if $a + b = 0$, $l_2 = 0$; if $a + b = n$, $u_2 = 1$. Similarly for l_3 and u_3.

10. Square-and-add method using score intervals but continuity corrected $\hat{\phi}$: as method 8 above, but with the numerator of $\hat{\phi}$ replaced by max $(ad - bc - n/2, 0)$ if $ad > bc$.
11. Symmetrised version of method 10: as method 10, but if a and d are unequal, they are first replaced by $(a + d)/2$.
12. Score interval: the most convenient formulation (Tango 1999) is that the limits for θ are the two roots of the equation $b - c - n\theta = \pm z\sqrt{\{n(2q + \theta(1-\theta))\}}$ where, in the absence of zero cells, $q = \{\sqrt{(y^2 - 4nc\theta(\theta - 1))} - y\}/(2n)$ and $y = (2n - b + c)\theta - b - c$. As pointed out in Appendix 3 of Tango (1998), correct implementation necessitates heeding special cases where b or c is zero.

8.5.2 Examples

Tables 8.2 and 8.3 show 95% limits for θ calculated by the 12 methods, for chosen combinations of $a + d$, b and c. Although these examples were selected to cover all possible cases rather than to be representative of what is encountered in practice, they show how methods 5 to 7 and 10 to 12 obviate the aberrations of the traditional ones.

In practice the most important boundary case is where the c (or b) cell is zero. For example, Plugge et al. (2009) reported a questionnaire study of a representative sample of 217 women in two prisons in the United Kingdom, which studied changes in drug-taking in the first month following imprisonment. Results were presented for daily use of any drug and for several specific drugs. Numbers of women taking amphetamines, ecstasy and hallucinogens reduced from 7, 4 and 1 to zero following imprisonment. For the reduction in amphetamine use, (7 – 0)/217 = 0.0323, the 95% Tango interval is from 0.0143 to 0.0651. 0.0323 is of course simply the proportion of women taking amphetamines before imprisonment, 7/217. A 95% Wilson interval for this proportion is from 0.0157 to 0.0651. Thus the Tango interval for (7 – 0)/217

TABLE 8.2

95% Limits for the Paired Difference of Proportions for Selected Combinations of $a + d$, b and c

	Cell Frequencies						
Method	$a + d = 36$, $b = 12, c = 2$	$a + d = 36$, $b = 14, c = 0$	$a + d = 2$, $b = 97, c = 1$	$a + d = 2$, $b = 98, c = 0$	$a + d = 0$, $b = 29, c = 0$	$a + d = 0$, $b = 30, c = 0$	$a + d = 36$, $b = 12, c = 2$
Wald intervals							
1 Without CC	0.0642, 0.3358	0.1555, 0.4045	0.9126, >1[a]	0.9526, >1[a]	0.8049, >1[a]	1.0[a], 1.0	0.0[a], 0.0[a]
2 With CC	0.0442, 0.3558	0.1355, 0.4245	0.9026, >1[a]	0.9426, >1[a]	0.7715, >1[a]	0.9667, >1[a]	−0.0185, 0.0185
Conditional intervals							
3 Exact	0.0402, 0.2700	0.1503, 0.2800[a]	0.8711, 0.9795	0.9076, 0.9800[a]	0.6557, 0.9983	0.7686, 1.0	0.0[a], 0.0[a]
4 Mid-p	0.0575, 0.2662	0.1721, 0.2800[a]	0.8834, 0.9790	0.9210, 0.9800[a]	0.6928, 0.9967	0.8099, 1.0	0.0[a], 0.0[a]
Unconditional intervals							
5 Exact	0.0497, 0.3539	0.1619, 0.4249	0.8752, 0.9916	0.9132, 0.9976	0.6557, 0.9983	0.7686, 1.0	−0.0660, 0.0660
6 Mid-p	0.0594, 0.3447	0.1691, 0.4158	0.8823, 0.9900	0.9216, 0.9966	0.6928, 0.9967	0.8099, 1.0	−0.0540, 0.0540
7 Profile likelihood	0.0645, 0.3418	0.1686, 0.4134	0.8891, 0.9904	0.9349, 0.9966	0.7226, 0.9961	0.8760, 1.0	−0.0349, 0.0349
Square-and-add intervals							
11 Symmetrised	0.0562, 0.3290	0.1441, 0.3963	0.8737, 0.9850	0.9171, 0.9916	0.6666, 0.9882	0.8395, 1.0	−0.0351, 0.0351
Score interval							
12 Tango	0.0611, 0.3447	0.1747, 0.4167	0.8698, 0.9866	0.9068, 0.9945	0.6666, 0.9882	0.7730, 1.0	−0.0664, 0.0664

Note: Calculated using nine methods which disregard the $a{:}d$ split. CC = continuity correction.

[a] Boundary aberrations.

TABLE 8.3

95% Limits for the Paired Difference of Proportions for Selected Combinations of *a*, *b*, *c* and *d*

a	*b*	*c*	*d*	No Continuity Correction (Method 8)	Conventional Continuity Correction (Method 9)	Continuity Correction to $\hat{\phi}$ (Method 10)
36	12	2	0	0.0569, 0.3404	0.0407, 0.3522	0.0569, 0.3404
20	12	2	16	0.0618, 0.3242	0.0520, 0.3329	0.0562, 0.3292
18	12	2	18	0.0618, 0.3239	0.0520, 0.3327	0.0562, 0.3290
36	14	0	0	0.1528[a], 0.4167	0.1360[a], 0.4271	0.1528, 0.4167[a]
35	14	0	1	0.1573[a], 0.4149	0.1435[a], 0.4249	0.1461, 0.4175[a]
18	14	0	18	0.1504[a], 0.3910	0.1410[a], 0.3989	0.1441, 0.3963[a]
2	97	1	0	0.8721, 0.9854	0.8589, 0.9887	0.8721, 0.9854
1	97	1	1	0.8737, 0.9850	0.8610, 0.9885	0.8737, 0.9850
2	98	0	0	0.9178, 0.9945	0.9064, 0.9965	0.9178, 0.9945
1	98	0	1	0.9174, 0.9916	0.9063, 0.9933	0.9171, 0.9916
0	29	1	0	0.6666, 0.9882	0.6189, 0.9965	0.6666, 0.9882
0	30	0	0	0.8395, 1.0	0.8001, 1.0	0.8395, 1.0
54	0	0	0	−0.0664, 0.0664	−0.0827, 0.0827	−0.0664[a], 0.0664[a]
53	0	0	1	−0.0640, 0.0640	−0.0758, 0.0758	−0.0729[a], 0.0729[a]
30	0	0	24	−0.0074, 0.0074	−0.0079, 0.0079	−0.0358[a], 0.0358[a]
29	0	0	25	−0.0049, 0.0049	−0.0053, 0.0053	−0.0354[a], 0.0354[a]
28	0	0	26	−0.0025, 0.0025	−0.0026, 0.0026	−0.0352[a], 0.0352[a]
27	0	0	27	0.0, 0.0	0.0, 0.0	−0.0351[a], 0.0351[a]

Note: Calculated using three methods based on applying squaring and adding to score intervals for the single proportion. The entry for method 10 in the final line of each block is the method 11 interval as in Table 8.2.

[a] Non-monotonic behaviour as the *a*:*d* split alters.

extends slightly further to the left than the Wilson interval for 7/217; both methods give identical upper limits.

Table 8.2 shows intervals for several datasets from Newcombe (1998c) calculated by methods 1 to 7, 11 and 12 which disregard the *a*:*d* split. Only methods 1 and 2 are capable of violating the [−1, +1] boundaries, in which case the resulting interval is truncated. In this table, asterisks indicate overshoot and tethering aberrations. As noted above, point estimate tethering is a particularly frequent issue here for Wald and conditional methods.

Table 8.3 presents corresponding intervals for methods 8 to 10, showing the effect of varying the *a*:*d* split. In this table, asterisks are used to indicate where non-monotonic behaviour occurs as the *a*:*d* split varies. The only tethering that occurs here is that methods 8 and 9 (but not 10) produce a zero-width interval when $b = c = 0$ and $a = d$.

8.5.3 Evaluation

The main evaluation follows Newcombe (1998c), but also includes two additional methods. It involves exactly calculated coverage of nominal 95% intervals based on a large number of randomly generated parameter space points. The choice of an appropriate set of PSPs is rather complex here: very different choices are appropriate for two related problems, a paired difference of proportions and the mean of a variable taking the values 0, 1 and 2 as in Chapter 9. To be plausible for the paired case, a pseudo-prior distribution must ensure $\phi \geq 0$, although occasional instances of $\hat{\phi} < 0$ are permissible. In the present evaluation, pseudo-priors which constrain either ϕ or the inverse tetrachoric odds-ratio $\eta = \pi_2\pi_3/\pi_1\pi_4$ to range on (0,1) achieve this.

For each of n = 10, 11,..., 100, one hundred triples (ϕ, ν, ξ) were obtained, with ϕ, ν and ξ sampled independently from U(0,1), U(0.5, 1) and U(0.5, 1), respectively. Given these parameter values, the corresponding $\gamma \in (0, 1)$ and $\theta \in (0, \gamma)$ are derived by solving the relevant equations in Table 8.1. Their distributions are highly skewed, very rarely approaching the theoretical limit of +1, and having means 0.220 and 0.117, medians 0.191 and 0.068, in the chosen sample of 9100 PSPs.

For each method, the probabilities of mesial and distal non-coverage, boundary violation and inappropriate tethering were calculated exactly for each PSP by summation, then average and worst values over the 9100 PSPs obtained. Newcombe (1998c) also reported mean and minimum coverage probabilities for nominal 90% and 99% intervals for methods 1 to 10, and some results for large sample sizes but small to moderate discordant cell frequencies. Expected interval width was calculated exactly for 95% intervals by each method, truncated to lie within [−1, +1] where necessary, for 12 selected PSPs.

8.5.4 Results of the Evaluation

Table 8.4 shows results for coverage and location for the 12 methods. The coverage probability for 95% intervals, averaged over the 9100 PSPs, ranges from 0.764 (method 4) to 0.977 (method 5). Methods 1, 4 and even 3 (despite being based on an exact interval for ξ) are grossly anticonservative on average, and right non-coverage, due to a ZWI at 0, occurs with probability which tends to 1 as $\gamma \to 0$, holding $\xi > 0$ fixed. Even for $50 \leq n \leq 100$, these methods have mean coverage below 0.9. The maximum CP for method 1 was only 0.957. Method 2 is very much better, somewhat conservative on average, but occasionally incurs very high non-coverage either to the right or to the left. For these four methods interval location is too mesial.

Method 5 has rather similar coverage characteristics to the Clopper–Pearson method for the single proportion, with a minimum CP of 0.955, appropriate to the exact paradigm, and consequently a rather high mean CP.

TABLE 8.4

Coverage and Location for 95% Intervals for the Paired Difference of Proportions Calculated by 12 Methods

Method	Coverage		Mesial Non-Coverage		Distal Non-Coverage		MNCP/NCP
	Mean	Minimum	Mean	Maximum	Mean	Maximum	
Wald intervals							
1 Without CC	0.8543	0.0006	0.0194	0.3170	0.1262	0.9994	0.133
2 With CC	0.9690	0.6542	0.0091	0.3170	0.0219	0.3458	0.294
Conditional intervals							
3 Exact	0.7816	0.0006	0.0106	0.0829	0.2079	0.9994	0.049
4 Mid-*p*	0.7637	0.0006	0.0196	0.1188	0.2166	0.9994	0.083
Unconditional intervals							
5 Exact	0.9766	0.9546	0.0117	0.0263	0.0117	0.0239	0.500
6 Mid-*p*	0.9657	0.9332	0.0170	0.0372	0.0173	0.0465	0.496
7 Profile likelihood	0.9488	0.8539	0.0242	0.0590	0.0270	0.1387	0.473
Square-and-add intervals							
8 Without CC	0.9505	0.6388	0.0150	0.0474	0.0345	0.3610	0.303
9 CC to score limits	0.9643	0.6388	0.0094	0.0277	0.0263	0.3610	0.263
10 CC to $\hat{\phi}$	0.9672	0.9031	0.0114	0.0285	0.0214	0.0960	0.348
11 Symmetrised	0.9614	0.8606	0.0113	0.0323	0.0272	0.1387	0.293
Score interval							
12 Tango	0.9636	0.9310	0.0197	0.0529	0.0167	0.0537	0.541

Note: Based on 9100 parameter space points with $10 \le n \le 100$, $0 < \gamma < 1$, $0 < \theta < \gamma$. NCP = non-coverage probability; CC = continuity correction; MNCP = mesial non-coverage probability.

However the MNCP exceeded 0.025 slightly for 18 of the 9100 PSPs, suggesting that the existence of PSPs for which CP < $1 - \alpha$ cannot be ruled out. Method 6 yields a somewhat conservative mean CP (in line with the findings for the mid-*p* method for a single proportion), and a respectable minimum of 0.933. Method 7 is anticonservative, to a slight degree on average. These three methods tend to achieve highly symmetrical coverage as assessed by MNCP/NCP. Accordingly, in terms of both overall coverage probability and interval location, methods 5 and 6 are highly appropriate to the exact and mid-*p* interpretations of $1 - \alpha$, respectively, but are computationally intensive.

The much simpler score-based method 8 achieves a mean CP of almost exactly 0.95, at the cost of many PSPs, not only isolated examples, yielding a CP lower than this. Mean coverage was examined for several zones of PSPs defined by each of the parameters n, θ, $n\theta$, γ, $n\gamma$, ξ, ν and ϕ. Method 8's mean coverage was lower than $1 - \alpha$ for many of these zones.

The effect of a conventional continuity correction (method 9) is to increase the overall mean CP, and produce a mean CP over 0.95 in all except one of the zones examined. But the minimum at 0.639 is unaltered. Applying the continuity correction to ϕ instead (method 10) results in coverage only slightly inferior to method 6, by eliminating the most severe dips: as for methods 5 and 6, the mean CP exceeded 0.95 for each of the zones of the parameter space examined. Method 11, the symmetrised version of method 10, has slightly less favourable coverage properties than method 10. These four intervals derived from Wilson intervals by squaring and adding err towards mesial location.

As noted by Newcombe (2003a), the true score interval developed by Tango (1998) has a very favourable performance in terms of both coverage and location, similar to that of the mid-*p* interval. It has the advantage that it can be programmed in closed form, although the programming is quite tricky. It should be regarded as slightly superior to method 10, and distinctly superior to method 11, thus when the *a* and *d* cells are not separately identifiable, as in Newcombe (2003a), the Tango interval is clearly the superior choice.

For the unpaired difference of proportions (Chapter 7), the square-and-add Wilson interval was if anything slightly better than the Miettinen–Nurminen score interval. For the paired difference, the score interval (Tango) is slightly better than the interval derived by squaring and adding Wilson intervals, then correcting for non-independence. The rationale for method 10 here is very much *ad hoc*, not theoretically based. And the symmetrised version, method 11, is a further step in the same direction.

Only methods 1 and 2 can yield calculated limits outside [–1, +1]. With positive values for θ, the incidence of truncation at –1 was naturally extremely low, less than 10^{-6}. Truncation at +1 occurred with probability 0.0007 for method 1, and 0.0020 for method 2. These are much lower than for the single proportion and unpaired difference cases, because high values of θ are barely compatible with the plausibility constraint $\phi > 0$ here. Nevertheless some parameter combinations gave a high overshoot probability for method 1, up to 0.76. For method 2, the overshoot probability becomes arbitrarily close to 1 as γ and $\xi \to 1$.

The extreme configurations 2, 3 and 4 listed in Section 8.3, which can lead to inappropriate tethering or degeneracy, can each occur with arbitrarily high probability for suitable points in the parameter space. Tables with one of the off-diagonal cell frequencies zero, leading to unilateral tethering for methods 3 and 4, occur often, with probability around 0.26 in this evaluation. ZWIs at zero occur with frequency 9.5% for methods 1, 3 and 4, but only 0.1% for methods 8 and 9. Method 1 produces ZWIs at ±1 much more rarely. These frequencies apply irrespective of the $1 - \alpha$ chosen.

Variation in expected interval width (Table 8.5) between different methods is most marked when $n\gamma$ is low. The width is then least for methods 1, 3 and 4, largely on account of the high ZWI probability. Generally, width is greatest for method 5, followed by methods 12 and 6, as expected. Method 10 led to lower expected width than method 12 for all twelve PSPs, in some instances much lower.

TABLE 8.5

Average Width of 95% Intervals for the Paired Difference of Proportions Calculated by 10 Methods, for Selected Parameter Space Points

n	10	10	10	10	10	10	100	100	100	100	100	100
π_1	0.49	0.30	0.20	0.05	0.04	0.01	0.49	0.30	0.20	0.05	0.04	0.01
π_2	0.01	0.20	0.55	0.70	0.91	0.94	0.01	0.20	0.55	0.70	0.91	0.94
π_3	0.01	0.20	0.05	0.20	0.01	0.04	0.01	0.20	0.05	0.20	0.01	0.04
π_4	0.49	0.30	0.20	0.05	0.04	0.01	0.49	0.30	0.20	0.05	0.04	0.01
γ	0.02	0.40	0.60	0.90	0.92	0.98	0.02	0.40	0.60	0.90	0.92	0.98
θ	0.00	0.00	0.50	0.50	0.90	0.90	0.00	0.00	0.50	0.50	0.90	0.90
Wald intervals												
1 Without CC	0.125	0.950	0.926	1.002	0.398	0.399	0.556	0.992	0.997	1.009	0.930	0.921
2 With CC	0.478	1.212	1.165	1.160	0.564	0.555	0.780	1.072	1.084	1.073	1.071	1.023
Conditional intervals												
3 Exact	0.068	0.816	0.878	1.118	1.033	1.108	0.385	1.030	0.792	1.038	0.765	1.018
4 Mid-p	0.066	0.759	0.780	1.017	0.877	0.961	0.363	0.967	0.729	0.984	0.668	0.931
Unconditional intervals												
5 Exact	1.119	1.153	1.141	1.133	1.085	1.115	1.037	1.043	1.045	1.035	1.041	1.025
6 Mid-p	0.947	1.051	1.038	1.050	0.956	0.977	0.901	1.005	1.005	1.005	0.975	0.970
7 Profile likelihood	0.669	0.976	0.936	1.006	0.748	0.785	0.748	0.996	0.997	1.003	0.946	0.945

(*continued*)

TABLE 8.5 (Continued)

Average Width of 95% Intervals for the Paired Difference of Proportions Calculated by 10 Methods, for Selected Parameter Space Points

Square-and-add intervals												
8 Without CC	0.315	0.835	0.866	0.965	0.800	0.839	0.559	0.973	0.985	0.996	0.989	0.990
9 CC to score limits	0.362	0.951	0.998	1.111	0.961	1.007	0.587	1.022	1.042	1.054	1.095	1.096
10 CC to $\hat{\phi}$	0.699	0.912	0.926	0.967	0.806	0.839	0.729	0.986	0.995	0.996	0.996	0.990
11 Symmetrised	0.633	0.900	0.918	0.962	0.795	0.836	0.724	0.985	0.995	0.996	0.993	0.989
Score interval												
12 Tango	1.000	1.000	1.000	1.000	1.000	1.000	1.000	1.000	1.000	1.000	1.000	1.000
Actual average width for method 12	0.5659	0.7629	0.7273	0.8720	0.6125	0.6448	0.0892	0.2483	0.2310	0.3111	0.1365	0.1687

Note: Expressed relative to average width for Tango score interval. CC = continuity correction.

9

Methods for Triads of Proportions

9.1 Introduction

In this chapter, we consider three methods that may be applied to data on a trinomial scale. In some situations, the three scale points constitute an equally spaced ordinal scale. Examples include procedure rates based on paired organs, and gene frequency without the assumption of the Hardy–Weinberg equilibrium (HWE). The mean of such a variable is equivalent to a paired difference based on trinomial data, accordingly a confidence interval may be derived from one of the recommended methods for paired differences introduced in Chapter 8. However, the regions of the parameter space that are plausible for these contexts are quite different to that for the difference between proportions. When evaluated over appropriate regions of the parameter space, these approaches turn out to have much better performance than methods that analyse the data as if from a Gaussian model, or methods that disregard the paired structure of the data. The chosen method, based on the Tango interval, is readily implemented in MS Excel. Two extensions are discussed, to clinical trials with paired organ related outcomes, and to estimation of the degree of first-move advantage in chess.

Sometimes, the three scale points are identified only as being in a particular order, without any implication that they are equally spaced. The observed distribution across the three categories may be compared to a pre-specified standard, using a two-dimensional generalisation of the one-sided p-value. A mid-p formulation based on aggregated tail probabilities is most appropriate here. An alternative graphical representation including confidence regions may also be formulated.

When there is no inherent ordering, the three proportions may be represented by a point within an equilateral triangle. A confidence region may be drawn to illustrate the impact of sampling imprecision. This display may be less suitable when a meaningful ordering applies, as this property is disregarded.

A fourth application, to the ratio of sizes of two overlapping groups, is developed in Section 10.8. This is derived from an interval for a ratio of paired proportions.

9.2 Trinomial Variables on Equally Spaced Scales

The research described here, which is published as Newcombe (2003a), was prompted by a request from a colleague for a confidence interval for a procedure rate based on paired organs (Hood 1999; Edwards et al. 1999). It turns out that this can easily be derived from an interval for a paired difference between proportions. Two further contexts in which an equally spaced 3-point ordinal scale arise are described in Sections 9.2.2 and 9.2.7. However, the regions of the parameter space that are plausible for these contexts are quite different to that for the difference between proportions. It is thus necessary to validate that good coverage properties extend throughout the parameter space.

Throughout this chapter, it is assumed that the three scale points are denoted by 0, 1 and 2. This is most often the natural representation. Naturally, the methods are equally applicable to a scale represented by {–1, 0, +1}, {0, 0.5, 1} or any other similar characterisation.

9.2.1 Procedure Rates on Paired Organs

Table 9.1 gives results for 350 women undergoing breast investigation (Edwards et al. 1999). Each woman has to undergo at least one procedure of

TABLE 9.1

Procedure Rates in n = 350 Women Undergoing Breast Investigations

	Number of Women With			
Procedure	**No Procedure** n_0	**One Procedure** n_1	**Two Procedures** n_2	**Point Estimate of Procedure Rate per Woman** $\bar{X} = (n_1 + 2n_2)/350$
Cyst aspiration	297	50	3	0.1600
Diagnostic biopsy	340	10	0	0.0286
Fine needle biopsy	254	94	2	0.2800
Mammogram	101	29	220	1.3400
Ultrasound	164	133	53	0.6829
Wide bore biopsy	329	20	1	0.0629

Source: Data from Edwards, A. G. K. et al. 1999. *British Journal of General Practice* 49 (447): 811–812.

some kind to be included in the series, but clearly many women have more than one type of procedure. Some procedures such as mammography are rarely performed unilaterally, whereas others such as fine needle biopsy are seldom bilateral. We could clearly estimate trinomial probabilities, and for each type of investigation represent the probabilities of no use, unilateral and bilateral use by a point within an equilateral triangle, as described in Section 9.4, surrounded by a confidence region. But this would disregard the ordered, indeed numerical, nature of the scale. We seek to estimate the rate of performance of each procedure, for economic and other reasons, with a sensible confidence interval.

9.2.2 Gene Frequency

Table 9.2 gives genotype frequencies for angina patients and controls in the Diet and Reinfarction Trial (DART) study (Burr et al. 1989) in relation to the methylene tetrahydrofolate reductase allele (Section 1.1). For a gene with two alleles, three genotypes can occur. Estimation of gene frequency based on n subjects would often be performed by calculating a confidence interval for a proportion with denominator $N = 2n$. But this amounts to assuming that the two alleles of the same individual are independent (i.e., that HWE applies). While almost all Mendelian populations examined have been found to approximate to HWE, there are many possible causes of departure in either direction, towards deficiency or excess of heterozygotes (Cavalli-Sforza et al. 1971). Thus it is appropriate to apply interval estimation methods that do not rely on assuming HWE.

An obvious difference between the paired organs and allele frequency situations is that we would usually express the procedure rate per woman, but the allele frequency per chromosome. While these differ by a factor of 2, this is purely a matter of how the results are finally expressed, the processes and interpretation in the two cases are equivalent.

TABLE 9.2

Methylene Tetrahydrofolate Reductase Gene Frequency in 624 Angina Cases and 605 Controls in the DART Trial

Group	Total Number of Subjects n	Number of Subjects with Genotype CC n_0	CT n_1	TT n_2	Point Estimate of Gene Frequency per Chromosome $p = (n_1 + 2n_2)/2n$
Angina	624	274	272	78	0.3429
Controls	605	290	242	73	0.3207

Source: Data from Burr, M. L. et al. 1989. *Lancet* 2: 757–761.

9.2.3 General Derivation of the Approach

For any of the breast procedures listed in Table 9.1, the natural estimate for the procedure rate per woman is $\bar{X} = (n_1 + 2n_2)/n$. Here n_2, n_1 and n_0 are the numbers of women in the sample who have the procedure applied to 2, 1 and 0 breasts, and $n = n_0 + n_1 + n_2$ is the total number of women. Then

$$\bar{X} = (n_1 + 2n_2)/n = (n_0 + n_1 + n_2 - (n_0 - n_2))/n = 1 - (n_0 - n_2)/n. \quad (9.1)$$

This is the complement of the difference of the two proportions n_0/n and n_2/n, which are both based on the same sample of n subjects. The essence of the proposed approach is that confidence limits for $(n_0 - n_2)/n$ may be calculated, using one of the methods developed in Chapter 8, and complemented to produce limits for the procedure rate $\bar{X}$.

For example, $n_2 = 3$ women underwent cyst aspiration to both breasts, $n_1 = 50$ to just one breast, and $n_0 = 297$ did not undergo this procedure. Thus the overall procedure rate is 56/350 = 0.16 procedures per woman. Here $(n_0 - n_2)/n$ is (297 − 3)/350 = 0.84 (≡ 1 − 0.16), with 95% interval 0.7939 to 0.8762 using the Tango score method. Thus a 95% interval for the paired organ procedure rate is 0.1238 to 0.2061 per woman. Equivalently, we could report a rate of 0.08 procedures per breast, with 95% interval 0.0619 to 0.1030.

Thus a good interval estimator for such a mean may be obtained using a good interval estimator for the difference in proportions from paired data. In the notation of Table 8.1, the theoretical cell proportions π_i, $i = 1\ldots4$ are reparametrised in terms of $\theta = \pi_2 - \pi_3$ (which is one of three possible representations of the parameter of interest) and a nuisance parameter $\gamma = \pi_2 + \pi_3$. Given n_0, n_1 and n_2, the frequencies of scoring 0, 1 and 2, with $n_0 + n_1 + n_2 = n$, we obtain the difference of interest by substituting $b = n_0$ and $c = n_2$.

For the original issue of interval estimation for the mean of a variable X on {0, 1, 2}, let ζ_i, $i = 0, 1, 2$ denote the theoretical proportions of individuals scoring 0, 1 and 2, respectively, with of course $\zeta_0 + \zeta_1 + \zeta_2 = 1$. Then the theoretical mean of this variable is $\mu = 2\zeta_2 + \zeta_1 = 1 + (\zeta_2 - \zeta_0)$. Just as we identify n_0 with b and n_2 with c, so also we identify ζ_0 with π_2 and ζ_2 with π_3, hence $\mu = 1 - \theta$. Several of the methods introduced in the next section obtain point and interval estimates for $\mu = E\bar{X}$ by calculating point and interval estimates for θ and complementing. Furthermore two methods, which would be natural to consider in the gene frequency estimation context, are primarily methods that produce interval estimates for $\pi = \mu/2$, the gene frequency per allele, rather than μ, the gene frequency per individual. Thus each of the approaches developed here can be regarded as estimating any of the three equivalent parameters, θ (the paired difference of proportions), μ (e.g., the procedure rate per woman, as in Table 9.1) and π (e.g., the gene frequency as in Table 9.2, or the procedure rate per breast). These parameters are simply interrelated by $1 - \theta = \mu = 2\pi$, accordingly have equivalent estimation properties.

In Chapter 8 we saw that most, but not all, good confidence intervals for θ are based on *b*, *c* and *n* alone, and disregard the distinction between the double positive and double negative cells *a* and *d*. Methods 8 to 10 based on squaring and adding Wilson score intervals for single proportions treat *a* and *d* separately, whereas given merely b ($\equiv n_0$) and c ($\equiv n_2$), these cell frequencies are not separately identifiable. The simplest choice is to take $a = d = n_1/2$, which is the origin of method 11 of Chapter 8. The resulting non-integer cell frequencies in the event of odd n_1 are not a problem for these methods, because they are not defined in terms of tail areas. The symmetrised interval that results is usually but not always narrower than for other choices of *a* and *d* which sum to n_1.

9.2.4 Eight Methods and Examples

The eight methods are most naturally expressed as intervals for μ (method 1), π (2 and 3) and θ (4 to 8) as follows.

1. Wald t-interval for the mean (Section 2.1)

   The interval for μ is $\bar{X} \pm t \times \text{SE}$ where

$$\text{SE} = \sqrt{[\{n_1 + 4n_2 - (n_1 + 2n_2)^2/n\}/\{n(n-1)\}]} = \sqrt{[\{n_1(n_0 + n_2) + 4n_0n_2\}/\{n^2(n-1)\}]} \quad (9.2)$$

   and *t* denotes the two-sided $1 - \alpha$ point of the central *t*-distribution with $n - 1$ degrees of freedom.

2. Wald interval applied to overall proportion positive

   The interval for π is $p \pm z\sqrt{\{p(1-p)/N\}}$ where *z* denotes the two-sided 1–α point of the standard normal distribution.

3. Wilson score interval applied to overall proportion positive

   The interval for π is delimited by

$$[2Np + z^2 \pm z\sqrt{\{z^2 + 4Np(1-p)\}}]/\{2(N + z^2)\}. \quad (9.3)$$

   The remaining methods are formulated as intervals for the paired difference θ and are defined in Section 8.5.1.

4. Wald interval for paired difference (method 1 in Section 8.5.1).
5. Symmetrised square-and-add Wilson interval for paired difference (method 11 in Section 8.5.1).
6. Exact profile likelihood interval for paired difference (method 5 in Section 8.5.1).
7. Mid-p profile likelihood interval for paired difference (method 6 in Section 8.5.1).
8. Tango score interval for paired difference (method 12 in Section 8.5.1).

In special cases with n_0, n_1 or n_2 zero, methods 2 and 3 reduce to obvious linear functions of Wald and Wilson intervals for binomial proportions with denominator $N = 2n$, and method 4 to linear functions of Wald intervals for proportions with denominator n. Furthermore, methods 8, 6 and 7 usually reduce to corresponding linear functions of the Wilson, mid-p and exact intervals, except where boundary cases affect their formulation.

Examples of these calculations are given in Table 9.3. Ten illustrative datasets $\{n_0, n_1, n_2\}$ are chosen, taken from previous studies. They have been selected to exemplify a range of sample sizes and to include overflow and zero width aberrations for methods 1, 4 and 2.

Furthermore, for given n, the single parameter of interest (whether scaled as μ, π or θ) is not sufficient to identify all three parameters ζ_0, ζ_1 and ζ_2 which are constrained to sum to 1. We reparametrise in terms of two parameters, π, and a second parameter κ, which plays an essential role in the subsequent evaluation, and is designed to be orthogonal to π. At $\kappa = -1$, π is constrained to be 0.5, but as κ increases, a wider range of π values is available, symmetrical about 0.5, and extending to the whole of [0,1] when $\kappa = +1$.

Figure 9.1 shows the relationship between κ, π and ζ_i, $i = 0, 1, 2$. The five curves correspond to five zones of the parameter space identified by the letters S to W, with $\kappa = -0.75, -0.5, 0, +0.5$ and $+1$. Similarly, the 10 illustrative datasets in Table 9.3 incorporate a range of κ values from -1 to nearly $+1$ and hence approximate all of zones S to W. In this table and the subsequent evaluation zones are also numbered 1, 2 or 3 to denote small, medium or large sample size. An approximate zone is indicated for each example, although in some cases n is intermediate between two zones. κ cannot be estimated for example J, but such datasets tend to come from zones U, V and especially W.

The reparametrisation follows Bloch and Kraemer (1989):

$$\zeta_0 = (1 - \pi)^2 + \pi\,(1 - \pi)\kappa \tag{9.4}$$

$$\zeta_1 = 2\pi\,(1 - \pi)\,(1 - \kappa) \tag{9.5}$$

$$\zeta_2 = \pi^2 + \pi\,(1 - \pi)\kappa \tag{9.6}$$

or equivalently $\kappa = 1 - \zeta_1/(2\pi\,(1 - \pi))$, estimated by $\hat{\kappa} = 1 - n_1/(2n\hat{\pi}\,(1 - \hat{\pi}))$ where $\hat{\pi} = 1/2(1 - (n_0 - n_2)/n)$. Strictly, this parameter is the version of the kappa statistic introduced by Scott (1955), in which marginals are homogenised. This parameter is not identical to Cohen's kappa (1960), but this is one of several contexts in which Scott's parameter is widely but confusingly referred to as kappa (Newcombe 1996). Note also that the sign of $\hat{\kappa}$ is counter-intuitive: for the ordinary 2×2 table corresponding to example 1, we would normally

TABLE 9.3

Confidence Limits for the Mean of a Variable Taking the Values 0, 1 or 2 by 8 Methods: 10 Exemplar Datasets

Example	$n_0 = b$	$n_1 = a + d$	$n_2 = c$	n	Source	$\hat{\kappa}$	Approx. Zone	$\bar{X}$	Confidence Limits for μ Calculated by Method 1	2	3	4	5	6	7	8
A	0	54	0	54	a	−1.00	S2	1	1*	0.811	0.815	1*	0.965	0.934	0.946	0.934
									1*	1.189	1.185	1*	1.035	1.066	1.054	1.066
B	0	43	1	44	b, c	−0.96	S2	1.023	0.977	0.814	0.817	0.979	0.962	0.935	0.948	0.941
									1.069	1.232	1.226	1.067	1.082	1.120	1.107	1.118
C	12	36	2	50	a, d	−0.50	T2	0.800	0.659	0.608	0.619	0.664	0.671	0.646	0.655	0.655
									0.941	0.992	0.996	0.936	0.944	0.950	0.941	0.939
D	254	94	2	350	e	−0.12	U3	0.280	0.231	0.229	0.233	0.232	0.237	0.233	0.234	0.235
									0.329	0.331	0.335	0.328	0.334	0.332	0.331	0.331
E	98	2	0	100	a	−0.01	U2	0.020	<0*	<0*	0.005	<0*	0.008	0.002	0.003	0.006
									0.048	0.048	0.071	0.047	0.083	0.087	0.078	0.093
F	297	50	3	350	e	+0.03	U3	0.160	0.119	0.120	0.124	0.119	0.125	0.121	0.123	0.124
									0.201	0.200	0.205	0.201	0.208	0.206	0.205	0.206
G	290	242	73	605	f	+0.08	U3	0.641	0.586	0.589	0.590	0.587	0.588	0.587	0.588	0.588
									0.696	0.694	0.695	0.696	0.698	0.698	0.697	0.697
H	97	2	1	100	a	+0.49	V2	0.040	<0*	0.001	0.016	<0*	0.015	0.008	0.010	0.013
									0.088	0.079	0.101	0.087	0.126	0.125	0.118	0.130
I	101	29	220	350	e	+0.81	W3	1.340	1.246	1.270	1.269	1.246	1.243	1.242	1.244	1.243
									1.434	1.410	1.408	1.434	1.430	1.432	1.431	1.430
J	30	0	0	30	a	–	–	0	0	0	0	0	0	0	0	0
									0*	0*	0.120	0*	0.161	0.231	0.190	0.227

* = Boundary anomalies–overshoot and zero width intervals.
a = Newcombe, R. G. 1998c. *Statistics in Medicine* 17: 2635–2650.
b = Tango, T. 1998. *Statistics in Medicine* 17: 891–908.
c = Miyanaga, Y. 1994. *Japanese Journal of Soft Contact Lenses* 3: 163–173; Tango, T. 1998. *Statistics in Medicine* 17: 891–908.
d = Armitage, P., Berry, G. and Matthews, J. N. S. 2002. *Statistical Methods in Medical Research*, 4th Edition. Blackwell Science, Oxford.
e = Edwards, A. G. K. et al. 1999. *British Journal of General Practice* 49(447): 811–812.
f = Burr, M. L. et al. 1989. *Lancet* 2: 757–761.

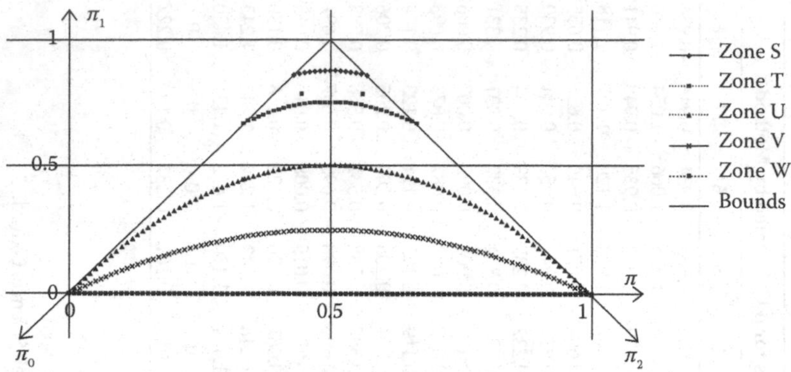

FIGURE 9.1
Parameter space for a variable taking values from {0, 1, 2}. This can be viewed as parametrised by (π, π_1) or by (π_0, π_2). The permissible region of the space is bounded by the isosceles triangle. Zones S to W and points corresponding to the mean for the paired difference case evaluation (Chapter 8) are marked.

TABLE 9.4

Estimated Mean Coverage Probabilities for Nominal 95% Intervals for the Mean of a Variable Taking the Values 0, 1 or 2 by 8 Methods

	Method							
Zone	**1**	**2**	**3**	**4**	**5**	**6**	**7**	**8**
S1	0.8438	0.9988	0.9996	0.8238	0.9747	0.9985	0.9944	0.9934
T1	0.9380	0.9884	0.9941	0.9078	0.9525	0.9873	0.9727	0.9720
U1	0.8852	0.8712	0.9524	0.8600	0.9381	0.9804	0.9681	0.9600
V1	0.8564	0.7964	0.9000	0.8335	0.9400	0.9754	0.9627	0.9555
W1	0.8348	0.7275	0.8463	0.8157	0.9457	0.9800	0.9643	0.9534
S2	0.9461	0.9998	0.9999	0.9401	0.9531	0.9677	0.9537	0.9494
T2	0.9482	0.9933	0.9943	0.9426	0.9490	0.9614	0.9501	0.9487
U2	0.9329	0.9301	0.9515	0.9276	0.9467	0.9638	0.9544	0.9522
V2	0.9225	0.8660	0.8926	0.9173	0.9467	0.9609	0.9522	0.9512
W2	0.9187	0.8056	0.8369	0.9135	0.9506	0.9670	0.9545	0.9520
S3	0.9479	0.9999	0.9999	0.9470	0.9488	0.9573	0.9496	0.9493
T3	0.9490	0.9944	0.9946	0.9478	0.9490	0.9551	0.9494	0.9491
U3	0.9461	0.9454	0.9504	0.9448	0.9488	0.9566	0.9512	0.9509
V3	0.9431	0.8840	0.8910	0.9421	0.9498	0.9553	0.9508	0.9507
W3	0.9418	0.8278	0.8343	0.9410	0.9500	0.9580	0.9507	0.9502

calculate $\hat{\kappa} = +1$, not -1. The explanation is that here, it is the *b* and *c* cells, not the *a* and *d* cells, that are of primary interest.

9.2.5 Evaluation

The main evaluation of the eight methods for the mean of a variable on {0, 1, 2} (Newcombe 2003a) involves evaluating their performance in 15 parameter space zones defined by five levels of κ and three ranges of sample size, 10 to 20 (zone 1), 50 to 100 (zone 2) and 250 to 500 (zone 3). Different applications correspond to different κ zones. Thus for paired organ procedure rates, κ values near to 0 or positive (zones U, V and W) are plausible. For gene frequency, κ is expected to be close to zero, and performance in zone U is most relevant. For the original context of a paired difference between proportions, zones S and T are plausible. Thus in the evaluation in Chapter 8, the means for γ and θ (or rather $|\theta|$) were 0.220 and 0.117, corresponding to $\zeta_0 = \pi_2 = 0.1685$, $\zeta_2 = \pi_3 = 0.0515$ or vice versa, with $\kappa = -0.58$. These points are marked on Figure 9.1.

Criteria for evaluation are the same as in other chapters: degree and symmetry of coverage, expected interval width, and avoidance of anomalous behaviour at or near boundaries. We calculate confidence intervals for π by the eight methods, and examine mean coverage and interval width based on an appropriate sample of parameter space points, and mean mesial and distal non-coverage. The frequencies with which methods 1, 2 and 4 produce boundary abnormalities of zero width intervals and overshoot are counted. Methods 3 to 8 avoid tethering abnormalities.

Coverage properties are evaluated by simulation. As usual, we avoid restricting to rounded values for sample sizes and proportions, which may be atypical. In each zone, we generate 100,000 PSPs as follows. We first choose the total sample size *n* from an integer uniform distribution spanning the relevant range for the zone. Here and elsewhere, all sampling is random and independent. We then choose a single set of three hypothetical proportions (ζ_i, $i = 0, 1, 2$) as follows. Without loss of generality we restrict to $\pi \le 1/2$, because all methods evaluated are equivariant. We choose $\pi \sim U[\pi_{min}, 1/2]$ where $\pi_{min} = 0$ if $\kappa \ge 0$, $\kappa/(\kappa - 1)$ if $\kappa < 0$, thus π_{min} takes values 3/7 and 1/3 in zones S and T. Then ζ_0, ζ_1 and ζ_2 are calculated from κ and π as in Section 9.2.4. One random dataset is then sampled per PSP, by sampling one set of cell frequencies (n_i, $i = 0, 1, 2$) randomly using *n* and (ζ_i, $i = 0, 1, 2$). We calculate 95% intervals for this dataset by the eight methods, determine whether mesial or distal non-coverage or any boundary aberration occurs, and as a by-product calculate the interval width after truncation of overshoot. We then summarise across the 100,000 runs for the zone, including mean coverage and width for each PSZ.

Tables 9.4 and 9.5 give the results of the main evaluation for the {0, 1, 2} mean estimation problem. Table 9.4 shows the mean coverage probability for 95% intervals calculated by eight methods for the 15 zones. Table 9.5 shows

TABLE 9.5
Estimated Mean Widths of Nominal 95% Intervals for the Parameter π by 8 Methods

	Method							
Zone	**1**	**2**	**3**	**4**	**5**	**6**	**7**	**8**
S1	0.180	0.361	0.339	0.157	0.193	0.276	0.242	0.246
T1	0.271	0.354	0.333	0.237	0.246	0.314	0.282	0.276
U1	0.298	0.269	0.275	0.262	0.273	0.331	0.303	0.299
V1	0.356	0.264	0.272	0.315	0.318	0.373	0.346	0.339
W1	0.399	0.258	0.269	0.356	0.365	0.427	0.389	0.373
S2	0.081	0.162	0.160	0.079	0.082	0.094	0.087	0.088
T2	0.114	0.159	0.157	0.111	0.112	0.122	0.115	0.115
U2	0.129	0.126	0.127	0.126	0.127	0.135	0.129	0.129
V2	0.157	0.126	0.126	0.153	0.153	0.162	0.156	0.156
W2	0.181	0.125	0.126	0.177	0.177	0.190	0.179	0.178
S3	0.036	0.072	0.072	0.036	0.036	0.038	0.037	0.037
T3	0.051	0.071	0.071	0.050	0.050	0.052	0.051	0.051
U3	0.057	0.057	0.057	0.057	0.057	0.059	0.057	0.057
V3	0.070	0.057	0.057	0.070	0.070	0.071	0.070	0.070
W3	0.081	0.057	0.057	0.080	0.080	0.083	0.081	0.080

the corresponding mean interval width for π. Methods 5 to 8 derived from good methods for the paired difference of proportions θ performed very well whilst methods 1 to 4 generally performed poorly. Only in zone U3 do all 8 methods perform similarly and reasonably well, with mean CP close to 0.95, ranging from 0.945 (methods 4 and 2) to 0.957 (method 6) and mean width 0.059 for method 6 else 0.057.

Of the four good methods, generally 7 and 8 aligned mean CP best with the nominal $1 - \alpha = 0.95$. In most zones mean coverage and width were highest for method 6, followed by 7 and 8, and lowest for method 5. For all four methods the highest mean CP occurred in zone S1 and was very conservative (method 5 0.975, others >0.99). The lowest mean CP was at 0.955 for method 6 (zone T3), 0.949 for 7 (T3) and 8 (T2), and 0.938 for 5 (U1).

For all four of these methods, the interval location tended to be too mesial in zones T1 to W1. For other zones, methods 6 and 7 gave acceptably symmetrical coverage. Method 5 also resulted in predominantly distal non-coverage in zones S2 to W2 and S3; elsewhere interval location was satisfactory. Method 8 was too mesial in zones U1 to W1 and U2 to W2, too distal in zones S1, T1 and S2, otherwise satisfactory.

For the four good methods, there is the usual trade-off between mean coverage and narrowness, but this does not apply to the poorer methods. Thus in zone W1, method 1 has mean width 0.399, higher than that for methods 7 and 8 (0.389 and 0.373), but the mean CP is only 0.835, as against 0.964 and 0.953.

The *t*-based method 1 was very anticonservative for small n and extreme κ (mean CP 0.844 for zone S1, 0.835 W1), although only slightly in zone T1 (0.938), and remained slightly anticonservative for large n (mean CP 0.942 to 0.949 for zones S3 to W3). Similarly method 4 was very anticonservative for small n and extreme κ, and remained slightly anticonservative for large n.

Method 2 goes from being very conservative in zones S and T to very anticonservative in zones V and W, with both mesial and distal non-coverage probabilities large. This occurs even for large n (mean CP 0.99986 in zone S3, 0.828 W3). The average width corresponds. Method 3 is a little better: in zone U1 method 2 is anticonservative, with mean CP 0.872, due to predominantly mesial non-coverage, but method 3 has a very satisfactory mean CP of 0.952.

As in related contexts, only the methods that produce symmetrical intervals, here 1, 2 and 4, can produce aberrations. These three methods all produced overshoot in zones T1 to W1, U2 to W2 and U3 to W3, maximally in zone W1 for method 1 (frequency 0.393) and 4 (0.334), and in zone U1 for method 2 (0.204). They also produce a ZWI at the boundary, here $\pi = 0$, when $n_1 = n_2 = 0$. Furthermore, methods 1 and 4 (but not 2) produce a ZWI at the midpoint, $\pi = 1/2$, when $n_0 = n_2 = 0$. ZWIs were most frequent for small n. A midpoint ZWI occurred with frequency 0.134 in zone S1. A ZWI at the boundary occurred with frequency 0.130 in zone W1. In zone U1, all three methods produced a ZWI at $\pi = 0$ for 6746 out of 100,000 runs, and also methods 1 and 2 produced a ZWI at $\pi = 1/2$ for two runs.

The properties of the eight methods may be linked to the rationale of their derivation. Thus of all the methods evaluated, it is method 1 that most obviously corresponds to the idea of estimating the mean of a variable on {0, 1, 2}, and hence to the problem of estimating a paired organ procedure rate, or a gene frequency, for which zones U, V and W are *a priori* plausible. Correspondingly, the coverage of this method is reasonable for zone W3, in accordance with its known asymptotic robustness, but poor for zone W1.

Similarly, methods 2 and 3 are what might be contemplated in estimating gene frequency, with the assumption of HWE, and could also be considered for paired organ procedure rates, disregarding the pairwise non-independence. Here, zone U is the most plausible, with V and W also possible. In zone U1, method 2 performs badly, but method 3 performs well, in line with the known small-sample behaviour of these methods as described in Chapters 3 and 5.

Method 4 is derived from the simplest method for the paired difference of proportions and might be expected to perform well in zones S and T. It performed anticonservatively in zones S1 and T1, but reasonably in zones S3 and T3, again in line with previous findings as in Chapter 8.

Methods 1 and 4 are closely related, despite their very different algebraic derivation and rationale, method 1 corresponding to zones U–W and method 4 to zones S and T above. They differ only in that the sample variances imputed to $\bar{X}$ and its complement $\hat{\theta}$ differ by a factor $1 - 1/n$, and use of the *t*-distribution instead of *z*. On both counts method 1 results in a wider interval than method 4, and this is reflected by their mean coverage and width characteristics, as well as the examples in Table 9.3.

Methods 5 to 8 are greatly preferable to methods 1 to 4. From the standpoint of aligning mean coverage with the nominal $1 - \alpha$, Tango's score method 8 performs excellently. Its mean CP never fell below $1 - \alpha$ by more than sampling error, and was only too high in zone S1 (0.994, in common with other good methods except 5), T1 (0.972) and the original evaluation following the lines of Newcombe (1998c) (0.964). In all other zones mean coverage was always between 0.949 and 0.960, although the interval location can be slightly misplaced for small to moderate sample sizes. It is relatively simple to program, and can be highly recommended for use in all parts of the parameter space.

Methods 6 and 7 which involve a more complex program result in a slightly more appropriately located interval; method 7 had coverage similar to method 8 but slightly more conservative, whilst method 6 was the most satisfactory from a strictly conservative standpoint, with mean CP above $1-\alpha$ in all zones.

Method 5 also has approximately correct coverage, although often slightly anticonservative, apparently resulting from the need to symmetrise *a* and *d*. The issue of the effect of pooling double positive and double negative cells (*a* and *d*) or keeping them separate relates only to the paired difference of proportions problem—for the mean of a variable on {0, 1, 2} we usually cannot meaningfully identify these separately. Of the 12 methods evaluated in Chapter 8, only methods 8, 9 and 10 distinguish these cells. Of these three, only method 10 is satisfactory, much better than methods 1 through 4 and 7, although slightly less good than the computationally more complex methods 5 and 6 and the Tango method 12. Symmetrising method 10 is quite unnecessary in the paired difference context, but method 10 can only be applied to the {0, 1, 2} problem after first symmetrising, which lowers the minimum coverage considerably. Differences in coverage properties between methods are not a direct consequence of whether they symmetrise the *a* and *d* cells, although doing so unnecessarily causes the coverage properties to deteriorate.

The Tango method is implemented in the spreadsheet MEAN012.xls.

9.2.6 Clinical Trials

Newcombe (2003a) suggested that the approach outlined above can easily be extended to produce an interval for the difference between two independent sample means $\bar{X}_1$ and $\bar{X}_2$, each based on a variable taking only the values 0, 1 and 2, by applying the squaring and adding or MOVER approach as in Section 7.3. Method 5 is the obvious candidate for this process, being already of this form, and doing so is virtually equivalent (except for the symmetrisation of *a* and *d* cells) to the estimation of an interaction effect for paired

TABLE 9.6

Results of a 14-Day Course of Antibiotic Treatment in Children Presenting with Bilateral Otitis Media with Effusion

Number of Ears with Adequate Resolution of Symptoms	Antibiotic	
	Amoxicillin	Cefaclor
0	15	14
1	3	9
2	13	21
Total	31	44

Source: Data from Mandel, E.M. et al. 1982. *Pediatric Infectious Disease Journal* 1: 310–316.

data as described in Section 14.2. Nevertheless, method 8 is the preferred approach here, as explained in Section 9.2.5.

These suggestions were explored further by Pei et al. (2012) and Tang, He and Tian (2012). See also Tang, Qiu, Tang et al. (2012). Table 9.6 presents the results of a randomised trial of a 14-day course of antibiotic treatment in children presenting with bilateral otitis media with effusion (Mandel et al. 1982). The proportions of ears in which symptoms were adequately resolved were 29/62 = 0.4677 on amoxicillin and 51/88 = 0.5795 on cefaclor. The corresponding 95% Tango limits are (0.3012, 0.6400) and (0.4418, 0.7079). The difference in efficacy between the two antibiotics is 0.5795 – 0.4677 = 0.1118 in favour of cefaclor. Using the square-and-add approach to combine the intervals for the two groups, the 95% interval for the difference in efficacy is –0.0966 to +0.3079.

9.2.7 The Degree of First-Move Advantage in Chess

While issues relating to games of chance have played a large role in the development of statistical thinking, it has often been regarded as unfair that chance should have a large impact on outcome. In some instances, the impact of the chance element can be greatly reduced. For instance, competitive contract bridge is played using a duplicate system. Two teams of four players compete. The 52 cards are dealt into four packs of 13 cards each, in the usual manner, at table 1; an identical deal is then placed at table 2. Members of team A play north–south at table 1 and east–west at table 2, whilst members of team B play north–south at table 2 and east–west at table 1, with an obvious link to split-unit experimental designs such as the Battenberg design used in a split-mouth study by Kinch et al. (1988) as described in Section 2.3.

Readers may wonder how statistical issues could possibly arise in relation to totally "fair" games of a deterministic nature, such as chess, checkers (draughts) or Go (碁), known in Chinese as **weiqi** (围棋). For such games, the player who moves first—traditionally, white in chess—may be at an advantage. Essentially, there are two hypotheses, somewhat analogous to the null and alternative hypotheses in statistical hypothesis testing:

H_0: With best play, the second player can always achieve a draw
H_1: With best play, the first player can always achieve a win

A third hypothesis—that with best play there can be an advantage to moving second—has sometimes been mooted, but this seems highly implausible, for chess at any rate, even though it is black who can mate in two moves with the queen and in three moves with a knight.

In 2007 Schaeffer et al. demonstrated by computer that for checkers, best play always leads to a draw. The computer time to solve chess and Go in this way is reckoned to be prohibitive. The best available evidence bearing on the issue comes from outcomes of series of games between the same two players, although it is conceded that such evidence cannot be conclusive concerning these hypotheses. A Wikipedia site (Anonymous 2011b) describes the history of this issue, and presents evidence that the average score for the opening player (scoring 1 for a first-player win, 0.5 for a draw and 0 for a second-player win) has historically been of the order of 0.55. It is obviously straightforward to compute this figure for a large series of games, with a confidence interval calculated using one of the methods described in this chapter. These calculations would be appropriate for a contest comprising several games between identical copies of the same chess software. But no two human opponents can be assumed to be identically strong players. The calculations require to be adjusted in some way for differences in playing strength. One possibility is to use established rating scales for player strength such as the Elo system. A more attractive proposition is to base inferences on a substantial series

TABLE 9.7

Results of 79 Chess Games between Viswanathan Anand and Veselin Topalov, Up to and Including the 2010 World Chess Championship

	White Win	Draw	Black Win	Total
Prior to 2010 championship				
Anand (white) vs. Topalov (black)	13	16	4	33
Topalov (white) vs. Anand (black)	9	21	4	34
2010 championship				
Anand (white) vs. Topalov (black)	2	4	0	6
Topalov (white) vs. Anand (black)	2	3	1	6
Total				
Anand (white) vs. Topalov (black)	15	20	4	39
Topalov (white) vs. Anand (black)	11	24	5	40
All 79 games	26	44	9	79

of games between the same two players, which allows the adjustment to be made using evidence internal to the series.

For example, in April–May 2010 Viswanathan Anand defeated challenger Veselin Topalov by 6½ : 5½ to retain the title of World Chess Champion. Prior to the 2010 match, Anand and Topalov had played 67 games against each other at classical time control. Table 9.7 summarises results of the 79 games. Based on all 79 games, the simplest estimate of the average white score is $(26 + 44 \times 0.5)/79$ or 0.6076. A 95% interval using method 8 as described in Section 9.4 is from 0.5358 to 0.6757.

However, this analysis is distorted by the fact that Topalov played white in more games than Anand (albeit minimally). Based on the 39 matches in which Anand played white, the estimate is 0.6410, with 95% interval from 0.5339 to 0.7359. Based on the 40 matches in which Topalov played white, the estimate is 0.575, with 95% interval from 0.4756 to 0.6696. In close analogy to the Hills–Armitage (1979) analysis for the two-period crossover trial, a better overall estimate of the average white score is $(0.6410 + 0.575)/2 = 0.6080$. A 95% interval is obtained from the above intervals by an obvious extension of the square-and-add approach as 0.5349 to 0.6750. While for these data, the more refined analysis is practically identical to the crude one, it would not always necessarily be the case that the numbers of games played as white and black would be so evenly balanced. This technique could be applied to much more extensive data on many pairs of players, then combined as described in Section 7.3.3 in a meta-analysis.

Needless to say, all these analyses are dependent on the usual assumption of independence, which in this context would require that the outcome of one game between a pair of contestants should not depend in any way on the outcome of preceding matches between them. This assumption is highly questionable here. Notwithstanding this argument, the above crossover-type analysis should be regarded as preferable to analyses that either disregard player strength or adjust for it using external ratings of skill, which are also subject to the same criticism.

9.3 Unordered Trinomial Data: Generalising the Tail-Based *p*-Value to Characterise Conformity to Prescribed Norms

Generally, in the research setting, *p*-values quoted are two-sided, for the usual reasons. An even-handed approach to the data is an important principle of interpretation, and if a one-sided test is chosen, it is far from clear what one should report when there is a large difference in a direction opposite to anticipation. Furthermore, normally two-sided intervals are used, to give a meaningful margin of error on either side of the observed result, and correspondence between *p*-values and confidence intervals is generally desirable. Conversely, in the audit paradigm, often 1-sided *p*-values and confidence limits are appropriate. For example, the final protocol for the Comparative Effectiveness of Magnetic Resonance Imaging in Breast Cancer (COMICE) trial (Turnbull et al. 2010) evaluating the usefulness of magnetic resonance imaging (MRI) in staging breast cancer quotes historical data indicating that a prescribed standard was not met for an important binary outcome: "Currently 14.2% of 50- to 65-year-old women with a C5/B5 pre-operative diagnosis undergo more than one operation for primary breast cancer (British Association for Surgical Oncology 2002), although the quality assurance standard for the NHS breast screening programme is <10% re-operation rate for incomplete tumour excision (NHSBSP 1996)." A one-sided *p*-value expressing the probability of observing a re-operation rate of 14.2% or higher in a sample of this size based on a population 10% re-operation rate would appropriately summarise the evidence that the observed re-operation rate indicated that the 10% target is not being met. In a complementary manner, a lower confidence limit for the observed proportion could be calculated, and inferences based largely on whether the resulting interval excludes the target value of 10%. In the following sections a related context involving an ordered trinomial scale is described in which the concept of calculating precise probabilities of all possible outcomes and aggregating them can be applied in a slightly more elaborate way. Some results are also shown for the simpler, more familiar binomial case in order to draw out parallels and differences between these two applications.

9.3.1 The Trinomial Case: An Example from Dental Radiology

Newcombe and Farrier (2008d) presented some results from a randomised controlled trial (Farrier et al. 2009) in which patients requiring intra-oral periapical radiology to a single tooth were randomly allocated between two different types of digital system, A, a charged coupled device (Sidexis, Sirona Dental Systems GmbH, Bensheim, Germany) and B, a photostimulable storage phosphor image plate system (Vistascan, Dürr Dental GmbH, Bissingen, Germany). Randomisation was stratified by six areas of the dentition. The primary interest was image diagnostic quality; however, non-diagnostic parameters, patient comfort and acceptability were also assessed as these too are important for successful implementation of any radiographic system. If a repeat exposure is necessary to produce a high-quality diagnostic image, this not only uses additional resources but more importantly increases the radiation dose received by the patient; this may offset advantages such as a reduced patient radiation dose per exposure and faster acquisition time and substantially affect the balance of advantages and disadvantages of the two systems.

The study gave clear evidence that the device B was superior to device A in terms of several outcomes, including diagnostic image quality and patient acceptability. However, based on the manufacturer's recommended dose for each system, the mean radiation dose for a single intra-oral periapical radiograph using system B was 2.94 times that for system A; this ratio reduces to 2.45 when the need for a repeat radiograph in taken into account, as in Section 14.3.

Quality assurance in dental radiography seeks to ensure consistently adequate diagnostic information is obtained from radiographs, whilst radiation doses to patients and staff are controlled to be as low as reasonably practicable. Ideally all radiographs must be of high diagnostic quality, and it is important to avoid those exposures which have no merit. A simple subjective quality rating system has been suggested to monitor image quality of dental radiographs (National Radiological Protection Board (NRPB) 1994). Quality standards for specific radiographs provide a means of comparing radiographs against an ideal (Office for Official Publications of the European Communities (OOPEC) 2004). Using the criteria given for ideal film quality each radiograph can be allocated a quality score: 1 (excellent), 2 (diagnostically acceptable) and 3 (unacceptable). In no sense should this be regarded as a meaningfully equally spaced scale: the labels 1, 2 and 3 are convenient and arbitrary, and for practical purposes the difference between 2 and 3 is more crucial than the difference between 1 and 2. According to these quality ratings, performance targets have been recommended, albeit not evidence-based. Whilst there can be no tolerable level of "unacceptable" radiographs, it has been specified that as a minimum target, no greater than 10% should be of unacceptable quality (NRPB 1994; OOPEC 2004). Thus the minimum targets for radiographic quality were set at ≥70% excellent and ≤10% unacceptable. (A target of ≤20% for the acceptable category was also specified, but this is clearly an inappropriate additional requirement, in that a 70: 30: 0 split is clearly preferable to a 70: 20: 10 one, whereas the former has over 20% classed as acceptable and the later does not. For present purposes we will define the target in terms of proportions classed as excellent and unacceptable only.) In recognition that these ultimate targets may not be initially achievable for reasons such as new technology, the developing process or operator training or familiarity, an interim target was set at ≥50% excellent and ≤10% unacceptable (NRPB 1994).

While the primary objective of the trial was to compare the two systems, it is also of interest to quantify how the performance of each system should be judged relative to the above standards. Any possible outcome can be represented by a point in an outcome space in which X represents the proportion of radiographs classed as excellent and Y represents the proportion judged unacceptable. This outcome space may be plotted as the unit square, restricted to the triangle $X + Y \leq 1$. Vertical and horizontal lines may be plotted to represent the preset standards for proportions excellent and unacceptable, with observed (x, y) points superimposed.

Given the total number of cases, N, the outcome space comprises $\sum_{n_1=0}^{N}\sum_{n_2=0}^{N-n_1} 1 = \sum_{n_1=0}^{N}(N - n_1 + 1) = (N+1)(N+2)/2$ points.

We partition these into several zones:

U (upper tail): $X \geq x$ and $Y \leq y$, with at least one inequality strict
O (observed): $X = x$ and $Y = y$
Q (questionable): Q_1: $X > x$ and $Y > y$ or Q_2: $X < x$ and $Y < y$
L (lower tail): $X \leq x$ and $Y \geq y$, with at least one inequality strict

Given π_E, π_A and π_U, the proportions excellent, acceptable and unacceptable prescribed as standard, the probability of each zone based on these proportions is readily calculated as a sum of trinomial probabilities. Here, U is the aggregate probability of observing an outcome (X, Y) strictly better than the observed (x, y) assuming the standard trinomial proportions. Similarly, L is the probability of observing an outcome (X, Y) strictly worse than the observed (x, y). The calculations are illustrated using the hypothetical small dataset in Table 9.8 compared to the ultimate standard of 70% excellent, 10% unacceptable.

The probability of the observed outcome is O = $(6!/3!\ 2!\ 1!)\ 0.7^3\ 0.2^2\ 0.1^1$ = 0.08232.

The probability of a questionable outcome with $X > x$ and $Y > y$ is

$$Q_1 = (6!/4!\ 0!\ 2!)\ 0.7^4\ 0.2^0\ 0.1^2 = 0.036015.$$

Probabilities of other outcomes are obtained by summating similar trinomial probabilities.

When the observed x and y are identical to the standards, neither L nor U is low. A high value of U (with a low value of L) indicates that the observed

TABLE 9.8
Comparison of Radiograph Quality Results on a Three-Point Ordinal Scale with a Preset Standard; Definitions of Zones U, O, Q and L for a Simple Hypothetical Example with $N = 6$

		Number of Radiographs Judged Excellent, x						
		0	1	2	3	4	5	6
Number of radiographs judged unacceptable, y	6	L						
	5	L	L					
	4	L	L	L				
	3	L	L	L	L			
	2	L	L	L	L	Q_1		
	1	L	L	L	O	U	U	
	0	Q_2	Q_2	Q_2	U	U	U	U

Source: From Newcombe, R. G. and Farrier, S. L. 2008d. *Statistical Methods in Medical Research* 17: 609–619.

Note: Numbers of radiographs judged excellent, acceptable and unacceptable are 3, 2 and 1, respectively.

TABLE 9.9

Image Quality for Two Digital Periapical Radiography Systems, in Relation to Two Published Standards

	System A	System B	Ultimate Standard	Interim Target
Excellent	36 (37%)	77 (71%)	70%	50%
Diagnostically acceptable	40 (41%)	25 (23%)	(20%)	(40%)
Unacceptable	22 (22%)	6 (6%)	10%	10%
Total (*N*)	98	108		

Source: From Newcombe, R. G. and Farrier, S. L. 2008d. *Statistical Methods in Medical Research* 17: 609–619.

results fall short of the standard, whereas a high L (and low U) indicate results that surpass the standard.

Table 9.9 shows the results obtained, alongside these norms. It is clear that system B achieves the desired standard, whereas system A falls short of even the interim one. However, these obvious conclusions are based on the observed proportions alone, and take no account of sampling uncertainty. We develop appropriate tail-based probabilities to quantify this.

Table 9.10 gives the aggregated outcome probabilities for the categories U, O, Q and L defined above, for systems A and B relative to the ultimate standard and the interim target. These figures are derived straightforwardly from Table 9.9 but can be regarded as adding value in much the same way that a conventional *p*-value adds value to the informal comparison of an observed binomial proportion to some hypothesised value. Thus for system A in relation to the ultimate target, U = 0.9999 and L < 0.0001 indicate, based on the sample of 98 radiographs, very strong evidence that the performance of system A falls short of the ultimate standard. Similarly, there is strong evidence

TABLE 9.10

Comparison of Radiographic Image Quality Obtained Using Two Digital Systems against Prescribed Ultimate and Interim Norms

Standard Radiography System	Ultimate A	Interim A	Ultimate B	Interim B
Upper tail	0.9999	0.9969	0.0568	<0.0001
Observed	<0.0001	<0.0001	0.0032	<0.0001
Lower tail	<0.0001	<0.0001	0.6390	0.9649
Questionable	0.0001	0.0031	0.3009	0.0351
U + O/2	0.9999	0.9969	0.0585	<0.0001
L + O/2	<0.0001	<0.0001	0.6406	0.9649

Source: From Newcombe, R. G. and Farrier S. L. 2008d. *Statistical Methods in Medical Research* 17: 609–619.

Note: Aggregated tail probabilities.

that its performance falls short of the interim target. Conversely, system B is apparently slightly better than the ultimate standard, and much better than the interim target. There is substantial evidence that system B surpasses the interim target, with U < 0.0001 and L = 0.9649. However, we cannot be quite so sure that it reaches the ultimate standard, for which U = 0.0568 and L = 0.6390. Q is high here, because the actual proportions in excellent and unacceptable categories are both very close to their respective target values.

Table 9.10 also includes probabilities for U + O/2 and L + O/2, which can be regarded as generalising mid-*p* *p*-values. Thus in a mid-*p* sense, assuming true probabilities of 0.7, 0.2 and 0.1 of excellent, diagnostically acceptable and unacceptable image quality, an outcome at least as good as that for system B, 77 : 25 : 6 would be observed with probability 0.0585, whereas an outcome similar to this or poorer would occur with probability 0.6406. These figures are certainly consistent with the ultimate target, but do not constitute clear evidence at a conventional level of statistical significance to reject the possibility that the true probabilities were 0.7, 0.2 and 0.1 (or poorer) but the play of chance resulted in an outcome better than this. In view of the ambiguity regarding where to place the probability of the outcome as observed—which arguably could equally reasonably be aggregated with either U or L—it seems most appropriate to share it out and quote this pair of mid-*p* aggregate probabilities.

The generalised *p*-values shown here are apparently inherently one-sided, it does not seem to make sense to seek to construct any kind of two-tailed *p*-value here. Corresponding asymptotic methods could in principle be developed, although on account of the important roles of the O and Q categories the present approach based on exactly computed probabilities of all outcome space points seems clearer, especially for the trinomial case.

9.3.2 The Binomial Case: An Example from Cancer Surgery

Similar calculations can be performed for the simpler situation in which an observed proportion of cases in which some criterion is (or is not) satisfied is compared against some prescribed normative value. For breast cancer, the Quality Assurance Guidelines for Surgeons (NHS Breast Screening Programme 1996) states that, "At least 90% of operations carried out with a proven pre-operative diagnosis of cancer (*in situ* and invasive) should not require a further operation for incomplete excision." The results in Table 9.11 relate to the data described above (British Association of Surgical Oncology 2002), in which 833 (14.2%) of 5866 patients required re-operation on account of incomplete excision—a performance which falls well short of the guideline.

Table 9.11 presents proportions requiring re-operation for the United Kingdom as a whole, and for a selection of five of the 16 regions. No region performed significantly better than the guideline, for most the performance fell significantly short. Two-sided 95% Clopper–Pearson exact intervals for these proportions are reported, for compatibility with the tail probabilities quoted. We also give upper tail, observed and lower tail probabilities—the

TABLE 9.11

Proportion of Breast Cancer Patients Who Required More Than One Operation, in Comparison to NHS Breast Screening Programme Target of (under) 10%

						Tail Probabilities				
		More Than One Operation		Clopper–Pearson 95% Limits		Exact			Mid-*p*	
Region	Total Cancers	Number	Proportion	Lower	Upper	Upper Tail	Observed	Lower Tail	U + O/2	L + O/2
Whole of UK	5866	833	0.1420	0.1332	0.1512	>0.9999	<0.0001	<0.0001	>0.9999	<0.0001
Mersey	182	41	0.2253	0.1668	0.2929	>0.9999	<0.0001	<0.0001	>0.9999	<0.0001
Northern	282	40	0.1418	0.1033	0.1881	0.9844	0.0058	0.0098	0.9873	0.0127
N. Thames	689	85	0.1234	0.0997	0.1503	0.9736	0.0065	0.0199	0.9769	0.0231
Wales	346	36	0.1040	0.0739	0.1411	0.5732	0.0679	0.3589	0.6072	0.3928
N. Western	326	25	0.0767	0.0502	0.1111	0.0629	0.0286	0.9085	0.0772	0.9228
Hypothetical	300	15	0.0500	0.0283	0.0811	0.0006	0.0007	0.9987	0.0009	0.9991

Note: Whole of UK, 1 April 1997 to 31 March 1998, and selected regions.

"questionable" category cannot arise in this simpler case. A high upper (or lower) tail probability indicates performance is significantly worse (better) than the guideline according to the usual exact binomial test. We also give the mid-*p* versions, U + O/2 and L + O/2, which as a consequence are complementary.

Nationwide, the 95% interval for the re-operation rate lies well above the target value of 0.1. The lower tail probability is very small, indicating that the probability of observing a re-operation rate as high as 14.2% on this large sample size is very low if the true rate is 10%. Re-operation rates in the 16 regions range from 7.7% to 22.5%. For the Mersey region, the interval also lies well to the right of 0.1 and the lower tail probability is very small. For the Northern region, the interval lies just to the right of 0.1, and the upper tail probability is a little above 97.5%. For the North Thames region, the interval just includes 0.1, and the upper tail probability is just below 97.5%. For Wales, the re-operation rate is just above the target, and neither upper nor lower tail probabilities are small. The North Western region has the lowest re-operation rate, but the interval does not exclude the target value. The lower tail probability here is high, but does not reach 0.975. Finally, we show results for a fictitious region with a re-operation rate significantly below i.e., surpassing the target. (These figures were not included in the total, which relates to all the actual patients.) Here, the upper limit is below the target, and the lower tail probability is greater than 0.975.

Of course, here U + O/2 and L + O/2 are simply the two possible mid-*p* *p*-values for the data. These figures are shown in detail largely in order to illustrate the fact that in the binomial case, the probability of the outcome actually observed can be relatively large, in particular when the observed proportion is close to the standard. In contrast, in the trinomial case, it is generally very small, but the "questionable" category often has a substantial probability. In both cases, it is recommended to consider both the upper and lower tail probabilities.

9.3.3 An Alternative Approach

Another approach is to weight the outcomes to produce an overall score. Radiographs judged excellent, acceptable and unacceptable are awarded scores of 1, λ and 0, for some chosen λ satisfying $0 < \lambda < 1$. Figure 9.2 shows the mean scores for systems A and B, alongside those corresponding to the ultimate standard and interim target, plotted as functions of the parameter λ. Whatever value of λ within this range is chosen, system B is superior to the ultimate standard whilst system A falls short of the interim target. This is a direct consequence of the fact that both the relevant probabilities for system B surpass the ultimate standard whereas those of system A fall short of the interim target. However, this representation of the data gives no indication of the impact of sampling variation on this conclusion.

We can extend this approach to address this issue in two ways. For any given λ between 0 and 1, a tail-based hypothesis test could be constructed, using either an exact or mid-*p* aggregation of tail probabilities based on

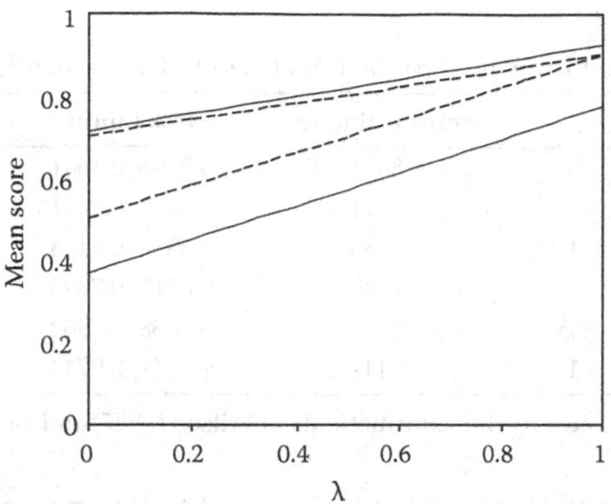

FIGURE 9.2
Mean quality scores for two radiography systems A and B (lower and upper solid lines) as functions of a weighting parameter λ representing the intermediate (acceptable) category. Reference lines (broken) corresponding to interim and ultimate quality norms are also shown, for comparison. (Based on Newcombe, R. G. and Farrier, S. L. 2008d. *Statistical Methods in Medical Research* 17: 609–619.)

the resulting distribution of scores, very much as for the single proportion case. Alternatively, the graph may be enhanced by incorporating confidence bounds as follows. Mean scores plotted at $\lambda = 0$ are identical to proportions judged excellent. The proportion of radiographs judged excellent for system B was 77/108 or 0.7130. A 95% Wilson interval for this proportion is 0.6215

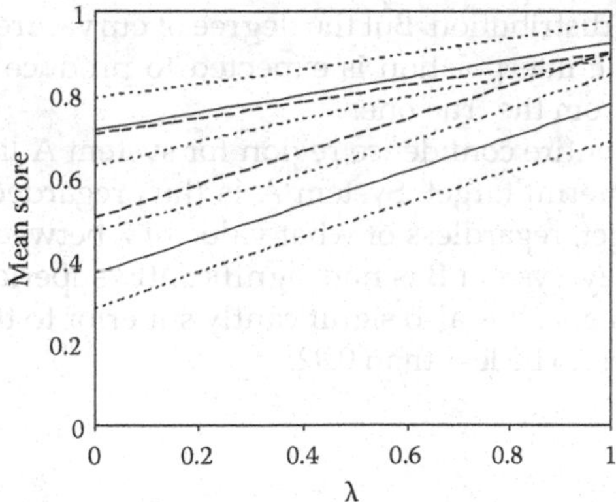

FIGURE 9.3
Mean quality scores for two radiography systems A and B (lower and upper solid lines) as functions of a weighting parameter λ representing the intermediate (acceptable) category, with 95% confidence regions delimited by the dotted curves. Reference lines (broken) corresponding to interim and ultimate quality norms are also shown, for comparison. (Based on Newcombe, R. G. and Farrier, S. L. 2008d. *Statistical Methods in Medical Research* 17: 609–619.)

TABLE 9.12

Confidence Limits for Proportions Used to Construct Figure 9.3

System	Λ	Point Estimate	95% Limits	Method
A	0	0.3673	0.2786, 0.4661	Wilson
	0.5	0.5714	0.4952, 0.6441	Tango
	1	0.7755	0.6834, 0.8468	Wilson
B	0	0.7130	0.6215, 0.7898	Wilson
	0.5	0.8287	0.7658, 0.7862	Tango
	1	0.9444	0.8841, 0.9743	Wilson

Note: Calculated by the score methods of Wilson (1927) and Tango (1998).

to 0.7898. Similarly, the proportion judged not to be unacceptable (plotted at $\lambda = 1$) was 102/108 = 0.9444, with 95% interval 0.8841 to 0.9743. In Figure 9.2 the line joining the points (0, 0.7130) and (1, 0.9444) represents system B. Figure 9.3 shows the same four straight lines as in Figure 9.2, and also includes approximate confidence bounds for the lines representing the two devices, much as in the sensitivity-specificity plot in Section 12.5.

These were obtained as follows. When $\lambda = 0.5$, we have a variable that takes three equally spaced values, 0, 0.5 and 1, consequently the methods introduced in the first half of this chapter are applicable. Table 9.12 gives score intervals for both systems for $\lambda = 0$ and 1 (Wilson 1927) and $\lambda = 0.5$ (Tango 1998). The concave confidence regions plotted in Figure 9.3 were determined by three-point quadratic interpolation. In principle, with substantial algebraic and programming effort, the Tango approach could be extended to any λ between 0 and 1, or equivalently any linear function of the three parameters of a trinomial distribution. But the degree of curvature here is sufficient slight that quadratic interpolation is expected to produce a curve visually indistinguishable from the true one.

In Figure 9.3 the entire confidence region for system A lies below the line representing the interim target. System A is thus regarded as significantly inferior to that target, regardless of what value of λ between 0 and 1 is chosen. In the same way, system B is non-significantly superior to the ultimate target for all values of λ. It is also significantly superior to the interim target, provided λ is chosen to be less than 0.92.

9.4 A Ternary Plot for Unordered Trinomial Data

A graphical characterisation is available for trinomial proportions, which does not make the assumption that the three groups have an inherent ordering. Suppose we sample n individuals or units, of which r_i are classed as belonging to group i, for $i = 1, 2, 3$. Let π_i and $p_i = r_i/n$ denote the theoretical and observed proportions, for $i = 1, 2, 3$, so that $\Sigma r_i = n$ and $\Sigma \pi_i = \Sigma p_i = 1$. A natural characterisation of the triad of trinomial proportions (p_1, p_2, p_3) is as a point on a triangle. The most appealing characterisation treats the three proportions interchangeably, using an equilateral triangle with all sides 1 unit in length. It is equivalent, but less intuitive, to plot in (p_1, p_2) space, restricted to the triangle defined by $0 \le p_1 \le 1$, $0 \le p_2 \le 1$ and $0 \le p_3 = 1 - p_1 - p_2 \le 1$. In the equilateral triangle plot, p_1, p_2, p_3 denote the perpendicular distances of any point P from the three sides of the triangle. Then $p_1 + p_2 + p_3 = 1$, a result known as Viviani's theorem. So any triad of trinomial proportions may be represented by a point in this space.

In the population genetics literature the equilateral triangle plot was often called the Streng diagram (see e.g., Edwards (1992), p. 139–142). Edwards' illustrative example relates to ABO gene frequencies estimated from frequencies of the four phenotypes, AA or AO, BB or BO, AB and OO. Nowadays it is most commonly used in the earth sciences to characterise compositional data such as the relative quantities of three substances in a mineral sample, where it is called the ternary diagram (see Weltje (2002)). Watson and Nguyen (1985) and Watson (1987) describe the construction of a goodness-of-fit confidence region. This comprises all triads $(\pi_i, i = 1, 2, 3)$ satisfying $1 + \frac{z^2}{n} = \sum_{i=1}^{3} \frac{p_i^2}{\pi_i^2}$, heeding the special case where any of the observed proportions is zero. Alternatively, a likelihood-based confidence region could be constructed.

10

Relative Risk and Rate Ratio

10.1 A Ratio of Independent Proportions

As explained in Chapter 1, a strong case can be made for basing the interpretation of data primarily on an appropriate measure of effect size with a confidence interval, rather than a p-value. For the comparison of two proportions, an issue arises that is more profound than the choice of which test to use. Several effect size measures are commonly used, the principal ones being the absolute difference, the ratio of proportions, and the odds ratio. These three measures have distinctive characteristics, and all three play important roles, which are considered in Sections 10.2 and 11.1. Fagerland et al. (2011) give an excellent current review of confidence intervals for these three measures.

In this chapter, we consider specifically the ratio of two proportions. We examine its applicability vis-à-vis the absolute difference, which was discussed in detail in Chapter 7. We examine confidence interval methods for the ratio of independent proportions, and the closely related rate ratio. For both measures, an asymptotic (delta) interval based on the standard error of the log-transformed measure takes a simple form. But this interval has limitations similar to those for the binomial proportion Wald interval. Better methods are available, in particular a score interval developed by Miettinen and Nurminen (1985). This is not of closed form; nevertheless it may be implemented in spreadsheet software using a fixed number of interval bisections. We defer detailed discussion of the more complex, less intuitive odds ratio to Chapter 11, including several additional arguments which repudiate the tendency to regard the odds ratio as the gold standard for comparison of proportions.

The ratio of two proportions is commonly known as the relative risk or risk ratio (RR). Following Fleiss, Levin and Paik (2003), in Chapters 10 and 11 we denote the ratio of two independent proportions, $\pi_1 \div \pi_2$ by θ, and the odds ratio, $\pi_1/(1-\pi_1) \div \pi_2/(1-\pi_2)$ by ω (Table 10.1).

The empirical estimate of θ is $p_1 \div p_2$ which can take positive values from 0 to ∞. It is important which proportion is divided by which, just as for the difference. But for the ratio, there is also the issue of which outcome is labelled "positive" and which is labelled "negative". And use of the word "risk" is

TABLE 10.1

Notation for Ratio Measures Comparing Independent Proportions

2 × 2 contingency table representation:

	Observed frequencies		Observed proportions		Theoretical proportions	
	Group 1	Group 2	Group 1	Group 2	Group 1	Group 2
Positive	a	b	p_1	p_2	π_1	π_2
Negative	c	d	$1-p_1$	$1-p_2$	$1-\pi_1$	$1-\pi_2$
Total	m	n	1	1	1	1

Parameter of interest: Relative risk $\theta = \frac{\pi_1}{\pi_2}$

Odds ratio $\omega = \frac{\pi_1/(1-\pi_1)}{\pi_2/(1-\pi_2)}$

appropriate only if the positive outcome is a detrimental one. In that situation, $1 - p_1/p_2$ may be termed the relative risk reduction (RRR), although this nomenclature is sensible only if $p_1 < p_2$. Moreover for a non-significant difference, the ensuing negative lower limit requires careful explanation to non-statisticians.

The issue of labelling is, unfortunately, compounded by the way these quantities are calculated by widely used software. The SPSS Crosstabs routine produces clearly labelled contingency tables with two (or more) rows or columns. For the 2 × 2 table, the associated risk subcommand produces three derived statistics, one odds ratio and two relative risks, each with a confidence interval calculated by the default delta method. Depending on which variable is used to define rows and which defines columns, and the numerical order of the codes representing positive and negative states of the two variables, risk can calculate two different odds ratios and no fewer than eight different relative risks! Obviously careful checking is required to ensure that the data are set up in such a way that the intended ratio is produced.

10.2 Three Effect Size Measures Comparing Proportions

The *Cochrane Handbook for Systematic Reviews of Interventions* (Higgins and Green 2008) includes a most helpful section, 9.4.4.4, on choice of comparative measures for dichotomous outcomes. Here Deeks et al. (2008) observe that no single measure is uniformly best, so the choice inevitably involves a compromise. Empirical evidence suggests that relative effect measures tend to be more consistent than absolute ones (Engels et al. 2000; Deeks 2002), but absolute measures of effect are believed to be more easily interpreted by clinicians than relative effects (Sinclair and Bracken 1994).

When the two proportions are both small, the absolute difference must be small, but a ratio measure comparing them can be small or large. The interpretation of what would amount to an importantly large difference, in the context of the application, may be substantially different for absolute and ratio measures. An absolute difference in risk of a particular size may be regarded as more important when the risks are close to 0 (or 1), such as 0.010 and 0.001, than when they are near the middle of the range, for example 0.410 and 0.401.

In a helpful expository paper Schechtman (2002) recommended that both a relative and an absolute measure should always be reported, with appropriate confidence intervals (as in Table 10.3). Table 10.2 illustrates the absolute difference, relative risk and odds ratio for four quite different comparisons of proportions. The first three cases have the same relative risk (and RRR), while case D is very different. However, the absolute risk reduction (and hence NNT) and odds ratio are substantially different in the three cases studied. For odds ratios, case B is different from cases A and C, which are similar. Cases A and D have the same absolute risk reduction and odds ratio, but very different relative risk.

Schechtman also considered a comparison between the incidence of dyskinesia after ropinirole (17/179, 0.095) and levodopa (23/89, 0.258) in a prospective study in early Parkinson's disease (Rascol et al. 2000). Here, the risk difference, risk ratio and odds ratio are –0.163, 0.368 and 0.302. All measures show that ropinirole is associated with much less frequent dyskinesia compared to levodopa, and all three correspond to the same highly significant difference with $\chi^2 = 12.5$, $p < 0.001$. Here, the results are best summarised by stating that on ropinirole the risk of developing dyskinesia is only about one-third as high as when levodopa is used, a reduction by about 16% in absolute terms.

This example presupposes that patients either are or are not susceptible to developing the adverse event of interest. When there is a time element, the

TABLE 10.2

Measures of Effect Size for Four Comparisons of Proportions

	Case A	Case B	Case C	Case D
Intervention group	0.05	0.14	0.017	0.70
Control group	0.30	0.84	0.10	0.95
Absolute difference	0.25	0.70	0.083	0.25
Relative risk	0.167	0.167	0.167	0.737
Odds ratio	0.123	0.031	0.156	0.123

Source: Data from McQuay, H. J. and Moore, R. A. 1997. *Annals of Internal Medicine* 126: 712–720.

absolute difference may well widen as the duration of follow-up increases. But the relative risk is likely to remain constant—which is of course the basis for the widely used proportional hazards model used for survivorship data with censoring (Cox 1972); see Section 11.5.5.

A similar issue arises for paired binary data. In the study of drug use by women prisoners (Plugge et al. 2009) considered in Section 8.5.2, the proportion who reported current daily use of any drug reduced from 111/217 (51.2%) before imprisonment to 31/217 (14.3%) when in prison. This was reported as an absolute reduction of 36.9% (95% interval 29.3% to 43.9%) using method 10 of Chapter 8. But it is equally relevant to report that the prevalence of drug use was only 28% as high following incarceration as it was before arrest, with 95% interval 20% to 39% using the Bonett–Price method introduced in Section 10.7. These two ways to express the comparison give complementary information.

In health-related research, traditionally comparative measures such as the risk difference and risk ratio have been produced primarily for dissemination to a clinical and scientific readership. Increasingly such measures are communicated to the wider public, including individuals both with and at risk of disease, in the contexts of health promotion and shared decision making. Indeed, a major requirement of the IPDASi scoring system for rating the quality of patient decision aids (Elwyn et al. 2009) is quantitative communication of the risk on different options. Schechtman also pointed out that a patient's perception of benefit from an intervention may be greatly influenced by whether the benefit is framed in relative or absolute terms, pointing to results of a widely cited study by Malenka et al. (1993). Four-hundred and seventy patients agreed to complete a questionnaire while waiting for their clinic visit. They were asked to choose between two equally efficacious medications for the management of a hypothetical serious disease. The benefit of one medication was stated in relative terms, the other in absolute terms. Patients could choose either medication alone, indicate indifference to the choice of medication, or choose not to answer. Fifty-seven percent of the patients chose the medication whose benefit was expressed in relative terms; 15% chose the medication whose benefit was expressed in absolute terms; and only 15% expressed indifference to the choice of medication. The preference for the medication with benefit expressed in relative terms applied across a wide range of ages and educational levels. Further questioning suggested that the patients thought benefit was greater when expressed in relative terms because they ignored the underlying risk of disease and assumed it was 100%—which could be an artefact of using a counterfactual, hypothetical scenario rather than a real-life decision the patient faces. These results also imply that the degree to which patient choice is swayed by framing varies greatly between patients. Much subsequent work has sought to determine the relative impact of different ways to present risks and benefits to patients. It is fair to say that the observation that both a relative and an absolute measure should always be reported applies equally when measures of risk are communicated to the patient population, although it is a moot point whether these should be accompanied by confidence intervals expressing their statistical imprecision.

The above considerations are now incorporated in the CONSORT guidelines for reporting clinical trials (Schulz et al. 2010). Item 17b states, "For binary outcomes, presentation of both absolute and relative effect sizes is recommended ... When the primary outcome is binary, both the relative effect (risk ratio (relative risk) or odds ratio) and the absolute effect (risk difference) should be reported (with confidence intervals), as neither the relative measure nor the absolute measure alone gives a complete picture of the effect and its implications. Different audiences may prefer either relative or absolute risk, but both doctors and lay people tend to overestimate the effect when it is presented in terms of relative risk. The size of the risk difference is less generalisable to other populations than the relative risk since it depends on the baseline risk in the unexposed group, which tends to vary across populations. For diseases where the outcome is common, a relative risk near unity might indicate clinically important differences in public health terms. In contrast, a large relative risk when the outcome is rare may not be so important for public health (although it may be important to an individual in a high risk category)."

10.3 Ratio Measures Behave Best on a Log Scale

Differences between means or proportions belong to the class of difference measures. When two quantities are equal, their difference is 0, consequently for hypothesis testing, the usual null hypothesis is that the difference is zero. The sampling distribution of the difference of two Gaussian means is Gaussian, hence symmetrical; the sampling distribution of the difference of two proportions is approximately symmetrical, provided the proportions are not extreme and the sample sizes are not small. Consequently, confidence intervals for difference measures are either symmetrical or approximately symmetrical on an untransformed, additive scale. Interchanging the roles of the two groups simply reverses the signs of the difference and its confidence limits.

In contrast, the relative risk and the odds ratio belong to the class of ratio measures, along with ratios of arithmetic or geometric means. When two quantities are equal, their ratio is 1, consequently for hypothesis testing, the usual null hypothesis is that the ratio is 1. Interchanging the roles of the two groups has the effect of inverting the ratio, not reversing its sign. However, a key feature of log transformation is that it turns ratios into differences, because $\log(a/b) = \log(a) - \log(b)$. Hence interchanging the two groups reverses the signs of log RR and its confidence limits.

Relative risks and odds ratios are generally analysed on the log scale because the distributions of the log ratios tend to be those closer to normal than of the ratios themselves (Altman and Bland 2003). Confidence intervals for log-transformed ratio measures are either symmetrical or approximately symmetrical. The interval for a ratio measure is either symmetrical on a ratio scale (i.e., of the form $\hat{\theta} \div k$ to $\hat{\theta} \times k$), or approximately so.

Suppose we calculate a relative risk as 3.0, with standard error 0.51 on an untransformed scale, leading to a symmetrical 95% interval from 2.0 to 4.0. If we interchange the roles of the two groups, we get a relative risk of 0.33. The corresponding 95% interval is 0.25 to 0.50, which is far from symmetrical. The point estimate, 0.33, is at the harmonic mean, which is always lower than the arithmetic mean of the lower and upper limits.

However, a relative risk of 1.6, with ratio scale symmetrical limits 1.6 ×/÷ 2.0 (i.e., from 0.8 to 3.2) inverts to a relative risk of 0.625, which also has ratio scale symmetrical limits 0.625 ×/÷ 2.0 (i.e., from 0.3125 to 1.25). Here, 1.6 is the geometric mean of 0.8 and 3.2, and 0.625 is the geometric mean of 0.3125 and 1.25. This, rather than additive scale symmetry, is the natural shape for a confidence interval for a ratio measure.

10.4 Intervals Corresponding to the Empirical Estimate

10.4.1 Delta Interval

In the notation of Table 10.1, the relative risk we aim to estimate is defined as $\theta = \pi_1/\pi_2$. The empirical estimate is $\tilde{\theta} = \frac{p_1}{p_2} = \frac{a/m}{b/n}$. The variance of ln θ estimated from the sample is

$$V = \frac{1}{a} + \frac{1}{b} - \frac{1}{m} - \frac{1}{n}. \tag{10.1}$$

So the simple asymptotic or delta limits for θ are

$$L = \frac{p_1}{p_2} \div \exp\left(z\sqrt{\frac{1}{a} + \frac{1}{b} - \frac{1}{m} - \frac{1}{n}}\right) \text{ and } U = \frac{p_1}{p_2} \times \exp\left(z\sqrt{\frac{1}{a} + \frac{1}{b} - \frac{1}{m} - \frac{1}{n}}\right). \tag{10.2}$$

Here, as usual, z denotes the appropriate quantile of the standard normal distribution, 1.960 for the usual two-sided interval with 95% nominal coverage. The delta interval is closed form and easily calculated. Because $\sqrt{LU} = \tilde{\theta}$, it is symmetrical on a ratio scale.

Sometimes this procedure results in an apparently reasonable interval. Comparing the incidence of dyskinesia after ropinirole (0.095) and levodopa

(0.258) in Section 10.2, the risk ratio is 0.368, with 95% delta interval 0.207 to 0.652. The interpretation here is well expressed in terms of the relative risk reduction: by using ropinirole instead of levodopa, we expect to reduce dyskinesia by 1 – 0.368 = 0.632 (i.e., by about 63%) in relative terms. Ninety-five percent limits for this relative risk reduction are 35% (derived from 1 – 0.652) and 79% (from 1 – 0.207).

However, the delta method fails to produce an interval when either a or b is zero, because the Gaussian approximation is inadequate: in fact the calculated standard error is infinite. When $a = 0$ and $b > 0$, $\tilde{\theta} = 0$. In this situation, a good method should produce a lower limit at zero, and a finite upper limit. If we interchange the roles of the two groups, b becomes 0 and a becomes positive. The point estimates and confidence limits should be the reciprocals of those for the case $a = 0$ with $b > 0$. So here, $\tilde{\theta}$ should be reported as ∞. A good method should report a finite lower limit and an infinite upper "limit." The delta method is not capable of producing such intervals in boundary cases. These properties are explored further in Section 13.4. Miettinen and Nurminen (1985) pointed out that even when $p_1 = p_2 = 1$, a zero width interval at 1 is produced, unnecessarily, and showed that the coverage of the delta interval is poor.

10.4.2 Miettinen–Nurminen Score Interval

Miettinen and Nurminen (1985) developed closely interrelated score intervals for the difference of proportions (Sections 7.4.1 and 7.5.1), the ratio of proportions and the odds ratio (Section 11.3.2). These intervals may be regarded as natural extensions of the Wilson score interval to measures comparing two independent proportions. The Miettinen–Nurminen score interval for the ratio of two proportions comprises all θ satisfying

$$|p_1 - \theta p_2| \le z\sqrt{V} \tag{10.3}$$

where

$$V = \lambda\left\{\frac{r_1(1-r_1)}{m} + \theta^2\frac{r_2(1-r_2)}{n}\right\} \tag{10.4}$$

Here, the quantities r_1 and r_2 are constrained estimates of π_1 and π_2:

$$r_1 = \frac{B - \sqrt{B^2 - 4\theta(m+n)(a+b)}}{2(m+n)} \tag{10.5}$$

and $r_2 = r_1/\theta$, with $B = (m + b)\theta + a + n$ and $\lambda = \dfrac{m+n}{m+n-1}$.

A slightly different score interval was formulated by Koopman (1984).

The three Miettinen–Nurminen intervals are not of closed form as such. Nonetheless lower and upper limits may be calculated by a single-depth iterative process, consequently these methods can be implemented in spreadsheet software. Use of 40 interval bisections estimates a parameter with bounded support such as [0, 1] or [–1, 1] to ample precision. However, ratio scale parameters usually range from 0 to ∞. So it is convenient to rescale such parameters to have a finite range, preferably from –1 to +1. Log transformation would result in a range from –∞ to ∞, which is even worse. For the odds ratio (Section 11.7), rescaling to $Q_{OR} = (OR - 1)/(OR + 1)$ (Yule 1900) does the trick. For the ratio of proportions, we use an analogous measure $Q_{RR} = (RR - 1)/(RR + 1)$. Both Q_{OR} and Q_{RR} range from –1 to +1.

Miettinen and Nurminen (1985) showed some results for left and right non-coverage for the ratio of proportions as well as the difference of proportions, although not for the odds ratio. They evaluated their proposed method alongside the delta and profile likelihood intervals. The evaluations involved generating 10,000 random 2 × 2 tables for 15 parameter space points, seven with $\pi_1 = \pi_2$ and eight with $\pi_1 > \pi_2$. When $\pi_1 = \pi_2$, the usual distinction between mesial and distal non-coverage is inappropriate; only mesial non-coverage can occur. For these PSPs, left and right non-coverage were balanced when $n_1 = n_2$, of course, but could be unequal when $n_1 \neq n_2$. For those PSPs with $\pi_1 > \pi_2$, the left (mesial) non-coverage of the Miettinen–Nurminen interval for the ratio was usually but not invariably lower than the right (distal) non-coverage, suggesting that the interval usually errs towards being too mesial. The Miettinen–Nurminen interval for the ratio gave excellent mean coverage, ranging from 95.0% to 96.5%. Coverage was slightly more conservative than the corresponding mean coverage for the difference, ranging from 94.7% to 96.1%; conversely, the likelihood interval was more anticonservative for the ratio than for the difference. As a consequence of the very favourable performance of the Miettinen–Nurminen interval here, this is the recommended interval corresponding to reporting the simple, unshrunk empirical estimate p_1/p_2, and also by extension, for the odds ratio.

The 95% Miettinen–Nurminen interval for the dyskinesia example is from 0.209 to 0.649, very similar to the delta interval. Once again, we would report a relative risk reduction of 63%, with 95% interval 35% to 79%.

Of course, there are situations in which the delta and Miettinen–Nurminen intervals are very different, or where the delta interval cannot be calculated. In the Eastern Co-operative Oncology Group study (Section 7.2), 14 subjects were randomised to treatment A and 11 to treatment B. For the primary outcome, tumour shrinkage by at least 50%, no successes were observed on either treatment. In this situation, the empirical estimates of both θ (if construed as the ratio of success rates, 0/14 ÷ 0/11) and ω are undefined, and neither delta nor Miettinen–Nurminen intervals can be computed. For the ratio of failure rates, 14/14 ÷ 11/11, the delta interval cannot be computed, but the Miettinen–Nurminen interval is 0.778 to 1.364.

Furthermore, the estimated relative risk for life-threatening treatment toxicity on treatment A compared to treatment B was 2/14 ÷ 1/11 or 1.57. The delta and Miettinen–Nurminen 95% intervals are 0.22 to 11.7 and 0.16 to 15.2. While both intervals correctly indicate that the data are consistent with a large risk ratio favouring either treatment, they are less similar than for the dyskinesia example above.

Goodfield et al. (1992) presented results of a randomised placebo-controlled trial evaluating the effect of the antimycotic agent terbinafine on toenail and fingernail infections. While there was a clear difference in efficacy in favour of the active treatment, respiratory adverse effects were noted in five (9%) of 56 patients on terbinafine, but none of the 29 who took the placebo. Various confidence intervals for the difference between these proportions appear in column (d) of Table 7.2. Here, we consider the corresponding ratios. We would normally express the risk of adverse effects on the active treatment relative to that on placebo, leading to a relative risk of (5/56) ÷ (0/29) which is infinite. The Miettinen–Nurminen method yields a finite lower limit here, 0.72, so we would report a 95% interval from 0.72 to ∞. Equivalently, we could report a zero relative risk on placebo, compared to that on active treatment, with 95% interval from 0 to 1.39. The delta method fails to produce an interval in both instances.

10.5 Infinite Bias in Ratio Estimates

When we seek to make inferences about one proportion, π_1 relative to another, π_2, it is normally expected that these two parameters will have possible values satisfying $0 < \pi_i < 1$ for $i = 1$ and 2. Here, the two inequalities are strict—we do not anticipate that the true, population value of either parameter would actually be zero. Indeed, it is taken for granted that both either $a = 0$ or $b = 0$ can occur. A study of what range of possibilities can occur, as opposed to what usually happens, would normally be framed in the qualitative paradigm, not the quantitative one, with no expectation of estimating proportions.

The constraint that π_2 is strictly less than 1 has a disquieting consequence. The probability of observing $b = 0$, while usually very small, is not zero. This is no problem for the RD, but has serious repercussions for the RR and the OR. Whenever $b = 0$, the observed ratio $\tilde{\theta} = \frac{p_1}{p_2}$ is infinite—provided of course that a is not zero, but so long as $\pi_1 > 0$, a will generally not be zero. The event $\tilde{\theta} = \infty$ occurs with finite probability. Consequently the expectation of $\tilde{\theta}$ will be infinite. This means that $\tilde{\theta}$ is a seriously biased estimate of the parameter θ—the degree of bias is infinite. The same applies to the odds ratio ω, which becomes infinite whenever either b or c is zero.

In such situations, exemplified by the ECOG trial data, we need to consider not only what confidence interval methods are appropriate, but also what point estimates are sensible. It is desirable to use alternative, less biased estimates for θ or ω, so as to yield finite values even when $p_2 = 0$. Infinite or very high values of θ or ω should be shrunk in a mesial direction, towards 1. Logically, this should be done irrespective of whether a zero cell occurs or not, otherwise we are using a hybrid method, with unknown coverage properties and the potential for unsmooth or even non-monotonic behaviour when the data are altered minimally. Confidence limits should then correspond to the shrunk point estimate—otherwise the upper limit, at least, should also be infinite.

10.6 Intervals Based on Mesially Shrunk Estimated Risks

For the RR and the OR, several approaches are available which are based on mesially shifted estimates for the two proportions, and thereby avoid any possibility of dividing by zero. Agresti and Min (2005a) considered Bayesian intervals for the RR (after Koopman 1984) and the OR. As usual, tail-based intervals are preferable on grounds of invariance, although of course HPD intervals are shorter. Coverage of tail-based intervals for RD, RR and OR with B(ψ,ψ) priors for ψ = 0.5, 1, 1.5 and 2 and logit-normal priors was evaluated. For RD, the mean CP is exactly 0.95 for ψ = 1, lower for other values, although the Jeffreys prior is better for a distance criterion |CP-0.95| and the proportion of PSPs with coverage below 0.93. Use of more informative priors usually results in a great reduction in coverage. Agresti and Min found that the score interval formulated by Mee (1984) performs at least as well in terms of coverage, at expense of being slightly wider. Likewise for the RR and OR, the Jeffreys prior led to the best coverage properties, with mean CP close to 0.95 for several pseudo-priors. Again, the score interval performed very well. Agresti and Min also discussed use of joint priors for the two proportions. They concluded by recommending that the main reason for choosing Bayes intervals here was in order to produce shrinkage in a mesial direction, which several frequentist methods also do.

Price and Bonett (2008) developed an easily computed approximation to the Bayes interval with excellent performance. Specifically, in the notation of Table 10.1, upper and lower limits are given by

$$\frac{p_1}{p_2} \times \exp\left(\pm z \sqrt{\left[a+\beta_1-1+\frac{(a+\beta_1-1)^2}{c+\beta_2-1} \right]^{-1} + \left[b+\beta_1-1+\frac{(b+\beta_1-1)^2}{d+\beta_2-1} \right]^{-1}} \right) \tag{10.6}$$

Performance was found to be optimised by the choices $\beta_1 = 1.25$ and $\beta_2 = 2.5$.

Pseudo-frequency methods also achieve mesial shift and thereby avoid boundary anomalies. To estimate the odds ratio in tables with zero cells, Haldane (1955) and Gart and Zweifel (1967) suggested adding $\psi = 0.5$ to each cell. The RevMan software for meta-analysis developed by the Cochrane Collaboration adds 1/2 to each cell whenever point or interval estimates for RR or OR would not otherwise be computable (Higgins and Green 2008, Section 9.2.2.2). An approach of this nature is essential in this context, as it would be very unsatisfactory for the algorithm to fail whenever a zero cell frequency occurs.

But it is not essential for all cells to be incremented by the same quantity. Jewell (1986) proposed nearly unbiased estimators $\theta^* = \frac{a/m}{(b+1)/(n+1)}$ for the RR and $\omega^* = \frac{ad}{(b+1)(c+1)}$ for the OR, and gave variance estimators designed for the small sample case.

A deterministic bootstrap interval for the risk ratio based on median unbiased estimates is described in Section 11.4, alongside similar intervals for the odds ratio and risk difference.

10.7 A Ratio of Proportions Based on Paired Data

We have seen how, in a parallel-groups study, it is natural to seek to characterise the contrast between the two groups by ratio as well as difference measures. This applies equally to studies with individual-level pairing, such as longitudinal studies in which the same group of individuals are assessed for the presence or absence of a binary characteristic on two occasions.

For the comparison of two proportions based on paired data, we revert to the notation of Table 8.1. The parameter of interest is the ratio of the marginal proportions, $\lambda = \frac{\mu_1}{\mu_2}$ where μ_1 and μ_2 denote the marginal proportions, $\mu_1 = \pi_1 + \pi_2$ and $\mu_2 = \pi_1 + \pi_3$. The obvious empirical estimate is the ratio of the marginal frequencies, $\tilde{\lambda} = \frac{a+b}{a+c}$.

Desu and Raghavarao (2004) describe asymptotic limits

$$\tilde{\lambda} \times \exp\left[\pm z\sqrt{\frac{b+c}{(a+b)(a+c)}}\right] \tag{10.7}$$

which may be derived by back-transforming Wald intervals for ln λ.

This interval may be regarded as combining Wald intervals for μ_1 and μ_2 in a particular way. Bonett and Price (2006) showed that a similar process may be used to combine Wilson intervals for μ_1 and μ_2, leading to a Wilson-like boundary-respecting interval delimited by

$$\frac{2(a+b)+\zeta^2 \pm \zeta\sqrt{\zeta^2+4(a+b)\left(1-\dfrac{a+b}{a+b+c}\right)}}{2(a+c)+\zeta^2 \mp \zeta\sqrt{\zeta^2+4(a+c)\left(1-\dfrac{a+c}{a+b+c}\right)}} \tag{10.8}$$

where

$$\zeta = z\frac{\sqrt{(b+c+2)/\{(a+b+1)(a+c+1)\}}}{\sqrt{1/(a+b+1)-1/(a+b+c+2)}+\sqrt{1/(a+c+1)-1/(a+b+c+2)}}. \tag{10.9}$$

These expressions are of closed form, consequently implementation is straightforward. Both the Desu-Raghavarao and Bonett-Price intervals are symmetrical about $\tilde{\lambda}$ on a ratio scale. Bonett and Price show how, for tables with zero cells, the intervals reduce to ones derivable directly from Wilson score intervals.

An alternative boundary-respecting interval is described by Nam and Blackwelder (2002). This involves solving a quartic equation, which may be done iteratively. Nam (2009) gave closed-form expressions to calculate the resulting limits. An infrequent special case with $a = 0$ requires a different formulation; nevertheless this is a single, unified method regardless of the cell frequencies. Shi and Bai (2008) describe a Bayesian interval.

10.7.1 Estimation of False-Negative Rates in Sentinel Node Biopsy

In Section 7.7 we encountered the ALMANAC study (Mansel et al. 2006), a randomised clinical trial in several centres in the United Kingdom, which yielded definitive evidence establishing the benefits of sentinel node biopsy in the staging of clinically early breast cancer. In the New Start programme (Goyal et al. 2009) this technology was then rolled out in a carefully controlled manner to a large number of centres in the United Kingdom. In each series, both radioisotope and blue dye were used as sentinel node localising agents. Table 10.3 shows results on localisation for 1730 node-positive cases in the New Start series with unequivocal information on the method of localisation. For this series, we can identify how many cases would have been false negatives or failed localisations had only one localising medium been used.

The interpretation of the results is tricky, because of the complex relationship between false-negative status and failed localisation. In the upper panel, the top left-hand cell is a structural zero, because the analysis is restricted to cases with unequivocal information on localisation method.

TABLE 10.3

Detection of Axillary Positivity in 1730 Node-Positive Breast Cancers by Blue Dye and Radioisotope in the New Start Series

	Blue Status, All Nodes			
	No Blue Nodes	Blue Nodes, All Negative	One or More Blue and Positive	Total
Hot Status, All Nodes				
No Hot Nodes	0	7	75	82
Hot Nodes, All Negative	11	140	42	193
One or More Hot and Positive	120	38	1297	1455
Total	131	185	1414	1730

False-Negative Rate	Rate	As %	95% Limits[a]	Risk Ratio	95% Limits[b]	Difference	95% Limits[c]
Actual	158/1730	9.1%	7.9%, 10.6%	1		0	
If only isotope used	193/1730	11.2%	9.8%, 12.7%	1.22	1.14, 1.33	2.0%	1.3%, 2.9%
If only dye used	185/1730	10.7%	9.3%, 12.2%	1.17	1.09, 1.28	1.6%	0.8%, 2.4%

Source: Data from Goyal, A. et al. 2009. Results of the UK NEW START sentinel node biopsy training program: A model for future surgical training. 32nd Annual San Antonio Breast Cancer Symposium. *Cancer Research* 69: 538S.

[a] Wilson (1927).

[b] Nam and Blackwelder (2002); Nam (2009).

[c] Tango (1998, 1999).

Using both media resulted in a total of 7 + 11 + 140 = 158 false negatives. If isotope alone were used, we would only identify 1455 cases as sentinel node positive. There would then be 193 false negatives. The remaining 82 cases are not false negatives but are additional failed localisations. Since these would proceed to axillary clearance, they are of less concern than the false negatives which would be undertreated relative to current standards of care. Similarly, if blue dye alone were used, 1414 cases would be identified as sentinel node positive, with 185 false negatives and 131 additional failed localisations.

The lower panel of Table 10.3 gives analyses comparing the actual false negative rate when both localising agents are used with the (counterfactual) false negative rate when a single localising medium is used. The set of 158 actual false negatives is not a subset of the set of 193 or 185 false negatives with a single medium; there is partial overlap. Thus methods for paired differences or ratios of proportions are indicated here.

10.8 A Ratio of Sizes of Overlapping Groups

Bonett and Price (2006) noted that the widths of the Desu–Raghavarao and Nam–Blackwelder intervals have no contribution from either the table total (n) or the double negative (d) cell of Table 8.1, but depend only on the three cells a, b and c. This property applies equally to $\tilde{\lambda}$ and the Bonett–Price limits. The corollary is that these methods are also applicable to the marginal ratio in a paired 2 × 2 table in which the double negative d cell is either structurally zero or unobservable (Newcombe 2007b). This ratio may then be regarded as the ratio of the sizes of two overlapping groups in a simple Venn diagram, such as Figure 10.1. While all published work in this area has been conceptualised as relating to a marginal ratio in a paired 2 × 2 table, the next sections describe applications to the comparison of two overlapping groups.

10.8.1 Police Under-Recording of Assaults

Only a limited proportion of assault victims presenting at hospital accident and emergency (A&E) departments are reported in police records. The most satisfactory appraisal of the relative numbers of assaults contributing to A&E caseloads and police records comes from series in which records from both sources for the same time period are ascertained and matched systematically. In most series the interpretability of the results is seriously compromised by the fact that A&E catchment areas are not clearly definable and are generally not coterminous with the geographical areas used to record police data. The results presented in Table 10.4 and Figure 10.1 relate to a study of assaults in Funen, Denmark during the period 1991–2002 (Faergemann 2006).

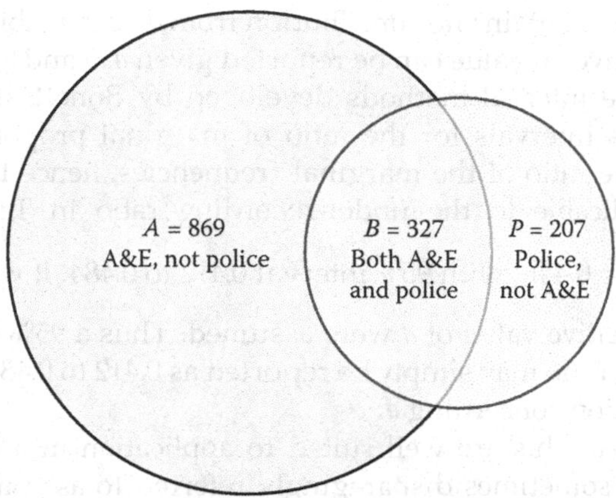

FIGURE 10.1
Venn diagram illustrating the ratio of the sizes of two overlapping groups. Assaults in Funen Island. Comparison of numbers recorded by accident and emergency services and numbers in police records. The number of assaults that are not reported to either agency is poorly defined and unidentifiable, nevertheless an unequivocal confidence interval for the underreporting ratio $(B + P)/(B + A)$ may be calculated. (Data from Faergemann, C. 2006. Interpersonal violence in the Odense municipality, Denmark 1991–2002. Ph.D. thesis, Odense, University of Southern Denmark.) A&E = accident and emergency.

Because Funen is an island, the key assumption of coterminosity may be regarded as reasonable.

The degree of under-recording in police figures compared to A&E departments may be characterised by the ratio $(a + b)/(a + c)$ which is 0.446 here. The fourth cell of the table, d, is the number of assaults missed by both methods of ascertainment. Unless a third source of data is available, it is unobservable. It is displayed as zero here, but is expected to be positive, as the more minor assaults are unlikely to contribute to either series—indeed d could well be larger than $a + b + c$ in this context. The under-recording ratio as defined

TABLE 10.4
Assaults in Funen Island, Denmark, 1991–2002

	Known to A&E Departments?		
	Yes	**No**	**Total**
Known to Police?			
Yes	$a = 327$	$b = 207$	534
No	$c = 869$	$d = 0$	869
Total	1196	207	$n = 1403$

Source: Data from Faergemann, C. 2006. Interpersonal violence in the Odense municipality, Denmark 1991–2002. Ph.D. thesis, Odense, University of Southern Denmark.

Note: Comparison of numbers of assaults recorded by hospital accident and emergency services and numbers in police records.

here, $(a + b)/(a + c)$, contains no contribution from d or the table total n, which is why an unequivocal value can be reported given a, b and c but not d.

The confidence interval methods developed by Bonett and Price (2006) are presented as intervals for the ratio of marginal proportions, but this is identical to the ratio of the marginal frequencies, hence the same methodology is applicable to the under-recording ratio in Table 10.4. Here, $\tilde{\lambda} = \frac{a+b}{a+c} = \frac{534}{1196} = 0.446$, with 95% interval 0.412 to 0.484. It would be exactly the same if a positive value of d were assumed. Thus a 95% interval for the under-recording ratio may simply be reported as 0.412 to 0.484 without making any assumption concerning d.

Methods such as this are well suited to application in a field of events-related research sometimes disparagingly referred to as "numerator epidemiology." Notwithstanding the correct emphasis that statistical analyses should be linked to a clearly defined population, in real-world applications often 'numbers at risk' are not clearly definable. An example involving a quite different statistical analysis relates to deliberate grass fires in Section 14.13.

10.8.2 Validation of a Cough Monitoring System

A major clinical feature of idiopathic pulmonary fibrosis is frequent coughing, with little preventive treatment available. A study by Kilduff (2009) aims to validate a system for monitoring coughing frequency in affected patients. The Leicester Cough Monitor detects movement, and is designed to be very specific to coughing movements. A European Union directive recommends establishing the sensitivity and specificity of a monitoring system relative to the number of coughs actually recorded by the patient in a period of at least 1 hour. In the resulting paired 2×2 table, the d cell counting coughs detected by neither the patient nor the device would not be identifiable; consequently strictly the sensitivity and specificity also cannot be identified. Here, unlike in the police under-reporting example, it is likely that few coughing episodes would be missed by both the patient and the device. Nevertheless, here too, the marginal ratio would be the natural measure to quantify the appropriateness of the device for monitoring coughing frequency in these patients. Naturally, several patients should be studied, not just one; moreover it would be preferable to use two or more monitoring periods per patient. Both aspects of the study would result in multi-level data.

10.9 A Ratio of Two Rates

Closely related to the ratio of two proportions is the ratio of two rates. For example, in Section 6.1.2 we considered breast cancer incidence in two groups of women (Rothman and Greenland 1998). Among women who underwent X-ray fluoroscopy during treatment for tuberculosis, $a = 41$ cases of breast cancer developed in $M = 28{,}010$ person-years, a rate of 1.46 per 1000 person-years. In a control series comprising women who had not had X-ray fluoroscopy, there were $b = 15$ cancers in $N = 19{,}017$ person-years, a rate of 0.79 per 1000 person-years. Here, a and b denote the numerators, which have the same interpretation as in Table 10.1. But the denominators M and N are displayed in upper case, in recognition that the interpretation is not the same as for the denominators in the binomial case. The 95% intervals for these two rates separately, based on the Poisson distribution and using the mid-p method as in Section 6.2, are 1.05 to 1.99 and 0.44 to 1.30 per 1000 person-years.

In this situation, interest naturally centres on the ratio of the two rates. The empirical estimate is simply calculated by division. For this example, the observed rate ratio is $\dfrac{a/M}{b/N} = \dfrac{1.46}{0.79} = 1.856$. Several confidence intervals are available. The delta limits are based on the asymptotic standard error of the log rate ratio in the usual way:

$$\frac{a/M}{b/N} \times \exp\left(\pm z\sqrt{\frac{1}{a} + \frac{1}{b}}\right) \tag{10.10}$$

For the X-ray fluoroscopy example, the delta interval is 1.027 to 3.353.

A conditional interval may be obtained by calculating limits for $a/(a + b)$ using a suitable method from Chapter 3, turning these into limits for a/b by substitution, then dividing these by M/N. For the X-ray fluoroscopy example, the 95% Wilson limits for 41/56 are 0.604 and 0.830. These transform into limits $0.604/(1 - 0.604) = 1.526$ and $0.830/(1 - 0.830) = 4.897$ for a/b. Dividing by $M/N = 1.473$ leads to the interval (1.036, 3.325) for the rate ratio. As a consequence of the logit-scale symmetry of the Wilson interval, this is symmetrical on a log scale about the empirical estimate 1.856. Observation of the log-scale symmetry property of the Wilson conditional interval for the rate ratio led to the discovery, 67 years after Wilson's publication, that the score interval for the binomial proportion is symmetrical on the logit scale—with several repercussions including the relationship between score and delta logit intervals for both proportions and rate ratios as described below, and the development of the Box–Cox index of symmetry in Section 4.4.3.

Alternatively, the MOVER-R algorithm (Section 7.3.4) may be used to combine suitable intervals for the two Poisson rates, a/M and b/N. For this example, the resulting interval based on mid-p Poisson intervals for the two numerators is 1.041 to 3.246.

In principle, a still better approach is to use a purpose-built interval for the rate ratio. Graham et al. (2003) describe a score interval. In contrast to the score intervals for the difference and ratio of two proportions and the odds ratio described by Miettinen and Nurminen (1985), this has the great advantage that it is of closed form: the upper and lower limits are simply

$$\frac{N}{2Mb^2}\left\{2ab+z^2(a+b)\pm z\sqrt{(a+b)\left(4ab+z^2(a+b)\right)}\right\} \tag{10.11}$$

Graham et al. suggested adding a pseudo-frequency 0.5 whenever $b = 0$.

Although it is not apparent from the formulae, there is in fact a very close relationship between the Graham and delta intervals for the rate ratio. This relationship is closely analogous to a similar relationship between two intervals for the binomial parameter π, the Wilson score interval and the delta logit interval introduced in Section 3.4.2 (Newcombe 2001a). In the notation of Chapter 3, the delta logit interval for a binomial proportion π estimated by $p = r/n$ is obtained by calculating limits for logit π as logit $p \pm h$ where

$$h = z\sqrt{\frac{1}{r}+\frac{1}{(n-r)}} \tag{10.12}$$

then transforming back to the scale of π. Like the Wilson and other good intervals, the delta logit interval cannot produce limits for π outside the interval [0, 1]. The delta logit interval for π is a considerable improvement on the Wald interval $p \pm z\sqrt{(pq/n)}$, conservative instead of anticonservative, but both methods unnecessarily fail to produce informative bounds in the extreme cases $r = 0$ or $r = n$.

The property common to the Wilson and delta logit intervals for π is symmetry on the logit scale. In response to a letter by Rindskopf (2000), Agresti and Coull (2000) observed that for an extensive series of possible non-extreme outcomes, the delta logit interval always contained the Wilson interval. Newcombe (2001a) demonstrated that this property always holds. In fact, the quarter-widths of these two intervals for the binomial parameter π on the logit scale are related by an inverse sinh transformation: the Wilson limits for logit π can be expressed as

$$\text{logit}\ p \pm 2\sinh^{-1}\left(\frac{1}{2}h\right) \tag{10.13}$$

Here, as usual, $\sinh^{-1}$ stands for the inverse hyperbolic sine function, $\sinh^{-1}(x) = \ln(x + \sqrt{(x^2 + 1)})$. Because $\sinh^{-1}(x) < x$ for all $x > 0$, this property implies that the Wilson interval is always contained strictly within the delta logit interval.

Exactly the same inverse sinh relationship links the quarter-widths on the log scale of the two corresponding intervals for the Poisson parameter, the inversion interval (Section 6.2) and the delta log interval delimited by $\exp(\ln r \pm z/\sqrt{r})$.

The delta intervals for the ratio of two proportions (Section 10.4.1) and the odds ratio (Section 11.3.1) take a similar form to the delta logit interval for π: in each case, the variance on the transformed scale involves the sum of the reciprocals of cell frequencies. Newcombe (2001a) suggested that intervals obtained in an analogous way for the ratio of proportions and the odds ratio, derived from the delta intervals by applying a similar inverse sinh transformation to the quarter-width on a log scale, would be worthy of consideration as closed-form methods. For a table with a single zero cell entry, the non-extreme limit is simply calculated by substituting z^2 for the zero.

Exactly the same property links the Graham score interval for the ratio of two rates and the corresponding delta interval: the Graham interval is always contained within the delta interval, and when a or b is zero, the non-extreme limit is simply calculated by substituting z^2 for the zero (Newcombe 2003b). The Graham limits are then conveniently programmed as

$$\frac{a/M}{b/N} \times \exp\left(\pm 2 \sinh^{-1}\left(\frac{z}{2}\sqrt{\frac{1}{a}+\frac{1}{b}} \right) \right). \tag{10.14}$$

For the X-ray fluoroscopy data, the Graham interval is 1.036 to 3.325. This is slightly narrower than the delta interval, as expected. The Graham and conditional Wilson intervals are in fact identical, as is apparent by comparing Equation 10.14 with Equations 10.12 and 10.13. This indicates that for the rate ratio, unlike the paired difference of proportions (Section 8.2), nothing is lost by a conditional approach.

The suggestion that inverse sinh transformation might improve the delta interval for the ratio of two proportions was explored by Price and Bonett (2008). Once again, when either numerator is zero, the non-extreme limit is simply calculated by substituting z^2 for the zero. The resulting interval was superior to the delta interval but was less good than the Koopman (1984) score interval and the non-iterative approximate Bayesian method (Equation 10.6). The same transformation would be expected to improve the Desu–Raghavarao interval for the marginal ratio (Equation 10.7), but to be inferior to the Bonett–Price and Nam–Blackwelder intervals.

10.10 Implementation in MS Excel

As noted in Section 10.4.2, Miettinen–Nurminen intervals for the ratio of proportions and the odds ratio are not of closed form; nevertheless implementation is practicable using a fixed number of interval bisections, as in the spreadsheet ODDSRATIOANDRR.xls. This spreadsheet also calculates the closed-form Graham interval for the rate ratio, using Equation 10.14, heeding the special cases with a or b zero.

Bonett and Price give Gauss code to implement their closed-form interval. This is readily translated into other languages. The spreadsheet BONETTPRICE.xls calculates a confidence interval for the ratio of marginals, λ starting from the a, b and c cell frequencies, or equivalently for the ratio of sizes of overlapping groups, $(a + b)/(a + c)$.

11

The Odds Ratio and Logistic Regression

11.1 The Rationale for the Odds Ratio

The odds ratio is widely used in health-related research as an alternative measure of relative risk. It has several advantages and several disadvantages compared to the more intuitive ratio of proportions, including two novel drawbacks explained in Section 11.2.

In the notation of Table 10.1, the odds ratio for the standard unpaired 2 × 2 table is defined as $\omega = \dfrac{\pi_1/(1-\pi_1)}{\pi_2/(1-\pi_2)}$. The usual empirical estimate $\tilde{\omega} = \dfrac{p_1/(1-p_1)}{p_2/(1-p_2)} = \dfrac{a/c}{b/d}$ is often expressed as the cross-product odds ratio, ad/bc. This formulation has three advantages: it is more convenient, more memorable and can still be calculated if $d = 0$.

The odds ratio comparing two small proportions approximates closely to the relative risk, and may then be regarded as a proxy for it. Why, then, is the odds ratio so often used, rather than the arithmetically and conceptually simpler relative risk? There are four arguments favouring the odds ratio as a measure of effect size for the 2 × 2 table.

1. A particular value for the odds ratio is always meaningful, irrespective of the baseline level of risk. In contrast, analyses involving the RD or RR may encounter ceiling effects. Thus, starting from a baseline risk of 0.4, say, an absolute increase by 0.7 or multiplication by a factor of 3 is unachievable. Even when this issue does not apply, it is often found that odds ratios are similar across subgroups of a dataset that have quite different inherent levels of risk.
2. The odds ratio is the natural measure in a study such as a retrospective case-control study, which aims to elucidate disease aetiology by comparing exposure between a group of cases who have the disease and a control series who do not exhibit the disease. It is important to ensure that the controls are as similar as possible to the cases in relation to potential confounders. With this in mind, the controls

may or may not be matched individually to the cases. In both situations, sampling is not naturalistic; consequently neither the risks in the exposed and unexposed groups nor the RR and RD are directly estimable.

3. The odds ratio arises naturally in logistic regression (Section 11.5): its natural logarithm is identical to the regression coefficient, and both lead to equivalent simple asymptotic intervals.
4. When comparing proportions that are not small, it may be arbitrary whether we take the ratio of proportions who respond "yes" or the ratio of proportions who respond "no"; the OR gets around this issue neatly by incorporating contributions from both. The relative risk classes two proportions of 99.9% and 99% as practically identical, whereas the corresponding odds differ by a factor of practically 10.

Figure 11.1 gives some results for four binary outcomes from several randomised trials of patient decision support interventions (PDSIs). Data were extracted from a Cochrane review (O'Connor et al. 2009) and re-analysed (Joseph-Williams et al. 2012) to demonstrate the degree of correlation between efficacy as shown by randomised trials and PDSI quality as assessed by the IPDASi instrument (Elwyn et al. 2009). The meta-analyses included in the Cochrane review were based on the relative risk. In the re-analysis it was considered that a relative risk criterion was less appropriate here than using the odds ratio. Many of the proportions considered were in the middle of the range: one could equally well use the ratio of proportions with a positive response, or the ratio of proportions with a negative response. Neither of these options is totally satisfactory here, as the two approaches could well lead to quite different conclusions (Deeks et al. 2008). The odds ratio obviates the ambiguity neatly by incorporating contributions from both these ratios.

In meta-analyses using the customary Cochrane tools, methods based on the RD, RR and OR are all available, as well as choice between fixed- and random-effects methods, but the RR makes sense only as a measure for comparing low proportions, except in the special application to derive confidence intervals for the projected positive and negative predictive values (Section 14.9). Deeks et al. suggest that in the meta-analysis context it is often sensible to estimate a summary odds ratio or RR pooling information from the several sources available, then convert the resulting point and interval estimates to absolute measures such as the RD or NNT using an assumed baseline rate. If the latter is a rounded figure based informally on existing evidence, intervals for the RD or NNT are readily obtained by substitution. However, if an original baseline rate estimate is available, based on a clearly defined sample size, it is preferable to take account of the sampling imprecision of both the relative risk measure and the baseline rate. In some

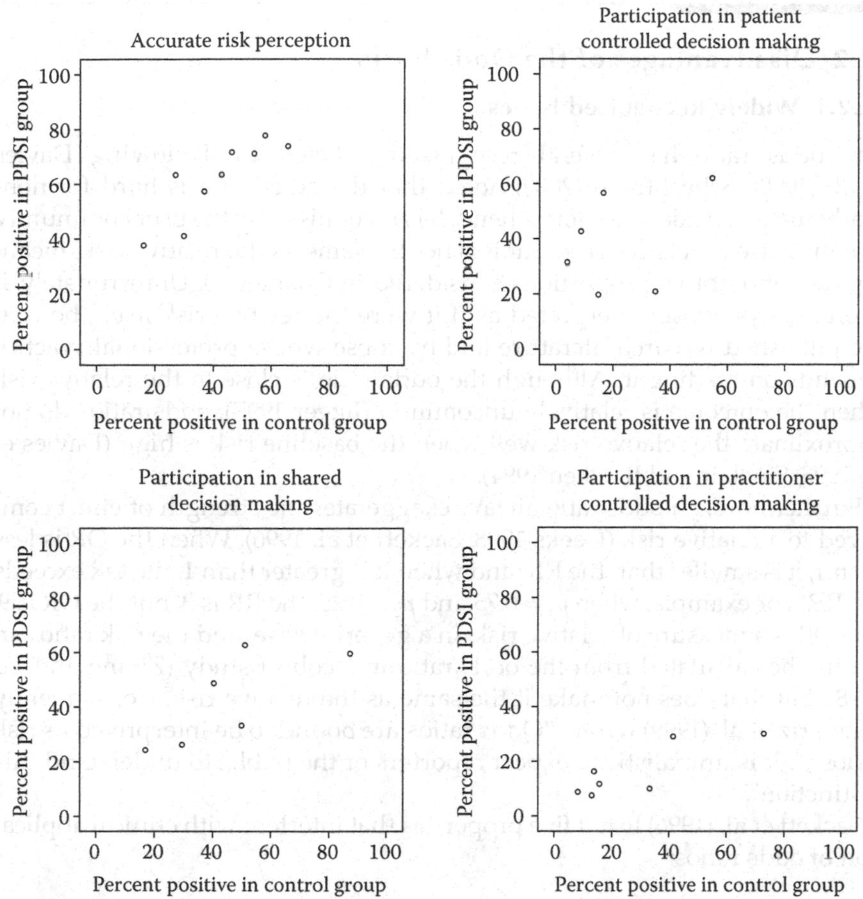

FIGURE 11.1
Some results from several randomised trials of patient decision support interventions, in relation to four binary outcome criteria. (Data from O'Connor, A. M., Bennett, C. L., Stacey, D. et al. 2009. Decision aids for people facing health treatment or screening decisions (Review). *The Cochrane Library*, Issue 3; and Joseph-Williams, N., Newcombe, R. G., Politi, M. et al. In preparation, 2012. Toward minimum standards for the certification of patient decision aids: A correlation analysis and modified Delphi consensus process. In submission to *Journal of Clinical Epidemiology*.) Each plotted point represents the proportions of consultations regarded as positive for the criterion in question in the intervention and control groups of a single trial. Many of these proportions are middling or high, indicating that the odds ratio is preferable to the relative risk as a measure of effect size here.

related situations this may be implemented using MOVER, as in Section 14.4. But intervals for the RD or NNT to incorporate the uncertainty of both baseline risk and OR cannot be obtained in this way. This is the application that prompted the development of the PropImp algorithm described in Section 13.2.

11.2 Disadvantages of the Odds Ratio

11.2.1 Widely Recognised Issues

The odds ratio has several recognised deficiencies. Following Davies et al. (1998), Schechtman (2002) noted that the odds ratio is hard for non-mathematical readers to comprehend. It is recognised by the user community as *a* measure of relative risk. But it is not *the* same as the relative risk, meaning the ratio of two proportions considered in Chapter 10. Unfortunately, it is often erroneously interpreted as if it were the relative risk itself, both in the published research literature and by those whose professional practice depends on reading it. Although the odds ratio is close to the relative risk when the outcome is relatively uncommon (Egger 1997), odds ratios do not approximate the relative risk well when the baseline risk is high (Davies et al. 1998; Sinclair and Bracken 1994).

Furthermore, an odds ratio always exaggerates the strength of effect compared to a relative risk (Deeks 1998; Sackett et al. 1996). When the OR is less than 1, it is smaller than the RR, and when it is greater than 1, the OR exceeds the RR. For example, when $p_1 = 0.75$ and $p_2 = 0.25$, the RR is 3, but the OR is 9. The OR is a measure of relative risk, in a generic sense, and the risk ratio can readily be calculated from the odds ratio in a cohort study (Zhang and Yu 1998), but that does not make it the same as the relative risk. Consequently Schwartz et al. (1999) wrote: "Odds ratios are bound to be interpreted as risk ratios … It is unrealistic to expect reporters or the public to understand this distinction".

Sackett et al. (1996) listed five properties that interfere with clinical application of odds ratios:

1. "Because very few clinicians are facile at dealing with odds and relative odds, odds ratios are not useful in their original form at the bedside or in the examining room."
2. In many studies, odds ratios are very dissimilar to risk ratios.
3. Odds ratios cannot be used in exactly the same way as risk ratios to calculate the rate difference or number needed to treat.
4. Two treatments may yield the same relative risk reduction but different odds ratios, and vice versa.
5. Odds ratio and relative risk reduction may order treatments differently.

King and Zeng (2002) argued similarly: "The key advantage of OR is that it has been easier to estimate than the other quantities … when OR is not the quantity of interest then its 'advantages' are not sufficient to recommend its use … some statisticians seem comfortable with OR as the ultimate quantity of interest, but that is not common ... we have found no author who claims to be more comfortable communicating with the general public using an odds ratio ... (it) has been used 'largely because it serves as a link between results obtained from follow-up studies and those obtainable from case-control studies.'"

The non-equivalence of the RR and OR also has repercussions for the planning stage of a study, and for interpretation of the degree of clinical or epidemiological importance of the findings. Suppose that clinicians or epidemiologists decide that in a particular context they would regard a risk ratio of, say, 2 or 0.5 as representing a "clinically relevant" size of effect. It is incorrect to then regard the same figure as a clinically relevant size of effect on the odds ratio scale also, unless the prevalence is very low.

One of the advantages often claimed for the OR is invariance. Agresti (2002, page 45–46) drew attention to a multiplicative invariance property: on multiplying entries in one row of Table 10.1 by a constant, the OR remains unaltered but the RR does not. Edwards (1966) claimed two properties for the OR, approximate invariance over choice of dichotomy point for two overlapping, normal distributions, and approximate proportionality to the corresponding standardised difference. However, Fleiss (1970) showed that these properties apply more closely to other measures than to the OR itself.

11.2.2 A Novel Paradox

When used to compare chained or conditional probabilities between two independent groups, odds ratios can behave paradoxically, by failing to show a reinforcement effect that is expressed very clearly when risk ratios are used (Newcombe 2006c). This novel yet simple paradoxical effect can arise when odds ratios are used with the unrealistic expectation that they will behave in the same manner as risk ratios. This property is curious only because many non-statisticians conceptualise the odds ratio just as if it was the risk ratio—but this practice is widespread.

The 2002 Welsh Health Behaviour in Schools Survey (Mckeown 2005) is part of a major multinational survey including many self-reported behaviours, including sexual behaviour in year 11 (age 15–16) pupils, in relation to predisposing factors. Respondents were asked, "Have you ever had sex?", and if they replied positively were then asked, "Last time you had sex, was a condom used?". Many explanatory variables were also elicited, including self-assessed school performance.

Comparing reported sexual risk behaviour between lowest and highest school performance groups yielded the results in Table 11.1. These results may validly be summarized by either risk ratios or odds ratios, as in Table 11.2.

Clearly, the risk ratios are simply related by multiplication. The risk ratio for outcome 3, $RR_3 = 4.57$ is identical to the product of those for outcomes 1 and 2, $RR_1 = 3.11$ and $RR_2 = 1.47$, and appropriately indicates a stronger

TABLE 11.1

Self-Reported Sexual Behaviours in Relation to Self-Assessed School Performance

	School Performance	
	Very Good	**Below Average**
Total	212	65
Never had sex	168	23
Ever had sex	44	42
Last sex protected	29	21
Last sex unprotected	15	21

Source: Data from the 2002 Welsh Health Behaviour in Schools Survey (Mckeown 2005).

effect than either of the component risk ratios 3.11 and 1.47. Here, the fact that $RR_3 > RR_1$ is specifically because $RR_2 > 1$.

Considering the corresponding odds ratios, $OR_3 = 6.26$ is not the same as the product of $OR_1 = 6.97$ and $OR_2 = 1.93$, as is obvious from how these figures are derived. What is paradoxical is that the odds ratio for outcome 3, 6.26, is lower than that for outcome 1, 6.97, even though 1.93 is greater than 1. The correct conclusion from the data is that in respect of both having sex at all and using protection if they do so, the high performers engage in less risky behaviour than the low performers. This inference is summarised appropriately by the risk ratios. The odds ratios present a less cogent message, suggesting incorrectly a weaker association of school performance with outcome 3 compared to outcome 1. It is straightforward to demonstrate that for three chained outcome proportions such as those in Table 11.2, when both associations are positive in direction so that OR_1 and OR_2 are both greater than 1,

TABLE 11.2

Risk Ratios and Odds Ratios for Sexual Behaviours Comparing "Below Average" against "Very Good" School Performers

Outcome	**Risk Ratio**	**Odds Ratio**
1. Any sex	$RR_1 = \frac{42}{65} / \frac{44}{212} = 3.11$	$OR_1 = \frac{42}{23} / \frac{44}{168} = 6.97$
2. Latest sex unprotected, given any sex	$RR_2 = \frac{21}{42} / \frac{15}{44} = 1.47$	$OR_2 = \frac{21}{21} / \frac{15}{29} = 1.93$
3. Latest sex unprotected as proportion of total respondents	$RR_3 = \frac{21}{65} / \frac{15}{212} = 4.57$	$OR_3 = \frac{21}{44} / \frac{15}{197} = 6.26$

Source: Data from the 2002 Welsh Health Behaviour in Schools Survey (Mckeown 2005).

the resulting OR_3 lies between $\min(OR_1, OR_2)$ and $OR_1 \times OR_2$, with equality in boundary cases, but it can be either larger or smaller than $\max(OR_1, OR_2)$.

It is easily shown that the paradoxical behaviour described above applies in exactly the same way to the Jewell (1986) OR estimator introduced in Section 10.6, but not to the corresponding RR estimator.

The simplest possible simulation was carried out, in which the four proportions p_i, i = 1,…, 4 were randomly and independently sampled from U[0, 1], but with the constraint RR_1 and $RR_2 > 1$. This confirmed the above property, and indicated that configurations such as the above in which $OR_3 < \max(OR_1, OR_2)$ would frequently occur—indeed, occurred in 5839 of 10,000 runs, with this prior of admittedly limited plausibility.

Thus to the drawbacks of the odds ratio listed in Section 11.2.1 may be added the paradoxical effect demonstrated here. In the above context, there was no particular reason to calculate odds ratios rather than risk ratios, nevertheless in line with common practice the investigator had calculated the odds ratios shown in Table 11.2. The greater clarity of interpretation of the RR here, compared to the OR, has a parallel in that survival curves are based on chained products of proportions, and thus the hazard ratio used in the Cox proportional hazards model is analogous to the RR rather than the OR.

The paradox described here may be regarded as a novel kind of non-transitive dominance paradox, complementary to the well-known Condorcet paradox which applies to the Mann–Whitney test statistic and measures of effect size derived from it (θ = U/mn, the area under the ROC curve, or equivalently Somers' D; see Section 15.2.3), albeit for different reasons. The Condorcet paradox relates to a single variable and pairwise comparisons of three groups, and can lead to either an apparent attenuation of effect size similar to that shown here, or in extreme cases reversal of the expected direction of the difference. The odds ratio paradox relates to three chained proportions compared between the same two groups, and involves the magnitude, but not the direction of the effect.

In conclusion, when applicable, the risk ratio is more transparent than the odds ratio—in any case—and especially where chained or conditional probabilities are involved.

11.2.3 A Second Novel Paradox

There is a quite different situation in which odds ratios, if lifted out of their context, could provide conflicting interpretations of the same data (Newcombe 2008b). The results in Table 11.3 are taken directly from a case-control study by Zhu et al. (2008) which examined the genotype and allele distribution of two IL-1B polymorphisms in nasopharyngeal carcinoma patients and a control series. Results are presented for the polymorphism –511C/T.

What is striking about these figures is that when the data are summarised at allele level, the odds ratio (c) for the T allele relative to the C allele is

TABLE 11.3

Genotype and Allele Distributions of the IL-1B Polymorphism –511C/T in Nasopharyngeal Carcinoma Patients and a Control Series

Genotypes	NPC Patients	%	Controls	%	Odds Ratio (95% CI)	*p*-Value
CC	20	17.7	44	30.6	1 (reference)	–
CT	54	47.8	62	43.1	(a) 1.92 (1.01 to 3.64)	0.046
TT	39	34.5	38	26.4	(b) 2.26 (1.13 to 4.51)	0.020
Total	113	100	144	100		
Alleles						
C	94	41.6	150	52.1	1 (reference)	–
T	132	58.4	138	47.9	(c) 1.53 (1.07 to 2.17)	0.018
Total	226	100	288	100		

Source: Data from Table 1 of Zhu, Y., Xu, Y., Wei, Y. et al. 2008. *Clinical Oncology* 20: 207–211.
Note: NPC = nasopharyngeal carcinoma.

substantially lower than for either the CT (a) or the TT (b) genotype relative to the CC genotype. Which of these should be quoted as the most appropriate measure of association? The authors unfortunately state that, "the –511T allele carriers were associated with a significantly increased risk of nasopharyngeal carcinoma as compared with the non-carriers (odds ratio = 1.53, 95% CI 1.07 – 2.17, $p = 0.018$)"—falling into the trap of interpreting an association at allele level as if it were at patient level.

Consideration of the nature of the contrast between TT and CC genotypes suggests the main explanation: the figure of 1.53 is per allele, so the difference between having two alleles and none would be expected to be the square of this figure. However, this does not explain why odds ratio (a), 1.92 is also considerably higher than (c), 1.53 here—and indeed data may be constructed for which odds ratio (a) is higher than both (b) and (c).

The authors state that the genotype distributions among the controls and cases were in agreement with the Hardy–Weinberg equilibrium. Table 11.4

TABLE 11.4

Genotype and Allele Distributions of the IL-1B Polymorphism-511C/T in Nasopharyngeal Carcinoma Patients and a Control Series

Genotypes	NPC Patients	%	Controls	%	Odds Ratio
CC	19.5	17.3	39.1	27.1	1 (reference)
CT	54.9	48.6	71.9	49.9	1.53
TT	38.5	34.1	33.1	23.0	2.33
Total	113	100	144	100	

Source: Data from Table 1 of Zhu, Y., Xu, Y., Wei, Y. et al. 2008. *Clinical Oncology* 20: 207–211, modified to correspond exactly to the Hardy–Weinberg equilibrium.

shows an analysis at genotype level for a series of 113 cases and 144 controls, with the same allele frequencies as in Table 11.3, but with cells populated using the binomial frequencies expected under HWE. While the genotype distribution in cases was very close to HWE, that for the controls was less close, with overdispersion (i.e., fewer heterozygotes than HWE would predict). These binomial data give odds ratios of 1.526, identical to (c), for genotype CT relative to CC, and $1.526^2 = 2.330$ for TT relative to CC. The overdispersion in controls has driven (a) upwards relative to the HWE value of 1.526, and conversely has lowered (b).

The intervals reported for each odds ratio were evidently calculated by the default delta method (Section 11.3.1), as implemented in SPSS, the package used by the authors. Odds ratios (a) and (b) conform to standard practice for a $k \times 2$ table with $k > 2$, in which CIs are calculated separately for genotype categories CT and TT relative to the reference category, CC. These are valid, of course, subject to the usual caveat that they are not statistically independent of each other— if genotype CC happens to be underestimated in cases, both odds ratios will be higher as a consequence.

What is of greater concern here is the assumption of independence underlying the application of the ordinary CI to odds ratio (c). This assumption is equivalent to HWE. It is always of concern when the sample size in an analysis is greater than the total number of individuals—whether the explanation relates to paired organs, or as here, to alleles. In both situations it is recommended that interval estimates should allow for the potential for non-independence in an appropriate way, as in Chapter 9, using the methods developed in Newcombe (2003a). Based on the Tango (1998) score interval for the paired difference of proportions, 95% intervals for the –511T allele frequency in patients and controls are 0.518 to 0.647 and 0.418 to 0.541. Combining these intervals using MOVER-R (Section 7.3.4) gives a 95% interval 1.060 to 2.198 for the odds ratio at allele level. The correct inference is then that that the odds for disease increase by a factor 1.526 per T allele, with 95% interval 1.060 to 2.198. Squaring all these figures, the fitted risk for the TT genotype is represented by an odds ratio of 2.330 relative to the CC genotype, with 95% interval 1.124 to 4.329. Alternatively, the method of propagating imprecision (PropImp) may be used (Chapter 13), giving a 95% interval 1.061 to 2.196 for the odds ratio at allele level. Both intervals are similar to that reported by Zhu et al. but do not rely on the assumption of HWE.

The *p*-values quoted by Zhu et al. correspond to their confidence intervals in that each compares the data in the row in question against the referent category. Neither *p*-value is ideal; it is preferable to have an analysis of the data as a whole. For the 3×2 table, $X^2 = 5.91$, $p = 0.052$, which includes a linear trend component with $X^2 = 5.08$, $p = 0.024$. All analyses lead to the conclusion of a marginal level of statistical significance. Nevertheless it is striking that the lowest, the most extreme of their three *p*-values corresponds to the lowest, i.e., least extreme odds ratio—no doubt because of the inappropriately inflated sample size on which it is based.

The paradox expounded by Newcombe (2006c) relates specifically to the odds ratio. The relative risk could equally well have been used, and cannot show such paradoxical behaviour. By contrast, in the example above, the odds ratio is the natural effect size measure—because it is a case-control study the relative risk is not applicable. Moreover, even if it were, this would not particularly obviate the problematic behaviour seen here.

11.3 Intervals Corresponding to the Empirical Estimate

11.3.1 Delta Interval

Like the relative risk, the logarithm of the odds ratio has a simply computed standard error which may be used to calculate an asymptotic interval often referenced to Woolf (1955). With the odds ratio ω and its empirical estimate $\tilde{\omega}$ defined as in Section 11.1, the variance of ln ω is estimated from the sample as $\frac{1}{a}+\frac{1}{b}+\frac{1}{c}+\frac{1}{d}$. So the delta limits for ω are $\frac{ad}{bc} \div \exp\left(z\sqrt{\frac{1}{a}+\frac{1}{b}+\frac{1}{c}+\frac{1}{d}}\right)$ and $\frac{ad}{bc} \times \exp\left(z\sqrt{\frac{1}{a}+\frac{1}{b}+\frac{1}{c}+\frac{1}{d}}\right)$. Here, once again, z denotes the appropriate quantile of the standard normal distribution. This closed-form interval is readily calculated. The formula has an extremely appealing symmetry of form. The derivative of the function ln x is $1/x$. Imputing a variance $1/a$ to ln a, and similarly for the other three cells, leads directly to the delta method variance for ln ω and hence the above interval.

Just as for the relative risk, sometimes this procedure results in an apparently reasonable interval. We considered a comparison between the incidence of dyskinesia after ropinirole (0.095) and levodopa (0.258) in a prospective study of patients with early Parkinson's disease (Rascol et al. 2000; Schechtman 2002). Here, the odds ratio is 0.301, with 95% delta interval 0.151 to 0.600. In Chapter 10 we saw how the corresponding risk ratio of 0.368, with 95% delta limits 0.207 and 0.652, may be interpreted in terms of the relative risk reduction. No such intuitive interpretation applies to the odds ratio.

When any of the four cells are zero, the delta method fails to produce an interval for the odds ratio, because the calculated variance is infinite. When $ad = 0$ and $bc > 0$, $\tilde{\omega} = 0$. In this situation, a good method should produce a lower limit at zero, and a finite upper limit. If we interchange the roles of the two groups, or of positive and negative outcomes, bc becomes 0 and ad becomes positive. The point estimates and confidence limits should be the reciprocals of those for the case $ad = 0$ with $bc > 0$. So here, $\tilde{\omega}$ should be reported as ∞. A good method should report a finite lower limit and an infinite upper 'limit.' The delta method cannot produce such intervals in boundary cases. In Section 13.4 we examine two general methods to derive an interval for measures such as the RR and OR that can be expressed in ratio form, MOVER and PropImp. These methods still produce interpretable intervals in some but not all of these situations.

The delta interval for the odds ratio from the standard 2 × 2 table is identical to the usual asymptotic interval obtained from a logistic regression model, described in Section 11.5.

11.3.2 Miettinen–Nurminen Score Interval

The Miettinen–Nurminen (1985) score interval for the odds ratio comprises all ω satisfying

$$|a - mr_1| \le z\sqrt{V}. \tag{11.1}$$

Here, the variance V takes the remarkable form

$$V = \lambda / \left\{ \frac{1}{mr_1(1-r_1)} + \frac{1}{nr_2(1-r_2)} \right\} \tag{11.2}$$

where the quantities r_1 and r_2 are constrained estimates of π_1 and π_2:

$$r_2 = \frac{-B + \sqrt{B^2 - 4n(\omega - 1)(a+b)}}{2n(\omega - 1)} \tag{11.3}$$

and

$$r_1 = \frac{\omega r_2}{1 + r_2(\omega - 1)}, \tag{11.4}$$

with

$$B = m\omega + n - (\mathrm{a} + \mathrm{b})(\omega - 1) \tag{11.5}$$

and

$$\lambda = \frac{m+n}{m+n-1}. \tag{11.6}$$

The 95% Miettinen–Nurminen interval for the dyskinesia example is from 0.152 to 0.596, very similar to the delta interval. But, just as for the relative risk, there are situations in which delta and Miettinen–Nurminen intervals are very different, or where the delta interval cannot be calculated. In Sections 7.2 and 10.4.2 we considered some results from a very small ECOG clinical trial cited by Parzen et al. (2002). The odds ratio comparing the risk of life-threatening treatment toxicity between treatments A and treatment B is 1.67. The delta and Miettinen–Nurminen 95% intervals are 0.18 to 14.9 and 0.13 to 21.2. As before, while both intervals correctly indicate that the data are consistent with a large odds ratio favouring either treatment, they are less similar than for the dyskinesia example above.

For the comparison of the proportions 5/56 versus 0/29 from Goodfield et al. (1992) (Section 10.4.2), the odds ratio expressing the risk of adverse effects on the active treatment relative to that on placebo is ∞, with 95% interval from 0.70 to ∞. Equivalently, we could report a zero odds ratio for the risk on placebo relative to that on active treatment, with 95% interval from 0 to 1.44. Just as for the relative risk, the delta method fails to produce an interval.

Another example with a zero cell comes from a study of the clinical usefulness of surface swabs in comparison to tissue biopsies in clinically non-infected leg wounds (Davies et al. 2007). Table 11.5 shows the association between presence of Gram-positive and Gram-negative aerobes in surface swabs from 66 patients. Remarkably, no patient was negative for both organism types. A 95% Wilson interval for the proportion of cases harbouring neither organism is 0 to 5.5%. Normal practice is to characterise the relationship between two such binary variables by the odds ratio. The delta method cannot be used for this zero odds ratio; the Miettinen–Nurminen 95% interval is 0 to 0.42. Both these analyses, but particularly the latter, emphasise that there is certainly no guarantee that such wounds would always carry at least one type of organism. Even when such an analysis is performed *post hoc*, it is worthwhile to counteract the natural inference that the double-negative combination would seldom if ever occur.

TABLE 11.5

Gram-Positive and Gram-Negative Aerobes in Surface Swabs from 66 Patients with Chronic Leg Wounds

	Gram-Negative Aerobes		
	Present	**Absent**	**Total**
Gram-Positive Aerobes			
Present	29	26	55
Absent	11	0	11
Total	40	26	66

Source: Data from Davies, C. E., Hill, K. E., Newcombe, R. G. et al. 2007. *Wound Repair and Regeneration* 15: 17–22.

11.4 Deterministic Bootstrap Intervals Based on Median Unbiased Estimates

The methods considered in this section use a very different approach based on mesially shifted estimates for the two proportions. They were developed by Stuart Lipsitz and colleagues in three papers relating to the odds ratio (Parzen et al. 2002), the risk difference (Lin et al. 2009) and the relative risk (Carter et al. 2010). The motivation for development of this approach is that finite point estimates and confidence limits can be produced for all these measures, even in situations in which other methods encounter zero division problems. Accordingly, the potential advantage in using these methods relates principally to the two ratio measures, rather than to the risk difference.

It is beyond the scope of this book to describe in detail all the possible applications of bootstrapping to inferences relating to proportions. A deterministic bootstrap procedure as described here has the obvious advantage of being fully reproducible. Deterministic bootstrapping could be used with any estimators of p_1 and p_2, but can only be recommended when some kind of mesially shrunk point estimate is chosen. When empirical estimates are used, behaviour in boundary cases is inappropriate: if the observed proportion is 0 or 1, no resampling procedure is capable of shifting away from that value. The median unbiased estimate (MUE) is preferable, being a mesially shrunk estimate of a proportion, which is intermediate between the posterior mean and median when a Jeffreys prior is used.

The bootstrapping approach used here is labelled "exact" in Parzen et al. (2002), "fully specified" in Lin et al. (2009) and "deterministic" in Carter et al. (2010). The methodology is essentially the same in all three situations; "deterministic" seems the most appropriate description here.

In group 1, the probability density for the number of cases with positive outcome is

$$\Pr[A = a \mid \pi_1] = \binom{m}{a} \pi_1^a (1-\pi_1)^{m-a}. \tag{11.7}$$

We define π_1^L to be the lowest value of π_1 such that $\Pr[A \geq a \mid \pi_1 = \pi_1^L] = 0.5$.

Similarly, we define π_1^U to be the highest value of π_1 such that $\Pr[A \leq a \mid \pi_1 = \pi_1^U] = 0.5$.

Then the MUE of π_1 is defined as $(\pi_1^L + \pi_1^U)/2$, with a corresponding definition for group 2.

These quantities may be obtained using built-in functions for the incomplete beta integral, as described by Parzen et al. (2002). The MUEs for the special cases $a = m$ and $a = 0$ reduce to $0.5(1 \pm 0.5^{1/m})$.

The point estimate for the comparative measure—whether the RD, the RR or the OR—is then calculated from the MUEs for π_1 and π_2. Parzen et al. point out that while an MUE of the odds ratio itself can be derived from the conditional non-central hypergeometric distribution as described by Hirji et al. (1989), this does not adequately obviate the zero cell problem, being undefined when both proportions are 0 or 1.

The resulting method is of closed form. The method of construction of the bootstrap distribution is best illustrated in Tables I and II of Lin et al. (2009), which relate to the risk difference. There are $(m + 1)(n + 1)$ possible outcomes defined by $0 \le a \le m$ and $0 \le b \le n$. These are tabulated together with their probabilities based on the actual MUEs for π_1 and π_2 and Δ, the difference between the MUEs corresponding to a/m and b/n. These entries are then sorted by Δ, and the cumulative frequency distribution computed. The desired values delimiting a $1\text{-}\alpha$ interval for Δ are then obtained as the $\alpha/2$ and $1\text{-}\alpha/2$ quantiles of this distribution, estimated by linear interpolation. Exactly the same method is used for the RR or the OR; while interpolation on a log scale would theoretically be more satisfactory in these situations, in practice this makes little difference to the calculated limits.

The illustrative examples in Parzen et al. (2002) and Lin et al. (2009) relate to the ECOG trial (Sections 7.2 and 10.4.2), in which proportions of 2/14 and 1/11 arose for the secondary endpoint of safety and 0/14 and 0/11 for the primary endpoint of achieving at least 50% tumour shrinkage. The illustrative examples in Carter et al. (2010) relate to three interim analyses from a small trial evaluating intensive glycaemic control in renal transplantation patients, in which the data and safety monitoring board stipulated an intense schedule of interim analyses for safety reasons. The relevant pairs of proportions experiencing severe hypoglycaemia were 0/3 versus 0/4, 1/9 versus 0/11 and 1/12 versus 1/15 at the first three interim analyses. The MUE bootstrap approach was developed precisely with such data in mind, and produces ostensibly appropriate results in all these situations. For all three comparative measures, simulation results are presented indicating favourable coverage characteristics. Thus for the toxicity endpoint of the ECOG trial, the two proportions are 2/14 and 1/11. The corresponding MUEs are 0.152 and 0.105, leading to an estimated odds ratio of 1.53, with 95% interval 0.104 to 12.4. For the primary endpoint of tumour shrinkage, the two proportions are 0/14 and 0/11. The MUEs are $0.5(1 - 0.5^{1/14}) = 0.024$ and $0.5(1 - 0.5^{1/11}) = 0.031$ respectively, leading to an estimated odds ratio of 0.786, with 95% interval 0.104 to 5.60.

Table 11.6 shows some results comparing the MUE bootstrap with several other methods for the risk ratio and the odds ratio. As well as the Miettinen–Nurminen interval, we consider the use of MOVER-R to combine intervals for the two proportions by several methods. While Wilson intervals have been the usual choice so far, Bayes intervals with the usual uninformative conjugate priors have the advantage here that they shrink point and interval estimates of proportions mesially towards 0.5. We compare 2/14 versus 1/11 and 0/14 versus 0/11, as in the ECOG trial. These represent cases NZ and RZ

TABLE 11.6

Point Estimates and 95% Confidence Limits for the Risk Ratio and Odds Ratio by Several Methods, for Three Comparisons Based on an Eastern Cooperative Oncology Group Trial

	Risk Ratio	95% Limits	Odds Ratio	95% Limits
Case NZ: 2/14 vs. 1/11				
MOVER-R Wilson	1.57	0.230, 11.2	1.67	0.165, 15.7
Miettinen–Nurminen	1.57	0.225, 11.7	1.67	0.176, 14.9
MUE bootstrap	1.45	0.122, 9.46	1.53	0.104, 12.4
MOVER-R Jeffreys	1.33	0.192, 17.8	1.40	0.150, 22.8
MOVER-R Uniform	1.22	0.223, 9.84	1.27	0.169, 13.1
Case OZ: 0/14 vs. 1/11				
MOVER-R Wilson	0	0, 4.15	0	0, 4.99
Miettinen–Nurminen	0	0, 2.97	0	0, 3.18
MUE bootstrap	0.231	0.085, 2.70	0.212	0.057, 2.82
MOVER-R Jeffreys	0.267	0.0003, 4.81	0.241	0.0002, 5.19
MOVER-R Uniform	0.406	0.011, 4.05	0.367	0.009, 4.55
Case RZ: 0/14 vs. 0/11				
MOVER-R Wilson	N/A	N/A	N/A	N/A
Miettinen–Nurminen	N/A	N/A	N/A	N/A
MUE bootstrap	0.791	0.124, 4.48	0.786	0.104, 5.60
MOVER-R Jeffreys	0.800	0.0008, 770	0.793	0.0008, 798
MOVER-R Uniform	0.813	0.020, 32.0	0.800	0.018, 35.3

Source: Data from Parzen, M., Lipsitz, S., Ibrahim, J. et al. 2002. *Journal of Computational and Graphical Statistics* 11: 420–436.

as defined in Section 7.4.2. We also compare 0/14 versus 1/11, to represent the intermediate case OZ. For each of these comparisons, MOVER-R applied to Wald or Agresti–Coull intervals for the two proportions would fail to produce intervals for the risk ratio and the odds ratio.

Results for the five methods are arranged in increasing order of degree of mesial shrinkage to the point estimate. The pattern is very similar for the risk ratio and the odds ratio. For the NZ example, 2/14 versus 1/11, all five methods produce reasonable intervals. For the OZ example, 0/14 versus 1/11, the MOVER-R Wilson and Miettinen–Nurminen methods give a point estimate and lower limit at zero, with a positive upper limit, which is appropriate for a method which does not incorporate mesial shrinkage. The MUE bootstrap shrinks the point estimate less strongly towards 1 than the MOVER-R Bayes intervals; nevertheless it produces narrower intervals.

For the RZ example, 0/14 versus 0/11, the MOVER-R Wilson and Miettinen–Nurminen methods produce neither meaningful point estimates nor confidence limits. Once again, the MUE bootstrap shrinks the point estimate less strongly towards 1 than the MOVER-R Bayes intervals; nevertheless it produces narrower intervals, indeed to a far more striking degree than was the case for the OZ example. The Jeffreys prior is associated with particularly wide intervals here.

Given that the MUE bootstrap approach has been demonstrated to have favourable coverage characteristics, the above results suggest that this is an excellent choice of method for very small samples, when there is a substantial chance of zero cells. The discreteness of the bootstrap distribution does not appear to be a great problem when constructing an interval for the risk difference, relative risk or odds ratio, since in general there are $(m + 1)(n + 1)$ usually distinct values, and interpolation to obtain the required $\alpha/2$ and $1 - \alpha/2$ quantiles is reasonably satisfactory. However, when a single proportion $p = r/n$ is being estimated, the bootstrap distribution has only $n + 1$ points, leading to large steps in the cumulative distribution function; consequently this approach is not recommended for this application.

Another application in which the MUE approach is highly appropriate is described in Section 14.3.

11.5 Logistic Regression

Many hypothesis tests and confidence intervals have their roots in mathematical models that can be fitted to data. In the simplest applications, the analyses that are usually performed lead more directly to the same inferences as would be obtained by fitting the model. The strengths of the modelling approach are that additional explanatory variables may be included in the model, and additional interpretable outputs such as mean squares (for linear models) and deviances (for logit models) are produced.

Linear regression, the model used for continuous outcome variables, is inappropriate for a binary outcome, primarily because it does not impose the constraint $0 < p < 1$. When fitted to binary outcomes, linear models are liable to lead to meaningless fitted probabilities outside this range, which cast doubt on the validity of the model. Several alternative models incorporating the above constraint have been developed for the case where the outcome Y is binary. Logistic regression, the commonest of these models, is widely available in statistical software. In the simplest situation, there is a single explanatory variable, X, which may be binary or continuous. The logistic regression model is then

$$\begin{aligned} p = \Pr[Y = 1 \mid X = x] &= \exp(\alpha + \beta x)/(1 + \exp(\alpha + \beta x)) \\ &= 1/(1 + \exp(-\alpha - \beta x)). \end{aligned} \tag{11.8}$$

Equivalently, logit(p) is modelled by

$$\text{logit}(p) = \ln\,(p/(1 - p)) = \alpha + \beta x. \tag{11.9}$$

As p ranges from 0 to 1, the logit function can take any value, positive or negative, and so is suitable for modelling as a linear function of the explanatory variable.

When the explanatory variable X is binary, the slope parameter β is the natural log of the odds ratio. The default interval for $\exp(\beta)$ is identical to the delta interval for the odds ratio, and the default p-value is the same as that for the chi-square test. When X is continuous, $\exp(\beta)$ represents the factor by which the odds of getting $Y = 1$ are multiplied for each unit increase in X—very much like the odds ratio of 1.526 per allele in Section 11.2.3.

As in other modelling methods, the model can include multiple explanatory variables. If there are k explanatory variables, the model then takes the form

$$\text{logit}(p) = \ln\left(p/(1-p)\right) = \alpha + \beta_1 x_1 + \beta_2 x_2 + \ldots + \beta_k x_k \tag{11.10}$$

Explanatory variables may be continuous, binary or categorical. A binary variable is normally coded so that 1 indicates the factor is present and 0 indicates that it is absent. The quantity $\exp(\beta)$ then represents the odds ratio associated with the presence of the factor. A categorical explanatory variable with c categories is represented in the model by a set of $c - 1$ dummy binary variables. Automatic variable selection algorithms are available, similar to those for linear regression models.

Interaction effects may also be incorporated to characterise effect modification. An alternative linear interaction effect is considered in Section 14.2, which is often appropriate, but there is a caveat regarding boundedness, as in Section 14.2.4.

Several extensions are possible, including to ordinal and nominal level outcome variables. Section 11.5.3 describes a further extension of the binary logistic regression principle.

11.5.1 An Illustrative Example with Binary Explanatory Variables

Table 11.7 summarises the results of a cross-sectional study which examined the relationship of hypertension to snoring, smoking and obesity in 433 men aged 40 and above (Norton and Dunn 1985). Table 11.8 shows three possible binary logistic models expressing the dependence of hypertension on these three factors. All of these are main effects models: no interaction terms are included, although models including interactions could also be considered.

Model A includes all three explanatory variables. The coefficient in the model expressing the dependence of hypertension on snoring is +0.872, with an asymptotic (Wald) standard error 0.398, leading to a critical ratio test z = +2.19 and a two-sided p-value 0.028. The odds ratio expressing the strength of relationship of hypertension to snoring is then $e^{0.872} = 2.39$. The 95% limits are $e^{0.872 \pm 1.960 \times 0.398}$ i.e., 1.10 and 5.21. Similarly, the odds ratios associated with smoking and obesity in this model are 0.93 and 2.00. Obesity and snoring

TABLE 11.7

Prevalence of Hypertension in Men Aged 40 and over in Relation to Snoring, Smoking and Obesity

Snoring	Smoking	Obesity	Prevalence of Hypertension
No	No	No	5/60 (8%)
No	No	Yes	1/8 (13%)
No	Yes	No	2/17 (11%)
No	Yes	Yes	0/2 (0%)
Yes	No	No	35/187 (19%)
Yes	No	Yes	15/51 (29%)
Yes	Yes	No	13/85 (15%)
Yes	Yes	Yes	8/23 (35%)
Total			79/433 (18%)

Source: Data from Norton, P. G. and Dunn, E. V. 1985. *British Medical Journal* 201: 630–632.

appear to both influence the risk of hypertension independently—the data do not suggest the effect of one variable is mediated via the others.

Model B relates to the effect of smoking only. It might be supposed that smoking would adversely affect the risk of hypertension, but this does not appear to be the case, for this series at any rate. From Table 11.7, the prevalence of hypertension was 23/127 (18.1%) in smokers and 56/306 (18.3%) in

TABLE 11.8

Logistic Regression Analysis of Data on Hypertension and Snoring

	Regression Coefficient b	Standard Error	z	p-Value	Odds Ratio	95% Confidence Limits
Full model A						
Constant	−2.378	0.380				
Snoring	+0.872	0.398	+2.19	0.028	2.39	1.10, 5.21
Smoking	−0.068	0.278	−0.24	0.81	0.93	0.54, 1.61
Obesity	+0.6953	0.285	+2.44	0.015	2.00	1.15, 3.50
Model B including smoking only						
Constant	−1.496	0.148				
Smoking	−0.013	0.274	−0.05	0.96	0.99	0.58, 1.69
Model C omitting smoking						
Constant	−2.392	0.376				
Snoring	+0.865	0.397	+2.18	0.029	2.38	1.09, 5.17
Obesity	+0.6954	0.285	+2.44	0.015	2.00	1.15, 3.50

Source: Data from Norton, P. G. and Dunn, E. V. 1985. *British Medical Journal* 201: 630–632.

non-smokers. The odds ratio based on these figures is 0.99, which is identical to that obtained by this univariate binary logistic model.

Because smoking is clearly contributing very little to this model, we also consider a third model C, which includes snoring and obesity as explanatory variables, but not smoking. In this example, omitting smoking from the model results in very small alterations to the coefficients and corresponding odds ratios for the other factors, although sometimes omitting a variable has a much larger effect than this.

How can we say which model is preferable? Output from statistical software usually includes p-values associated with each factor, which compare the goodness of fit of the model as shown and the model based on the remaining factors only. Thus it is clear that smoking may be omitted from the model without impairing model fit, but this does not apply to either of the other factors. Alternatively, a deviance criterion based on minus twice the log of the maximised likelihood is used to compare goodness of fit of logistic models. The deviances associated with the three models are as follows

Three-factor model	398.916
Smoking only	411.422
Smoking excluded	398.976

The effect of dropping smoking from the model is assessed by referring 398.976 – 398.916 = 0.060 to the chi-square distribution with 1 degree of freedom. The resulting *p*-value of 0.807 is the same as that shown for the significance of smoking in the three-factor model in Table 11.8.

Conversely, a model including smoking only is a much poorer fit to the data than the three-factor model. This may be demonstrated by calculating 411.422 – 398.916 = 12.505. This is then compared to the chi-square distribution with 2 degrees of freedom, since we have dropped two explanatory variables from the model. The resulting *p*-value is then $e^{-12.505/2} = 0.002$.

Comparing deviances of two models in this way presupposes that one model is nested in the other—in other words, we are examining the effect of dropping one or more factors from a model. It is incorrect to subtract 398.976 from 411.422 in an attempt to compare models B and C. A deviance comparison is especially useful when considering a model reduction by 2 or more degrees of freedom, which would occur when a set of variables is dropped from the model—in particular, when a categorical variable is represented by $c - 1$ dummy variables.

11.5.2 An Illustrative Example with Continuous Explanatory Variables

Returning to the wound microbiology study (Davies et al. 2007), Table 11.9 shows summary statistics for the log-transformed number of colony forming units (CFUs) per gram at surface (swab) and tissue (biopsy) sites for

TABLE 11.9

Summary Statistics for Log-Transformed (to Base 10) Numbers of Colony-Forming Units per Gram in 20 Patients Whose Wounds Had Healed Completely within 6 Months and 46 Who Had Not

		Mean Log CFU/g	SD	*t*-Test
Tissue	Healed	4.43	1.91	$t = -1.75$
	Not healed	5.27	1.48	$p = 0.090$
Swab	Healed	5.78	1.68	$t = -2.47$
	Not healed	6.87	1.54	$p = 0.019$

Source: Data from Davies, C. E., Hill, K. E., Newcombe, R. G. et al. 2007. *Wound Repair and Regeneration* 15: 17–22.
Note: CFU = colony-forming units.

20 patients whose wounds had healed completely within 6 months and 46 who had not. As in Section 2.4.2, such data displays extreme positive skewness, which is greatly improved by log transformation. The logit function is defined in the usual way, as the natural log of $p/(1 - p)$. However, it is convenient to use log transformation to base 10 for the bacterial counts, as this leads to a more readily interpretable effect size measure, namely the odds ratio associated with a tenfold change in bacterial count. Here, higher bacterial counts are associated with poor healing rates. The difference is larger in absolute terms for surface compared to tissue, and only the surface contrast reaches the conventional level of statistical significance. Thus the surface CFU/g appears rather more discriminatory than the tissue CFU/g.

Table 11.10 shows the results from logistic regression analyses exploring the dependence of wound healing on tissue and surface bacterial counts simultaneously. For the first analysis, using tissue CFU/g levels only, the regression coefficient of −0.342 means that the log odds on healing are reduced by

TABLE 11.10

Logistic Regression Analysis of Data on Wound Healing and Bacterial Counts

CFU/g in Sample From	Regression Coefficient *b*	Standard Error	*z*	*p*-Value	Odds Ratio	95% Confidence Limits
Model using log CFU/g in tissue sample						
Tissue	−0.342	0.184	1.85	0.064	0.71	0.49, 1.02
Model using log CFU/g in surface sample						
Surface	−0.423	0.178	2.38	0.017	0.66	0.46, 0.93
Model using log CFU/g in both tissue and surface samples						
Tissue	−0.221	0.203	1.09	0.28	0.80	0.54, 1.19
Surface	−0.352	0.190	1.86	0.063	0.70	0.48, 1.02

Source: Data from Davies, C. E., Hill, K. E., Newcombe, R. G. et al. 2007. *Wound Repair and Regeneration* 15: 17–22.
Note: Constant terms are not shown. CFU = colony-forming units.

0.342 for each tenfold increase in bacterial count. The corresponding odds ratio of 0.71 indicates that the odds on healing are reduced by a factor of 0.71 (i.e., by 29%) for each tenfold increase in bacterial count.

The second model indicates a rather stronger dependence on surface bacterial count. Here, the odds on healing are reduced by 34% for each tenfold increase in bacterial count. The *p*-values for these two models are similar to those produced by the corresponding *t*-tests.

When we model on both factors simultaneously, the significance of each factor is reduced, because they are substantially positively correlated with each other. Indeed, neither factor makes a statistically significant contribution to the model when the other one is included—a familiar phenomenon when two explanatory variables are closely interrelated. Similarly the odds ratios become less extreme. We may evaluate whether performing a biopsy adds significantly to the information produced by the surface bacterial count, by calculating the difference between the two relevant deviance values, 74.745 – 73.513 = 1.232, which gives $p = 0.27$ when referred to the χ^2 distribution on 1 df. Thus based on this dataset, the surface bacterial count per gram gives statistically significant (albeit limited) prognostic information, and the additional invasiveness and effort of performing a biopsy does not significantly enhance prediction.

11.5.3 Logistic Regression with a Plateau

This example is included here primarily because it illustrates the flexibility of the logistic regression approach. Here, the classic model is extended to include a plateau below 100%. This is sometimes referred to as a ceiling, but it is preferable to reserve the latter term for an impassable upper bound at 100% as in Section 14.2.4.

The case series (Lewis 2010) comprised 147 consecutive patients with a chronic disease who were maintained on a licensed drug for between 2 months and 16 years. This drug is known to have a very specific adverse effect (AE), but previous evidence has been insufficient to determine what proportion of patients would develop the adverse effect as a result of long-term use. All patients were examined for this adverse effect by the same investigator on one occasion only.

The appropriate family of models for the progression of incidence of AE (p) with increasing exposure to the drug (x) is logistic regression. In addition to the standard logistic regression model A (Equation 11.8 or 11.9), a plateau model B

$$p = k/(1 + \exp(-\alpha - \beta x)) \tag{11.11}$$

was evaluated. Here, k denotes a plateau value with $0 < k \leq 1$, and α and β are intercept and slope parameters. In this model the risk of the AE approaches k not 1, at the highest exposures. In the standard model $k \equiv 1$. The plateau parameter k represents the proportion of patients who develop the AE following a high degree of exposure to the drug. A further refinement C was formulated, which also constrains p to be 0 when $x = 0$:

$$p = k \times \frac{1/(1+\exp(-\alpha-\beta x)) - 1/(1+\exp(-\alpha))}{1-1/(1+\exp(-\alpha))} \tag{11.12}$$

All models are fitted by maximum likelihood using Excel's Solver facility. Obtaining confidence intervals for parameters is less straightforward. The practicable way to do this is by the profile likelihood method. We have already encountered this approach when constructing intervals for unpaired and paired differences of proportions. The method is recognised to be generally anticonservative, but only to a slight degree; it is a practicable approach to use in such problems. While procedures such as NLIN in SAS or CNLR in SPSS can fit the plateau model, as suggested by Rindskopf (2010), this has the disadvantage that only symmetrical intervals can be produced. Obviously, we would be concerned to get the right shape of interval for the parameter k because it is of such crucial interest here. In this instance, boundary anomalies are not expected to affect estimation of the slope, intercept and ceiling parameters, since there would be no reason to fit a plateau model in a boundary or near-boundary case.

For the model Equation 11.11, to obtain an upper confidence limit for k, we seek k_U between $\hat{k}$ and 1 such that with $\alpha = \alpha_k$ and $\beta = \beta_k$, the log likelihood is $z^2/2 = 1.92$ lower than for the maximum likelihood model. Here, α_k denotes the profile estimate of α; that is, the value of α that maximises Λ given that particular value of k, and similarly β_k denotes the profile estimate of β. Similarly, to get the lower limit for k, we seek k_L between 0 and $\hat{k}$ with the same property. This two-stage iterative process is actually practicable in Excel: the inner stage to get the profile estimates is done automatically using Solver and the outer stage is done by trial and error, which converges quite rapidly to produce estimates to four decimal places. A similar procedure may be used to obtain confidence intervals for the risk at 1, 2, … units exposure, by using a re-parametrisation.

Overall, 87 (59%) of the patients exhibited the AE. Figure 11.2 shows the risk of developing the AE according to duration of drug use (upper panel) and cumulative dose (lower panel). In each panel, the 10 points represent decile groups defined by the relevant exposure, and show the proportions of patients in each exposure group who were classed as having the AE, plotted against the median exposure for the group. The smooth curves are fitted by a logistic regression model incorporating a plateau. The middle curve shows the estimated cumulative risk of AE at each exposure. The lower and upper curves represent 95% confidence limits.

Table 11.11 shows the relevant parameter estimates. The 95% limits are shown for the plateau k, the slope parameter β and the corresponding odds

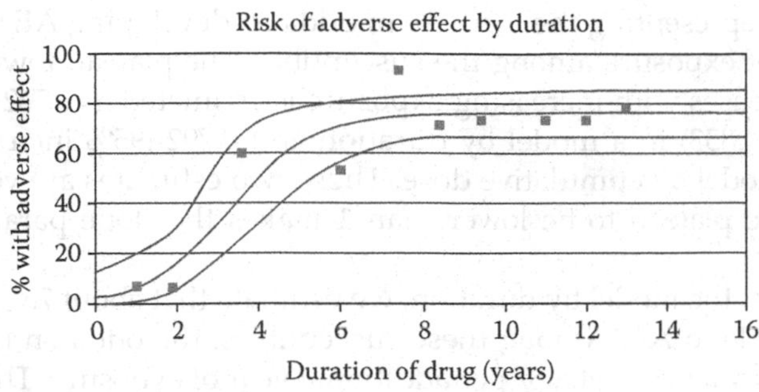

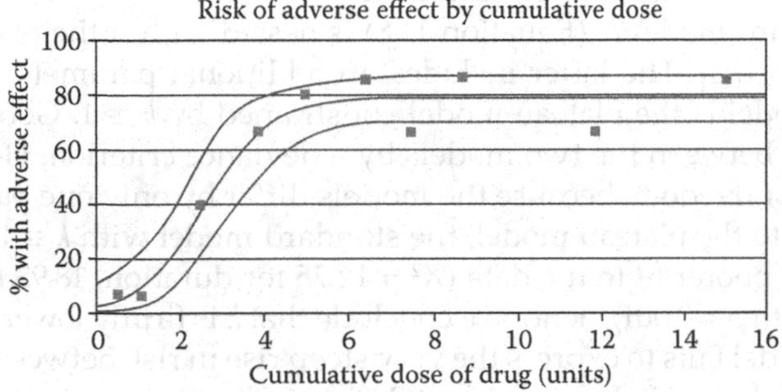

FIGURE 11.2
Risk of developing an adverse effect according to duration (upper panel) and cumulative dose (lower panel) of a drug. (Data from G. Lewis, 2010, personal communication.). In each panel, the 10 points represent decile groups defined by the relevant exposure, and show the proportions of patients in each exposure group who were classed as having AE, plotted against the median exposure for the group. The smooth curves are fitted by a logistic regression model incorporating a plateau. The middle curve shows the estimated cumulative risk of AE at each exposure. The lower and upper curves represent 95% confidence limits.

TABLE 11.11
Parameter Estimates for Plateau Logistic Regression Models for Risk of Developing an Adverse Event by Duration and Cumulative Dose of a Drug

	Model by Duration (Years)	Model by Cumulative Dose (Units)
Intercept α	−3.28	−3.36
Slope β	0.98 (0.48, 2.38)	1.29 (0.71, 2.28)
Odds ratio per unit exposure, e^{β}	2.67 (1.61, 10.79)	3.63 (2.03, 9.74)
Plateau k	0.762 (0.665, 0.853)	0.792 (0.699, 0.872)

Source: Data from G. Lewis, 2010, personal communication.
Note: The plateau parameter k may be regarded as the proportion of the caseload of patients treated with the drug who would develop the adverse event following a high degree of exposure to the drug. The 95% limits are shown for the plateau k, the slope parameter β and the corresponding odds ratio e^{β} representing the increase in odds on developing the adverse event for an additional unit of exposure among the susceptible.

ratio exp(β) representing the increase in odds on developing AE for an additional unit of exposure among the susceptible. The plateau towards which the risk increases with increasing exposure is estimated as 0.762 (95% interval 0.665 to 0.853) in a model by duration and 0.792 (95% interval 0.699 to 0.872) in a model by cumulative dose. These two estimates are very similar. Allowing the plateau to be lower than 1 makes the slope parameter β far greater.

To interpret the model by duration, we estimate that about 76% of patients are susceptible to AE. Among these susceptibles, the odds on manifesting AE increase by a factor of 2.67 per additional year of exposure. The model by cumulative dose is interpreted similarly.

The standard model A (Equation 11.8) is nested within the plateau model B (Equation 11.11). The latter includes an additional parameter, k, and the standard model is the plateau model constrained by $k \equiv 1$. Goodness of fit is compared between the two models by a deviance criterion. Here, there is one degree of freedom, because the models differ by only one parameter. In comparison to the plateau model, the standard model with $k \equiv 1$ is a highly significantly poorer fit to the data ($X^2 = 12.75$ for duration, 18.93 for cumulative dose, both $p < 0.001$), hence we conclude that k is firmly lower than 1. The standard model fails to express the very steep rise in risk between exposures of about 2 and 6 units. It fits a risk of 20% to 25% at zero exposure, rising to over 90% for the highest exposures in this series, both of which are much higher than the observed risks. The observed risk at minimal exposure is very low as in Figure 11.2, but not zero: one patient developed the AE after taking 0.2 units of the drug over 9 weeks.

A plateau model including both exposure variables together fits the data significantly better than a model by duration alone ($X^2 = 9.02$, $p = 0.003$), but not appreciably better than a model by cumulative dose alone ($X^2 = 0.09$, $p = 0.77$), suggesting that the risk of developing the AE is determined by cumulative dose rather than duration.

However, these plateau models have two drawbacks, one theoretical and one practical. The theoretical drawback is the point noted above, that the fitted risk at zero exposure is small and positive, not zero. But the model Equation 11.11 actually fits the data slightly better than the more complex model Equation 11.12 which was designed to represent zero exposure by zero risk. This occurs because occasionally someone with very low exposure develops the AE. The practical drawback is that when we allow the plateau to be flexible, the resulting interval for the slope parameter β becomes very wide.

Seventy-one (48%) of the patients were receiving the drug at the time of examination. The AE was marginally significantly commoner in these patients (48/71, 68%) than in those who had discontinued treatment (39/76, 51%; $p = 0.045$). However, the patients who were currently taking the drug had significantly longer duration of use (mean 8.6 versus 6.2 years, $p < 0.001$) and significantly greater cumulative dose (7.3 versus 5.5 units, $p = 0.017$) than those who were not. Including an additional term for whether currently on the drug in the plateau model did not lead to an appreciably better fit (X^2 = 0.46, p = 0.50 for duration; X^2 = 0.04, p = 0.84 for cumulative dose). Thus the results do not indicate any difference in risk by current treatment status once exposure is taken into account, hence give no support to the hypothesis that the AE may be reversible on discontinuation of treatment.

The best presentation of the findings includes three elements, listed in increasing order of persuasiveness:

- A graph such as Figure 11.2
- A p-value comparing the fit of plateau and non-plateau models
- The estimated plateau, with a confidence interval

In the graph, a feature that is characteristic of a plateau below 100% is the irregular pattern at high exposure—if the true incidence asymptotes to 100%, the degree of variation would become very low at high exposures.

The findings reported here relate to a relatively large series of patients, many of whom had high levels of cumulative exposure to the drug. Eighty-five (58%) of the 147 patients had received the drug for more than 7 years and 47 (32%) for more than 10 years. This study was the first to report an increase in the frequency of AE to clinically unacceptable levels based on a substantial number of patients with such a degree of exposure. The compelling findings were that the frequency of AE rises steeply up to around 6 years or a cumulative dose of 5 units, then levels out at a plateau of around 75% to 80%. Duration and cumulative dose of the drug were highly correlated, with very similar patterns of increasing frequency of AE by increasing duration and cumulative dose. The degree of increase was substantially greater than that previously reported: the plateau level is much higher than any figures suggested by previous series with lower degrees of exposure, yet it is firmly well below 100%. The findings at low exposures are concordant with previous reports.

A conventional logistic regression model for the risk of AE by drug exposure leads to a seriously misleading conclusion. It would be crass to infer that the risk approaches 100% with increasing duration simply because the mathematical model fitted constrained this to be the case. Goetghebeur and Pocock (1995) demonstrated that, among people with no history of heart disease, the alleged increase in risk of cardiovascular events at very low centiles of the distribution of blood pressure is purely an artefact of using a quadratic model to fit a curvilinear relationship; a more flexible model is needed, such as one including fractional polynomials. There is a serious risk of wrong conclusions being drawn whenever a model is chosen from a family that is not sufficiently adaptable to model the observed data.

In Chapter 15, the index U/mn, equivalent to the area under the ROC curve is developed as a widely applicable characterisation of the degree of separation of two samples. However, when the issue is dose-response, quoting U/mn is less informative than fitting a logistic regression model which identifies interpretable parameters. This point applies more strongly in this example, as U/mn based on the whole dataset gives no indication whatever that a plateau is needed. Also U/mn depends on where we sample from the distribution of exposure X: a plot of U/mn restricted to cases with $X > x_0$ against x_0 would soon asymptote to 1/2.

11.5.4 A Contrasting Example

In a study similar to that of Absi et al. (2011) reported in Section 2.1, Bailey et al. (2012) studied the gain in dental personnel's knowledge of disinfection and decontamination issues following a statutory refresher course. There were 1180 participants who attended 18 one-day courses given over the period from October 2008 to April 2010 and completed matching pre- and post-course 20-item best-of-five multiple choice instruments. Question 16 involved identifying the correct label for the new standard in relation to decontamination in dentistry as HTM 01-05. For this question, it was noted that the pre-training percentage correct improved markedly, from 10% or below for the first three courses to over 50% for the last five, as this became wider knowledge among this professional group. This raises a question: Can we say whether the proportion correct is likely to continue to rise towards 100%, or to plateau at little above 50%? We would like a more objective way to determine where the proportion correct is likely to level out, that does not depend on an arbitrary choice of which cohorts should be regarded as having reached the plateau level of knowledge.

The results for the 18 cohorts of attendees are shown in Figure 11.3, with curves representing the increase over time fitted using the ordinary (A) and modified (B) logistic regression models (Equations 11.8 and 11.11). The dates of the courses are represented by day numbers, with 1 representing 1 January 2008. The two models fit nearly identically well here, and the two curves are barely distinguishable within the range of the data, but diverge greatly when extrapolated forwards.

For these data, model B does not fit the data significantly better than model A: the null hypothesis, $k = 1$ is not rejected by the data. Using model B, the plateau is estimated as 0.660, with 95% interval 0.496 to 1. The apparent explanation is that the five most recent cohorts were very close together in time, a fact that is less evident when they are simply arranged in date order and the actual dates are disregarded. The lower limit, just below 0.5, means that it is possible that the true plateau value is around 50%, but the five most recent cohorts were just "lucky" ones that all achieved slightly higher levels of pre-course knowledge than this. Here, the importance of fitting an asymmetrical interval is still greater than in the adverse effect example in Section 11.5.3.

On model A, the proportion of correct responses to question 16 continues to increase towards an asymptote at 1. The proportion correct post-training

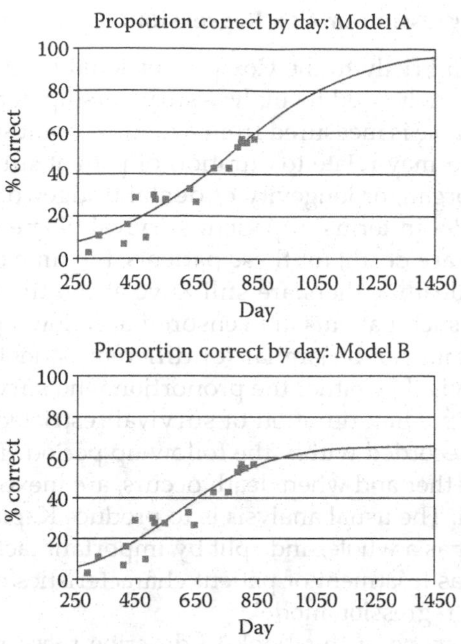

FIGURE 11.3
Proportions of correct answers to a question on identifying the current standard for decontamination in dentistry, in 18 cohorts of trainees about to take continuing professional dental education courses in Wales in 2008–2010. (Data from Bailey, S. E. R., Bullock, A. D., Cowpe, J. G. et al. 2012. Continuing dental education: Evaluation of the effectiveness of a disinfection and decontamination course. *European Journal of Dental Education,* in press.) Models A (ordinary logistic regression; Equation 11.8) and B (modified to incorporate a plateau; Equation 11.11) are indistinguishable in terms of goodness-of-fit to the data, but subsequently diverge.

was 1124/1180 = 0.9525, which is sufficiently close to 1 not to negate the applicability of model A. Nevertheless the main message here is that, however well our model fits the data to hand, extrapolation must not be regarded as reliable. In this example, we simply cannot tell whether future behaviour will conform to model A or model B. Similarly, the odds ratio representing the risk of Down syndrome is understood to double for every 2 years increase in maternal age. But we cannot determine whether the risk would plateau at older ages or asymptote to 100%, since there are practically no births at ages substantially beyond the usual menopausal age. These two examples contrast with the adverse event example, in which no extrapolation is involved—before the data presented in Section 11.5.3 became available, it was pure speculation whether the long-term risk of the adverse event would asymptote to around one-third, to 100%, or to some intermediate level.

11.5.5 Logistic Regression or Cox Regression?

Section 10.2 alluded briefly to the Cox proportional hazards model. This is the method most widely used to analyse survivorship data or time-to-event data. The time variable is measured from a defined point such as randomisation or surgery, and may relate to duration of patient survival, functioning of a transplanted organ, or longevity of dental bridgework. It is convenient to explain the model in terms of patient survival. Some patients reach the endpoint—in this case death; for these patients, the time on test is the interval from entry to death. Others are still alive at the time of follow-up; the survival times for such patients are censored at follow-up. The duration of follow-up may or may not be similar for different patients. It is inadequate to characterise survival by either the proportion who survived to follow-up, or by the mean or median duration of survival restricted to those patients whose deaths are recorded within the follow-up period. The answers to the two questions, whether and when death occurs, are inextricably interlocked in data of this kind. The usual analysis is to produce Kaplan–Meier survival plots for the cohort as a whole, and split by important factors. The influence of covariates such as treatment or patient characteristics on survival is then modelled by a Cox regression model.

It is beyond the scope of this book to describe these methods in detail, which are presented in many statistical texts and widely implemented in software. But it cannot be emphasised too strongly that it is crucial to choose the correct model for a dataset such as the one described in Section 11.5.3. This is one area in which it is sometimes extremely difficult for a statistician, when brought in only at the analysis phase, to elicit clear information from the clinical investigators. The adverse effect in this instance is one of which the patient and the prescribing physician may or may not be aware, and may be detected only following a meticulous examination by a specialist observer. The history of this collaboration is that the analysis was performed in three phases. In phase 1, the dataset was described in such a way as to strongly suggest each patient was assessed on just one occasion to determine whether the adverse effect was detectable. This scenario indicates conventional analyses for dose-response, culminating in logistic regression. Following discussion of the resulting analyses with the investigators, the author became convinced that he had misunderstood the data, that this was in fact time-to-event data, in which the survival time recorded for cases who had developed the adverse effect was the actual time at which this occurred. This led to a second phase of analyses, using the Cox model. On discussing these analyses with the investigators, it was far from clear that this re-analysis was justified. The author pressed the investigators for a crystal-clear explanation of what had taken place during the study, both for his own benefit as analyst and also for clarity of exposition in the resulting paper. It became clear that the original decision, to use a logistic regression based model, was correct. Following clarification of a few data queries the dataset was re-analysed on the same basis as in phase 1, resulting in the analyses reported above.

How much difference does choice of model make here? Using the modified logistic regression model above, we estimate the plateau as approximately 75% to 80%. A standard logistic regression model without a plateau asymptotes towards 1, of course, and leads to an estimated risk of having developed the side effect by 15 units of exposure (years or units) of well over 90%. Had the choice to analyse by Cox regression been the correct one, this would lead to an estimated risk approaching 100% at the highest exposures studied. All these analyses agree that the long-term risk is very much higher than previously reported figures which were based on much shorter periods of exposure. But it matters a great deal whether the risk asymptotes to 100% or to 75% to 80%. In the latter situation, a plausible explanation is that 25% to 30% of the patient population are resistant to the side effect, possibly on a genetic basis, which could lead to elucidation of the mechanism of action and modification of the medication to avoid this adverse effect.

In the above example, the (incorrect) Cox regression model would indicate that the risk asymptotes to practically 100%, because there were a very few subjects with failure times reported in the range 10–15 years. Nevertheless, a survivorship model is also capable of having a pattern of risk that does not approach 100% at the highest exposures. King et al. (2012) studied longevity of a large series of resin-retained dental bridges. The maximum duration of follow-up was of the order of 15 years, very similar to that for the drug adverse effect dataset discussed above. In this series, actual times of failure were recorded; non-failed bridges were checked for soundness at a follow-up visit in 2010. Survivorship analyses indicated that around 20% of bridges failed during the first few years after fitting, with extremely few failures thereafter, even when followed up for around 15 years. These results indicated a great improvement relative to an earlier series (Djemal et al. 1999) in which the proportion of bridges surviving decreased linearly during follow-up.

11.6 An Odds Ratio Based on Paired Data

As explained in Section 11.1, a major use of odds ratios is in retrospective case-control studies which aim to elucidate disease aetiology. It is advantageous for controls to be individually matched to cases with regard to obvious factors such as gender and age, and sometimes also other specified factors that could otherwise distort the relationship. Here, we consider in particular the case of a binary exposure. Most often a single control is matched to each case, leading to the data layout shown in Table 11.12. This is similar to Table 8.1, except that "First Classification" and "Second Classification" are replaced

TABLE 11.12

Notation for Comparison of Proportions from an Individually Matched Case-Control Study

Frequencies		Exposure status of control		
		Positive	Negative	Total
Exposure status of case	Positive	a	b	$a+b$
	Negative	c	d	$c+d$
	Total	$a+c$	$b+d$	n

by "Exposure Status of Case" and "Exposure Status of Control". The cell frequencies are labelled a, b, c and d, just as in Table 10.1, but the interpretation is quite different.

Just as in the unpaired case, the natural expression of effect size is the odds ratio. The simplest estimate for the data in Table 11.12 is the ratio $\tilde{\omega}_M = b/c$, based purely on the two discordant cells of the paired 2 × 2 table. This estimate may be derived as the MLE conditional on the observed value of $b + c$, or the Mantel–Haenszel (1959) estimate when the dataset is regarded as consisting of n strata, each relating to one case-control pair.

A conditional interval for this ratio is derived simply from the corresponding interval for the proportion $b/(b + c)$ using substitution, as in the illustrative example below. When a Wilson interval is used, its logit scale symmetry translates into an intuitively appealing multiplicative scale symmetry for the odds ratio. Interval location, in the usual sense of the degree of balance between mesial and distal non-coverage, also maps across to the odds ratio scale, so the Wilson-based interval errs on the side of being too mesial, relative to the natural centre of symmetry of the ratio scale, namely 1.

For example, Schlesselman (1982) presented results for an individually matched study relating endometrial cancer to oral conjugated oestrogen use. In the notation of Table 11.12, $a = 12$, $b = 43$, $c = 7$ and $d = 121$, whence $\tilde{\omega}_M = 6.14$. For these data, $b/(b + c) = 43/50 = 0.86$, with 95% Wilson limits 0.738 and 0.930. These transform into limits $0.738/(1 - 0.738) = 2.82$ and $0.930/(1 - 0.930) = 13.4$ for the odds ratio. The resulting interval is symmetrical on a multiplicative scale, since the limits can be written as $6.14 \div 2.18$ and 6.14×2.18.

Liang and Zeger (1988) discussed the relative properties of the above "matched" estimate $\tilde{\omega}_M$ and the "pooled" estimate $\tilde{\omega}_P = \dfrac{(a+b)(b+d)}{(a+c)(c+d)}$ (equivalent to $\tilde{\omega}$ in Section 11.1) which disregards the individually matched structure. Usually, $\tilde{\omega}_P$ will be less extreme than $\tilde{\omega}_M$, although not invariably, because in the (admittedly highly implausible) case where $a = d = 0$, $\tilde{\omega}_P$ reduces to $(b/c)^2$. Liang and Zeger noted that while $\tilde{\omega}_M$ may be less subject to bias (on the log scale) than $\tilde{\omega}_P$, it may suffer from decreased precision when the proportion of discordant pairs is small. For the endometrial cancer and oestrogen example, the "matched" and "pooled" estimates are 6.14 and 3.71, although the latter has a much narrower confidence interval. A mesially shrunk estimate $\tilde{\omega}_S = \frac{(a+b)(b+d)+bn(ad/bc-1)}{(a+c)(c+d)+cn(ad/bc-1)}$ was proposed, which takes the value 5.34 here. A closed-form confidence interval is available from a mixed model, derived by the delta method.

When cases are scarce, it may be advantageous to use $k > 1$ cases per control. Fleiss, Levin and Paik (2003, pp. 384–387) describe how the resulting data may be analysed by regarding each set of $k + 1$ individuals comprising a single case and his or her matched controls as a stratum, then constructing a Mantel–Haenszel estimate of the common odds ratio. An estimate of the variance of its log is available. Conversely, occasionally there can be more cases than controls, if the condition is a common one and determination of exposure status requires an unpleasant invasive procedure which is part of the regular care of cases but would be unacceptable to most healthy people. The same analysis methods are also applicable in this situation.

11.7 Implementation

The Miettinen–Nurminen (1985) score interval for the odds ratio for unpaired data (Section 11.3.2) may be obtained using the spreadsheet ODDSRATIOANDRR.xls introduced in Section 10.10. The four bold cells for the numerator and denominator of the first and second proportions are to be replaced by the appropriate figures for the user's dataset. The calculated limits then appear in place of those for the exemplar dataset in the usual way.

As explained in Section 10.4.2, the iterative process is based on Q_{OR} = (OR – 1)/(OR + 1). In the event that the current trial value of ω is exactly 1, r_2 becomes infinite and no further iteration is possible. To obviate this problem, when this occurs the pooled proportion $\frac{a+b}{m+n}$ is substituted for r_2. The next trial value will not be 1, so the iteration bypasses this potential singularity.

When considering a ratio of two independent proportions, it is generally very clear which two marginal totals are conditioned on, which then become m and n in the notation of Table 10.1, and are used as the two denominators in ODDSRATIOANDRR.xls. When the odds ratio is the measure of interest, sometimes it is not obvious which pair of marginal totals to use as m and n. Fortunately, for the odds ratio, it makes no difference which is chosen, as may be seen by replacing the exemplar dataset by the corresponding figures with rows and columns interchanged—namely, comparing 32/49 versus 118/245.

Doing so alters the relative risk and its confidence limits, but not the odds ratio and its interval.

SAS code is available to compute MUE bootstrap intervals for the odds ratio and risk ratio. Alternatively, for small samples it is reasonably straightforward to construct the bootstrap distribution using Excel, copying values and sorting to produce the required cumulative distribution function. In the interests of stability, very low probabilities in the tails of the distributions for a and b should be replaced by zero.

To obtain a Wilson interval for the odds ratio estimate b/c for an individually matched case-control study as described in Section 11.6, the spreadsheet CIPROPORTION.xls may be used. The first block of the spreadsheet is used to obtain the Wilson limits L and U for the conditional proportion $b/(b + c)$. The corresponding limits for the paired odds ratio are then L/(1 – L) and U/(1 – U).

12

Screening and Diagnostic Tests

12.1 Background

Epidemiology is the study of issues relating to health and health care in a population context. Statistical issues are of fundamental importance in all branches of epidemiology. One important area of epidemiology concerns the performance of screening tests and diagnostic tests. The aim is to characterise the relationship between an approximate test procedure T and a gold standard criterion G which is used to define which patients do and which do not have the disease of interest—in other words, the actual final diagnosis. Familiar and widely studied examples include the use of mammography to screen women for breast carcinoma and prostate specific antigen (PSA) measurement to detect prostate cancer in men. In general, both T and G may be single-test procedures or more complex algorithms.

In some situations, it is practicable to use the definitive test as it stands. In other contexts, it is preferable to use an approximate test first, and only use the gold standard test later for subjects with positive findings from the approximate test. The gold standard test may be invasive and dangerous, and is usually substantially more expensive than the approximate test. Sometimes the gold standard classification would only become available at post-mortem examination. In many contexts, especially the detection of malignancy, histopathology is regarded as the gold standard, even though it is recognised that this is not really definitive but is subject to considerable observer variation.

There are important differences between the screening and diagnosis paradigms. Screening, sometimes termed case finding, is applied to populations of symptom-free, apparently healthy individuals, in order to detect covert early disease, with a better prospect of cure (e.g., for cancers) or effective control of the disease process and its complications (e.g., for maturity-onset diabetes). Following the approximate screening test, individuals are classed as either abnormal (suggesting disease may be present) or normal (suggesting little current cause for concern). Screening of newborns for innate conditions such as phenylketonuria obviously takes place only once, whereas screening for adult-onset diseases may take place at regular intervals (e.g., mammography at 3-year intervals).

In contrast, the diagnosis paradigm often really relates to differential diagnosis. The patient has a health problem, but it is not obvious which of two (or more) diseases best fits the patient's presentation. For example, a patient presenting as an emergency with acute abdominal pain may or may not have appendicitis requiring surgical removal. The clinical signs relating to acute appendicitis and non-specific abdominal pain overlap considerably; nevertheless there is substantial discriminatory evidence from the presence or absence of clinical signs such as guarding (De Dombal and Horrocks 1978). The simplest choice involves two disease categories which are exclusive and exhaustive—we are trying to determine whether the patient has disease A or B. In practice, the diseases A and B may co-exist in some patients, and there may be more than two possible diagnoses, including no abnormality detectable (NAD) which accounts for the patient's presentation.

The most basic diagnostic paradigm is pattern matching, in which the clinician simply chooses which of the possible diagnoses most closely matches the patient's immediate clinical presentation. Best practice goes substantially beyond this, and takes account of the background information, including the patient's demographic characteristics, family history and so forth as well as clinical signs and any tests carried out. These two approaches link to the likelihood and the posterior in Bayes' theorem.

Results reported by Stroud et al. (2002) and Newcombe et al. (2008a) imply that diagnosticians are, appropriately, influenced by background information as well as clinical signs, but to widely differing degrees. Many neurologically compromised patients have swallowing difficulty (dysphagia). In some patients this is benign while in others it involves aspiration, which is dangerous, but this is difficult for even the highly trained ear to assess. Twelve speech and language therapists were asked to assess, on two occasions, whether sound recordings of individual test swallows in 20 patients with dysphagia were indicative of aspiration. In one round the patient's correct clinical scenario was presented, in the other round an alternative scenario representing a very different degree of prior risk was given. The observers graded each swallow on an ordinal scale as normal, abnormal but not aspiration, or aspiration. Moving from a low-risk to a high-risk scenario resulted in a more severe grade of abnormality being given in 72 (30%) of instances, unchanged in 150 (63%), and a less severe grade in 17 (7%), a clear preponderance of changes in the predicted direction. Using ordinal logistic regression, the degree to which prior information swayed the diagnosis varied highly significantly ($p < 0.001$) between the 12 observers.

The presentation in this chapter relates specifically to the screening situation. Many concepts carry over to both differential diagnosis and also non-health-related contexts, although often with some modification. For example, the false-positive rate of a medical test is simply a proportion, with no specific time element involved. An alarm system has a false-positive rate per unit time.

The gold standard is assumed to be inherently binary. Some approximate tests yield a result which is intrinsically binary. For example, in Section 14.9 we consider some data from Mir (1990) in which a saliva test was evaluated 75 subjects, 17 of whom were identified as human immunodeficiency virus (HIV)-positive according to a definitive serum test. Some tests produce results on a continuous scale, such as fasting blood sugar level or PSA level. Others such as cervical smear specimens are classified using an ordinal scale. For the purpose of making a decision about how the patient should be treated, such tests need to be dichotomised into binary form, and several cut points are possible. In Sections 12.2 and 12.3 we consider how to characterise the performance of a test that is expressed in binary form. In Section 12.4 we consider what happens when there is a choice of possible cut points.

12.2 Sensitivity and Specificity

When both the gold standard and the approximate test are expressed in binary form, data relating to the performance of the test may be expressed in a familiar 2 × 2 table format, as in Table 12.1 which is closely analogous to Table 7.1. Naturally we assume that the relationship between approximate and gold standard tests is in the positive direction, so that the odds ratio $\tilde{\omega} = ad/bc > 1$. Indeed, a positive relationship is a minimum requirement here—to be useful, the sensitivity or specificity (or preferably both) need to be substantially greater than 1/2.

The natural quantities describing the performance of the test are the sensitivity and specificity. These are simple proportions, for which

TABLE 12.1

Notation for 2 × 2 Contingency Table Characterising the Performance of a Screening or Diagnostic Test

		Result of gold standard determination G		Total
		Positive	Negative	
Result of approximate test T	Positive	a	c	$a + c$
	Negative	b	d	$b + d$
	Total	$a + b$	$c + d$	t

confidence intervals may be calculated using the first block of the spreadsheet CIPROPORTION.XLS. Sensitivity and specificity are calculated as follows

$$\text{Sensitivity} = a/(a+b) \tag{12.1}$$

$$\text{Specificity} = d/(c+d) \tag{12.2}$$

The sensitivity measures how well the test detects the condition. It is the proportion of patients who really have the condition who are classified correctly by the approximate test T.

The specificity measures how well the test rules out the condition. It is the proportion of patients who really do not have the condition who are classified correctly by T.

An equivalent characterisation is based on the false-positive rate and the false-negative rate, which are the complements of the specificity and the sensitivity:

$$\text{False-positive rate } \alpha = c/(c+d) \tag{12.3}$$

$$\text{False-negative rate } \beta = b/(a+b) \tag{12.4}$$

There is a natural link between these quantities and the false-positive and false-negative rates of a hypothesis test. A hypothesis test screens for the possibility that the null hypothesis does not hold. The power of the test is then $1 - \beta$, which is equivalent to the sensitivity when viewed as a screening test. Consequently, the sensitivity and specificity of a screening test are sometimes denoted by $1 - \beta$ and $1 - \alpha$. The important limitation to this analogy is that when screening for disease, we are normally concerned to detect a deviation from the norm in a specified direction, such as lowered haemoglobin or raised intra-ocular pressure, whereas statistical hypothesis testing should usually be two-sided.

12.3 Positive and Negative Predictive Values

The positive predictive value (PPV) and negative predictive value (NPV) are quantities related to the sensitivity and specificity, which express what happens when the test is applied to a population with a particular prevalence. They may be calculated directly from the original dataset as simple proportions. In the notation of Table 12.1:

$$\text{PPV} = a/(a+c) \tag{12.5}$$

$$\text{NPV} = d/(b+d) \tag{12.6}$$

Assuming that this table is the result of naturalistic sampling, starting with t individuals who were not selected on the basis of disease status, confidence intervals may be calculated using the first block of the spreadsheet CIPROPORTION.XLS.

An equivalent characterisation is in terms of the post-test probabilities. Before we have the information from the approximate test, the prior risk that an individual has the disease is simply the prevalence of disease, which in this situation is estimated as $(a + b)/t$. The test divides the series into a higher risk group, comprising $a + c$ subjects, with risk $a/(a + c)$, and a lower risk group, comprising $b + d$ subjects, with risk $b/(b + d)$. These two posterior risks are the post-test probabilities for positive and negative test results, with

$$\frac{b}{b+d} < \frac{a+b}{t} < \frac{a}{a+c}. \tag{12.7}$$

Segregating into higher and lower risk groups in this way is known as triage when done in the emergency care context, but the same concept applies in all areas of health care.

It is often important to estimate what would happen if the test were applied to a population with a quite different prevalence. *If* we are prepared to make the assumption that the sensitivity and specificity will carry across, the projected PPV and NPV may be calculated using Bayes' theorem. Letting π, *se* and *sp* denote the prevalence, sensitivity and specificity, these quantities are

$$\text{PPV} = \frac{\pi \times se}{\pi \times se + (1-\pi) \times (1-sp)} \tag{12.8}$$

$$\text{NPV} = \frac{(1-\pi) \times sp}{\pi \times (1-se) + (1-\pi) \times sp} \tag{12.9}$$

In particular, often the original series comprises hospital in-patients or out-patients, and the application of the test to screening asymptomatic populations is being considered. The prevalence of disease is likely to be much lower in the community setting; consequently the PPV is greatly reduced: the odds are stacked against screening.

For example, the results in the first panel of Table 12.2 relate to 300 men admitted to hospital with retention of urine who were investigated for carcinoma of the prostate. Among the tests performed by the urologist was a digital rectal examination. Each patient subsequently underwent needle biopsy, which is assumed to be definitive in assessing the presence or absence of prostate cancer.

TABLE 12.2

Relationship between Results of Digital Examination and Needle Biopsy

Urology ward		Biopsy for prostate carcinoma			**Community**		Prostate carcinoma		
		Positive	Negative	Total			Present	Absent	Total
Digital examination	Positive	48	25	73	Digital examination	Positive	35	1077	1112
	Negative	21	206	227		Negative	15	8873	8888
	Total	69	231	300		Total	50	9950	10,000

Note: The first panel relates to 300 men admitted to a urology ward with retention of urine.The second relates to the application to screen men aged 50 and over in the community, with prevalence 0.5%.

Directly from these results, the estimated sensitivity and specificity are 70% and 89%, with PPV = 66%, NPV = 91%, all reasonably high. However, a test that has adequate sensitivity and specificity for use in a specialist clinic or ward where the prevalence of the condition is high may not be useful as a screening test for a population in which the prevalence is low.

For instance, consider the suggestion that men aged 50 years and over should be screened for prostate cancer by digital examination, on a regular, annual basis, regardless of the presence or absence of symptoms. Suppose that each year 1 in 200 men aged over 50 develop a prostate malignancy. Then, if we assume that the sensitivity and specificity calculated above apply also in this situation, and calculate the equivalent table for a notional 10,000 men, we get the figures shown in the second panel of Table 12.2.

These figures are derived as follows. Of the total of 10,000 men, 0.5%, or 50, have a new carcinoma of the prostate. The 70% sensitivity implies that 35 of these will be detected by digital examination and the remaining 15 will be false negatives. With a specificity of 89%, 8873 of the 9950 unaffected subjects will be correctly identified by the test and the remaining 1077 will be false negatives. The PPV is then only 3.1%; only 35 of the 1112 men who tested positive have the condition. With 1 in 9 men referred for further investigation each year, diagnostic facilities will be subjected to greatly increased demands, for little yield. Large numbers of men will be subjected unnecessarily to both the physical hazards of further diagnostic procedures and unnecessarily raised anxiety levels. Nevertheless some tests with low PPV, such as mammographic screening for breast cancer (with PPV about 1 in 6) can result in a marked beneficial impact on the disease-specific death rate in future years, as demonstrated definitively in randomised trials comparing groups offered and not offered the test, analysed by intention to treat. Furthermore, although the NPV looks impressively high at 99.8%, this may be of little consequence in the general population if the level of concern about this particular disease is low. The sensitivity of 70% still means that 15 (30%) of the affected subjects are not picked up by the test. So even from the standpoint of excluding disease, the test is not particularly reassuring.

When such calculations are performed, it is sometimes assumed that the sensitivity and specificity are subject to sampling imprecision, but the assumed prevalence is a notional, usually rounded value, not based on a specific sample, without any particular sampling uncertainty. Confidence intervals for the projected PPV and NPV may then be derived from intervals for ratios of proportions, as described in Section 14.9, and a plot of PPV and NPV versus prevalence produced, including confidence regions, as in the spreadsheet PPVNPV.xls.

While such calculations are fairly simple to perform, the assumption that the sensitivity and specificity will carry across to a different context is highly questionable (Kraemer 1992, p. 101). Notwithstanding, it remains true that when a test is transferred to a population with a much lower intrinsic risk of disease, the PPV will be commensurately reduced.

12.4 Trade-Off between Sensitivity and Specificity: The ROC Curve

If the discriminatory variable is not dichotomous but ordinal or continuous, a choice of cut points is available. There is then a trade-off between sensitivity and specificity—as the cut point is moved in one direction, sensitivity increases while specificity decreases, if it is moved in the opposite direction there is a gain in specificity at the expense of sensitivity. This is best displayed by plotting a receiver operating characteristic (ROC) curve which is a plot of sensitivity against 1-specificity for different cutoff values of the variable. The ROC curve originated in signal detection theory, but is now widely used to provide a graphical display of the degree of separation between two samples for a continuous or ordinal variable, commonly in relation to medical tests. Similar issues apply in the ordinal and continuous cases, the main difference being the number of possible dichotomy points.

Table 12.3 shows some analyses for a radiology dataset (Hanley and McNeil 1982), relating to 109 patients with neurological problems. Computed tomography images are classed on a five-point ordinal scale ranging from "definitely normal" to "definitely abnormal." Subsequently, 51 of the patients are identified as abnormal by a definitive test.

When a scale with $k = 5$ points is used, there are $k - 1 = 4$ possible dichotomy points. The table shows the sensitivity and specificity corresponding to each of these, and also, for completeness, the extreme rules "class all patients as abnormal" and "class all patients as normal". The ROC curve then consists of points representing the resulting $k + 1 = 6$ (1-specificity, sensitivity) pairs, as in Figure 12.1. In practice ROC curves are generally displayed with the points joined either by line segments or smoothed curves.

A diagonal reference line is usually displayed. When the horizontal and vertical axes have equal scales, the plot is square and the diagonal line is at 45°. A point on this line corresponds to sensitivity + specificity = 1, which would imply a worthless test. Discriminatory tests lie in the upper left-hand triangular area, well towards the upper boundary (which corresponds to perfect sensitivity) and/or the left-hand one (corresponding to perfect specificity). A simple measure of the performance of a test is the area under the ROC curve (AUROC). This is considered in depth in Chapter 15, in its other guise as a generalisation of the Mann–Whitney test statistic, including a confidence interval with favourable coverage properties.

TABLE 12.3

Sensitivity and Specificity for Various Cut Points for a Radiology Dataset[a]

		True Status		Cumulative Frequencies						
		Normal	**Abnormal**	**Normal**	**Abnormal**	**Spec%**	**100-Spec%**	**Sens%**	**0.5 × Sens% + 0.5 × Spec%**	**0.75 × Sens% + 0.25 × Spec%**
Classification of computed tomography image	Definitely normal	33	3	0	51	0	100	100	50	75
	Probably normal	6	2	33	48	57	43	94	76	**85**
	Questionable	6	2	39	46	67	33	90	79	84
	Probably abnormal	11	11	45	44	78	22	86	**82**	84
	Definitely abnormal	2	33	56	33	97	3	65	81	73
	Total	58	51	58	0	100	0	0	50	25

Source: Data from Hanley, J. A. and McNeil, B. J. 1982. *Radiology* 143: 29–36.

Note: Spec = specificity; Sens = sensitivity.

[a] Including sensitivity and 1-specificity for ROC analysis.

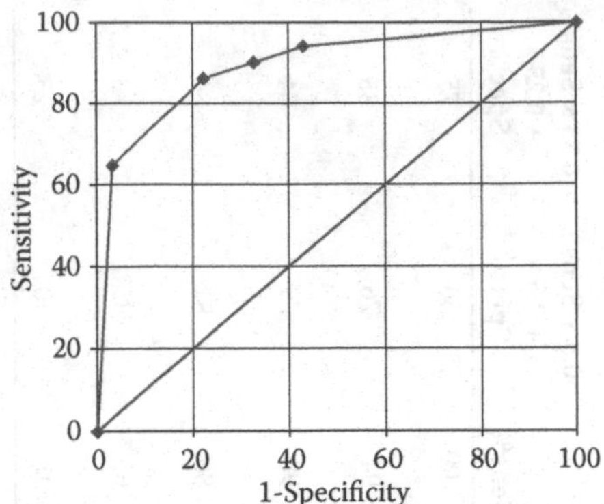

FIGURE 12.1
Example of receiver operating characteristic (ROC) curve displaying the trade-off between sensitivity and specificity. (Data from Hanley, J. A. and McNeil, B. J. 1982. *Radiology* 143: 29–36.)

12.4.1 Choice of Dichotomy Point

Given the trade-off between sensitivity and specificity, how should the "best" dichotomy point be chosen? This issue arises in several contexts, not only screening and diagnostic tests. Particularly when the sample size is large, it is natural to choose the point on the curve that is furthest above the diagonal reference line representing sensitivity = 1-specificity; in other words, furthest "north-west" on the ROC. This is probably reasonable for an ordinal variable. However, for a continuous variable, to literally do this (e.g., in Kyle et al. 1996) is logically flawed due to over-fitting to the vagaries of fine structure of the training dataset.

A more reasonable approach involves identifying the exact optimal point, then rounding this cutoff value, although this requires to be done very carefully as such rounding may greatly impair or completely destroy the discriminatory ability of the measurement. King et al. (2012) studied longevity of resin-retained dental bridges. A Cox regression model was envisaged, involving several dichotomous variables, and also two continuous ones, date of fitting and age of the patient at fitting, which it is convenient to dichotomise in the model. Because virtually all failures occurred early, a simple analysis treating the outcome as binary would not be greatly misleading; consequently this approach was appropriate for preliminary analyses. For these two variables, optimal cut points were identified as the furthest north-west points on the ROCs (Figure 12.2). For age at fitting, this criterion identifies a single optimum point, with sensitivity 74.3% and specificity 44.6%, corresponding to age 29.21 years. This was rounded to 30 years, resulting in sensitivity = 70.4% and specificity = 45.7%.

For date of fitting, the point furthest above the diagonal reference line has sensitivity 44.1% and specificity 71.1%, corresponding to a cutoff date of

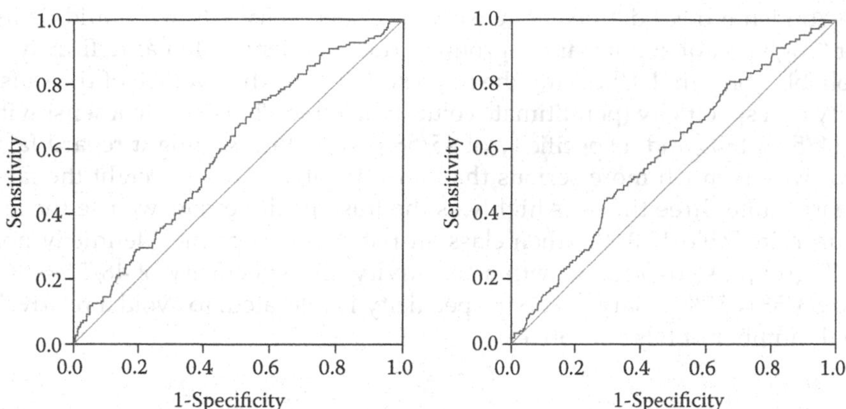

FIGURE 12.2
ROC curves for age (first panel) and date at fitting (second panel) of resin-retained dental bridges in relation to whether failure occurred. Data from King, P. A., Foster, L. V., Yates, R. J. et al. 2012. Survival characteristics of resin-retained bridgework provided at a UK Teaching Hospital. In preparation for *British Dental Journal*.)

1 May 1997. The height above the reference line is only very slightly lower than this for later dates in 1997, but the curve falls away sharply towards the diagonal for earlier cutoff dates. Rounding the dichotomy point down to the nearest new year, 1 January 1997, destroys the predictive ability of this variable, with sensitivity 28.9% and specificity 76.1%. It is preferable to round up here, to 1 January 1998, leading to a sensitivity 52.6% and specificity 59.1%, indicating still substantial although not over-optimised discrimination.

12.4.2 North by North-West or Sense and Sensitivity: The Asymmetry of Interpretation of ROC Curves

A deficiency of the AUROC is that it disregards the relative importance of the two types of errors. It may be inappropriate to do so when screening for disease; it may be preferable to regard false negatives as more serious than false positives, in which case sensitivity will be more important than specificity. For example, Charbit et al. (2007) studied the usefulness of low fibrinogen levels in identifying the risk of severe blood loss due to post-partum haemorrhage. Here, there is clearly a greater loss associated with a false negative (substantial risk of maternal death) than with a false positive (unnecessary use of donated blood). In this situation we do not want to choose the point furthest "north-west," but rather somewhere intermediate between north and north-west, hence "north-by-north-west."

Returning to the computed tomography data (Table 12.3 and Figure 12.1), in order to optimise sensitivity + specificity (or equivalently, their average, or the Youden index J defined as sensitivity + specificity – 1), we would define our "high risk of abnormality" group as those patients rated as definitely or probably abnormal. Choosing the highest value for the average of the sensitivity and specificity (penultimate column of Table 12.3) leads to a sensitivity of 44/51 = 86% and a specificity of 45/58 = 78%. But we might regard false negatives as much more serious than false positives. If we weight the false negative rate three times as highly as the false positive rate, we use the last column in Table 12.3. We then class all patients except the "definitely normal" group as suspicious, with a sensitivity and specificity of 48/51 = 94% and 33/58 = 57%. A large drop in specificity is tolerated, to avoid a relatively small number of false negatives.

12.5 Simultaneous Comparison of Sensitivity and Specificity between Two Tests

Often investigators seek to determine which of two diagnostic or screening tests T_1 and T_2 is the better surrogate for a gold standard test G, the criterion that is regarded as the definitive diagnostic classification. All three classifications, T_1, T_2 and G are assumed to be binary. Two study designs are available. In the usual and more satisfactory paired design, the same set of N subjects undergo all three tests. Either an unselected series of N subjects are recruited, or else M subjects who are known to be affected and N-M who are known not to be affected are studied. In the unpaired design N_1 subjects are given tests T_1 and G, and N_2 are given tests T_2 and G. The paired design has the obvious advantage that tests T_1 and T_2 are compared directly on the same subjects, whereas the unpaired design is profligate in its use of both patients and the gold standard test G which may well be expensive. An unpaired comparison is particularly unsatisfactory if retrospective in nature, as test G may well be applied quite differently in two independently performed studies, and patient characteristics may be different. Consequently the paired design is preferable whenever it is practicable. The unpaired design can only be recommended if for some reason tests T_1 and T_2 cannot be applied to the same individual or unit.

In the paired design, every individual undergoes all three tests T_1, T_2 and G. Tests T_1 and T_2 may be compared for sensitivity in the usual way. In the notation of Table 12.4, the two sensitivities are $\eta_{1.}$ and $\eta_{.1}$, where dots are used to indicate summation with respect to the relevant suffix. The null hypothesis $\eta_{1.} = \eta_{.1}$ may be tested using a McNemar test. The difference in sensitivity between the two tests is $\theta_1 = \eta_{1.} - \eta_{.1} = \eta_{12} - \eta_{21}$. The interval estimation methods introduced in Chapter 8 may be used to calculate a confidence interval for θ_1. But these analyses disregard both any observed difference in

TABLE 12.4

Notation for Comparison of Two Tests in the Paired Design

True proportions

Affected		Test T_2			Unaffected		Test T_2		
		Positive	Negative	Total			Positive	Negative	Total
Test T_1	Positive	η_{11}	η_{12}	$\eta_{1.}$	Test T_1	Positive	ζ_{11}	ζ_{12}	$\zeta_{1.}$
	Negative	η_{21}	η_{22}	$\eta_{2.}$		Negative	ζ_{21}	ζ_{22}	$\zeta_{2.}$
	Total	$\eta_{.1}$	$\eta_{.2}$	1		Total	$\zeta_{.1}$	$\zeta_{.2}$	1

Difference in sensitivity $\theta_1 = \eta_{1.} - \eta_{.1} = \eta_{12} - \eta_{21}$

Difference in specificity $\theta_2 = \zeta_{2.} - \zeta_{.2} = \zeta_{21} - \zeta_{12}$

Observed frequencies

Affected		Test T_2			Unaffected		Test T_2		
		Positive	Negative	Total			Positive	Negative	Total
Test T_1	Positive	a_{11}	a_{12}	$a_{1.}$	Test T_1	Positive	b_{11}	b_{12}	$b_{1.}$
	Negative	a_{21}	a_{22}	$a_{2.}$		Negative	b_{21}	b_{22}	$b_{2.}$
	Total	$a_{.1}$	$a_{.2}$	M		Total	$b_{.1}$	$b_{.2}$	$N-M$

specificity and its imprecision due to sampling variation. Similar analyses may be performed to compare the specificities of the two tests, $\eta_{2.}$ and $\eta_{.2}$. But running these analyses separately for sensitivity and specificity fails to take account of the known trade-off between sensitivity and specificity.

Lu and Bean (1995) considered the issue of testing equivalence of sensitivity. They commented that equivalence should focus on both sensitivity and specificity, and suggested that equivalence in specificity should be assessed by applying the same methods to a series of normal subjects. They noted that the relative importance of sensitivity and specificity varies according to clinical context.

In recognition of this, Vach et al. (2012) identify strong, liberal and weak criteria to define the success of a diagnostic study. The liberal criterion is based on a weighted mean of sensitivity and specificity, and permits a compensation between sensitivity and specificity to a clinically relevant degree. A similar weighted mean approach underlies the graphical methods developed in Sections 12.5.2 to 12.5.6.

A naive approach to the comparison of two tests focuses on the overall accuracy, $(a + d)/t$, of each test. This index is flawed on two counts. It fails to incorporate weightings for the relative seriousness of the two types of errors and the prevalence. Also, it presents both tests in an unduly favourable light—by failing to take into account the degree of agreement expected by chance, it incurs the problems which (in the rather different context of test reliability) Scott's pi (1955) and Cohen's kappa (1960) were designed to obviate.

12.5.1 Positive and Negative Likelihood Ratios

Clearly, the conditions $se_1 \geq se_2$ and $sp_1 \geq sp_2$ taken together (with at least one inequality strict) are sufficient to imply that test T_1 is superior in performance to test T_2—although of course there could be other factors to take into consideration such as prohibitive cost or invasiveness. But they are not necessary conditions. Biggerstaff (2000) pointed out that the relevant comparison here is in terms of two derived quantities, and suggested a graphical method of comparison on the ROC plane. The positive and negative likelihood ratios (PLR and NLR) of a test with a given sensitivity (se) and specificity (sp) are defined as se/(1 − sp) and sp/(1 − se). For the first panel of Table 12.2, the PLR and NLR are 6.43 and 2.93. The projected PPV and NPV (Equations 12.8 and 12.9) are derived from these quantities and the prevalence, and may be written

$$\text{PPV} = 1/\{1 + (1/\pi - 1)/\text{PLR}\} \tag{12.10}$$

$$\text{NPV} = 1/\{1 + (1/(1/\pi - 1))/\text{NLR}\} \tag{12.11}$$

The PLR and NLR derived from the resulting table (second panel) are 6.47 and 2.97, which only differ from the original values due to rounding.

Biggerstaff gave an example in which for two tests A and C, $se_A = 0.6$, $sp_A = 0.8$, $se_C = 0.85$ and $sp_C = 0.75$. Test C is much more sensitive than test A, at the cost of slightly lower specificity. Nevertheless $PLR_A = 3$, $PLR_C = 3.4$, $NLR_A = 2$, $NLR_C = 5$, indicating that for any assumed prevalence π, both the projected PPV and NPV will be superior for test C. This implies that test C should be regarded as superior to test A, unequivocally—as far as point estimates are concerned.

Biggerstaff noted that, in practice, the sampling imprecision of each estimated sensitivity and specificity should also be taken into account. Section 12.5.6 describes how this may be done.

12.5.2 Sensitivity-Specificity Plot for a Paired Comparison

Newcombe (2001b) presented a different graphical method to compare two tests, which does take sampling uncertainty into account. A weighted mean f of the differences in sensitivity and specificity between the two tests is calculated, to represent the difference in utility between the two tests. Then f is plotted against a mixing parameter λ, which is allowed to range from 0 to 1. The parameter λ represents the prevalence in the population to which the test is to be applied, together with the relative seriousness of false-negative and false-positive decisions. The plot of f against λ is simply a straight line. For each value of λ, a confidence interval for f is obtained by a simple extension of an interval for the difference between proportions. This results in a concave confidence region, which resembles but should not be confused with the concave confidence bounds around a regression line. All computations are of closed form. The original graphs in Newcombe (2001b) were produced by a Minitab overlay plot macro. More conveniently, Excel spreadsheets are available.

In the paired design, the confidence region is derived from the interval for a paired difference of proportions. As noted above, the difference in sensitivity between the two tests is $\theta_1 = \eta_{1.} - \eta_{.1} = \eta_{12} - \eta_{21}$. Similarly the difference in specificity is $\theta_2 = \zeta_{2.} - \zeta_{.2} = \zeta_{21} - \zeta_{12}$. The simple approach developed here enables simultaneous comparison of sensitivity and specificity. It is assumed that tests T_1 and T_2 are applied independently. This method is not appropriate if T_1 and T_2 arise as two different dichotomisations of the same continuous or ordinal variable.

12.5.3 Examples of Paired Comparisons

Faecal occult blood tests (FOBTs) are potentially useful in screening asymptomatic populations to facilitate detection of subclinical colorectal neoplasia. High sensitivity and specificity are a prerequisite to usefulness for a screening programme, but are not readily achieved. The tests work by detecting substances present in small quantities of blood that may be released from lesions. Bleeding may arise not only from carcinomas but also from adenomatous polyps, which are not strictly malignant but have the potential for malignant transformation. A FOBT may also yield a false-positive result due to residues of certain dietary items, even though subjects are instructed to avoid these in preparation for the test. Low sensitivity arises primarily because not all early lesions bleed continuously.

It is convenient (although not ideal) to assess the performance of FOBTs by studying series of patients whose clinical presentation indicates more invasive methods are required. Hope et al. (1996) applied three FOBTs, Monohaem (MH), Hemoccult II (HOII) and BM-Test Colon Albumin (BMCA) to a series of patients who subsequently underwent colonoscopy, which is here regarded as definitive. Results of MH and HOII were available for 160 patients, 24 of whom were affected; BMCA results were available for 108 of them, including 16 affected, as shown in Table 12.5.

Table 12.6 gives results for the ability of two tests, the inactive urinary kallikrein:creatinine ratio (IUK:Cr) and the angiotensin sensitivity test (AST) to predict pre-eclampsia in pregnancy (Kyle et al. 1996).

TABLE 12.5

Comparison of Three Faecal Occult Blood Tests[a]

	Sensitivity		Specificity		Positive LR	Negative LR
MH	14/24	58%	131/136	96%	15.9	2.31
HOII	9/24	37%	118/136	87%	2.83	1.39
MH	7/16	44%	87/92	95%	8.05	1.68
HOII	4/16	25%	81/92	88%	2.91	1.17
BMCA	4/16	25%	82/92	89%	2.30	1.18

Source: Results from Hope, R. L. et al. 1996. *Gut* 39: 722–725.
[a] 160 subjects underwent tests MH and HOII, 108 also underwent BMCA.

TABLE 12.6

Comparison of Two Tests to Predict Pre-Eclampsia

Test	Sensitivity		Specificity		Positive LR	Negative LR
Outcome any pre-eclampsia						
IUK:Cr	42/63	67%	294/395	74%	2.61	2.23
AST	14/63	22%	334/395	85%	1.44	1.09
Outcome proteinuric pre-eclampsia						
IUK:Cr	16/20	80%	311/438	71%	2.76	3.55
AST	5/20	25%	368/438	84%	1.56	1.12

Source: Results from Kyle, P. M. et al. 1996. *British Journal of Obstetrics and Gynaecology* 103: 981–987.

12.5.4 The Plot

We set up a simple display, which weighs the clinical costs of false negatives and false positives. Economic costs of the two test procedures may alter greatly with the passage of time and are not readily incorporated. Define a loss function taking the value c_1 in the event of a false-negative result, c_2 for a false positive, and 0 for a correctly classified subject. This may be regarded as representing misclassification regret: the loss associated with a true positive is set at zero, even though this is a serious outcome, because this result should lead to optimal management. For any condition sufficiently serious to warrant considering a screening programme, it is reasonable to postulate $c_1 > c_2 > 0$. Let π denote the prevalence of the condition in any population to which application of test 1 or 2 is envisaged; this may be very different to M/N, the prevalence in the training set, and will often be low. Then the expected loss due to imperfect performance of test T_1 in relation to the gold standard G is

$$E_1 = \pi\,(1 - \eta_{1.})\,c_1 + (1 - \pi)\,(1 - \zeta_{2.})\,c_2. \tag{12.12}$$

The corresponding loss for test T_2 is

$$E_2 = \pi\,(1 - \eta_{.1})\,c_1 + (1 - \pi)\,(1 - \zeta_{.2})\,c_2. \tag{12.13}$$

Then

$$E_2 - E_1 = \pi\,c_1\,\theta_1 + (1 - \pi)\,c_2\,\theta_2. \tag{12.14}$$

Thus we consider a simple weighted mean of θ_1 and θ_2,

$$f = \lambda\,\theta_1 + (1 - \lambda)\,\theta_2 \tag{12.15}$$

where

$$\frac{\lambda}{1-\lambda} = \frac{c_1\pi}{c_2(1-\pi)}\ \text{i.e., } \lambda = \frac{1}{1+\dfrac{c_2(1-\pi)}{c_1\pi}}. \tag{12.16}$$

The mixing parameter $\lambda \in [0,1]$ represents an appropriate weighting of sensitivity and specificity, taking into account both the relative seriousness of the two types of errors and the prevalence. It depends on and represents the balance between c_2/c_1 and π, enabling the person applying the findings to appraise the implications for the intended application. If the penalty associated with a false negative is the dominant consideration, then $c_2 \ll c_1$. As $c_2/c_1 \to 0$, with π held non-zero, $\lambda \to 1$, and f reduces to θ_1. Conversely, if the prevalence is sufficiently low, the penalty associated with false positives may be the dominant consideration: as $\pi \to 0$, with c_2/c_1 held non-zero, $\lambda \to 0$, and f reduces to θ_2. The linear function f eliminates dependence on the arbitrary unit of the cost scale, and appropriately expresses the comparison between tests 1 and 2 for any chosen λ directly on a proportion scale close to that used for sensitivity and specificity themselves. It is not the intention that the reader would actually supply a value of λ. Rather, the plot is designed to be interpreted as a whole, showing how the comparison between the tests depends on a running parameter that reflects both c_2/c_1 and π.

Then, for $i = 1$ and 2, let (L_i, U_i) denote a $100(1-\alpha)\%$ interval for θ_i, calculated by method 10 defined in Section 8.5.1. For any $\lambda \in [0,1]$, we may apply the square-and-add process again, this time with $w_1 = \lambda$, $w_2 = 1-\lambda$, to obtain $100(1-\alpha)\%$ limits for f as

$$\hat{f} - \sqrt{\lambda^2(\hat{\theta}_1 - L_1)^2 + (1-\lambda)^2(\hat{\theta}_2 - L_2)^2} \text{ and } \hat{f} + \sqrt{\lambda^2(U_1 - \hat{\theta}_1)^2 + (1-\lambda)^2(U_2 - \hat{\theta}_2)^2}, \tag{12.17}$$

the positive value of each square root being taken. This reduces to (L_1, U_1) at $\lambda = 1$ and (L_2, U_2) at $\lambda = 0$.

A plot of f together with these limits, for λ ranging from 0 to 1, may be obtained using the spreadsheet SESPPAIRED.xls. This permits appraisal of the degree of preference for one test over the other, for any assumed prevalence and loss ratio. The reader can choose a value for λ to reflect both the prevalence (whether an actual local value or an assumed one) and their perceived relative seriousness of the two types of errors, and can discern how altering either of these would affect the preference for one test over the other.

It is easily shown that $\dfrac{\partial^2 L}{\partial \lambda^2} < 0 < \dfrac{\partial^2 U}{\partial \lambda^2}$, so the region enclosed by the plot is always concave. Usually $M < N/2$, in which case the interval for the difference in sensitivity (Δ sensitivity) θ_1 (plotted at $\lambda = 1$) is normally wider than the interval for the difference in specificity (Δ specificity) θ_2 (plotted at $\lambda = 0$).

12.5.5 Examples of Sensitivity-Specificity Plots

For the faecal occult blood test data (Table 12.5), Figures 12.3 to 12.5 are plots of f against λ comparing each pair of tests in turn, using a $1 - \alpha = 0.95$ confidence level.

Figure 12.3 shows a statistically significant superiority in specificity ($\lambda = 0$) of MH over HOII, although not in sensitivity ($\lambda = 1$), even though $\hat{\theta}_1 > \hat{\theta}_2$. The interval for θ_2 is narrower than that for θ_1, because M/N is well below 0.5; moreover, the estimated specificities $\hat{\eta}_{2.}$ and $\hat{\eta}_{2.}$ are relatively close to 1. Because of concavity, the whole of the lower limit curve lies above 0 precisely if both L_1 and L_2 do. This would be a rather stringent criterion for superiority, arising with frequency $(\alpha/2)^2$ when in reality $\theta_1 = \theta_2 = 0$ (to which should be added a further $(\alpha/2)^2$ chance of obtaining $U_1 < 0$ and $U_2 < 0$). The pattern shown in Figure 12.3 indicates considerable evidence that MH outperforms HOII: for any $\lambda \le 0.55$, the entire interval lies above 0. This contrasts

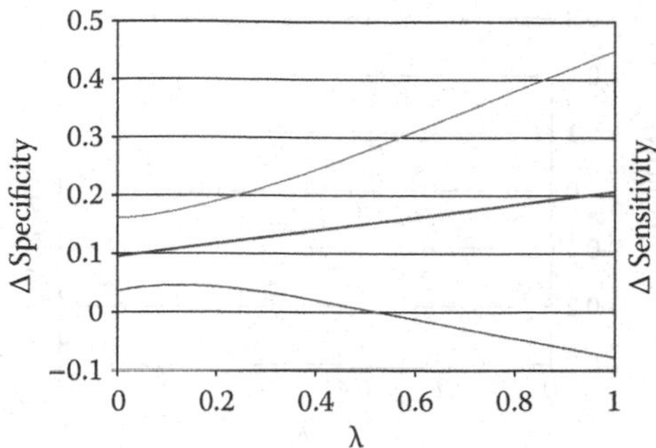

FIGURE 12.3
Comparison of sensitivity and specificity between faecal occult blood tests MH and HOII. (Data from Hope, R. L. et al. 1996. *Gut* 39: 722–725.)

somewhat with the conclusion from the positive and negative LRs, which suggest that MH is unequivocally superior to HOII. The explanation is that the comparison of positive and negative LRs disregards the sampling imprecision, which is particularly great for the sensitivity, as it often would be.

Figure 12.4 shows there is less firm evidence for superiority of MH over BMCA, based on the reduced sample size. There is no λ for which the interval for *f* excludes negative values, although the lower limit comes very close to 0. Here, comparison of positive and negative LRs implies that MH is unequivocally superior to BMCA.

HOII and BMCA obviously have nearly identical sensitivity and specificity based on the 108 subjects who had both these tests performed, with no clear preference from comparison of positive and negative LRs. In Figure

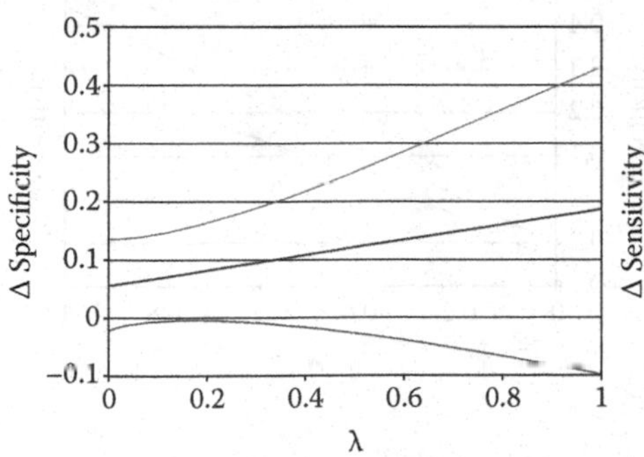

FIGURE 12.4
Comparison of sensitivity and specificity between faecal occult blood tests MH and BMCA. (Data from Hope, R. L. et al. 1996. *Gut* 39: 722–725.)

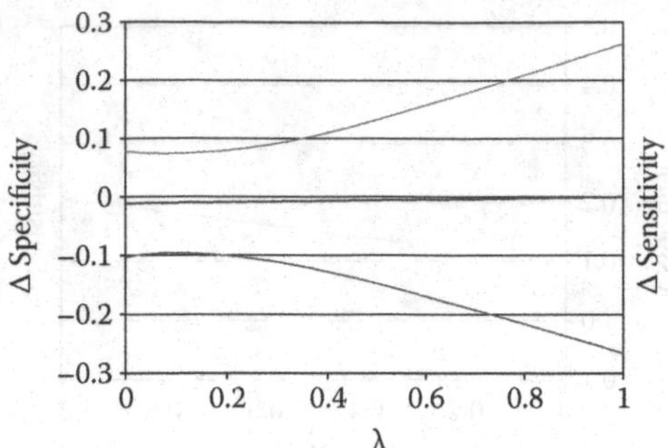

FIGURE 12.5
Comparison of sensitivity and specificity between faecal occult blood tests HOII and BMCA. (Data from Hope, R. L. et al. 1996. *Gut* 39: 722–725.)

12.5, f is subject to considerable imprecision, particularly at high λ. Using rejection of both $f = +0.1$ and $f = -0.1$ at a one-sided 0.025 level as a criterion of equivalence, then only for $0.001 \leq \lambda \leq 0.23$ can the results be regarded as demonstrating equivalence.

For the prediction of pre-eclampsia, Figure 12.6 shows that the much greater sensitivity of IUK:Cr (Table 12.6) outweighs the lower specificity provided $\lambda \geq 0.29$, although for very low $\lambda \leq 0.11$, it is inferior to AST. A similar plot is obtained when attention is restricted to proteinuric pre-eclampsia (Kyle et al. 1997). For both outcomes a comparison of positive and negative LRs implies clear superiority.

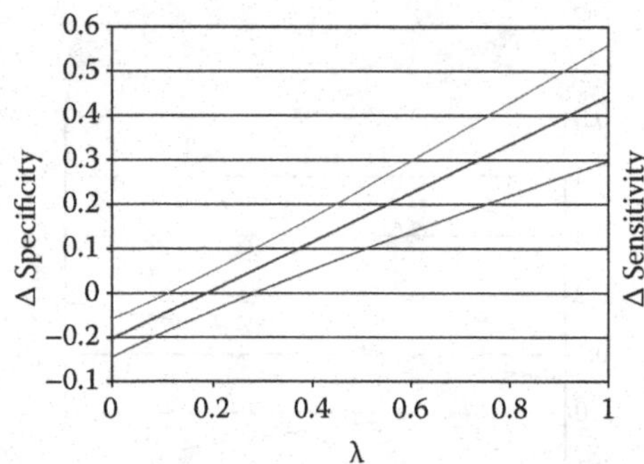

FIGURE 12.6
Comparison of sensitivity and specificity between the inactive urinary kallikrein:creatinine ratio and the angiotensin sensitivity test for prediction of pre-eclampsia. (Data from Kyle, P. M. et al. 1996. *British Journal of Obstetrics and Gynaecology* 103: 981–987.)

12.5.6 Comparisons for Unpaired Data

Sometimes it is not practicable to use all three tests on the same individuals. Some subjects then undergo tests T_1 and G, others undergo tests T_2 and G. It is obviously preferable that the groups of subjects who undergo these pairs of tests should be as similar as possible. This is best achieved by randomisation. A closely analogous plot may then be produced using the spreadsheet SESPUNPAIRED.xls with a confidence region derived from the interval for an unpaired difference of proportions.

Among patients attending accident and emergency departments, there is a high prevalence of alcohol misuse, so this setting affords a convenient opportunity for use of alcohol screening tests. Obviously the briefer the test, the greater the acceptability to both patients and staff. Hodgson et al. (2003) evaluated the Fast Alcohol Screening Test (FAST) (Hodgson et al. 2002), the Paddington Alcohol Test (PAT) (Smith et al. 1996) and the CAGE (Mayfield et al. 1974) against the Alcohol Use Disorders Identification Test (AUDIT) (Babor et al. 1992). Of the three brief tests, FAST had the best combination of sensitivity and specificity when tested against AUDIT as the gold standard.

In Hodgson et al. (2003), 2185 patients completed questionnaires at four A & E departments in different cities in the United Kingdom. Following informed consent and recording of demographic variables one randomly chosen short screening test, either CAGE, FAST or PAT was administered, followed by the full AUDIT questionnaire. The overall prevalence of alcohol misuse in the study population as defined by the AUDIT instrument was 843 out of 2175 (38.8%). Table 12.7 displays the sensitivity and specificity of the FAST, the CAGE and the PAT as predictors of a positive AUDIT, in all subjects receiving the test in question.

The above results indicate that FAST has high sensitivity and specificity. CAGE has low sensitivity but very high specificity. PAT has moderately high sensitivity and specificity. Although there was substantial evidence that sensitivity and specificity were affected by centre, age and gender, nevertheless the sensitivity and specificity of FAST were high in all localities, in men and women, and in those aged under and over 25 alike.

Figure 12.7 shows the comparison in sensitivity and specificity between FAST and CAGE as predictors of the AUDIT result. The difference in sensitivity is +52.5% in favour of FAST, with 95% interval +45.8% to +58.4%. The difference in specificity is 10.0% in favour of CAGE and appears at the left of the diagram as –10.0%, at $\lambda = 0$, with its 95% interval –13.5% to –6.8%. These intervals are calculated by method 10 of Newcombe (1998b) as in Section 7.5.1. Both intervals exclude zero, indicating that the two tests differ significantly in both sensitivity and specificity.

Here, FAST should be regarded as better than CAGE at values of λ for which the diagonal line is above 0, i.e., for $\lambda > 0.16$. For $\lambda > 0.21$, where the upper curve cuts the horizontal axis, FAST is significantly better than CAGE. Conversely, for $\lambda \leq 0.16$, CAGE is better than FAST, significantly so for $\lambda \leq 0.11$.

TABLE 12.7

Sensitivity and Specificity of FAST, CAGE and PAT Alcohol Abuse Screening Tests Relative to AUDIT as a Gold Standard

	Sensitivity		Specificity	
FAST	269/290	92.8%	404/461	87.6%
CAGE	122/303	40.3%	422/432	97.7%
PAT	175/250	70.0%	371/439	84.5%

	Positive LR	95% Limits	Negative LR	95% Limits
FAST	7.50	5.90, 9.62	12.1	8.09, 18.3
CAGE	17.4	9.42, 32.4	1.64	1.50, 1.81
PAT	4.52	3.59, 5.72	2.82	2.34, 3.44

	Positive LR ratio vs. FAST	95% Limits	Negative LR Ratio vs. FAST	95% Limits
CAGE	2.32	1.20, 4.54	0.135	0.088, 0.204
PAT	0.602	0.429, 0.843	0.233	0.147, 0.364

Source: Results from Hodgson, R. J. et al. 2003. *Addictive Behaviors* 28: 1453–1463.

Note: The 95% confidence limits for positive and negative likelihood ratios are calculated by the score method (Miettinen and Nurminen 1985). The 95% limits for ratios comparing PLRs or NLRs between two tests are derived from these by the MOVER-R algorithm.

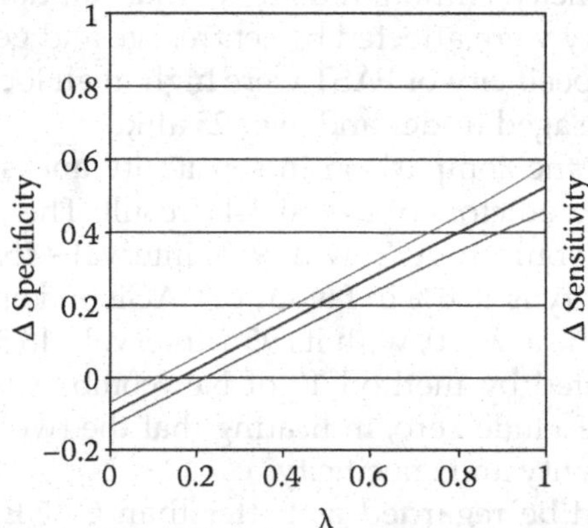

FIGURE 12.7

Comparison of sensitivity and specificity between FAST and CAGE tests for detection of alcohol abuse. (Data from Hodgson, R. J. et al. 2003. *Addictive Behaviors* 28: 1453–1463.)

In the current study, relatively high values of λ are expected. This is because prevalence is high when screening for hazardous and problem drinkers in a medical setting. Also a false negative is more problematic than a false positive. A false positive is probably a borderline case and potentially could benefit from a health promotion intervention. If we consider a false negative to be twice as serious as a false positive then, for a 10% prevalence, λ is 0.18. For 20% prevalence λ = 0.33 and for 30% prevalence λ = 0.46. These values of λ are all above the cut-off point at which FAST is better than CAGE.

Figure 12.8 shows the comparison in sensitivity and specificity between FAST and PAT as predictors of the AUDIT result. FAST has a large, highly significant advantage in sensitivity, and also an advantage in specificity that is somewhat short of statistical significance. FAST is better than PAT for all values of λ, and significantly better than PAT for any plausible value of 0.07 or above.

Table 12.7 includes an analysis of the positive and negative likelihood ratios for the three tests. The positive and negative LRs are ratios of proportions, for which the Miettinen–Nurminen score interval (Section 10.4.2) is appropriate. PAT has much lower PLR and NLR than FAST, indeed the 95% intervals do not overlap, leading to a very clear inference of inferiority. CAGE is clearly inferior to FAST in terms of NLR; conversely its PLR is superior, almost to the point of non-overlapping intervals. Confidence intervals for the ratio of the PLRs (or NLRs) of two tests may then be derived using the MOVER-R algorithm. These analyses confirm that PAT is unequivocally inferior to FAST, whereas CAGE is clearly superior to FAST in terms of PLR but inferior for NLR. The conclusions in this example are particularly clear—seldom would all the confidence intervals for PLR ratios and NLR ratios exclude 1 in this way. A caveat to interpreting such comparisons is that the

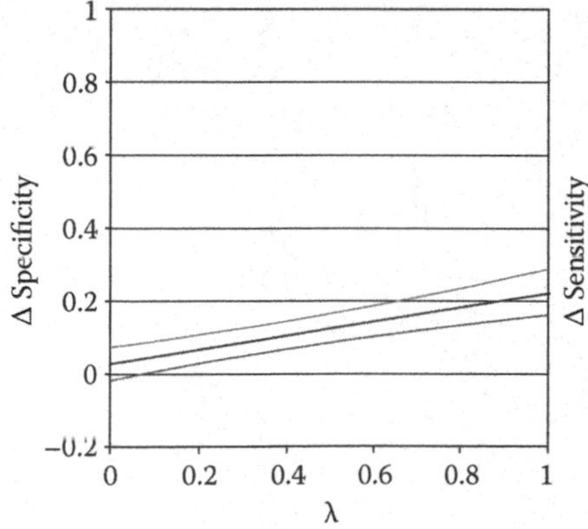

FIGURE 12.8
Comparison of sensitivity and specificity between FAST and PAT tests for detection of alcohol abuse. (Data from Hodgson, R. J. et al. 2003. *Addictive Behaviors* 28: 1453–1463.)

positive and negative LRs for a test are not statistically independent; each is derived from the same (sensitivity, specificity) pair.

The construction of a confidence interval for the ratio of two PLRs (or NLRs) described above presupposes that the two PLRs are derived from separate series of individuals. In the paired design, intervals for the PLR and NLR for a given test could be constructed just as above. But use of MOVER-R to derive intervals for ratios comparing these measures between tests would be invalid due to non-independence, and correct intervals would be complex to construct.

13

Widening the Applicability of Confidence Interval Methods: The Propagating Imprecision Approach

13.1 Background

Suppose we want to calculate a confidence interval for some quantity $F = f(p_1, p_2, \ldots, p_k)$ derived from k parameters (often but not necessarily proportions), each of which is estimated independently. In the simplest cases, $k = 2$, and f is some comparative measure such as the difference or ratio of two proportions or the odds ratio. In general, k may be greater than 2, and the relationship of f to its parameters may be non-linear, possibly even non-monotonic.

In the general case, computer-intensive methods are often resorted to, notably resampling (bootstrapping) or Bayesian methods such as MCMC. While these approaches are very useful and flexible, they have some disadvantages. They use special software that requires large amounts of computational time. They do not yield objectively reproducible results—because they involve generation of pseudo-random numbers, starting with a different seed may lead to a slightly different answer. Moreover, Bayesian methods, with their deliberate intrinsic shift towards the neutral point of the support space, and also bootstrapping without bias correction, do not correspond directly to the simple empirical estimate $\tilde{f} = f(\tilde{p}_1, \tilde{p}_2, \ldots, \tilde{p}_k)$ obtained by substituting the empirical estimates of the k parameters. Neither approach will be developed in detail here.

The method of propagating imprecision (PropImp) described by Newcombe (2011c) is a conceptually and computationally simpler, effective alternative approach. It is closely related to, and may be regarded as a generalisation of, two existing methods.

The substitution method (Section 6.3) starts with a confidence interval for a single parameter and calculates a derived interval on a transformed scale. We noted that in most applications, the measure for which we want a confidence interval depends on other parameters in addition to p, and in most instances it is inadvisable to disregard the sampling imprecision attaching to these.

In Section 7.3 we considered the square-and-add or MOVER-D algorithm, a computationally simple method to calculate a confidence interval for a difference between two quantities estimated from independent samples. This approach extends directly to arbitrary linear functions of two or more quantities. A further extension, MOVER-R (Section 7.3.4) produces confidence intervals for ratios. By combining these approaches, intervals may be produced straightforwardly for most linear and non-linear functions of the form $f(p_1, p_2,..., p_k)$ that are encountered; for example Bender (2001) describes an interval of this form for the number needed to treat. But when the p_i appear in the function in an interlocked manner, this approach cannot be used.

The PropImp procedure developed here may be regarded as further extending the applicability of these approaches. It is more general and flexible than either substitution or MOVER, as it also copes with functions in which the p_i are interlocked in f, and hence permits calculation of confidence intervals in situations in which it would be either impossible or very difficult to do so by any other method. PropImp is however computationally rather more complex and intensive, requiring iteration in k-1 dimensions. In the simplest case where $k = 2$ and intervals for p_1 and p_2 are obtained by closed-form methods, this approach is readily implemented for an automatic solution using spreadsheet software with a large, fixed number of interval bisections. Alternatively, a spreadsheet can be set up to enable dynamic optimisation, or a dedicated program in a language such as Fortran written.

In most applications f is monotonic in its arguments. All these approaches preserve the empirical estimate and preserve boundary-respecting properties provided obvious monotonicity criteria are fulfilled, and are therefore particularly suited to intervals relating to proportions. We later consider an example showing what can happen when f is not monotonic.

De Levie (2001, p. 54) identifies the concept of propagating imprecision with a simple formula involving partial derivatives. Such an approach assumes independent information on the contributory parameters—as does the PropImp algorithm described here. The De Levie method also assumes explicitly that asymptotic expansions are adequate to produce confidence intervals. This assumption is implicit in the MOVER approach, but to a lesser degree in PropImp, which should accordingly be regarded as of more general applicability.

13.2 The Origin of the PropImp Approach

Khan and Chien (1997) collated evidence on the potential benefit of magnesium sulphate as seizure prophylaxis in hypertensive pregnancies by combining parameter estimates from two independent sources. Information on baseline seizure risk (BR) was extracted from Burrows and Burrows (1995) for each of four categories of hypertensive disorder—gestational, chronic, pre-eclampsia and chronic with superimposed pre-eclampsia—and presented as simple proportions with Clopper-Pearson exact intervals. Separately, the odds ratio (OR) comparing risk of seizure activity between women treated with $MgSO_4$ and those on phenytoin was calculated in a meta-analysis of four published randomised trials as 0.26. Confidence intervals for both parameters were wide, the latter being 0.08 to 0.84, which is symmetrical on a log scale as expected.

The authors constructed an estimate of the rate difference to be expected for each category of hypertension:

$$RD = BR - 1/[1 + (1/BR - 1)/OR] \tag{13.1}$$

from which the number needed to treat (NNT) was calculated by reciprocation as usual:

$$NNT = 1/RD = 1/\{BR - 1/[1 + (1/BR - 1)/OR]\} \tag{13.2}$$

The NNT increases monotonically as the odds ratio increases over the range $0 < OR \le 1$. It displays an inverted U-shaped relationship to the baseline risk, but as the turning point is always at least 0.5, well outside the relevant working range of baseline risks here, behaviour is essentially monotonic.

The authors reported confidence intervals for the resulting estimated NNTs (Table 13.1) based solely on the imprecision of baseline risk, using

TABLE 13.1

Baseline Risk and Number Needed to Treat for Benefit of Magnesium Sulphate Seizure Prophylaxis for Four Categories of Hypertension in Pregnancy

	BR (%)		NNT		
				95% Interval	
Category of Hypertension	**Empirical Estimate**	**95% Interval (Clopper-Pearson)**	**Empirical Estimate**	**Khan and Chien**	**PropImp**
Non-proteinuric	0.1	0.003 to 0.75	1000	180 to 40000	170 to 41000
Proteinuric	4.3	2.5 to 6.9	32	20 to **55**	18 to **180**
Chronic	0.6	0.07 to 2.1	230	66 to 1900	62 to 2000
Superimposed	2.1	0.3 to 7.4	65	19 to 530	17 to 560

Source: Newcombe, R. G. 2011c. *Communications in Statistics—Theory and Methods* 40: 3154–3180.

Note: 95% intervals for the numbers needed to treat are shown, calculated by two methods: those reported by Khan and Chien (1997), which take account of sampling variation of the baseline risk only, and PropImp intervals, which properly incorporate the imprecision of the odds ratio (0.26; 95% interval 0.08 to 0.84) as well as the baseline risk. BR = baseline risk; NNT = number needed to treat.

essentially Daly's substitution method, by substituting in the above formulae the lower and upper Clopper–Pearson limits for the relevant baseline risk and the point estimate for the odds ratio. This procedure has the deficiency that it takes into account the sampling imprecision of the baseline risk, but disregards that of the odds ratio. In this instance, reassuringly, the baseline risk turned out to be the dominant source of variation for seven of the eight limits calculated. Nevertheless, confidence intervals that encapsulate only one of the sources of variation are potentially anticonservative. It is clearly preferable for the interval for the NNT to incorporate imprecision of both contributing parameters. Table 13.1 also includes 95% intervals calculated subsequently by the PropImp method described below, which is designed to do so.

A sensitivity analysis may be performed by instead using upper and lower limits for the odds ratio, combined with the point estimate for the baseline risk. Reassuringly, this indicated that when calculating lower limits for the NNTs, for all four hypertension groups the uncertainty in the baseline risk was the dominant source of imprecision. The same applied at the upper limit, with the exception of the analysis for proteinuric hypertension, for which the uncertainty in the odds ratio had a greater effect than that of the baseline risk. Thus the analyses used by Khan and Chien allow for the more important source of uncertainty in seven out of eight cases.

The PropImp intervals are necessarily at least as wide as those in which only one source of uncertainty is taken into account. Doing so makes little difference here, except for the upper limit for proteinuric hypertension.

In Table 13.1, all intervals for the baseline risk are calculated by the Clopper–Pearson exact method. A less conservative method could be used. Using the overconservative Clopper–Pearson method here tends to mitigate the anticonservatism resulting from disregarding the sampling imprecision of the odds ratio. But there is no particular reason why these should cancel out—it is better to use a purpose-built CI method for the NNT based on intervals for both the baseline risk and odds ratio that each have appropriate coverage properties.

In general, we require a method that will cope with a wide range of such functions of two or more independently estimated parameters. While the method described below is equally applicable to functions of three or more quantities, for clarity of exposition we consider the simplest case where $k = 2$. Normally we expect the function of interest, $F = f(X,Y)$, to be monotonic in both parameters, at least over their working ranges. If it is linear, its standard error may simply be derived from those of X and Y. But for proportions and related quantities, this approach is unlikely to be useful, as the most appropriate intervals for these quantities are not symmetrical on an additive scale.

Let $\tilde{X}$ and $\tilde{Y}$ denote empirical estimates for X and Y, and let (X_L, X_U) and (Y_L, Y_U) denote the corresponding 1-α (usually 95%) intervals. We would normally choose boundary-respecting intervals with appropriate coverage properties. Suppose for simplicity that all the above limits are positive. If F is an increasing function of both parameters X and Y, such as $X + Y$ or $X \times Y$, the obvious way to combine these intervals to construct an interval for F is to use $(f(X_L, Y_L), f(X_U, Y_U))$. For contrast measures such as $X - Y$ or $X \div Y$, where $\partial F/\partial X > 0$ but $\partial F/\partial Y < 0$, the corresponding interval for F is $(f(X_L, Y_U), f(X_U, Y_L))$. Such intervals are likely to be overconservative, as they incorporate the full sampling imprecision of both parameters. Lloyd (1990) developed a confidence interval for paired differences of proportions of this form, combining $1 - \alpha$ intervals for the off-diagonal proportion and the ratio of the off-diagonal cells, which he subsequently conceded was overconservative. Similarly, Rittgen and Becker (2000) developed an interval for a standardised mortality ratio which incorporates a correction factor to take into account a proportion of deaths for which the cause is uninformatively missing. Their method combines $1 - \alpha/2$ exact intervals for the two parameters, and thus would be anticipated to be grossly overconservative. Conversely, either of the intervals $(f(X_L, \tilde{Y}), f(X_U, \tilde{Y}))$ or $(f(\tilde{X}, Y_L), f(\tilde{X}, Y_U))$ tends to be anticonservative, since they reflect the imprecision of only one parameter and regard the other as known exactly. This is acceptable only if we know that the imprecision in F is almost entirely attributable to one of the two parameters. What is needed is an algorithm for an interval (f_L, f_U) with a clear rationale and appropriate performance.

When we seek an interval estimate for a measure comparing two independent proportions, there is a tenuous link to the suggestion by Tryon (2001, p. 374) and Tryon and Lewis (2009) to calculate inferential confidence intervals for the two proportions using a modified z-value of $1.9600/\sqrt{2} = 1.3859$, based on Goldstein and Healy (1995). Zou and Donner (2008a) refer to other papers that do something similar. The two intervals would be expected to overlap only if the proportions differ significantly at the conventional 5% α level. As explained in Section 7.8, a crucial drawback is that displaying ICIs for X and Y risks them being taken out of context and interpreted separately in a grossly anticonservative manner. However, in the applications envisaged here, interest centres on some kind of comparison between the two proportions, rather than intervals for the two proportions separately. We may obtain $(1 - \alpha^*)$ intervals for X and Y, for some α^* derived from α, not in order to display these two intervals but to use them to create an interval for $F = f(X,Y)$. We certainly want α^* to be greater than α, not equal to α (as in Lloyd 1990) or less than α (as in Rittgen and Becker 2000) corresponding to two options considered in Section 14.7. Using $1 - \alpha^* = 0.8342$ corresponding to $z = 1.9600/\sqrt{2}$ is reasonable if the two sources of imprecision affect F equally, but this is not always the case. If X is the dominant source of imprecision of $f(X,Y)$, the interval should approximate to $(f(X_L, \tilde{Y}), f(X_U, \tilde{Y}))$. Using $1 - \alpha^* = 0.8342$ for both X and Y leads to an interval that is narrower than this. This suggests we should partition the uncertainty between the two parameters in some appropriately data-driven way, using intervals for X and Y based on different α-values, α_1 and α_2, or equivalently different z-values, z_1 and z_2. In the Goldstein–Healy and Tryon approaches, the common value of z_1 and z_2 is taken to be $z/\sqrt{2}$, which ensures that

$$z_1^2 + z_2^2 = z^2 \tag{13.3}$$

This is the constraint we apply to z_1 and z_2. Accordingly, we choose z_1 and its corresponding z_2 so as to maximise the uncertainty expressed by the interval width, rather than arbitrarily constraining them to be equal. When this constraint is used, and the method is used to combine Wald intervals, it leads to the ordinary Wald interval for $p_1 - p_2$. If there is no uncertainty in one of the parameters, the method reproduces the base interval for the other one. Separate choices of (z_1, z_2) pairs are made for lower and upper limits.

13.3 The PropImp Method Defined

Thus the suggested solution is as follows. Suppose first that f is a monotonic increasing function of both its arguments. Let $\left(X_z^L, X_z^U\right)$ denote an interval for X with nominal coverage 1-2$Q(z)$, where Q denotes the standard Gaussian tail function, so that $\left(X_{1.96}^L, X_{1.96}^U\right)$ is the usual 95% interval.

Likewise, let $\left(Y_z^L, Y_z^U\right)$ be a 1-2$Q(z)$ interval for Y.

Then we choose $\theta = \theta_1 \in [0, \pi/2]$ to minimise the lower limit $L = f\left(X_{z\sin\theta}^L, Y_{z\cos\theta}^L\right)$.

Similarly, we choose $\theta = \theta_2 \in [0, \pi/2]$ to maximise the upper limit $U = f\left(X_{z\sin\theta}^U, Y_{z\cos\theta}^U\right)$.

We then use (L, U) as a 1-2$Q(z)$ interval for $f(X,Y)$.

If X is the dominant source of imprecision of $f(X,Y)$, we expect θ_1 and θ_2 to be close to $\pi/2$.

Generally, we do not expect θ_1 and θ_2 as defined here to be identical.

Equivalently, for each limit in turn, define $\lambda_1 = \sin\theta$, $\lambda_2 = \cos\theta = \sqrt{\left(1-\lambda_1^2\right)}$, and choose λ_1 iteratively to maximise the interval width.

This formulation extends in an obvious way to the case in which f is a decreasing function of one or both parameters. The case $\partial f/\partial X > 0 > \partial f/\partial Y$ is of particular importance, as it relates to contrast measures such as differences and ratios of similar parameters. Here,

$$L = \min_{0\le\theta_1\le\pi/2} f\left(X_{z\sin\theta_1}^L, Y_{z\cos\theta_1}^U\right) \text{ and } U = \max_{0\le\theta_2\le\pi/2} f\left(X_{z\sin\theta_2}^U, Y_{z\cos\theta_2}^L\right). \tag{13.4}$$

One justification for the square-and-add approach was that the interval for $p_1 - p_2$ obtained by calculating Wald intervals for p_1 and p_2 separately and squaring and adding is identical to the usual Wald interval for $p_1 - p_2$. Exactly the same applies to PropImp. Here, the PropImp and usual square-and-add limits are

$$p_1 - p_2 \pm z \max_{0 \le \theta \le \pi/2} \left(e_1 \sin\theta + e_2 \cos\theta\right) \text{ and } p_1 - p_2 \pm z\sqrt{e_1^2 + e_2^2} \tag{13.5}$$

where $e_i = \sqrt{\{p_i (1 - p_i)/n\}}$, $i = 1, 2$.

Putting $\psi = \tan^{-1} \frac{e_1}{e_2} \in [0, \pi/2]$, the equality of the PropImp and square-and-add limits follows from the fact that $\sin\theta \sin\psi + \cos\theta \cos\psi = \cos(\theta - \psi)$ attains its maximum value 1 when $\theta = \psi$.

The formulation also extends directly to a function of $k \ge 2$ parameters.

Here, let superscripts distinguish the k parameters, and let $F = f(X^1, X^2, \ldots, X^k)$.

Let $\left(L_z^i, U_z^i\right)$ denote a $1 - 2Q(z)$ interval for X_i, for $i = 1 \ldots k$.

Then, assuming F is an increasing function of all its parameters, the PropImp limits for F are

$$L = \min_{\{z_i\}:\Sigma z_i^2 = z^2} f\left(L_{z_1}^1, \ldots, L_{z_k}^k\right) \text{ and } U = \max_{\{z_i\}:\Sigma z_i^2 = z^2} f\left(U_{z_1}^1, \ldots, U_{z_k}^k\right). \tag{13.6}$$

Again, we interchange the roles of $L_{z_i}^i$ and $U_{z_i}^i$ for those parameters with $\partial f/\partial X_i < 0$.

To implement this for the two-parameter case, we must be able to calculate intervals for X and Y using arbitrary z_1 and z_2 between 0 and $z = 1.96$. The methods used to obtain the intervals for X and Y may or may not be well-specified. In the Khan–Chien example, the Clopper–Pearson exact interval was used for the baseline risk—as a member of the family of beta intervals this can readily be applied with arbitrary z when the raw data are known. The odds ratio and its confidence interval came from a meta-analysis. Here, and in general, we do not necessarily expect access to original data in order to reproduce the calculation of the interval for the odds ratio for arbitrary z, but sometimes we will be able to interpolate on some suitable transformed scale, in this instance log OR. This method of interpolation is chosen because in logistic regression the logs of the confidence limits for the odds ratio are linear in z. Wherever there is a scale on which the interval is symmetrical, such as the logit scale for the Wilson interval (Newcombe 1998a), such interpolation is feasible.

For the hypertension prophylaxis example, the relevant calculations may be performed in an interactive spreadsheet by a trial-and-error (TAE) approach as in Table 13.2. User input values, shown in bold, are required for the numerator and denominator of the baseline risk (proteinuric hypertension in this example), and point and interval estimates for the odds ratio. The 95% interval for the baseline risk is calculated by the Clopper–Pearson exact method, for compatibility with Khan and Chien; setting each $\lambda_1 = 1$

TABLE 13.2

Excel Spreadsheet to Calculate Confidence Intervals for the RD and NNT from BR and OR by the PropImp Method Using Clopper–Pearson Exact Interval for BR and Interpolation on Log Scale for OR

							95% Interval	
BR	**16**	out of	**372**	i.e.	0.043011	0.024781	to	0.068908
OR					**0.26**	**0.08**	to	**0.86**
				Point estimates:		RD	0.0314603	
						NNT	31.786	
Trial values of lambda1 and lambda2 for lower limit for RD:								
lambda1 = **0.065**					Lower limit for BR		0.040770	
lambda2 = 0.997885					Upper limit for OR		0.857827	
					Lower limit for RD		0.005593	
					Upper limit for NNT		178.81	
Trial values of lambda1 and lambda2 for upper limit for RD:								
lambda1 = **0.879**					Upper limit for BR		0.065647	
lambda2 = 0.476822					Lower limit for OR		0.148216	
					Upper limit for RD		0.055340	
					Lower limit for NNT		18.070	

Source: Newcombe, R. G. 2011c. *Communications in Statistics—Theory and Methods* 40: 3154–3180.

Note: User input values are shown in bold. Trial and error approach. Data for risk of proteinuric hypertension from Khan and Chien (1997). RD = rate difference; NNT = number needed to treat; BR = baseline rate; OR = odds ratio.

reproduces their published intervals. The Clopper–Pearson calculation used here is closed-form, based on the incomplete beta integral. For an arbitrary value of λ_1, the relevant $\lambda_1 z$ limit for the baseline risk is calculated by the Clopper–Pearson formula. For the corresponding value of λ_2, namely, $\sqrt{(1-\lambda_1^2)}$, the relevant $\lambda_2 z$ limit for the odds ratio is calculated by interpolation on the scale of log OR. The interval reported by Khan and Chien for the odds ratio is symmetrical on a log scale, but only to within rounding error, consequently interpolation on the log OR scale is performed separately for lower and upper limits.

To obtain the upper limit for the NNT, the user selects an initial value of λ_1 (bold cell), then modifies it keeping within the range (0, 1) and notes whether this decreases or increases the lower limit displayed. Given the monotonicity of the relationship of the rate difference to the baseline risk and odds ratio, the upper limit for the NNT has a typical inverted U-shaped relationship to λ_1, as displayed in Figure 13.1. Therefore, λ_1 can be adjusted sequentially to maximise the upper limit for the NNT. Similarly a separate value of λ_1 is chosen to minimise the lower limit for the NNT. Here, as in other applications involving proportions, generally optimising λ_1 to 3 dp is sufficient to optimise the corresponding limit for the rate difference to 6 dp.

For proteinuric hypertension, as for the other groups considered, the baseline risk is the dominant source of imprecision at the lower NNT limit, hence the λ_1 chosen here is close to 1. However, for this group only, the odds ratio is the dominant source of imprecision at the upper NNT limit, leading to a λ_1 close to 0 and a much larger upper limit than that reported by Khan and Chien (Table 13.1).

Interpolation on a log scale to obtain intervals with reduced z values for the odds ratio is what enables the PropImp method to be applied to the Khan–Chien example. No existing practicable method is available, and the formula for the NNT in terms of the baseline risk and odds ratio does not simplify in such a way as to make squaring and adding possible. For the Khan–Chien example, in principle an alternative approach, a kind of general Wilson method would be to derive the large sample standard deviation of the BR, then express it in terms of BR, and other nuisance parameter-like quantities which were functions of BR and OR. These nuisance parameters would be estimated, either directly with MLE, or by profile likelihood, or some other method. After that, an inversion procedure akin to the Wilson method could be used. The resulting intervals are likely to have good performance. The downside is that the amount of algebraic work involved would

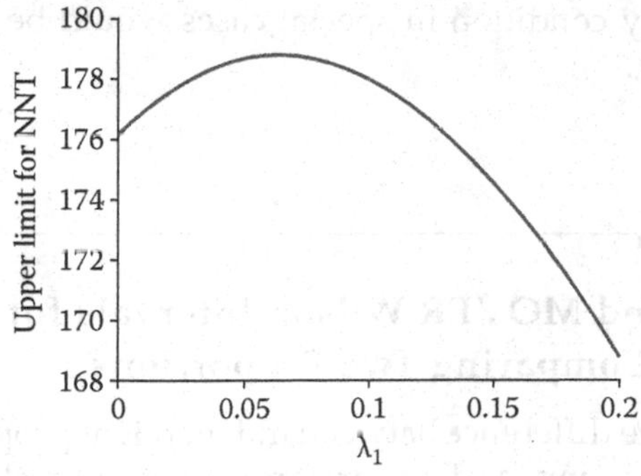

FIGURE 13.1
Optimisation of upper limit for NNT by choice of λ_1. (Data for risk of proteinuric hypertension from Khan, K.S. and Chien, P. F. W. 1997. *British Journal of Obstetrics and Gynaecology* 104: 1173–1179.)

be prohibitive in such an example. So this "conventional" approach is not a practical option here. Furthermore, the MCMC approach cannot be used here without recourse to the actual 2 × 2 tables for the four studies contributing to the meta-analysis. So, for this example, the PropImp method really provides the only accessible solution short of reworking the meta-analysis—it represents a high level of sophistication in post-processing others' results.

In Chapter 14 we examine several other applications of the PropImp algorithm. In many instances, but not all, a MOVER-R solution is available by using log-transformed scales for the parameters, after Zou and Donner (2008a). This permits a closed-form solution and generally gives similar results. However, this approach cannot be applied to the Khan–Chien problem, because the formula for the rate difference or NNT in terms of the baseline rate and odds ratio does not split into separate components relating to these two parameters.

An alternative perspective on how this method works may be described as follows. Let us set aside the question of combining parameters as if they were random variables, and all the Bayesian-like questions of principle that could be discussed. Suppose that using CI expressions as if they were probability statements we can write $p_1 = f(z_1)$, $p_2 = f(z_2)$, where z_1 and z_2 are independent standard normal variables. Suppose that a function $\theta = g(p_1, p_2)$ is a "new" parameter of interest. From these functional representations we can write $\theta = h(z_1, z_2)$, where the function h may be identified in principle. The PropImp method may be regarded as identifying certain points on curves of the form $z_1^2 + z_2^2 = z^2$, which are also tangent to the level contours of the function h. The approximation in the method, whether it is conservative or anticonservative, then depends on whether those contours are concave or convex, as well as of course on whether the base intervals used are conservative or anticonservative. If the contours are linear, then there is no implied approximation, suggesting that the method would work best in those cases where the contours were nearly linear. However, to evaluate the underlying convexity/concavity condition in special cases would be a formidable analytical undertaking.

13.4 PropImp and MOVER Wilson Intervals for Measures Comparing Two Proportions

When applied to the difference between independent proportions with Wald intervals, both PropImp and square-and-add approaches reproduce the familiar Wald interval for the difference. When a better method such as the Wilson score interval is used for each proportion, the resulting square-and-add and PropImp intervals for the difference of proportions have favourable properties. The same is true of PropImp applied to the relative risk and odds ratio. We study the effect of these methods in these simpler cases, with various base intervals, not so much to create new methods as to provide insight into what properties MOVER and PropImp intervals are likely to have in general. In these contexts, essentially, we are conditioning on two marginals *m* and *n* throughout.

The rate difference RD = $p_1 - p_2 = a/m - b/n$ is the simplest and, unsurprisingly, the least problematic case. Here, the square-and-add (MOVER-D) method is always applicable. In Section 7.4.2 we defined four cases NZ (no zero), OZ (one zero), RZ (row zero) and DZ (diagonal zero) according to the pattern of empty cells. In all these cases, both methods preserve the boundary-respecting properties of the Wilson interval. Both methods give point estimate tethering as defined in Section 7.5.4 only for case DZ, where it is appropriate anyway. In case RZ, $0/m - 0/n$, square-and-add is identical to PropImp, both methods give $(-U(0,n), U(0,m))$, simply because the PropImp limits are optimised by $\lambda_L = 0$, $\lambda_U = 1$. Similarly, in case OZ, $a/m - 0/n$, both square-and-add and PropImp methods give the same upper limit $U(a,m)$, because the upper limit is optimised by $\lambda_U = 0$, but slightly different lower ones.

Table 13.3 presents confidence intervals for five examples calculated using 11 methods, for which we later evaluate coverage and so forth in Section 13.6. The methods are defined in greater detail in Section 13.6.1. Methods A to C can display boundary abnormalities, as in Newcombe (1998a). Relative width and location properties of different intervals are generally in line with the properties shown in the evaluation (Section 13.6).

For the relative risk RR = $p_1/p_2 = (a/m)/(b/n)$ there is the additional complexity that zero cells can lead to infinity entering the calculations, either by division or as log(0). We define four cases NoZ, NumZ, DenZ and Indet as in Table 13.4. Examples are shown in Table 13.5. The MOVER-DL method, like the default delta method, is unnecessarily restricted to case NoZ, whilst PropImp and MOVER-R, like the score method are unrestricted and give appropriate results in all four cases. PropImp and MOVER intervals can be identical (e.g., for 10/10 versus 20/20). For 29/29 versus 51/56, these methods give identical upper limits, but slightly different lower ones. The explanation is the same as for the rate difference.

For the odds ratio OR = $p_1 (1 - p_2)/p_2 (1 - p_1) = a(n - b)/b(m - a)$, again there are basically four cases. We label these NoZ, NumZ, DenZ and Indet just as for the relative risk. But they are defined in a more complex manner than for the relative risk, as zeros in any of the four cells are relevant (Table 13.4). Just as for the rate ratio, the MOVER-DL method for the odds ratio (like the default delta method introduced in Section 11.3.1) is unnecessarily restricted to case NoZ, whilst MOVER-R, PropImp and the score method are unrestricted and give appropriate results in all four cases. Table 13.6 gives illustrative examples.

TABLE 13.3

95% Confidence Limits for Selected Contrasts between Independent Binomial Proportions Calculated by 11 Methods

	56/70 – 48/80 (NZ)	6/7 – 2/7 (NZ)	5/56 – 0/29 (OZ)	0/10 – 0/20 (RZ)	10/10 – 0/20 (DZ)
Wald intervals					
A Without CC	+.0575, +.3425	+.1481, +.9947	+.0146, +.1640	0*, 0*	+1*, +1
B With CC	+.0441, +.3559	+.0053, >1*	−.0116, +.1901	−.0750, +.0750	+.9250, >1*
Square-and-add intervals					
C Wald with CC	+.0481, +.3519	+.0478, >1*	+.0039, +.1747	−.0559, +.0559	+.9441, >1*
D Wilson score without CC	+.0524, +.3339	+.0582, +.8062	−.0381, +.1926	−.1611, +.2775	+.6791, +1
E Wilson score with CC	+.0428, +.3422	−.0290, +.8423	−.0667, +.2037	−.2005, +.3445	+.6014, +1
F Clopper–Pearson exact	+.0439, +.3442	−.0366, +.8567	−.0442, +.1962	−.1684, +.3085	+.6485, +1
G Mid-*p*	+.0524, +.3369	+.0259, +.8425	−.0236, +.1868	−.1391, +.2589	+.7061, +1
PropImp intervals					
H Wilson score without CC	+.0530, +.3380	+.0530, +.8478	−.0383, +.1926	−.1611, +.2775	+.7225, +1
I Wilson score with CC	+.0392, +.3506	−.0895, +.9193	−.0750, +.2037	−.2005, +.3445	+.6168, +1
J Clopper–Pearson exact	+.0403, +.3506	−.0870, +.9248	−.0508, +.1962	−.1684, +.3085	+.6437, +1
K Mid-*p*	+.0528, +.3392	+.0226, +.8856	−.0227, +.1868	−.1391, +.2589	+.7276, +1

Source: Newcombe, R. G. 2011c. *Communications in Statistics—Theory and Methods* 40: 3154–3180.

Note: Asterisks denote boundary anomalies. As noted elsewhere, the Wald interval for a difference of independent proportions corresponds to combining Wald intervals for the two proportions by either squaring and adding or PropImp. CC = continuity correction.

TABLE 13.4

Special Cases Defined for RR and OR, and Applicability of MOVER and PropImp Intervals

Case	Definition for RR	Definition for OR	Point Estimate	Delta and MOVER-DL Intervals	MOVER-R, PropImp and Score Intervals
NoZ	Neither *a* nor *b* zero	None of the four cells zero	0 < Point estimate < ∞	$0 < L < U < \infty$	$0 < L < U < \infty$
NumZ	Numerator cell $a = 0$ with $b > 0$	$a(n - b) = 0 < b(m - a)$	Point estimate = 0	Cannot be used	$L = 0 < U < \infty$
DenZ	Denominator cell $b = 0$ with $a > 0$	$b(m - a) = 0 < a(n - b)$	Point estimate = ∞	Cannot be used	$0 < L < U = \infty$
Indet	Both *a* and *b* zero	Both $a(n - b)$ and $b(m - a)$ zero	Point estimate indeterminate	Cannot be used	$L = 0$ and $U = \infty$

Source: Newcombe, R. G. 2011c. *Communications in Statistics—Theory and Methods* 40: 3154–3180.

Note: RR = relative risk; OR = odds ratio.

TABLE 13.5

95% Confidence Limits for Selected Ratios of Independent Proportions Calculated by Four Methods

Case	Contrast	Delta Method	MOVER-DL Wilson	MOVER-R Wilson	PropImp Wilson	Score
NoZ	56/70 ÷ 48/80	1.077, 1.651	1.0784, 1.6640	1.0783, 1.6629	1.079, 1.671	1.079, 1.672
NoZ	6/7 ÷ 2/7	0.895, 10.06	1.119, 10.49	1.108, 10.46	1.099, 10.71	1.064, 11.15
NoZ	29/29 ÷ 51/56	1.012, 1.192	0.95878, 1.23848	0.95901, 1.23848	0.95880, 1.23848	0.964, 1.240
NoZ	10/10 ÷ 20/20	1*, 1*	0.722, 1.192	0.722, 1.192	0.722, 1.192	0.716, 1.199
NumZ	0/29 ÷ 5/56	–	–	0, 1.589	0, 1.572	0, 1.394
DenZ	5/56 ÷ 0/29	–	–	0.629, ∞	0.636, ∞	0.717, ∞
Indet	0/10 ÷ 0/20	–	–	0, ∞	0, ∞	0, ∞

Source: Newcombe, R. G. 2011c. *Communications in Statistics—Theory and Methods* 40: 3154–3180.

* = Boundary anomalies.

TABLE 13.6

95% Confidence Limits for Selected Odds Ratios Comparing Independent Proportions Calculated by Four Methods

Case	Contrast	Delta Method	MOVER-DL Wilson	MOVER-R Wilson	PropImp Wilson	Score
NoZ	56/70 vs. 48/80	1.276, 5.572	1.287, 5.524	1.292, 5.547	1.282, 5.545	1.282, 5.540
NoZ	6/7 vs. 2/7	1.031, 218.3	1.395, 161.3	1.432, 167.7	1.240, 181.5	1.148, 172.0
NumZ	0/29 vs. 5/56	–	–	0, 1.672	0, 1.636	0, 1.437
DenZ	5/56 vs. 0/29	–	–	0.598, ∞	0.611, ∞	0.696, ∞
Indet	0/10 vs. 0/20	–	–	0, ∞	0, ∞	0, ∞

Source: Newcombe, R. G. 2011c. *Communications in Statistics—Theory and Methods* 40: 3154–3180.

Note: MOVER-DL and MOVER-R are applied to $p_i/(1-p_i)$, $i = 1, 2$.

For such ratio measures, we expect intervals to be either exactly or approximately symmetrical on a log scale—the issue here is different to that for simple proportions for which two natural scales exist. Delta intervals for the RR and OR are exactly symmetrical on a log scale, by definition. For the RR, none of the other intervals shown in Table 13.5 has this property. However, for the OR, the Wilson PropImp and MOVER-DL intervals have this property, but the MOVER-R interval does not.

For the odds ratio only, we consider what happens when we condition on row totals (in the notation here, $a + b$ and $m + n - a - b$) instead of column totals m and n. Doing so alters both the PropImp and MOVER limits. (No alteration occurs if the table has marginal homogeneity, of course.) Reversing the conditioning cannot alter which case the data corresponds to, which depends purely on the pattern of zero cells, not the marginals.

For example, Table 13.7, reconstructed from Iwatsubo et al. (1998), shows history of exposure to asbestos in 75 women with mesothelioma and 77 control women. The 95% interval for the odds ratio obtained by applying PropImp to Wilson intervals is 4.61 to 76.3 with the correct conditioning and 4.66 to 75.4 with the alternative conditioning. Altering the conditioning in this way has a similar effect on the limits obtained by applying MOVER-DL to Wilson intervals, changing them from (4.66, 75.5) to (4.69, 74.9). The Woolf limits obviously do not depend on the conditioning and are (4.25, 82.7) regardless. It is not obvious whether we should regard this non-invariance on reversing the conditioning as a bug or a favourable feature of this conditioning. In any case, the effect is relatively slight, even in a table with gross marginal heterogeneity such as our example.

TABLE 13.7

Excel Spreadsheet to Calculate 95% Confidence Limits for the Odds Ratio by the PropImp Method, Based on Wilson Score Intervals for the Two Proportions

	Numerator	Denominator	Point Estimate	Lower Limit	Upper Limit
*p*1	25	75	0.3333	0.2371	0.4458
*p*2	2	77	0.0260	0.0072	0.0898
PropImp		QOR	0.8987	0.6433	0.9741
		OR	18.7500	4.6064	76.3198

Apply PropImp Algorithm for Lower Limit

Current Best μ_0	Corresp. $f(\mu_0)$	Current Scan Interval Width h	$\mu_1 = \mu_0 - h$	z_1^2 for μ_1	z_2^2 for λ_1	Lower z_1 Limit for p_1	Upper z_2 Limit for p_2	$f(\mu_1)$	$\mu_2 = \mu_0 + h$	z_1^2 for μ_2	z_2^2 for μ_2	Lower z_1 Limit for p_1	Upper z_2 Limit for p_2	$f(\mu_2)$	Best μ for Next Step	Best f for Next Step	Best h for Next Step
0.5000				1.9207	1.9207	0.2629	0.0649	0.6743									
0.5000	0.6743	0.5000	0.0000	0.0000	3.8415	0.3333	0.0898	0.6702	1.0000	3.8415	0.0000	0.2371	0.0260	0.8419	0.0000	0.6702	0.2500
0.0000	0.6702	0.2500	0.0000	0.0000	3.8415	0.3333	0.0898	0.6702	0.2500	0.9604	2.8811	0.2824	0.0778	0.6470	0.2500	0.6470	0.1250
0.2500	0.6470	0.1250	0.1250	0.4802	3.3613	0.2968	0.0839	0.6434	0.3750	1.4405	2.4009	0.2717	0.0715	0.6580	0.1250	0.6434	0.1250
0.1250	0.6434	0.1250	0.0000	0.0000	3.8415	0.3333	0.0898	0.6702	0.2500	0.9604	2.8811	0.2824	0.0778	0.6470	0.1250	0.6434	0.0625
0.1250	0.6434	0.0625	0.0625	0.2401	3.6014	0.3072	0.0869	0.6467	0.1875	0.7203	3.1212	0.2889	0.0809	0.6440	0.1250	0.6434	0.0313
0.1250	0.6434	0.0313	0.0938	0.3601	3.4813	0.3015	0.0854	0.6443	0.1563	0.6002	3.2412	0.2926	0.0824	0.6434	0.1563	0.6434	0.0313
0.1563	0.6434	0.0313	0.1250	0.4802	3.3613	0.2968	0.0839	0.6434	0.1875	0.7203	3.1212	0.2889	0.0809	0.6440	0.1563	0.6434	0.0156
. . .																	

Source: Newcombe, R. G. 2011c. *Communications in Statistics—Theory and Methods* 40: 3154–3180.

Note: Illustrative data based on Iwatsubo et al. (1998). Optimisation of lower and upper limits is performed by interval bisection on the scale of Q_{OR} = (OR – 1)/(OR + 1) (Yule 1900).

13.5 Implementation of the PropImp Method

For purposes of programming development and checking, three types of algorithms were produced: interval bisection and trial-and-error in Excel, and dedicated Fortran programming. Excel is adequate for the purpose, as it generally displays #DIV/0! for ∞ and #NUM! for non-computable values. The programming needs to heed special cases with zero cells in order to display a lower limit 0 for an observed zero relative risk or odds ratio.

- *Interval bisection:* For a monotonic function of two parameters, precise solution in Excel is satisfactory, using say 40 interval bisections for $\mu = \lambda_1^2$ for each limit in turn. The odds ratio example shown in Table 13.7 illustrates the process. Here, just as for the Miettinen–Nurminen intervals (Section 10.4.2), interval bisection is conveniently performed on a transformed scale with finite range, namely $Q_{OR} = (OR - 1)/(OR + 1)$: a similar transformation would be used for a relative risk. It is necessary to use a refined interval bisection algorithm here. Let μ_0 denote the current best choice of μ, and h the current interval scan width, both initially set at 0.5. Identifying f with Q_{OR} here, we examine $f(\mu_0)$ and $f(\mu_0 \pm h)$, always subject to the constraint $0 < \mu < 1$. If either $f(\mu_0 - h)$ or $f(\mu_0 + h)$ is better than $f(\mu_0)$, we shift to that μ, but keep the current h for the next iteration. If $f(\mu_0)$ remains optimal, we hold μ_0 fixed and halve h for the next step. For compact display, 4 dp are displayed here, but of course the full precision available is used in the branching algorithm.
- *Trial and error:* For a monotonic function of two or more parameters, Excel provides an approximate trial-and-error approach in a very natural way. This obtains upper and lower limits to a precision that is adequate except for evaluation of coverage. For $F = f(p_1, p_2, \ldots, p_k)$, we set up a spreadsheet resembling Table 13.2 in which the user chooses $\lambda_1, \lambda_2, \ldots, \lambda_{k-1}$ to optimise the lower limit. The remaining parameter λ_k is then displayed, calculated as $\sqrt{\{1 - \lambda_1^2 - \lambda_2^2 - \ldots - \lambda_{k-1}^2\}}$. The user cycles through these $k - 1$ parameters in turn, adjusting each in turn so as to minimise the resulting value of F. A similar adjustment process is performed separately to maximise the upper limit. As noted for the Khan–Chien example, the function to be optimised normally has a flat peak, such that adjusting $\lambda_1, \lambda_2, \ldots, \lambda_{k-1}$ to just 3 dp is usually sufficient to optimise U and L to about 6 dp. The trial-and-error algorithm is transparent, no transformation of scale being involved, and robust, provided monotonicity is established. The Excel SOLVER macro is useful, especially when $k > 2$.
- *Fortran:* In general, a dedicated program in a language such as Fortran is needed. The procedure becomes more complex if we relax

the monotonicity condition as then at each stage we need to scan the entire interval, not just take the value at a pre-chosen endpoint – this would normally lead to an extra level of iteration. A scale transformation is needed for the odds ratio or relative risk as described for the interval bisection method above.

13.6 Evaluation

As well as examining similarity of PropImp intervals to those produced by existing methods, we present a formal evaluation of coverage and related issues for PropImp intervals. We do so primarily in order to obtain some broad brush evidence on whether in general this approach is likely to preserve the coverage properties of the interval methods used for p_1, p_2 and so forth, or tends to introduce conservatism or anticonservatism. For the simplest application, the difference between independent proportions, we evaluate coverage, location and width properties of new methods obtained by applying MOVER-D and PropImp to Wilson, Wilson continuity-corrected, Clopper–Pearson exact and mid-*p* intervals, defined in Section 3.5. As in Section 4.4.2, interval location is characterised by the balance between mesial and distal non-coverage. The known performance of the various satisfactory existing methods provides a yardstick to help interpret the findings for the PropImp intervals here.

The evaluation proceeds exactly as in Section 7.5, using the same 9200 parameter space points, so performance measures reported for methods evaluated in that study remain unaltered. For each of 230 (m, n) pairs, 40 (ψ, θ) pairs are sampled uniformly from the triangle delimited by $0 < \psi < 1$ and $0 < \theta < 1 - |2\psi - 1|$, where $\psi = (\pi_1 + \pi_2)/2$ and $\theta = \pi_1 - \pi_2$.

13.6.1 Eleven Methods for a Difference of Proportions

The 11 methods are listed in Table 13.3. Methods A and B are familiar Wald intervals for the difference of proportions without and with continuity correction. Methods C to G are formed from intervals for p_1 and p_2 by squaring and adding. Methods H to K correspond to D to G but use the PropImp algorithm.

The basic interval methods for the single proportion $p = r/n$, combined as above, are defined in Section 3.5:

Wald limits without continuity correction: $p \pm z\sqrt{(pq/n)}$, where $q = 1 - p$.

Wald limits with continuity correction: $p \pm \{z\sqrt{(pq/n)} + 1/(2n)\}$.

Wilson score interval without continuity correction: {π: |p − π| ≤ z √ (π(1 − π)/n)}.

Wilson score interval with continuity correction: {π: |p − π| − 1/(2n) √ z ≤ (π(1 − π)/n)}.

Clopper–Pearson exact interval: [L, U], with $L \le p \le U$, such that for all π in the interval:

$$\text{(i) if } L \le \pi \le p, \qquad kp_r + \sum_{j:r<j\le n} p_j \ge \alpha/2, \tag{13.7}$$

$$\text{(ii) if } p \le \pi \le U, \qquad \sum_{j:0\le j<r} p_j + kp_r \ge \alpha/2 \tag{13.8}$$

$$\text{respectively, where} \qquad p_j = \Pr[R = j] = \binom{n}{j} \pi^j (1-\pi)^{n-j}, \tag{13.9}$$

$j = 0, 1,\ldots, n$, R denoting the random variable of which r is the realisation, and $k = 1$. As usual an empty summation is understood to be zero.

Mid-p interval: as Clopper–Pearson, but with $k = 1/2$.

Method C is obtained by squaring and adding Wald continuity-corrected intervals for p_1 and p_2:

$$\frac{a}{m} - \frac{b}{n} \pm \sqrt{\left\{z\sqrt{\frac{ac}{m^3}} + \frac{1}{2m}\right\}^2 + \left\{z\sqrt{\frac{bd}{n^3}} + \frac{1}{2n}\right\}^2}. \tag{13.10}$$

It is easily demonstrated geometrically that this interval is narrower than the more usual continuity-corrected Wald interval for the difference of independent proportions (method B):

$$\frac{a}{m} - \frac{b}{n} \pm \left\{z\sqrt{\frac{ac}{m^3} + \frac{bd}{n^3}} + \frac{1}{2m} + \frac{1}{2n}\right\}. \tag{13.11}$$

13.6.2 Coverage

In Table 7.3 the mean coverage probability for the square-and-add Wilson interval without continuity correction was 0.9602, similar to that of Wald with continuity correction, but with a much more acceptable minimum coverage. Table 13.8 extends Table 7.3 and shows coverage and location properties for our 11 intervals. Squaring and adding other intervals that seek to

TABLE 13.8

Estimated Coverage Probabilities for 95% Intervals for a Difference between Independent Binomial Proportions

Method	Coverage		Mesial Non-Coverage		Distal Non-Coverage		Mean MNCP ÷ Mean NCP
	Mean	Minimum	Mean	Maximum	Mean	Maximum	
Wald intervals							
A Without CC	0.8807	0.0004	0.0417	0.7845	0.0775	0.9996	0.350
B With CC	0.9623	0.5137	0.0183	0.4216	0.0194	0.4844	0.486
Square-and-add intervals							
C Wald with CC	0.9455	0.4336	0.0262	0.4499	0.0283	0.5658	0.481
D Wilson score without CC	0.9602	0.8673	0.0134	0.0660	0.0264	0.1327	0.337
E Wilson score with CC	0.9793	0.9339	0.0061	0.0271	0.0147	0.0661	0.293
F Clopper–Pearson exact	0.9781	0.9339	0.0074	0.0213	0.0145	0.0661	0.339
G Mid-p	0.9636	0.9215	0.0136	0.0354	0.0228	0.0661	0.375
PropImp intervals							
H Wilson score without CC	0.9616	0.8936	0.0166	0.0828	0.0218	0.1064	0.433
I Wilson score with CC	0.9856	0.9634	0.0056	0.0221	0.0089	0.0353	0.386
J Clopper–Pearson exact	0.9840	0.9634	0.0063	0.0184	0.0097	0.0244	0.392
K Mid-p	0.9643	0.9215	0.0155	0.0373	0.0202	0.0470	0.433

Source: Newcombe, R. G. 2011c. *Communications in Statistics—Theory and Methods* 40: 3154–3180.

Note: Calculated by 11 methods including square-and-add and PropImp based methods. Based on 9200 parameter space points with $5 \le m \le 50$, $5 \le n \le 50$, $0 < \psi < 1$ and $0 < \theta < 1 - |2\psi - 1|$ as in Section 7.5.2. CC = continuity correction.

align with 1 – α the minimum coverage (Clopper–Pearson or Wilson with continuity correction) or mean coverage (mid-*p*) likewise results in mean and minimum coverage similar to those for the base method for single proportions, as does PropImp applied to the Wilson interval without continuity correction or the mid-*p* interval. None of the square-and-add methods, not even when based on Clopper–Pearson intervals, ensures strict conservatism. Squaring and adding preserves boundary-respecting properties, and approximate mean coverage, but does not propagate "exactness." This realisation evidently prompted Fagan (1999) to apply an *ad hoc* widening to interval *F* in a Draconian attempt to ensure strict conservatism. Not surprisingly, the resulting method had conservative coverage properties, and greater width than the Wald interval, to a degree that was large at small sample sizes and became small for large samples. Nevertheless, the PropImp Clopper–Pearson interval had a minimum coverage well above the nominal 0.95 for all 9200 PSPs evaluated here.

Although the algebraic difference in definition between intervals B and C is subtle, it results in a substantial difference in mean coverage, B being conservative and C anticonservative on average.

13.6.3 Location

Intervals for the RD produced by all 11 methods erred on the side of being too mesially located, as judged by the MNCP/NCP criterion introduced in Section 4.4.2. Remarkably, methods B and C had nearly balanced mesial and distal non-coverage; the balance was also within tolerance for the PropImp methods based on less conservative base intervals, *H* and *K*.

Methods I and J have similar MNCP/NCP ratio, and similar mean and minimum coverage probabilities. Only method J keeps both the maximum mesial and distal non-coverage probabilities below 0.025, unlike methods I and F, both of which do so for MNCP but not DNCP.

Score intervals for the single proportion, without or with a continuity correction, tend to be too mesially located, whereas the corresponding tail-based (exact or mid-*p*) intervals have appropriate location properties (Newcombe 1998a, and Table 5.3). Contrastingly, considering intervals for the rate difference derived from these methods, what leads to best interval location is not so much the choice of tail-based in preference to score intervals as choosing to combine using PropImp rather than squaring and adding.

For RD and other functions combining two or more proportions, PropImp is applied to Wilson score intervals as these are easy to compute, boundary-respecting and align mean overall coverage closely with the nominal 1-α. Non-coverage of the Wilson interval for the single proportion is predominantly distal, suggesting it might be preferable to base on some other interval method for proportions that does not produce such a mesial shift. However, it is clear that both PropImp and MOVER approaches do not propagate interval location properties. Thus the Wald interval for the single proportion is grossly too distal, but the Wald interval for $p_1 - p_2$, which is also the PropImp Wald interval, is in fact overall somewhat mesially located, as was evident from Table 7.3, although this is not apparent for every possible dataset such as 6/7 – 2/7 (Table 13.3). In fact, mean mesial and distal non-coverage rates are 0.0417 and 0.0775, suggesting too mesial, but the mean of the MNCP/NCP ratios across 9200 PSPs was 0.517, suggesting appropriate location—the disparity being attributable to extreme skewness and ratios not preserving linear properties—but there is certainly no suggestion that it is too distal, therefore MOVER and PropImp do not propagate location properties. Thus it is apparent that the location properties of an interval for a function $f(p_1, p_2)$, based on an interval method for p_1 and p_2 with known location properties (determined either from a simulation study or, more efficiently, the Box–Cox index of symmetry) cannot reliably be predicted and would need to be determined *de novo* in a dedicated simulation study.

13.6.4 Interval Width

Table 13.9 extends Table 7.4 and shows average interval width for nine selected PSPs. The PropImp approach is designed to produce an interval which is wider than the corresponding substitution method interval that takes into account only one of the sources of uncertainty involved. Any such algorithm that works by maximising interval width subject to constraints has the potential for overconservatism. It is necessary to evaluate interval width as well as coverage properties in any case, but especially in view of this. In this evaluation, usually but not invariably intervals I and J were the widest, in line with having the most conservative coverage. For a similar reason, interval A was generally the narrowest, as in the examples in Table 13.3. Interval B was invariably wider than A, with C intermediate, in the same order as these intervals are nested. For $n = 10$ with $m = 10$ or 100, with $\pi_1 = \pi_2 = 0.5$, B and C were actually wider than J and I. All this is generally in line with coverage properties.

13.6.5 Comparison of PropImp and Square-and-Add Intervals

Comparing corresponding square-and-add and PropImp methods, the PropImp interval always has the better location. When Clopper–Pearson or continuity-corrected score intervals are used, PropImp intervals were wider than square-and-add for all nine parameter space points in Table 13.8; for mid-p and score intervals without continuity correction, PropImp intervals were usually wider than square-and-add ones. For Clopper–Pearson and continuity-corrected score intervals, the PropImp mean and especially minimum coverage probabilities are importantly higher than with square-and-add. For score intervals without continuity correction, the difference in mean

TABLE 13.9

Average Width for 95% Intervals for a Difference between Independent Binomial Proportions Calculated by 11 Methods, for Selected Parameter Space Points

	m	10	10	10	100	100	100	100	100	100
	n	10	10	10	10	10	10	100	100	100
	π_1	0.01	0.5	0.95	0.01	0.5	0.95	0.01	0.5	0.95
	π_2	0.01	0.5	0.05	0.01	0.5	0.05	0.01	0.5	0.05
Wald intervals										
A Without CC		0.125	1.148	0.504	0.193	1.138	0.567	0.551	1.019	0.940
B With CC		0.480	1.424	0.716	0.528	1.340	0.743	0.774	1.093	1.096
Square-and-add intervals										
C Wald with CC		0.367	1.343	0.639	0.442	1.318	0.688	0.692	1.071	1.049
D Wilson score without CC		1.000	1.000	1.000	1.000	1.000	1.000	1.000	1.000	1.000
E Wilson score with CC		1.234	1.138	1.203	1.227	1.127	1.196	1.185	1.050	1.106
F Clopper–Pearson exact		1.115	1.183	1.134	1.101	1.168	1.130	1.007	1.057	1.066
G Mid-*p*		0.942	1.085	0.992	0.934	1.078	0.990	0.869	1.012	0.982
PropImp										
H Wilson score without CC		1.005	1.072	0.903	1.002	1.023	0.998	1.014	1.009	0.970
I Wilson score with CC		1.251	1.300	1.217	1.243	1.180	1.219	1.255	1.082	1.124
J Clopper–Pearson exact		1.128	1.315	1.172	1.112	1.209	1.156	1.060	1.084	1.107
K Mid-*p*		0.946	1.137	0.943	0.934	1.092	0.995	0.878	1.017	0.976
Actual average width for method D		0.5627	0.7231	0.4773	0.3289	0.5430	0.3522	0.0895	0.2707	0.1264

Note: All widths are expressed relative to the average width for the square-and-add Wilson interval without continuity correction. CC = continuity correction.

coverage is smaller, and negligible for mid-*p*. This all suggests that in general PropImp is superior to the MOVER-D approach, although it risks being overconservative.

13.6.6 Boundary Anomalies

When boundary-respecting methods are used for the intervals for X and Y, and f is monotonic in both parameters, then the square-and-add and PropImp limits for $f(X,Y)$ cannot introduce any boundary aberrations afresh—the only anomalous behaviour possible in our examples is then the known singularity of the confidence interval for the NNT in the non-significant case. However, when the intervals for X and Y are not boundary-respecting, their anomalies are also liable to appear in the resulting interval for $f(X,Y)$ (Table 13.3). Violation of the boundaries at ±1 can occur for methods B and C as well as for method A, which also produces a degenerate interval in cases RZ and DZ. In our evaluation, method A produced a zero-width interval with frequency 0.0460. The boundary violation frequencies for methods A, C and B were 0.0270, 0.0476 and 0.0594, once again naturally in the same order as these intervals are nested.

13.7 The Thorny Issue of Monotonicity

For the square-and-add method as such, which is designed for linear functions only, there is no possibility of violation of monotonicity. For the substitution method, there is an issue, as the function can be arbitrary. As noted previously, when a square-and-add interval for $p_1 - p_2$ is turned into one for the NNT by substitution, it is the reciprocation step, not the squaring and adding, that introduces the potential for anomalous behaviour. In general terms, it is clear that, whenever f is monotonic in X and Y, then $g(f(X,Y))$ will have similar properties if the function g is also monotonic. For example, Chatellier et al. (1996) expressed the NNT as $1/\{p_1 \times \text{RRR}\}$, where p_1 and p_2 denote the risk or event rates in control and intervention series and RRR denotes the relative risk reduction $1 - p_2/p_1$ as in Section 10.1. In this context, $f(X,Y)$ represents the risk difference $p_1 - p_2$ and $g(f(X,Y))$ is the NNT. Away from the singularity, the reciprocal transformation g preserves the desired properties here. Thus the substitution method remains by far the most appropriate way to calculate a confidence region for the NNT, notwithstanding.

We may also consider what happens when a confidence interval is calculated for a function such as $f(p) = (p - 1/2)^2$ for $\hat{p}$ close to 1/2. When $\hat{p} = 1/2$, point estimate tethering occurs, the interval extends in only one direction away from this value. This is appropriate behaviour here, even although $\hat{p}$ is strictly within its support. When $\hat{p}$ is close to 1/2, the lower limit for f is 0, and the upper limit will reflect imprecision in $\hat{p}$ in just one direction.

Now consider the PropImp algorithm. Consider the non-monotonic function

$$F = f(p_1, p_2) = \max\,[5\{1 - (p_1 - 0.1)^2\}\{1 - (p_2 - 0.1)^2\},\\ 4\{1 - (p_1 - 0.6)^2\}\{1 - (p_2 - 0.6)^2\}] \quad (13.12)$$

for p_1 and p_2 ranging on [0, 1], which is continuous and has two peaks, a mesial, lesser peak at (0.6, 0.6) and a distal, higher peak at (0.1, 0.1) as in Figure 13.2. While such a function would be an implausible choice for an informative effect size measure, it serves to demonstrate a range of abnormalities of behaviour that can arise when F is not monotonic.

For the observed data $p_1 = 2/10$, $p_2 = 70/100$, we get $\hat{F} = 3.33$, with 95% interval 2.84 to 3.94. There is no problem of non-monotonicity just here, since close to (0.2, 0.7), f is increasing in one parameter, decreasing in the other. But this does not apply to all possible data. It is easy to take for granted that we can use lower limits for x and y to get a lower limit for f and upper limits for x and y to get an upper limit for f – but in fact such a simplification is a luxury of the monotonic case. When there is non-monotonicity, obviously we may have to look at either the upper or lower limit for x, whichever results in the wider interval for f. Indeed, even this is inadequate. Suppose that x and y were both a little above 0.1. Then the upper limit for f would be 5, using values for x and y that are exactly 0.1 (i.e., strictly within the intervals for x and y. So, for any non-monotonic function f, the issue is not one of deciding whether to use upper or lower limits for x and y. We need to scan the whole of the z_1 interval for x and the z_2 interval for y to locate the extremum. In other words, we look for the extreme values of f, both lower and upper, over a rectangular subset of $[0, 1]^2$ swept out by the z_1 interval for x and the z_2 interval for y. There are also computational issues involved in this search, where non-monotonicity occurs—we need a more systematic approach than

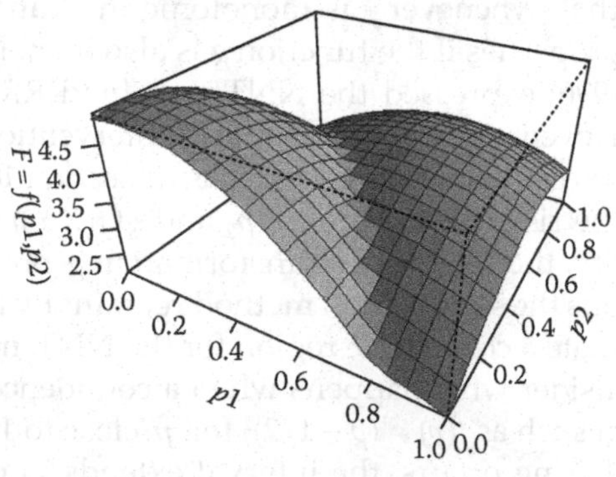

FIGURE 13.2
Non-monotonic function used to display possible behaviours of the PropImp algorithm where non-monotonicity applies.

starting at an arbitrary start point (e.g., $x = y = 0.5$) and looking for the largest increases in two directions, as here this would head towards the near peak at (0.6, 0.6, 4), not the taller, distant one at (0.1, 0.1, 5).

If we observe (0.1, 0.1), only a one-sided interval for f can be constructed. Point estimate tethering is normally an adverse aspect of the performance of a confidence interval method. Here, it is a direct consequence of non-monotonicity.

If we observe (0.6, 0.6), and consider 1-2$Q(z)$ intervals for f as z increases from 0 towards ∞, the upper limit for f remains constant at 4 for low values of z, then at some value of z, starts climbing towards 5, as we get past the contour that is level with the lower peak at 4, into the lower left part of the grid.

13.8 Some Issues Relating to MOVER and PropImp Approaches

The MOVER and PropImp approaches were designed to have wide applicability. The MOVER method depends on choice of base methods for the intervals being combined. PropImp depends on choice of methods for the intervals being combined or availability of interpolation on an appropriate scale. This contrasts with MCMC which requires starting with explicit individual or aggregate level data, and depends on choice of model and prior. The PropImp algorithm has nothing to contribute in the case of comparisons of proportions based on individually paired data including differences (Chapter 8) and ratios (Section 10.7), although MOVER methods can be modified to take non-independence into account, as in method 10 of Section 8.5.1 and Zou and Donner (2010a).

Copas (1997) suggested that the interval estimation methods for X and Y should be chosen with an eye to the form of f to be used. This suggestion closely parallels the choice of the Wilson interval as the obvious candidate for squaring and adding, rather than others such as Clopper–Pearson as in Fagan (1999). Because PropImp does not involve any specific formula for combining intervals, merely substitution (the formulaic part relates to the z_i), this issue appears to be more pertinent to the MOVER approach than for PropImp.

In conclusion, both MOVER and PropImp approaches are widely applicable—PropImp more so—it is even more flexible and has apparently better performance, but is computationally more complex and may sometimes err towards overconservatism.

14

Several Applications of the MOVER and PropImp Approaches

14.1 Introduction

In this chapter, we consider numerous further applications of the MOVER and PropImp algorithms. These include a method for calculating confidence intervals for interaction effects on an additive scale that has been comprehensively evaluated. Often, the quantity of interest $f(X, Y)$ can be expressed in the form $g(f_1(X) - f_2(Y))$ or $g(f_1(X)/f_2(Y))$ for some functions f_1, f_2 and g. In such situations the MOVER approach is easy to apply. This is the case for many but not all the examples considered here. In several situations this cannot be done, and the PropImp algorithm is needed.

14.2 Additive-Scale Interaction for Proportions

Newcombe (2001c) developed a simple, effective closed-form method to calculate a confidence interval for the difference between two differences of proportions. The method is based on Wilson score intervals for the proportions, and may be used to compare either unpaired or paired differences. It is equally applicable whether cell frequencies are large or small, and coverage properties are favourable. It is easily adapted to give a confidence interval for the treatment effect for a binary outcome in a two-period crossover trial.

For example (a) below from the Bristol antenatal care study (Jewell et al. 2000), this difference of differences is estimated as (17/65 – 17/75) – (18/72 – 16/65) = +0.0310. Simply applying the squaring and adding process twice to Wilson intervals produces a 95% interval from –0.1687 to +0.2343; there is no question of scale transformation here. Also, applying the PropImp algorithm introduced in Chapter 13 is feasible using the trial-and-error approach, adjusting 3 λs in turn, and gives –0.1717 to +0.2367, slightly wider than the square-and-add interval. The square-and-add approach appears to be perfectly adequate here; as in other examples involving linear functions of parameters, the additional complexity of PropImp adds little value.

This linear scale interaction is also readily applicable to the semi-paired case, in which changes in a binary variable following an intervention are compared between two independent groups, as in examples (b) to (d) below. The additive scale interaction is readily implemented in spreadsheet software, for both the case of four independent groups and the semi-paired case.

14.2.1 Four Applications

(a) The Bristol Antenatal Care Study (Jewell et al. 2000) sought to evaluate the effectiveness of routine antenatal visits by asking whether choice improves well-being. Women were randomised to either a traditional or a flexible schedule of antenatal care. As one of the outcomes (unpublished data), 277 women who developed a problem were asked whether their problem could have been recognised earlier than it was. Among nulliparous women, the proportions answering yes in the two groups were 17/65 and 17/75. Among multiparae, the proportions were 18/72 and 16/65. While clearly none of the differences between these four proportions are conventionally statistically significant, it is of interest to assess the degree to which the effect of the intervention differs between these two subsets of the parturient population, which it was anticipated could be quite different.

A factorial arrangement of data of this kind could also arise when subjects are randomised to four treatment regimes determined by all possible combinations of two binary factors, for example

1. Active A + Active B	2. Active A + Dummy B
3. Dummy A + Active B	4. Dummy A + Dummy B

In such situations, the outcome variable commonly has Gaussian distributional form. The correct analysis would then be to fit a 2 × 2 ANOVA model to model the main effect of factor A (active versus dummy, averaged over the two levels of factor B); the main effect of factor B (active versus dummy, averaged over the two levels of factor A); and an interaction term which indicates to what degree the effect of A is modified by factor B (or equivalently, the effect of B is modified by factor A).

When, as here, the outcome is binary in nature, often a logistic regression model is fitted to the 2 × 2 data structure (Section 11.5). This gives point and interval estimates for a coefficient representing the interaction on a logit scale; these would usually be transformed back by exponentiation to give an interaction estimate on an odds ratio scale and its associated confidence interval. Indeed, contemporary data analysis practice gives the impression that logistic regression is the only way to analyse data of this kind. But it is not. Sometimes, as here, there is particular interest in the interaction effect. Differences on a logit scale, without or with back-transformation to an odds ratio scale, can be difficult to grasp. The interaction effect on the logit scale embodies an additional layer of complexity. By contrast, the interaction on an additive scale is much simpler to explain: the chance of a problem being recognised in advance increased by 3.1% more following the intervention in primiparous women compared to their multiparous peers. The logit and linear scale interactions are obtained by different processes, are interpreted quite differently, and may even lead to different conclusions. Which way of looking at the data is most suitable often depends on the context.

(b) In a longitudinal study (McNamee 1999; related data published as Frank et al. 2008) the prevalence of wheezing was determined on two occasions 3 years apart in 446 children from households in which at least one parent smoked and 310 children from non-smoking households. The investigators sought to quantify the degree to which the change in wheezing prevalence over time differs between the two types of family.

(c) In a lifestyle intervention study (Oldroyd et al. 1999) the proportion of subjects who undertook vigorous physical activity more than three times per week was determined at baseline and after 6 months in 34 subjects in the intervention group and 32 subjects in the control group.

(d) In a crossover study of hospital versus home physiotherapy for chronic multiple sclerosis (Wiles et al. 2001), 31 out of 40 patients gave a positive response to treatment at home, whereas 25 patients gave a positive response to treatment in hospital. In Newcombe and Altman (2000), this example is quoted as a possible application of a good interval estimation method for a paired difference. However, just as in Section 2.3, it is preferable to compare period I minus period II differences between treatment groups, essentially as in Hills and Armitage (1979), primarily to obviate confounding with period differences.

In all these situations, we apply the usual squaring and adding approach a second time, to obtain an interval for the difference of two unpaired differences (example (a) above), or the difference of two paired differences (examples (b) to (d)). The methods used for illustration here are the ones based on squaring and adding Wilson intervals, method 10 of Chapter 7 (Newcombe 1998b) for the case of four independent groups in a factorial arrangement and method 10 of Chapter 8 (Newcombe 1998c) for the difference of two paired differences.

14.2.2 Specimen Calculations

(a) In the Bristol Antenatal Care Study, the estimated difference in nulliparous women was +0.0349, with 95% square-and-add Wilson limits (–0.1056, +0.1774). The corresponding difference in multiparae was +0.0038 (–0.1411, +0.1458). The interaction effect is estimated as +0.0310, with 95% limits (–0.1687, +0.2343) calculated directly from the above figures by applying the usual square-and-add or MOVER-D process.

(b) In the wheezing example, the estimated change was a reduction in prevalence from 119 to 96 out of 446 children from smoking families, a difference of –0.0516, with 95% limits (–0.0944, –0.0087) using method 10 of Chapter 8. The corresponding change in non-smoking families was from 82/310 to 69/310, a difference of –0.0419 (–0.0875, +0.0035). The difference between the changes in the two groups was –0.0096, with 95% limits (–0.0721, +0.0529) by applying the process described above. In this large-sample example, these are very similar to the limits (–0.0715, +0.0522) produced by squaring and adding Wald intervals.

(c) In the lifestyle intervention study, the estimated change was from five to 14 out of 34 in the intervention group, a change of +0.2647, with 95% limits (+0.0498, +0.4509). The change in the control group was (5–6)/32 = –0.0312 (–0.1680, +0.1020). The difference of the differences was +0.2960 (+0.0430, +0.5269). In this example with small numbers, this approach is greatly superior to the method based on Wald intervals.

(d) For the crossover study, the relevant results are shown in Table 14.1. As explained in Section 2.3, the treatment effect is best estimated as half the difference between the period differences in the two

TABLE 14.1

Crossover Trial of Home versus Hospital Physiotherapy for Multiple Sclerosis

Treatment Sequence	Number of Patients Benefiting From: Both	First	Second	Neither	Advantage to Home Treatment: Estimate	95% Limits
Home–Hospital	11	6	1	3	+0.2381	–0.0127, +0.4534
Hospital–Home	9	4	5	1	–0.0526	–0.3372, +0.2434
Difference					+0.2907	–0.0973, +0.6475
Halved					+0.1454	–0.0486, +0.3238

Source: Data from Wiles, C. M. et al. 2001. *Journal of Neurology, Neurosurgery and Psychiatry* 70: 174–179.

Note: Based on the treating physiotherapist's assessment of whether the patient benefited from either type of treatment.

treatment order groups. The resulting estimate, +0.1454 and 95% limits, (−0.0486, +0.3238), are here very similar to those obtained by direct application of method 10 of Newcombe (1998c), +0.1500 and (−0.0488, +0.3339).

14.2.3 Evaluation of Coverage

In Chapter 7, an interval for the difference between independent proportions p_1 and p_2 was derived by applying squaring and adding to Wilson limits for p_1 and p_2. In Chapters 7 and 13, we found that doing so preserves the boundary-respecting property of the Wilson interval, and approximately preserves its coverage properties. So it is plausible that the same would apply to intervals for a difference of differences derived by applying squaring and adding a second time, starting from Wilson intervals. Conversely, we would anticipate less favourable properties when the Wald interval is used as the base interval. It is straightforward to demonstrate that this is the case.

For the unpaired case, we evaluate coverage for the method described above, based on method 10 of Chapter 7. As a control, we also examine coverage of intervals produced by applying the squaring and adding process to Wald intervals (method 1 of Chapter 7). For the paired case, we evaluate coverage by the method described above, based on method 10 of Chapter 8. As a comparator, we also examine coverage of intervals produced by applying the squaring and adding approach to Wald intervals for the paired difference of proportions, method 1 of Chapter 8.

Criteria for evaluation are as in previous chapters. We examine mean coverage based on an appropriate sample of parameter space points, and mean mesial and distal non-coverage. As usual, the Wald methods are capable of producing boundary abnormalities of zero width intervals and overshoot. These properties can transfer to the interaction case but will be very infrequent on any plausible prior, and are not evaluated in detail.

Coverage is evaluated by a Monte Carlo method as in Newcombe (2001c). As usual, we avoid restricting to rounded values for sample sizes and proportions, which may be atypical. For the unpaired case, we first choose n_i, $i = 1\ldots4$ from an integer uniform distribution; here and elsewhere, all sampling is random and independent. We examine three ranges of sample size, zones A (10 to 20), B (50 to 100) and C (250 to 500). In each zone we choose 100,000 sets (n_i, $i = 1\ldots4$). For each of these, we sample a single PSP by choosing a set of four hypothetical proportions (π_i, $i = 1\ldots4$) as follows. To get π_1 and π_2, we choose ψ_1, θ_1 with $\psi_1 \sim U(0,1)$, $\lambda_1 \sim U(0,1)$ and $\theta_1 = \lambda_1\{1-|2\psi_1 - 1|\}$, then $\pi_1 = \psi_1 + \theta_1/2$, $\pi_2 = \psi_1 - \theta_1/2$, just as in Chapter 7. A further pair (ψ_2, θ_2) is chosen, leading to π_3 and π_4. Then the true interaction effect is $\theta = \theta_1 - \theta_2 = \pi_1 - \pi_2 - \pi_3 + \pi_4$. For each of $i = 1\ldots4$ we then sample one r_i from the binomial distribution $B(n_i, \pi_i)$. Then the empirical estimate of θ is $\hat{\theta} = \theta_1 - \theta_2 = \frac{r_1}{n_1} - \frac{r_2}{n_2} - \frac{r_3}{n_3} + \frac{r_4}{n_4}$.

We calculate the Wilson square-and-add interval for θ, as described above, and also the Wald interval which reduces to $\hat{\theta} \pm z\sqrt{\sum_{i=1}^{4}\{p_i(1-p_i)/n_i\}}$ and determine whether coverage or mesial or distal non-coverage occurs. We also examine the theoretical minimum coverage of both methods.

For the paired case, we again run 100,000 simulations for each zone using a similar approach. For i = 1 and 2 we choose n_i as above, and also additional parameters ϕ_i, ν_i and ξ_i, defined as in Chapter 8, from $U(0, 1)$, $U(0.5, 1)$ and $U(0.5, 1)$. We determine the corresponding hypothetical proportions for the four cells, sample cell frequencies randomly, calculate Wald and Wilson square-and-add intervals for the interaction, and determine coverage properties.

No evaluation of interval width was carried out here; it was assumed that width properties are expected to carry across from those for the corresponding two-sample case.

Table 14.2 shows mean coverage and mesial and distal non-coverage probabilities for low, medium and high sample sizes, for the unpaired and paired cases. For the unpaired case, for small n the Wilson method is somewhat conservative, and the interval is rather too mesially located. The Wald method is rather anticonservative, although much less so than the corresponding method for the unpaired difference (Newcombe 1998b). Within each sample size zone there is no clear relationship to the minimum of the expected frequencies across all eight cells. Both methods approach the desired coverage properties for large n.

TABLE 14.2

Mean Coverage, Mesial and Distal Non-Coverage Probabilities for 95% Intervals for Interaction Effects Based on Proportions, Using Wald and Wilson Intervals for the Single Proportion

	Wald				**Wilson**			
	CP	**MNCP**	**DNCP**	**MNCP/ NCP**	**CP**	**MNCP**	**DNCP**	**MNCP/ NCP**
Unpaired								
n 10 to 20	0.927	0.038	0.035	0.522	0.965	0.014	0.021	0.393
n 50 to 100	0.946	0.028	0.026	0.523	0.954	0.020	0.025	0.449
n 250 to 500	0.949	0.026	0.025	0.514	0.952	0.023	0.025	0.476
Paired								
n 10 to 20	0.891	0.030	0.078	0.280	0.979	0.008	0.013	0.382
n 50 to 100	0.942	0.026	0.032	0.444	0.962	0.017	0.021	0.451
n 250 to 500	0.949	0.024	0.026	0.480	0.954	0.021	0.025	0.463

Note: Estimated from 100,000 runs. CP = coverage probability; MNCP = mesial non-coverage probability; DNCP = distal non-coverage probability.

For the paired case, in zone A the Wald method is substantially anticonservative, with predominantly distal non-coverage. The Wilson method is rather conservative. Again, both methods approach the desired coverage properties for large *n*.

In the unpaired case, the same confidence interval for the interaction results whether we apply the process described here to compare 17/65 – 17/75 and 18/72 – 16/65 or 17/65 – 18/72 and 17/75 – 16/65 in example (a) above. Of course, this simple property does not apply to the paired case.

The method illustrated for the crossover study (d) above can be applied to any split-unit design in which the two halves are meaningfully identifiable apart from the study treatments. Occasionally, this cannot be done, and only direct application of the methods in Newcombe (1998c) or Tango (1998) is possible. For example, 125 articles submitted to a journal were sent to two reviewers, one of whom was randomised to be identified, the other to remain anonymous (Van Rooyen et al. 1999). In this study, 96 of the anonymous referees, but only 81 of the identified ones, agreed to perform the review. The 95% limits for the difference, using method 10 of Newcombe (1998c), are (+0.8%, +22.8%).

14.2.4 A Caveat

While the methods described above had good coverage properties in the evaluations that were performed, there are situations in which applying such methods may lead to highly misleading conclusions. For continuous outcome variables, it is well-known that analyses comparing simple incremental changes between two groups can be misleading when the two groups differ substantially at baseline. Analysis of covariance (ANCOVA) is the method of choice here, as it lets the data determine how much adjustment to make to the post-treatment difference per unit difference between the pretreatment means. The degree of adjustment is determined by the pooled within-groups regression slope *b*. An analysis based on increments is practically equivalent to assuming $b = 1$. But in practice, *b* is often much smaller than 1, consequently analyses based on increments tend to overadjust for baseline differences. While this argument does apply to randomised trials, this effect can make a much greater difference in observational studies, where baseline differences between groups can be very large.

This issue can also apply in the binary case. An additional feature that is prominent in the binary case is that there can be a ceiling effect. In the example below, however high the proportion of correct responses pre-training, the proportion of correct responses post-training cannot exceed 100%. This phenomenon is not unique to the binary case, however; often, continuous variables are subject to some kind of ceiling, even if not an impassable upper bound as in the binary case.

For example, in a non-randomised study Absi (2010) compared performance of two cohorts of trainees attending a refresher course on radiation protection in dentistry. The 2008 cohort comprised 272 trainees to whom teaching was delivered in a traditional didactic format. The 2010 cohort comprising 234 trainees received similar content, delivered in a more interactive manner including eliciting audience responses in real time using electronic response capture technology. According to educational theory, doing so should lead to better learning. However—at least from the standpoint of immediate post-course testing—there was no evidence of any benefit.

Both cohorts were tested for knowledge immediately before and immediately after training using identical 16-item best-of-five multiple choice instruments. Candidate performance is characterised simply by the total number of questions correctly answered, without negative marking. The main analyses are based on scores out of 16, with ANCOVA used to compare knowledge gains in the two cohorts. Overall, there was little difference in gain in score between the two cohorts. But there was also interest in performance on individual questions, which showed a wide heterogeneity in the apparent effect of the change of teaching style. The results in Table 14.3 relate to question 5, one of two questions on radiation doses. On the best analysis, this question showed a highly significant deterioration from the 2008 to the 2010 cohort. (There was an improvement, of similar magnitude, for the other question on doses.) Using an analysis based on within-subjects changes from pre-training to post-training, such as in Section 14.2.2, the conclusion would be a highly significant improvement from the 2008 to the 2010 cohort. Clearly, choice of analytic method here is crucial.

TABLE 14.3

Results for One Question in a Radiation Protection Training Course Study

Cohort	Response Pre-Training	Proportion with Correct Response Post-Training	Stage	Cohort	Proportion Correct
2008	Incorrect	121/137 (88.3%)	Pre-training	2008	0.496
	Correct	125/135 (92.6%)		2010	0.150
2010	Incorrect	147/199 (73.9%)	Post-training	2008	0.904
	Correct	31/35 (88.6%)		2010	0.761

Comparison	Difference 2010 Minus 2008	95% Limits for Difference	p-Value
Pre-training	–0.347		
Post-training	–0.144		
ANCOVA	–0.118	–0.186, –0.050	<0.001
Weighted mean difference	–0.109	–0.182, –0.045	<0.001
Logistic regression	[OR 0.415]		0.001
Incremental	+0.203	+0.105, +0.295	<0.001

Source: From E. G. Absi 2010, unpublished data.

The number who answered question 5 correctly increased from 135 (49.6%) to 246 (90.4%) among the 272 candidates in 2008, and from 35 (15.0%) to 178 (76.1%) among the 234 candidates in 2010. The simple gains in proportions with correct answers were 111/272 (40.8%) in 2008 and 143/234 (61.1%) in 2010, so from this standpoint, the 2010 trainees have the greater gain in knowledge due to the training. The resulting difference of +0.203 has 95% CI from +0.105 to +0.295 using the method described above.

However, the above analysis disregards the fact that, because of the ceiling at 100%, the improvement of 61.1% observed post-training in 2010 would be impossible starting from the 2008 baseline of 49.6%. If we disregard the pre-training information, the difference between the two cohorts is 0.761 – 0.904 = –0.144, i.e., post-training performance was 14.4% poorer in 2010 than in 2008. The corresponding pre-training difference was 0.150 – 0.496 = –0.347 (i.e., pre-training performance was 34.7% poorer in 2010 than in 2008). ANCOVA leads to an adjusted difference of –0.118; the degree of adjustment for the pre-training difference, as fitted by the model, is very small here, only $b = 0.074$ units for every unit difference pre-training. The values of b for the 16 questions range from 0.022 (for question 15) to 0.430 (for item 14). For the global score, b is 0.401, meaning that we only make 0.4 units adjustment for every unit difference pre-training; for all 16 items, the degree of adjustment is no greater than this, and often much less. This makes the adjusted difference of –0.118 look very illogical here. But in fact it is correct, because other analyses lead to practically identical conclusions. This applies to all 16 items in much the same way.

However, of course, ANCOVA is designed for continuous variables, whereas the pre-training and post-training scores at item level are binary. Generally, these two types of variables require totally different-looking methods of analysis. So we need to use methods that respect the fact that these variables are binary.

One possible approach is to restrict attention to those candidates who give an incorrect answer pre-training. For item 5, this analysis suggests that among such candidates, there was a fall in % correct of 14.5% from 2008 to 2010. 95% limits are easily calculated as (–22.3%, –5.9%) using method 10 of Newcombe (1998c). Like the ANCOVA, this analysis suggests a highly significant deterioration for this item. However, the above analysis has the deficiency that it disregards all information from candidates who gave the correct response pre-training. Such candidates "should" give the correct response post-training also—but that cannot be relied on, and indeed was not always the case. For item 5, the proportion incorrect post-training in such candidates was 10/135 (7.4%) in 2008, and 4/35 (11.4%) in 2010, so this group have also deteriorated slightly, by 4.0%.

A better analysis is based on a suitable weighted average of the changes from 2008 to 2010 in the two groups defined by whether they got question 5 correct pre-training. We simply use weights of 336 and 170, the combined numbers incorrect and correct pre-training. Using these leads to a weighted mean change from 2008 to 2010 of –0.109, i.e., a 10.9% deterioration in performance. This is very similar to the adjusted mean difference of –0.118 from ANCOVA. The square-and-add approach is applied a second time, to get a 95% interval for this weighted mean from –0.182 to –0.045. This is similar to the 95% interval from ANCOVA which is –0.186 to –0.050.

The square-and-add approach does not directly produce a *p*-value. But a *p*-value can be recovered, in this and other applications, by altering the confidence level until the interval just reaches zero. This is easily implemented using Excel's Solver facility. Here, the 99.901% interval just reaches zero. So this analysis yields a *p*-value of 0.00099. Once again, this is similar to the ANCOVA *p*-value, 0.00071.

For the remaining 15 questions, the results from this approach and from ANCOVA are very similar. This should be taken as confirming that ANCOVA is not misleading here. These weighted mean difference analyses should be regarded as the best ones for the data at question level.

Alternatively, logistic regression enables us to model the post-training score for any question on cohort and pre-training score. This, too, gives *p*-values for differences between the two cohorts that are very similar to those for both ANCOVA and the weighted mean difference analysis. However, it does not give an estimated difference on an additive scale—only on an odds ratio scale, which of course is not at all comparable. The estimated odds ratio comparing proportions of correct responses to question 5 post-training in the 2008 and 2010 cohorts adjusted for the corresponding pre-training response is 0.415, confirming that this analysis should be also regarded as indicating a deterioration from 2008 to 2010.

The upshot is that, even though the results for question 5 in particular look anomalous, when three very different methods are applied to the data, they give extremely similar results. Only an analysis based on simple changes in proportions correct would give a different conclusion. The latter analysis should be regarded as seriously flawed, as it takes no account of the presence of the ceiling at 100%, which has particularly serious consequences here as the two groups differ greatly at baseline.

14.3 Radiation Dose Ratio

For periapical radiology in dentistry, digital devices have many advantages over conventional ones, including an important decrease in patient radiation exposure. The results considered here come from a trial considered in section 9.3.1 (Farrier et al. 2009), in which patients requiring periapical radiology to a single tooth were randomly allocated between two devices. Randomisation was stratified by six zones of the mouth. The main concerns were image readability, and hence clinician-assessed need for a repeat radiograph, which not only uses additional resources but also increases the radiation dose to the patient. The study gave clear evidence that device B was superior to device A in terms of several outcomes, including image readability and patient acceptability. However, overall the mean radiation dose for a single radiograph using device B was 2.94 times as high as for device A. Radiograph readability is classed on a three-point ordinal scale as excellent, acceptable or unacceptable using established criteria. The latter category would normally prompt the clinician to request a repeat radiograph, although there are also other reasons to do so. After a second unacceptable result, a different technique is used, consequently the concern here is to estimate $(1 + p_1)/(1 + p_2)$ where p_1 and p_2 denote the proportions of unacceptable results produced by the two devices. We call this the ratio of dose units required (RDUR).

Only six of 108 (5.6%) radiographs using device B required repetition on account of unacceptable image quality, compared to 22 out of 98 (22.4%) using device A. While these figures suggest a clear preference for device B, this is outweighed by the much greater radiation dose it delivers per usage.

Consider for purposes of illustration the upper anterior zone. Here device B fared slightly less well than A, the proportions unacceptable were 1/20 for device A and 2/20 for device B. The resulting ratio in expected numbers of radiographs required, device A versus B, is then $(1 + 1/20) \div (1 + 2/20) = 0.955$. For this zone, the unit ratio—the radiation dose ratio per radiograph—is $0.05 \div 0.12 = 0.417$. So we estimate the adjusted dose ratio (ADR); that is, the expected radiation dose when device A is used for an upper anterior tooth divided by that for device B, as $0.955 \times 0.417 = 0.398$.

Here, we can use either the MOVER or the PropImp method to obtain an interval for RDUR, based as usual on Wilson intervals for the two proportions. The point estimate and confidence limits are then multiplied by the unit ratio to get corresponding figures for ADR. Farrier et al. reported MOVER-DL intervals. Tables 14.4 and 14.5 show confidence intervals calculated by five methods. The first block of Table 14.4 shows the raw data on proportions unacceptable and unit radiation doses (which are assumed to be without sampling uncertainty). The second block shows the empirical estimates and confidence intervals for the RDUR and ADR for each zone derived from Wilson intervals for $\ln(1 + p_i)$, $i = 1, 2$ by using MOVER-DL. The third block shows the corresponding results using the MOVER-R approach. All figures are shown to four decimal places here to illustrate the close similarity of the results obtained by these two methods. The first block of Table 14.5 shows corresponding analyses using the PropImp approach. In all instances, the upper limit for the RDUR or the ADR is derived from the upper limit for p_1 and the lower limit for p_2, and vice versa. The second block shows corresponding analyses using a deterministic bootstrap approach based on the median unbiased estimator as described in Section 11.4, but with a mid-p cumulative distribution function.

TABLE 14.4

Analysis of Radiation Dose Ratios Comparing Two Periapical Radiography Devices: Raw Data, and MOVER Analyses

Zone	Proportion Unacceptable		Unit Dose (Seconds at 5.98 mGy/sec)		
	Device A	Device B	Device A	Device B	UR A:B
Input Data					
Upper anterior	1/20	2/20	0.05	0.12	0.417
Upper premolar	2/20	0/20	0.06	0.16	0.375
Upper molar	6/20	1/20	0.06	0.25	0.240
Lower anterior	0/7	0/10	0.05	0.12	0.417
Lower premolar	5/13	1/18	0.06	0.16	0.375
Lower molar	8/18	2/20	0.08	0.25	0.320

Zone	RDUR = (1 + A)/(1 + B)		Dose Ratio A:B ADR	
	Empirical Estimate	95% Limits	Empirical Estimate	95% Limits
Analysis Using MOVER-DL: Square-and-Add Approach Applied to Wilson Limits for ln (1 + A) and ln (1 + B)				
Upper anterior	0.9545	0.8033, 1.1391	0.3977	0.3347, 0.4746
Upper premolar	1.1000	0.9336, 1.3010	0.4125	0.3501, 0.4879
Upper molar	1.2381	1.0070, 1.4540	0.2971	0.2417, 0.3489
Lower anterior	1.0000	0.7828, 1.3543	0.4167	0.3261, 0.5643
Lower premolar	1.3117	1.0331, 1.5669	0.4919	0.3874, 0.5876
Lower molar	1.3131	1.0498, 1.5353	0.4202	0.3359, 0.4913

Zone	RDUR = (1 + A)/(1 + B)		Dose Ratio A:B ADR	
	Empirical Estimate	95% Limits	Empirical Estimate	95% Limits
Analysis Using MOVER-R Applied to Wilson Limits for 1 + A and 1 + B				
Upper anterior	0.9545	0.8031, 1.1395	0.3977	0.3346, 0.4748
Upper premolar	1.1000	0.9332, 1.3010	0.4125	0.3500, 0.4879
Upper molar	1.2381	1.0068, 1.4543	0.2971	0.2416, 0.3490
Lower anterior	1.0000	0.7828, 1.3543	0.4167	0.3261, 0.5643
Lower premolar	1.3117	1.0329, 1.5674	0.4919	0.3874, 0.5878
Lower molar	1.3131	1.0496, 1.5358	0.4202	0.3359, 0.4914

Source: Data from Farrier, S. L. et al. 2009. *International Endodontic Journal* 42: 900–907.
Note: UR = unit ratio; RDUR = ratio of dose units required; ADR = adjusted for need to repeat.

TABLE 14.5

Analysis of Radiation Dose Ratios Comparing Two Periapical Radiography Devices: PropImp and MUE Bootstrap Analyses

	RDUR = (1 + A)/(1 + B)		Dose Ratio A:B ADR	
Zone	**Empirical Estimate**	**95% Limits**	**Empirical Estimate**	**95% Limits**
Analysis Using PropImp Approach				
Upper anterior	0.955	0.797, 1.148	0.398	0.332, 0.478
Upper premolar	1.100	0.928, 1.301	0.413	0.348, 0.488
Upper molar	1.238	1.007, 1.468	0.297	0.242, 0.352
Lower anterior	1.000	0.783, 1.354	0.417	0.326, 0.564
Lower premolar	1.312	1.033, 1.585	0.492	0.387, 0.594
Lower molar	1.313	1.051, 1.557	0.420	0.336, 0.498
	RDUR = (1 + A)/(1 + B)		**Dose Ratio A:B ADR**	
Zone	**Ratio of MUEs**	**95% Limits**	**Ratio of MUEs**	**95% Limits**
Analysis MUE Bootstrap Approach				
Upper anterior	0.956	0.842, 1.094	0.398	0.351, 0.456
Upper premolar	1.088	0.961, 1.233	0.408	0.360, 0.463
Upper molar	1.232	1.000, 1.431	0.296	0.240, 0.343
Lower anterior	1.013	0.866, 1.235	0.422	0.361, 0.514
Lower premolar	1.303	1.039, 1.583	0.489	0.390, 0.594
Lower molar	1.306	1.062, 1.564	0.418	0.340, 0.501

Source: Data from Farrier, S. L. et al. 2009. *International Endodontic Journal* 42: 900–907.
Note: RDUR = ratio of dose units required; ADR = adjusted for need to repeat; MUE = median unbiased estimator.

It is apparent that for each of the six zones of the dentition, all these methods produce appropriate intervals, despite numerators and denominators being very small or zero. In each instance the MOVER and PropImp intervals were rather wider than the MUE bootstrap intervals. MCMC intervals were also calculated; they tended to be intermediate between these.

14.4 Levin's Attributable Risk

In Section 6.3.2 we considered how Daly (1998) examined Levin's attributable risk in relation to the prevalence of a risk factor and its associated relative risk. This effect size measure may be expressed, in an obvious notation, as

$$LAR = \frac{1}{1 + 1/\{\mathrm{Prev} \times (RR - 1)\}} \qquad (14.1)$$

This relationship is ostensibly similar to those in the seizure prophylaxis example in Section 13.2, but there is an important difference. Here, both contributory parameters come from the same 2 × 2 table—in this respect paralleling Chatellier as considered in Section 13.7. Daly calculated a confidence interval for LAR by the substitution method which reflects the sampling imprecision of the relative risk using the delta method, but disregards the imprecision of the prevalence. Essentially, this amounts to conditioning on the observed prevalence, or equivalently, on the sizes of the high and low exposure groups. This approach would be incontrovertibly appropriate if the aim is to obtain point and interval estimates for the projected LAR corresponding to some hypothetical prevalence, say 5% or 8%. However, the prevalence here, 5215/72730 or 0.0717, is itself affected by sampling variation. Wald limits for this prevalence are 0.0698 and 0.0736. These limits are approximately 0.0717 × (1 ± 0.026), in contrast to the limits for the RR which are approximately 18.959 × (1 ± 0.12). It is clear that the RR is the dominant source of uncertainty here; nevertheless the imprecision of the prevalence should also be taken into account.

Another issue in this example is that the two parameters, the prevalence and relative risk, are being estimated from the same 2 × 2 table. The calculations for the relative risk are conditional on the marginal totals m and n, thus the uncertainty in the relative risk is expected to be approximately orthogonal to the uncertainty in the prevalence. This argument suggests that the application of this approach is reasonable here, but it might not be in other contexts in which the two parameters are estimated from the same dataset.

We reconstruct Daly's point estimate and 95% interval by the substitution method as 0.5629 and 0.5313 – 0.5938 to four decimal places, to enable comparison with MOVER-DL, PropImp and MCMC intervals. A straightforward interval for the LAR from intervals for the prevalence and relative risk may be constructed by applying the square-and-add approach to ln Prev and ln (RR – 1). Here, in contrast to the dose ratio example in the preceding section, the upper limit for the LAR comes from the upper limits of both the prevalence and the relative risk. The MOVER-DL interval for the LAR, using a Wald interval for the prevalence and a delta interval for the relative risk for consistency with Daly, is 0.5306 to 0.5944, just 2.1% wider than the interval calculated by Daly's method. The corresponding PropImp interval is identical to the MOVER-DL interval to four decimal places. MCMC with uniform priors gives a posterior median LAR of 0.5627, with 95% interval 0.5312 to 0.5938 conditioning on two margins, or median 0.5627, 95% interval 0.5313 to 0.5937 conditioning on the table total only. The latter interval is minutely narrower than the MOVER-DL or PropImp intervals here, suggesting that strictly these approaches are flawed here due to the two parameters coming from the same table.

We also consider what happens if we apply substitution and either MOVER-DL or PropImp to an example where it would be expected to make maximal difference, with 3000 exposed and unexposed, and risks 2/3 and 1/3. The substitution interval is 0.3080 to 0.3582. Both the MOVER-DL and PropImp intervals are 0.3074 to 0.3589, just 2.5% wider. Both examples suggest that the relative risk is robustly the dominant source of imprecision and that here a substitution interval taking account of only this source of uncertainty is only minimally anticonservative.

14.5 Population Risk Difference and Population Impact Number

When one source of variation predominates, use of either MOVER or PropImp results in an interval very similar to one which only incorporates the dominant source of variation. For example Bender and Grouven (2008) considered the population risk difference (PRD) and its reciprocal, the population impact number (PIN) which are derived from an estimated rate difference and exposure (Section 6.3.3). It turns out that the former is always the dominant source of variation here, and use of either MOVER or PropImp adds little value.

The rationale for the PIN is that it provides a measure for use in epidemiology which in some respects parallels the NNT measure used in clinical trials. The NNT may be regarded as the average number of exposed persons who need to be treated (by removal of exposure) for one person to benefit. It does not involve the prevalence of exposure. Within the limitations described in Section 7.6, it is a widely recognised measure for clinical trials to describe the effect of a treatment in a specific patient group. Paralleling the above, the PIN is the average number of persons of the population considered among whom one person benefits. It is dependent on the exposure prevalence. It may be used in public health research to describe the impact of an exposure or of an intervention that removes the exposure in the whole population, subject to the usual proviso for the non-significant case.

The PRD may be expressed either as $\Pr[D = 1] - \Pr[D = 1 \mid E = 0]$ or equivalently as $\Pr[E = 1] \times RD$, where D and E are binary variables representing disease and exposure, 1 and 0 denote positive and negative states, and the rate difference $RD = \Pr[D = 1 \mid E = 1] - \Pr[D = 1 \mid E = 0]$. Bender and Grouven suggested that for fixed exposure prevalence, interval estimation for the PRD, and hence for the PIN, can be based upon a suitable interval for the rate difference, in particular the square-and-add Wilson interval (Section 7.3.1). They noted that this approach is essentially conditional on two margins fixed, whereas in a study of this nature strictly it is the table total that should be conditioned on. This approach is essentially the substitution method, and disregards the sampling imprecision of $\Pr[E = 1]$.

Bender and Grouven presented analyses of data from Schräder, Grouven and Bender (2007), in which infection occurred in 135/8251 or 1.64% of knee replacement cases treated in hospitals with a low-volume caseload and in 735/102,098 or 0.72% of cases treated in hospitals with a high-volume caseload. For these data, the PRD is 0.000685 and the PIN is 1459.6. They reported a 95% interval 1098.4 to 2030.9 for the PIN based on the square-and-add Wilson interval. We need to consider carefully whether it is acceptable to disregard the sampling imprecision of Pr[E = 1] here.

A MOVER-DL interval for the PRD or PIN may be obtained from a Wilson interval for Pr[E = 1] and a Wilson square-and-add interval for the rate difference, via ln Pr[E = 1] and ln RD. This gives an interval from 1098.6 to 2033.3 for the PIN, very similar to the above interval. This may equally be regarded as a MOVER based on the three parameters Pr[E = 1], Pr[D = 1 | E = 1] and Pr[D = 1 | E = 0]. In this instance, the process of combining the intervals is done in two stages. The first step is to combine the intervals for the two infection rates in the usual way, without scale transformation. The second step combines the resulting interval for the rate difference and the Wilson interval for Pr[E = 1], both after log transformation.

Alternatively, we can apply the PropImp algorithm in two different ways here. If we use the square-and-add Wilson interval for the rate difference and the Wilson interval for Pr[E = 1], the resulting interval for the PIN is 1097.6 to 2032.1. Alternatively, the PIN may be expressed as a function of three binomial parameters, Pr[E = 1] × {Pr[D = 1 | E = 1] – Pr[D = 1 | E = 0]}, and PropImp applied accordingly, still using Wilson intervals throughout. The resulting interval for the PIN is 1097.6 to 2032.6.

So it is clear that for this example, the four approaches yield essentially identical intervals. For the two-parameter PropImp solution, the λ values for the contribution from the rate difference are 0.9984 and 0.9963 for the lower and upper limits, reiterating that the rate difference is the dominant source of variation here. It can be shown that the ratio of the variances of the rate difference and the exposure takes its lowest possible value, 4, in the (possibly unrealistic) case where Pr[D = 1 | E = 1], Pr[D = 1 | E = 0] and Pr[E = 1] are all 0.5, and even here, the rate difference is the dominant source of imprecision of the PRD and the PIN. Consequently, the approach suggested by Bender and Grouven is perfectly adequate to obtain confidence intervals for these measures if we are prepared to assume binomial sampling variation, and use of either MOVER or the PropImp algorithm is redundant here.

However, while there is nothing wrong *per se* with the methods developed above, none of them is really adequate for the scenario to which they were applied. Data comparing complication rates between hospitals which manage low and high volumes of the condition in question should be analysed by methods that heed the multilevel nature of the data. Any analyses that disregard this crucial feature of the data are likely to be seriously anticonservative. Analyses similar to those used in cluster randomised trials are more appropriate. For example, in a study by Pal et al. (2008) examining outcomes of major upper gastrointestinal surgery, a rate ratio comparing lower versus higher caseload hospitals was calculated, with a BCa bootstrap interval and a randomisation test *p*-value. These methods appropriately allow for both the marked overdispersion found in some parts of the data and the small numbers of procedures and deaths in many hospitals.

14.6 Quantification of Copy Number Variations

Dube et al. (2008) described a digital polymerase chain reaction (PCR) method to quantify copy number variations (CNVs) in a sample of DNA. An array comprising n chambers each of volume 6 nl is filled with the test material, where n is either 765 or a multiple of 765. Each chamber can be classed as positive (r chambers) or negative (n–r chambers) for the target DNA, but it is not possible to determine how many such molecules an individual chamber contains. If $p = r/n$ is small, it can serve as a reasonable approximation to λ, the concentration of target DNA per chamber (i.e., per 6 nl). Dube et al. suggested that if p is not small, an estimate based on the Poisson distribution with mean λ is preferable, namely $\hat{\lambda} = \ln(1 - p)$. An interval for λ may then be derived from a suitable interval for p by substitution. This interval expresses the precision of measurement of the concentration of the target DNA in one individual. Dube et al. used Wald intervals, which are indistinguishable from better intervals for the examples they display in which both r and n–r are very large.

Dube et al. also considered the comparison of two concentration estimates λ_1 and λ_2 from independent samples. They suggested that an interval for the ratio λ_1/λ_2 should be derived from intervals for λ_1 and λ_2 using the MOVER-R formula.

14.7 Standardised Mortality Ratio Adjusted for Incomplete Data on Cause of Death

In Section 6.1.3 we saw how a confidence interval based on the Poisson distribution may be used to calculate an interval for a standardised mortality ratio. The illustrative example, taken from Rittgen and Becker (2000), relates to mortality by cause in a cohort of male iron foundry workers. Unfortunately, information on cause was available for only 2896 of the 3972 deaths. Rittgen and Becker developed a method to estimate cause-specific standardised mortality ratios adjusted for incomplete data on cause of death. The adjusted SMR is simply calculated as SMR* = SMR/p, where p denotes the proportion of deaths in the series for which information on cause is available. Here, apart from for major, broad categories of cause of death such as all cancers or circulatory disease, the contribution from the imprecision of the correction factor will be minimal. Consequently, very much as in the Bender and Grouven example, one of the sources of imprecision may effectively be disregarded.

Table 14.6 is based on this dataset. An additional decimal place has been calculated for expected frequencies in such a way as to optimise closeness of

TABLE 14.6

95% Intervals for Cause-Specific Standardised Mortality Ratios for Lung and Liver Cancers in a German Cohort of Male Iron Foundry Workers, Adjusted for Incomplete Data on Cause of Death

	Lung Cancer	Liver Cancer
Observed deaths	322	28
Expected deaths	253.18	12.44
Crude SMR	127.2	225.0
SMR adjusted for missing data	174.4	308.6
1. Rittgen–Becker limits	150.1, 202.0	188.8, 477.7
2. Lloyd limits	153.0, 198.4	201.3, 454.8
3. PropImp limits	155.6, 194.9	204.9, 446.3
4. Daly limits	155.9, 194.6	205.1, 446.0
5. MOVER-R limits	155.7, 194.9	205.0, 446.3

Source: Based on Rittgen, W. and Becker, N. 2000. *Biometrics* 56: 1164–1169.

Note: All intervals are derived from exact intervals for the Poisson (number of deaths) and binomial (proportion of deaths with cause known) data. Five approaches are used to combine these intervals, described in Section 14.7. SMR = standardised mortality ratio.

fit to the derived figures published by Rittgen and Becker. The table shows confidence intervals for the adjusted SMRs for two causes of death, lung and liver cancers, calculated by five methods. The second and fourth methods here are labelled "Lloyd" and "Daly" because they are closely analogous to the approaches used in Lloyd (1990, discussed in Section 13.2) and Daly (1998).

1. Rittgen-Becker method uses 97.5% intervals for both parameters.
2. Lloyd method uses 95% intervals for both parameters.
3. PropImp uses z_1 and z_2 limits for the two parameters, such that $z_1^2 + z_2^2 = 1.960^2$, chosen to maximise interval width.
4. Daly method uses a 95% interval for the Poisson parameter and the point estimate for the correction factor.
5. MOVER-R applied to log-transformed exact intervals for the two parameters. (MOVER-DL produces practically identical intervals, with just one limit differing by 0.1.)

Rittgen and Becker's analysis used exact intervals for the two parameters involved, a Poisson parameter representing deaths accruing during the study period and a binomial parameter representing the proportion of deaths for which the cause is reported, which is used as a correction factor. To ensure comparability, all five methods used here utilise these as the base methods.

It is clear by considering the derivation of intervals 1 to 4 that each of them is nested within its predecessor. This is also apparent from Table 14.6. We would expect the Daly interval could err towards anticonservatism as a consequence of disregarding the uncertainty of the correction factor, conversely the Lloyd and especially the Rittgen–Becker approach are anticipated to be overconservative. In fact, the Daly, MOVER-R and PropImp intervals are very similar here. The PropImp analysis produced optimising λ_1 values of 0.986 and 0.984 for lung cancer and 0.9990 and 0.9983 for liver cancer, underlining the fact that the imprecision of the SMR is dominant, especially for the much rarer liver cancer. Consequently the simple substitution approach, termed the Daly approach here, is adequate in this instance, especially considering that the conservatism of the base method used counterbalances the neglect of the uncertainty of the correction factor.

All these analyses rely heavily on the assumption that missing data for cause of death is missing completely at random. Rittgen and Becker present an argument for the plausibility of this assumption, but this argument is far from convincing.

14.8 RD and NNT from Baseline Risk and Relative Risk Reduction

As described in Section 13.7, Chatellier et al. (1996) derived a formula and nomogram to obtain the NNT from the risk or event rate in the control series, p_1 and the relative risk reduction. Naturally, as with all such procedures, it is important to report a confidence interval as well as a point estimate. When p_1 and RRR both come from the same 2×2 table, it is inappropriate to use the substitution method to derive an interval for the rate difference or NNT from an interval for the RRR or RR and a point estimate for p_1, as this would amount to assuming p_1 to be without sampling variation, leading to a narrower interval than would normally be reported (Daly 1998). Nor is it appropriate to use squaring and adding or PropImp to take both sources of variation into account, because they are far from being estimated independently here. Suitable purpose-built methods for the RD exist, of course, as described in Sections 7.3 and 7.4.

But there are other situations in which such an approach has a greater potential usefulness. If the RR or RRR is available for some series, or from a meta-analysis of several series, with a confidence interval, then it is perfectly reasonable to use Chatellier's approach, in conjunction with substitution, to derive an interval for the projected RD = $p_1 \times$ RRR or NNT = $1/(p_1 \times \text{RRR})$ at any hypothetical value of the control series risk, or to produce a plot of the projected RD or NNT by p_1 ranging from 0 to 1, with confidence bounds.

Similarly, starting with point and interval estimates for an odds ratio, we may obtain a plot such as Figure 14.1, which shows both the projected RD

$$p_1 - 1/\{1 + (1/p_1 - 1)/\text{OR}\} \quad (14.2)$$

(corresponding directly to Equation 13.1) and similarly the projected RR

$$1/\{p_1 + (1 - p_1)/\text{OR}\} \quad (14.3)$$

as functions of the assumed baseline risk, with confidence regions.

Alternatively, if the RR or RRR and the control series risk are available from different sources, either the MOVER or PropImp approach may then be used to combine intervals for p_1 and RRR to construct an interval for the corresponding RD or NNT. Either source of variation could be dominant, depending on the relative sizes of the two series. The square-and-add approach is applicable to the Chatellier formula which expresses the RD or NNT in terms of p_1 and the RR or RRR, but not to the formulae underlying Figure 14.1, where the RD or RR is expressed in terms of p_1 and the odds ratio. The crucial difference is that the Chatellier formula factorises into separate components encapsulating the dependence on the two parameters; Equations 14.2 and 14.3 do not.

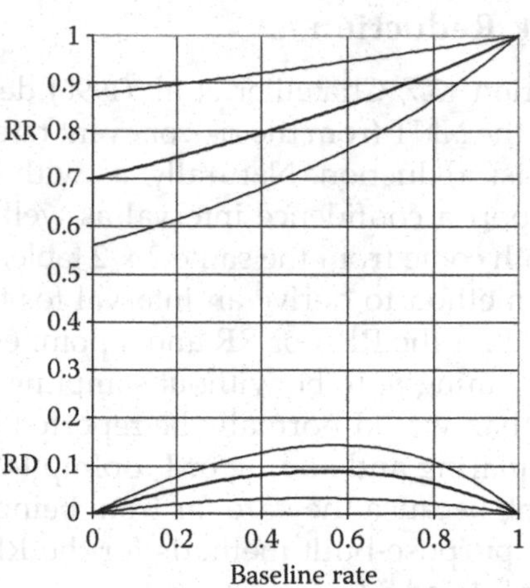

FIGURE 14.1
Relationship of the relative risk, RR (upper three curves) or rate difference, RD (lower three curves) to the odds ratio, for baseline rate varying from 0 to 1. The middle curve of each set plots the point estimate for the relative risk or rate difference against the theoretical baseline rate; the lower and upper curves delimit a confidence region. For the example shown, the odds ratio is 0.70, with 95% interval (0.56, 0.875).

14.9 Projected Positive and Negative Predictive Values

As we saw in Chapter 12, the screening context leads to several possible applications of the MOVER algorithms. Another application, quite similar to the pregnancy hypertension prophylaxis example, relates to the projected positive predictive value (PPV) at arbitrary prevalence p, based on the known sensitivity and specificity of a test. Mir (1990) found a sensitivity 17/17 (100%) and specificity 55/58 (95%) for the GACPAT salivary test for HIV. The PPV may be expressed as a function of the prevalence π and the PLR as in Equation 12.10. The substitution method can be used to display a confidence region for the projected PPV as a function of π, which encapsulates the imprecision of the PLR only. As in Section 12.5.6, the PLR is a ratio of proportions, for which the Miettinen–Nurminen interval (Section 10.4.2) is appropriate. However, if we have an estimate of disease prevalence based on an independent finite sample from a relevant population, intervals for $1/p - 1$ and RR may be combined by MOVER-R. This takes account of the uncertainty of both parameters and is therefore, rightly, wider than that obtained for the projected PPV based on a similar but hypothetical value for the prevalence. A similar approach may be used when interest centres on the projected NPV derived from π and the NLR as in Equation 12.11.

14.10 Estimating Centiles of a Gaussian Distribution

The examples in the next three sections are included to show how the MOVER and PropImp approaches may also be applied to parameters relating to continuous distributions.

Often investigators seek to determine a reference interval (Section 2.1) for some measurement Y based on ostensibly normal individuals. If the investigators are satisfied that the data conform to a Gaussian distribution, this is best achieved by calculating $\bar{Y} \pm z_c \times s_Y$. For the usual 95% interval, z_c, the z value to determine the centiles, is taken to be 1.96, which is often rounded to 2. Sometimes a transformation of scale, usually logarithmic, is required. When the distribution is not close to either Gaussian or log-Gaussian form, the relevant centiles may be estimated by an obvious non-parametric method.

In a method comparison study, two different measurement methods are performed on the same sample of patients (or specimens), resulting in two variables, X and Y. The primary analysis should seek to characterise the distribution of the individual level differences of the form $\Delta = Y - X$. While the Pearson correlation may be calculated, it is liable to be overinterpreted, as this may be both astronomically statistically significant and also close to 1, without implying that differences between the two methods when applied to the same subject are necessarily small. It is useful to estimate the mean difference, with a confidence interval, as in Section 2.3. But the most useful analysis describes the distribution of Δ, and also shows whether this is similar at low and high values. In a widely used graphical method (Bland and Altman 1986) the signed difference between the two measurements is plotted against their average. Often the cluster of points fans out at larger values of $(X + Y)/2$, indicating that variation is greater at higher values; this is frequently rectified by log-transforming the original measurements. The 95% limits of agreement are plotted as horizontal lines at $\bar{\Delta} \pm z_c \times s_\Delta$. Normally the great majority of observations will lie within these lines.

Constructs of the form $\bar{Y} \pm z_c \times s_Y$ or $\bar{\Delta} \pm z_c \times s_\Delta$ are themselves subject to sampling variation, of course. Bland (1995, p. 276–279) describes the calculation of confidence intervals for parametric and non-parametric centile estimates, and notes that the latter have considerably wider intervals. In programming intervals for parametric centile estimates, it is important to distinguish between z_c as defined above and z_i, the normal deviate used to construct 100(1 – α)% intervals: for the usual choice of 95% intervals, z_i will be 1.960, whereas it is often of interest to calculate centiles other than the 2.5 and 97.5 ones.

A parametric centile estimate of the form $\bar{Y} - z_c \times s_Y$ or $\bar{Y} + z_c \times s_Y$ has a variance which is derived from the variances of the two estimates $\bar{Y}$ and s_Y. This may be estimated as $s_Y^2\left(\frac{1}{n}+\frac{z_c^2}{2(n-1)}\right)$. Putting $z_c \approx 2$ for the usual 95% limits, this is approximately $\frac{3}{n}s_Y^2$. Obviously, the resulting confidence intervals for $\bar{Y} - z_c \times s_Y$ and $\bar{Y} + z_c \times s_Y$ are symmetrical. This may not be optimal as we would not normally use a symmetrical interval for a standard deviation estimate such as s_Y. The MOVER algorithm offers an alternative approach, to combine a symmetrical interval for $\bar{Y}$ based on the Gaussian or *t*-distribution and an asymmetrical interval for s_Y based on the chi-square distribution. The confidence limits for a variance estimate s^2 with ν degrees of freedom are $\nu s^2/\chi^2_{\nu,1-\alpha/2}$ and $\nu s^2/\chi^2_{\nu,\alpha/2}$ where $\chi^2_{\nu,q}$ denotes the q quantile of the relevant chi-square distribution.

Bland (1995) showed an example involving cord blood triglyceride measurements in 282 babies, which range from 0.15 to 1.66 mmol/l. The distribution is positively skew, with mean 0.51 and SD 0.22. The estimated 2.5 centile based on a Gaussian model would be $\bar{Y} - z_c \times s_Y = 0.08$ units, which is inappropriate as it is considerably lower than any of the values in the sample. Log_{10} triglyceride is much closer to Gaussian, with mean –0.331, SD 0.171. The resulting fitted 2.5 and 97.5 centiles for log triglyceride are –0.665 and +0.004, which transform back to 0.216 and 1.009 mmol/l on the original scale. 7 (2.5%) of the original observations were below 0.216 and 10 (3.5%) were above 1.009, confirming the satisfactory fit of the log-Gaussian model.

Using MOVER, the back-transformed 95% intervals for the 2.5 and 97.5 centiles are (0.199, 0.233) and (0.937, 1.097). These differ slightly from the intervals (0.196, 0.230) and (0.948, 1.109) reported by Bland, as a consequence of using $z_c = 1.960$ rather than 2 and using an asymmetrical interval for s_Y. For smaller sample sizes, choice of interval method for $\bar{Y} \pm z_c \times s_Y$ becomes more crucial as distributional skewness increases. While the sample size in the Bland example is considerably larger than is often used to establish reference ranges, Koduah et al. (2004) argued cogently for using a sample of size 500 to 1000 with non-parametric estimation, in recognition that sample sizes smaller than this do not give adequate reassurance of distributional form in the tails of a sample.

14.11 Ratio Measures Comparing Means

The need to choose between difference and ratio measures applies to means as well as proportions. Sometimes investigators feel more comfortable with the idea of expressing a difference in percentage terms, even when results are based on continuous distributions. For instance, in evaluations of oral hygiene products results are often summarised by statements such as "the mean plaque area (or index) was 20% lower on treatment A compared to treatment B." The fact that this is unit-free is appealing, but deceptive. It is unhelpful to express a power calculation in terms of the percentage difference, as this draws attention away from the need for anticipated means and SDs to be presented clearly. At the analysis stage, this kind of summary statement could be supported by an analysis on a log-transformed scale, leading to a geometric mean ratio. Alternatively, the actual ratio of arithmetic means may be reported.

Unfortunately, there is widespread misunderstanding of how ratio measures work. Studies evaluating toothpastes used to treat dentine hypersensitivity usually measure participants' responses to tactile and air blast stimuli applied to two teeth identified as sensitive. To measure tactile sensitivity, a force-sensing probe is drawn across the buccal surface, with force preset to 10g, which is increased in steps of 10g up to a maximum of 50g or until the respondent first indicates discomfort. Some minutes later, a standardised cold air blast is delivered to the tooth, and the participant's response is assessed on a 0–3 scale described by Schiff et al. (1994). For the tactile scale, a high score indicates a favourable response, whilst for the air blast scale, low scores indicate benefit. Ayad et al. (2009) demonstrated that both scales show a marked improvement immediately following a single dab-on application of a highly effective arginine-based toothpaste. In 41 participants who used this treatment, the mean tactile score increased from 11.46g to 33.17g, whilst the mean air blast score decreased from 2.90 to 1.26. These changes were presented as improvements of 189% and 57%. The casual reader would infer that the tactile test was much more sensitive to detect response than the air blast test. This is an incorrect conclusion here, as even if no subject had a positive air blast score for either tooth following treatment, the improvement would only be by 100%. Moreover, while the air blast scale is treated as equally spaced in the analysis, this is an arbitrary and questionable choice, and the percentage improvement would be rather different if responses were scaled in some other way, whether arbitrary or based on the frequencies of 0, 1, 2 and 3 responses in the sample using ranks or ridits (Bross 1958).

Two types of relative or scaled, unit-free measures are available for the comparison of means. The actual ratio of arithmetic means $\bar{y}_1/\bar{y}_2$ may be reported. The MOVER-R algorithm may then be used to derive a confidence interval for $\bar{y}_1/\bar{y}_2$ from intervals for $\bar{y}_1$ and $\bar{y}_2$. This may be appropriate when the two means are based on independent samples; it would be inapplicable in the Ayad study.

An alternative relative measure compares a mean, or a difference between means, to some kind of standard deviation. For an unpaired difference between means, the relevant quantity is the standardised difference, $(\bar{y}_1 - \bar{y}_2)/s$ where s is a standard deviation estimate, usually pooled, as described in Section 2.2. The corresponding construct for the paired difference case is $\bar{\Delta}/s_\Delta$. Confidence intervals for standardised difference measures are awkward as they use the non-central t-distribution, which is not widely available to users. A more convenient option is to use MOVER-R to combine intervals for the mean difference and the standard deviation estimate. An interval for $\bar{y}_1 - \bar{y}_2$ or $\bar{\Delta}$ is readily obtained using the methods already described. But confidence limits for s cannot be reconstructed for the results given by Ayad et al. who reported means and SDs pre-application and post-application, and a percentage change based on the two means, but no summary statistics for the amount of change as such. Note that in any case, the mean of the percentage changes within subjects is not the same as the percentage change based on the two means. In a closely related study (Schiff et al. 2009), such a reconstruction is possible, only for the tactile sensitivity scores, since at baseline all study teeth were sensitive to the lowest applied force, 10g. The mean provoking force immediately after fingertip application was 29.17g (SD 2.78, n = 84), so here $\bar{\Delta} = 19.17$ and $s_\Delta = 2.78$, leading to $\bar{\Delta}/s_\Delta = 6.90$, indicating that the degree of shift should be regarded as of an extreme degree of consistency. 95% limits for $\bar{\Delta}$ are 19.17 ± 1.989 × 2.78/√84; that is, (18.57g, 19.77g). 95% limits for s_Δ are obtained as 2.78 × √(83/110.09) = 2.41 and 2.78 × √(83/59.69) = 3.28. Thus, using MOVER-R as described in Section 7.3.3, 95% limits for $\bar{\Delta}/s_\Delta$ are (5.83, 7.97).

In the final Chapters 15 and 16, we will see how the results of both the unpaired and paired methods may be re-expressed to yield relative measures of effect size, U/mn and $\Delta T/\max$. Such non-parametric measures have more to recommend them than the standardised difference. Estimation properties of SDs or variances are highly dependent on assumed Gaussian distributional form. This impacts on intervals derived using the non-central *t*-distribution as well as those using MOVER-R.

Standardised difference measures are useful for sample size planning as in the Altman (1982) nomogram, and also can sometimes be reconstructed from published summary statistics in situations where the raw data are unavailable. But whenever the original data are available, the use of the appropriate non-parametric effect size measure is preferable as it obviates both of these issues.

14.12 Winding the Clock Back: The Healthy Hearts Study

In the Healthy Hearts study (Richardson et al. 2011) a cohort of adults identified as at increased risk of heart disease were followed up about 1.5 years after their initial visit, and changes in risk factors documented. We may compare the changes that occurred in the 738 participants who were followed up with an estimate of what would have happened to them purely as a result of ageing by $\bar{t} = 1.488$ years. A linear regression of each variable (after log transformation if indicated) on age, based on the baseline data, is used to produce a simple estimate of the effect of ageing. This is best restricted to those with follow-up data, so that the two estimates relate to exactly the same group of participants.

When this is done, we have two quantities that can be compared. The mean change in the parameter from baseline to follow-up, Δ, is obtained from the longitudinal aspect of the data. The regression coefficient per year of age, b, is derived by viewing the baseline data as cross-sectional. Standard errors for both these quantities are simply calculated in the usual way. These two quantities may be combined in two distinct ways.

The obvious quantity to construct is the adjusted difference, $\Delta - \bar{t} \times b$. It is adequate to regard $\bar{t}$ as a constant here, so the standard error of $\Delta - \bar{t} \times b$ is simply obtained from the standard errors of Δ and b. For example, for systolic blood pressure (recorded as usual in mm Hg), Δ was −2.271, with standard error 0.531, and b was +0.794, with standard error 0.114. Here, the adjustment increases the size of the difference by a little over 50% to become −3.543, with standard error $\sqrt{0.531^2 + (1.488 \times 0.114)^2} = 0.557$, and 95% interval −4.547 to −2.360. Thus our estimated benefit, of 2.3 units is enhanced to 3.5, and could be as small as 2.4 or as large as 4.5 units, within the uncertainty resulting from sampling variation in the two parameters Δ and b. Whilst strictly these two quantities are not quite independently estimated, the approximately uniform distribution of ages in the sample ensures that they are approximately orthogonal. For this simple linear function of the two parameters, each of which has a symmetrical sampling distribution, MOVER and PropImp approaches lead to exactly the same interval.

An alternative, complementary approach gives additional insight for some variables such as systolic blood pressure here. We may regard a reduction in systolic blood pressure of 2.271 units as equivalent to a reduction in age of $-\Delta/b$ = 2.271/0.794 or +2.86 years. Thus the apparent effect of the intervention is to wind the clock back (WTCB) by 2.86 years—whereas in reality 1.49 years have elapsed, so these participants are 4.36 years "younger" in terms of blood pressure than they would have been in the absence of the intervention. This expression of benefit may be linked to the notion of heart age or vascular age introduced by D'Agostino et al. (2008). For systolic blood pressure, a MOVER-DL interval for $-\Delta/b$ is readily derived from the usual intervals for ln $(-\Delta)$ and ln (b), as +1.47 to +4.72 years—to both of which figures we may add 1.49 years as above. MOVER-R gives a similar interval, from +1.48 to +4.73 years. Alternatively, the PropImp approach yields a 95% interval from +1.48 to +4.72 years, very similar to the MOVER intervals.

These analyses suggest a substantial beneficial effect on systolic blood pressure. The interpretation of this finding is of course subject to three limitations. A study of this nature is a limited proxy for a randomised trial comparing the complex intervention used to an appropriate control regime. Follow-up was far from complete: it is quite possible that those who returned for follow-up were preferentially those who felt the intervention had helped improve their well-being. Also, systolic blood pressure and Framingham score were the only two variables in this study for which this approach gives a satisfactory indication of unequivocal benefit.

The approach used above is sensible only when Δ and b are opposite in sign and the confidence interval for b is well away from zero, indicating that the slope with age is relatively well defined. If the latter condition does not hold, anomalous behaviour paralleling that of the number needed to treat (Section 7.6) can obviously occur. The only other variable recorded in the Healthy Hearts study satisfying both these conditions was the log-transformed Framingham score, with Δ = −0.0292 (SE 0.0168) corresponding to a 2.9% relative reduction in risk at follow-up, and b = +0.0530 (SE 0.0040) corresponding to a 5.2% increase in score per year based on a cross-sectional analysis of the baseline series. Here, the regression adjustment gives $\Delta - \bar{t} \times b = -0.1081$ (95% interval −0.1431 to −0.0730), corresponding to an adjusted relative reduction in risk at follow-up of 10.2%, with 95% interval from 7.0% to 13.3%. The WTCB approach gives $-\Delta/b$ = 2.924/5.295 or +0.55 years. Here, PropImp 95% limits may be obtained, with great care, as −0.07 and +1.20 years: the upper limit is derived from the upper limit for $-\Delta$ and the lower limit for b, but the lower limit is derived from the lower limit for both parameters. To all of these figures 1.49 years may be added as usual. Thus participants were actually 1.49 years older than at entry, but their risk profile corresponds to being 0.55 years younger, so it could be argued that at follow-up their risk was equivalent to retarding the progression of risk by 2.04 years, with 95% interval 1.42 to 2.69 years. The MOVER-DL approach should not be applied to the log-transformed Framingham score here, since the confidence interval for Δ includes zero. Nevertheless MOVER-R works straightforwardly in this situation, and produces 95% limits −0.07 and +1.20 years, identical to the PropImp limits here.

For the Framingham score, other perspectives for adjusting for ageing are possible, utilising the score's inbuilt age dependence, which obviously are not available for the other variables studied. The observed reduction of 0.0292 in the log-transformed Framingham risk score at follow-up at approximately 1.5 years means a 2.9% relative reduction in risk following intervention. These further analyses showed that the reduction in the score may be regarded as the resultant of two effects: an increase of 5.8 units due purely to the age component of the score being increased by 1.5 years, and a decrease of 8.7 or 8.8 units due to the other risk factors included in the Framingham score having improved at follow-up. These analyses are based on the formulae used to construct the Framingham score. Complementary to this, if we instead base the adjustment for age on the regression model fitted to the baseline data, the improvement in log-transformed Framingham score, adjusted for the effect of ageing by approximately 1.5 years, is 10.8 units. These two separate estimates of age-adjusted benefit, 8.7 or 8.8 and 10.8, are reasonably similar considering that they are derived in quite different ways. The intrinsic-adjusted estimate of benefit, 8.7 or 8.8, is probably the better estimate, inherently, and moreover had a slightly narrower confidence interval. But the closeness of these estimates does give a degree of reassurance that the process of using the regression coefficient from the baseline data to adjust for the effect of advancing age should be regarded as a robust one for the other variables, including systolic blood pressure, for which this is the only method of analysis available.

14.13 Grass Fires

Despite its rainy climate, Wales has a problem with fires on grassy hillsides started deliberately by adolescents. The peak incidence is in the spring, before the old season's growth is superseded by the more succulent new season's growth. In most years, the highest numbers of fires occur during the Easter school holiday fortnight. Researchers, in collaboration with the fire services, identified two areas in the South Wales valleys, Tonypandy and Aberdare, as having particularly large numbers of fires during several springs. An intervention targeting young people's behaviour was implemented in Tonypandy for the period leading up to Easter 2010 (Peattie et al. 2011). Aberdare served as an intervention-free control area.

While this non-randomised study cannot claim the same scientific rigour as a cluster randomised trial involving several areas, the choice of these two areas had several advantages. The two areas are topographically similar, and nearly but not quite adjacent. So they would be expected to have similar weather, but there is only a limited prospect that the intervention delivered in Tonypandy would influence the number of grass fires in Aberdare. Over the preceding 6 years, the two areas had closely similar numbers of grass fires, with exactly the same patterns of high and low incidence years.

The data analysed comprised numbers of fires in the intervention and control areas for a 6-week period spanning Easter 2010, and pooled figures for the corresponding periods for the preceding 6 years, as in Table 14.7.

For data of this kind, it is essential to have data from a control period, as well as a control area. This is because there is no possible way to justify the expectation that the two areas should have identical numbers of fires on the null hypothesis. This contrasts with the situation with patient-level health-related data, for which one would expect similar numbers of events amongst n patients allocated to an intervention and n allocated to control on H_0.

Accordingly, the simplest analysis of the results involves calculating the crude ratio for the intervention period, $a/b = 64/124$, then dividing it by an adjustment factor $c/d = 523/552$ representing the historical ratio of numbers of fires in the two areas. This produces the ordinary odds ratio calculated from the data as a 2 × 2 contingency table, 0.545. This is interpreted not as an odds ratio but as a relative risk, adjusted for the historic ratio of fires in the two areas across the preceding 6 years. The 95% limits for this ratio using the Miettinen–Nurminen formula developed for odds ratios (Section 11.3.2) are (0.394, 0.753). Accordingly, the investigators calculated a relative risk reduction of 45.5%, with 95% limits (24.7%, 60.6%). While clearly an analysis that interpreted the ratio of fires between intervention and control areas in 2010 in relation to how much it varied across preceding years

TABLE 14.7

Deliberate Grass Fires in an Intervention Area (Tonypandy) and a Control Area (Aberdare) During a 6-Week Period in Spring 2010, and Combined Figures for the 6 Years 2004–2009 in These Areas

	Deliberate Grass Fires In	
	Tonypandy	**Aberdare**
Intervention year 2010	$a = 64$	$b = 124$
Control years 2004–2009, merged	$c = 523$	$d = 552$

Source: Data from Peattie, S. et al. 2011. Social marketing to extinguish fire-setting behaviour. In I. V. Pereira (ed). Proceedings of the 10th International Congress of the International Association on Public and Nonprofit Marketing. Oporto, Portugal.

would in principle be preferable, the analysis performed here is the only practicable one with any possibility of demonstrating evidence for efficacy here.

However, interest centred not only on the ratio, but also on the absolute reduction in the number of fires in Tonypandy, compared to what would have been anticipated in the absence of intervention. It was estimated that each fire incurs a cost of approximately £2000 to the fire services. So to assess the cost-effectiveness of the intervention, we must estimate the absolute as well as the relative reduction attributable to it.

Then the expected number of fires in the intervention area during the study period is calculated as $b \times (c/d) = 124 \times (523/552)$ which is 117.5. The actual number of fires was 64. This is of course the same as 117.5 multiplied by 0.545, the adjusted ratio calculated above. So the intervention appears to have prevented $b \times (c/d) - a = 117.5 \times (1 - 0.545) = 53.5$ fires, and hence saved £107,000.

The simplest confidence limits for the number of fires prevented are calculated by the substitution method as $117.5 \times (1 - 0.753) = 29.0$ and $117.5 \times (1 - 0.394) = 71.2$. But it is unreasonable to treat the expected number of fires in the absence of intervention as if a predetermined constant.

Instead, we use PropImp to combine confidence intervals for the three quantities, *a*, *b* and *c*/*d*. The absolute numbers of fires in the two areas in the study period, *a* and *b*, may be modelled by Poisson distributions. An interval for *c*/*d* is readily obtained from an interval for the binomial proportion $c/(c + d)$. Using exact intervals for these parameters, the PropImp 95% interval for the number of fires prevented is 23.8 to 85.8. This is much wider than the substitution interval, suggesting that the latter is inappropriate.

A confidence interval for $bc/d - a$ may also be calculated using the MOVER algorithm twice. First, we calculate separate 95% exact intervals for *a*, *b* and $d/(c + d)$. The interval for $d/(c + d)$ is converted into an interval for *d*/*c*. Using the MOVER-R process, the intervals for *b* and *d*/*c* are combined to produce an interval for *bc*/*d*. Finally, the intervals for *bc*/*d* and *a* are combined by squaring and adding. This process leads to a 95% interval (24.3, 85.2) for fires prevented, very similar to the PropImp interval.

14.14 Incremental Risk-Benefit Ratio

Chen and Suryawanshi (2010) described a general approach for assessment of the balance between clinical risks and benefits in a comparative trial. Suppose treatment 1 is potentially more effective than treatment 2, but has a higher risk of an adverse effect. Let p_{1j} ($j = 1, 2$) denote the proportion of subjects receiving treatment *j* for whom the efficacy outcome is achieved, and let p_{2j} denote the proportion of subjects receiving treatment *j* experiencing the adverse effect. Lynd and O'Brien (2004) suggested displaying evidence from several studies by plotting the incremental risk $Y = p_{21} - p_{22}$ against the incremental benefit $X = p_{11} - p_{12}$ on a risk-benefit plane, the square defined by $|X| \leq 1$ and $|Y| \leq 1$. Chen and Suryawanshi suggested that this plot helps to interpret the outcome from a single trial, in relation to a prespecified risk-benefit threshold. Suppose we are willing to tolerate a μ% absolute increase in the incidence of the adverse event for 1% absolute increase in efficacy. Then the treatment is regarded as helpful if the point $(p_{11} - p_{12}, p_{21} - p_{22})$ lies below the line $Y = \mu X$. This is equivalent to determining whether the incremental risk-benefit ratio (IRBR) defined as $(p_{21} - p_{22})/(p_{11} - p_{12})$ is less than μ. This argument suggests that the IRBR may be a useful measure whenever there is an issue of balancing improved efficacy against adverse effects. One would not normally be concerned to calculate this measure when one treatment was superior to another in both respects. And, just as for the NNT, the division by a difference of proportions implies that confidence limits will be problematic whenever the difference in efficacy is not statistically significant. If neither risk nor benefit differs between the two treatments, such a ratio measure is uninterpretable—a positive value could mean positive differences in both parameters or negative differences in both of them.

Chen and Suryawanshi developed a simulation model to assess the probability that the active treatment was of net benefit, as a function of μ. This model takes account of the non-independence when evidence on efficacy and on adverse effects comes from the same trial. On their data the correlations between benefit and risk were practically negligible, −0.07 and −0.06 in groups 1 and 2. In this situation, and especially if the evidence on benefit and harm come from independent sources, a confidence interval for the IRBR may be obtained straightforwardly by the MOVER approach. Donner and Zou (2010) refer to the IRBR as a possible application of Fieller's theorem, but do not develop this further.

Table 14.8 shows the IRBR calculated for Chen and Suryawanshi's data, and for several modified versions of it, with MOVER-R and MOVER-DL limits. The data relate to a clinical trial of a drug treatment for obesity. The efficacy and safety endpoints were defined as 5% weight loss and occurrence of psychiatric adverse events. The differences in proportions attaining the efficacy and safety endpoints were 24.7% and 9.0%, resulting in IRBR = 0.37. If we disregard non-independence, either the MOVER-DL or MOVER-R algorithm may be used. For the observed data, both approaches yield the interval (0.10, 0.66).

In the second line of the table, we interchange the results for risk and benefit, resulting in an incremental benefit-risk ratio of 2.73. Whichever MOVER algorithm is used, the confidence limits are the reciprocals of those for the IRBR.

The third line shows what happens if the benefit and risk data are identical. Naturally, the point estimate is 1 here. Once again, the two methods lead to very similar intervals.

TABLE 14.8

Analyses for Incremental Risk-Benefit Ratio

	Risk		Benefit		Incremental Risk-Benefit Ratio	95% Confidence Limits	
	Active	**Placebo**	**Active**	**Placebo**		**MOVER-DL**	**MOVER-R**
Actual figures:							
Risk divided by benefit	219/811	37/206	397/811	50/206	0.366	0.100, 0.662	0.101, 0.665
Benefit divided by risk	397/811	50/206	219/811	37/206	2.729	1.510, 9.967	1.505, 9.889
Hypothetical figures:							
Benefit and risk data identical	219/811	37/206	219/811	37/206	1	0.256, 3.908	0.261, 3.825
Zero difference in risk	148/824	37/206	403/824	50/206	0	–	–0.265, 0.229
Zero difference in benefit	223/824	37/206	200/824	50/206	+∞	–	1.034, –1.042

Source: Based on Chen, M. and Suryawanshi, S. 2010. Application of methods for assessing risk-benefit in clinical trials. In JSM Proceedings, WNAR. Alexandria, VA: American Statistical Association, pp. 3858–3865.

When the observed risks in the two groups are identical, the IRBR is 0 (line 4). MOVER-R produces an interval which spans zero, which is appropriate here because the underlying difference in risk could be positive or negative. MOVER-DL fails to produce an interval.

When the observed benefits in the two groups are identical (line 5), the IRBR is +∞, because the risk is higher in group 1 with no compensating increase in benefit. Again, MOVER-DL fails to produce an interval. The MOVER-R limits, 1.034 and –1.042, are subject to the proviso we encountered for the NNT in Section 7.6. Here, the confidence region for the IRBR consists of two intervals, one from +1.034 to +∞ corresponding to increased risk on active treatment, the other from –∞ to –1.042 corresponding to decreased risk.

Thus, as in other contexts, MOVER-R is more widely applicable than MOVER-DL, but careful interpretation is required.

14.15 Adjustment of Prevalence Estimate Using Partial Validation Data

Most two-parameter applications we have considered permit a simple solution using the MOVER approach—a notable exception, of course, being the seizure prophylaxis example introduced in Section 13.2. In contrast, apart from very simple applications such as the additive scale interaction introduced earlier, applications involving more than two parameters generally combine them in a complex way. As a consequence the MOVER approach is no longer applicable, and PropImp leads to the most convenient solution. This is the case for the remaining three examples.

The examples in Sections 14.15 and 14.16 come from a prevalence survey of health care acquired infection covering England, Wales, Northern Ireland and the Republic of Ireland (Smyth et al. 2008). In each participating hospital following necessarily limited training, on a designated day in 2006 nursing staff completed a record form for each in-patient which captured basic demographic data and information on whether the patient currently had a health care acquired infection, and its nature and duration. This led to simple estimates of point prevalence in each country and overall, which being based on large numbers had narrow confidence intervals. However, there was concern that the resulting data might not conform so closely to the case definitions set out in the protocol as if data collection had been performed by the research team, for whom a much higher level of training in applying these definitions was feasible.

Accordingly, as a validation exercise, a member of the research team visited a small sample of the wards on the relevant day and completed a separate data sheet for each patient. While the data captured by hospital staff formed the basis of the figures reported for the study, the researchers also examined the degree of concordance between trust staff and the validator, and constructed analyses to assess the "true" prevalence in the entire series, regarding the validator's results as definitive. These analyses were restricted to the recording of the presence or absence of health care acquired infection, disregarding the type of infection, and other variables.

The figures reported here relate to England and Northern Ireland combined, because in these two countries validation was performed for a random sample of units. This contrasted with Wales, where the sampling was weighted towards apparently positive patients and the Republic of Ireland, where only apparently positive patients were reassessed. For England and Northern Ireland, the crude prevalence was 5010/62,419 or 8.026%. Adequate paired validation data was available for 4716 of the 62,419 patients. Assuming that the positive and negative predictive values observed in the validated series carry over to the remainder, the resulting calculations, set out in Table 14.9, result in an adjusted prevalence 4795/62,419 or 7.682%.

The calculations involve four sampled proportions:

p_1 = Pr[Trust staff +ve, validation series]	412/4716 = 8.74%
p_2 = Pr[Validator +ve | trust staff +ve]	334/412 = 81.07%
p_3 = Pr[Validator +ve | trust staff –ve]	55/4304 = 12.78%
p_4 = Pr[Trust staff +ve, non-validation series]	4598/57,703 = 7.97%

The adjusted prevalence is then expressed in terms of the p_i (i = 1…4) and the sizes of the validation series ($r + s$ and t in the notation of Table 14.9):

$$[(r + s)\{p_1p_2 + (1 - p_1)p_3\} + t\{p_4p_2 + (1 - p_4)p_3\}]/(r + s + t) \tag{14.4}$$

Confidence intervals for the adjusted prevalence were calculated by three methods, MCMC using uniform priors, bootstrap (conditioned on the validation series total $r + s$, not $a + c$ and $b + d$ separately), and PropImp based on Wilson intervals for the above proportions. The results obtained were as follows:

MCMC	median 7.692%	95% interval 7.23 to 8.165%
Bootstrap	median 7.68%	95% interval 7.22 to 8.15%
PropImp		95% interval 7.23 to 8.16%

These results indicate that PropImp can yield an interval that closely agrees with those calculated by the methods that would currently be most widely used for relatively complex constructs such as this.

TABLE 14.9

Calculation of Adjusted Prevalence of Health Care Acquired Infection in England and Northern Ireland Based on a Partial Validation Series

		Assessment by Hospital Staff		
		Positive	**Negative**	**Total**
Assessment by validator	Positive	a = **334**	b = **55**	r = 389
	Negative	c = **78**	d = **4249**	s = 4327
Validation series		$a + c$ = 412	$b + d$ = 4304	$r + s$ = 4716
Non-validation series		e = **4598**	f = **53,105**	t = 57,703
Total		$a + c + e$ = 5010	$b + d + f$ = 57,409	$r + s + t$ = 62,419
Crude prevalence		510/62,419 = 8.03%		
Corrected number positive:				
Validation series		r = 389		
Non-validation series		$a^*e/(a + c) + b^*f/(b + d)$ = 4406		
Total		4795/62,419 = 7.68%		

Source: Data from Smyth, E. T. M. 2008. *Journal of Hospital Infection* 69: 230–248.

The scenario described here is sometimes referred to as estimation using a double sampling scheme (Tenenbein 1970). This seems a poorer description of the scenario, in that often the results for the whole series ($a + c + e$ out of $r + s + t$) would be available first, with r+s participants subsequently reassessed. Tang, Qiu and Poon (2012) describe several alternative methods, which are applied both to the health care acquired infection data and also a much smaller dataset on graft-versus-host disease (Pepe 1992) in which (in the notation of Table 14.9) $a = 6$, $b = 3$, $c = 1$, $d = 8$, $e = 25$ and $f = 44$. Here, the adjusted proportion is 0.488. The PropImp 95% interval, 0.295 to 0.691, is similar to those produced by the methods of Tang et al.

14.16 Comparison of Two Proportions Based on Overlapping Samples

In the health care acquired infection study (Smyth et al. 2008), based on the data for all four countries together, the prevalence of healthcare acquired infection for patients with a urinary catheter was 2439/16,721 (14.6%), much higher than the prevalence for patients with a peripheral intravenous catheter, 2179/28,987 (9.4%). However, these two risk groups were not mutually exclusive but had a large overlap, as seen in Table 14.10.

The difference between the two prevalences above, 5.206%, is a simple linear function of the prevalences in the first three rows in Table 14.10, so the square-and-add Wilson approach, generalised to an arbitrary linear function of proportions as in Zou et al. (2009a), may be used. The resulting 95% limits are (4.760%, 5.662%). Alternatively, PropImp may be used to combine Wilson intervals for these three proportions. This leads to virtually identical limits, (4.759%, 5.662%). Consequently, for the difference between prevalences, the extra complexity of using PropImp is quite unnecessary.

TABLE 14.10

Prevalence of Health Care Acquired Infection in Patients with and without Urinary or Peripheral Intravenous Catheters

Urinary Catheter	Peripheral IV Catheter	Number of Patients at Risk	Number Infected	Prevalence of Infection
Yes	Yes	$n_{11} = 9507$	$r_{11} = 1434$	$p_{11} = 0.151$
	No	$n_{10} = 7214$	$r_{10} = 1005$	$p_{10} = 0.139$
No	Yes	$n_{01} = 19480$	$r_{01} = 1285$	$p_{01} = 0.066$
	No	$n_{00} = 39493$	$r_{00} = 2019$	$p_{00} = 0.051$

Source: Data from Smyth, E. T. M. 2008. *Journal of Hospital Infection* 69: 230–248.

The risk ratio is 14.6/9.4 or 1.555. In the notation of Table 14.10, this is

$$\frac{n_{11}p_{11}+n_{10}p_{10}}{n_{11}+n_{10}}\bigg/\frac{n_{11}p_{11}+n_{01}p_{01}}{n_{11}+n_{01}} \tag{14.5}$$

This is a function of three proportions, p_{11}, p_{10} and p_{01}: as in Section 10.8, the double-negative cell does not contribute. While MOVER is not applicable here, the PropImp approach is applicable just as for the difference between prevalences, and yields 95% limits (1.502, 1.610).

For such an application, great care is required to ensure that changes in the contributory parameters in the appropriate directions are considered. Generally, when the trial-and-error implementation is used, any such programming errors quickly become apparent, as the optimising value of the corresponding λ appears to be at the boundary. This occurrence is usually an indication that the programming requires careful checking.

14.17 Standardised Difference of Proportions

Sometimes the results of a placebo-controlled Phase II study are used to help plan a Phase III study, based on the observed efficacy from the Phase II study (Jia and Song 2010). The primary endpoint is the occurrence of a specified adverse outcome. In the Phase II study the observed event rates in the two groups were assumed to be 30/200 (15%) and 20/200 (10%). However, the Phase III study involves a different patient population, the event rate on placebo is anticipated to be about 7%. With the usual implicit assumption of equal sample sizes in the two groups, the authors sought to determine the most appropriate metric for the treatment effect to gauge the power for the intended Phase III study based on the Phase II study result.

Four effectsize measures were considered: the rate difference, the rate ratio, the odds ratio, and the standardised difference $(p_1-p_2)\big/\sqrt{p_1(1-p_1)+p_2(1-p_2)}$. For each measure, the power for the Phase III study was plotted against the event rate anticipated on placebo, for a range of group sizes. For the first three measures, the power altered grossly as the placebo event rate varied. But for the standardised difference, there was very little variation in power as the placebo event rate altered. This suggests that the standardised difference of proportions is a useful effect size measure to report, specifically as an adjunct to planning related studies.

Because the two proportions contribute to the standardised difference in an interlocked manner, the MOVER approach is not applicable here. The PropImp algorithm is applicable, and produces an appropriate interval.

Jia and Song reported a standardised difference 0.1072 with asymptotic standard error 0.071, leading to 95% limits (−0.0319, +0.2464). The PropImp limits are similar, (−0.0327, +0.2432).

Alternatively—particularly in an equivalence or non-inferiority study—the denominator used to standardise the difference could be based on pooled proportions, leading to $(p_1 - p_2)/\sqrt{2\overline{p}(1-\overline{p})}$ where $\overline{p} = (p_1 - p_2)/2$. The point estimate and PropImp limits become 0.1069 and (−0.0327, 0.2397). These are similar to the results obtained for the unpooled version. The two formulae differ substantially only if p_1 and p_2 differ widely. For example, if $p = 0.8$ and $p = 0.2$, the indices based on unpooled and pooled denominators are 1.06 and 0.85. The sample size for the new study based on such a gross difference would be unrealistically small. In this situation it is advisable to shade down the effect size from the precedent study, to arrive at a more reasonable sample size for the new study. Otherwise, the two versions of the measure do not differ greatly.

15

Generalised Mann–Whitney Measure

15.1 Absolute and Relative Effect Size Measures for Continuous and Ordinal Scales

In previous chapters we have considered binary data and absolute and relative measures for comparing proportions. Often, the estimated difference between two proportions and the confidence limits for the difference have a very natural, direct interpretation on the original scale of proportions. The difference of proportions and its reciprocal, the number needed to treat, are absolute measures. In some situations, the relative risk and odds ratio, which are relative or scaled measures, are more satisfactory measures to interpret.

In Chapter 2 we examined interval estimation for the mean of a continuously distributed variable, and for the absolute difference between two such means, for both unpaired and paired data structures. For example, a clinical trial comparing two antihypertensive drugs might report the mean reduction in systolic blood pressure was 5 mm Hg greater on drug A compared to drug B, with 95% confidence interval 2 to 8 mm Hg. This is an appealing expression of the findings, because clinicians are very familiar with this scale, and understand the clinical importance of a difference of 2, 5 or 8 units These summary statistics give a much more informative interpretation of the results than the *p*-value.

If desired—and especially in the event of seriously non-Gaussian distributional form, or for an ordinal outcome—the median difference may be estimated using the robust nonparametric Hodges–Lehmann estimator (Hodges and Lehmann 1963; Sen 1963; Hollander and Wolfe 1999). If the two datasets contain m and n data values, respectively, mn pairs of values (one from each set) can be formed and each pair gives a difference of values. The Hodges–Lehmann estimator for the median difference is the median of the mn signed differences. Like the mean difference, this is absolute, leading to point and interval estimates on the original scale of measurement. These mn differences also play a central role in the scaled effect size measure developed in this chapter. A related location estimate applicable to paired differences based on averages of pairs of observations is developed in Chapter 16.

In the unpaired case, a confidence interval for the median difference is described by Conover (1980), Hollander and Wolfe (1999) and Campbell and Gardner (2000). This too is derived from the *mn* differences of pairs. But this procedure has drawbacks, as explained in Section 15.2.1.

Sometimes, the binary variable is the outcome and the continuous one is the explanatory variable. In this situation the logistic regression coefficient can characterise the strength of relationship (Section 11.5). The units are then inverse to the original scale. For example, in a logistic regression of presence of disease by age (or duration of exposure) in a cross-sectional study, the regression coefficient might be 0.070 per year, interpreted as an increase in odds of presence of disease by a factor $e^{0.070} = 1.073$ per year.

All the above measures are absolute, and have units. Such measures are only helpful as a communication of the results if the scale of measurement is familiar to the relevant research community. Often this is not the case. Many comparisons involve outcomes such as visual analogue scales for symptom levels, or psychometric rating scales, albeit validated according to accepted criteria; in both instances, readers not involved in the original research may be at a loss to interpret a 1- or 10-unit difference for clinical importance. It is then more informative to quote a relative measure of effect size. In these final chapters, we consider alternative, relative effect size measures applicable to both continuous and ordinal data. In this chapter, we consider comparisons of two independent groups. Analogous methods for the paired case are developed in Chapter 16. For the homoscedastic Gaussian case, the natural relative measure is the standardised difference δ obtained by dividing the difference of means by the (pooled) standard deviation. However, it is more useful to construct an analogous relative effect size measure which does not embody parametric assumptions, which is then applicable to continuous and also ordinal data.

15.2 The Generalised Mann–Whitney Measure

An appropriate measure is the generalised Mann–Whitney measure

$$\theta = \Pr[Y > X] + \frac{1}{2}\Pr[Y = X] \tag{15.1}$$

Here X and Y denote independent random variables on the same support. Then θ is estimated by $\tilde{\theta} = U/mn$, the Mann–Whitney statistic U divided by the product of the two sample sizes m and n. (Generally, the superscript ~ (tilde) will be used to denote an empirical estimate, not ^ (circumflex or hat) as this would imply an MLE.) U is defined in the usual manner as

$$U = \sum_{i=1}^{m} \sum_{j=1}^{n} U_{ij} \tag{15.2}$$

where $U_{ij} = 1$, 0.5 or 0 when Y_j is greater than, equal to or less than X_i. Then U/mn serves as an obvious empirical estimate of θ. It expresses the degree of overlap (or conversely separation) between the two samples, and is applicable to both continuous and ordinal cases.

Both U/mn and the corresponding theoretical value θ range from 0 to 1, with values of 0 and 1 indicating no overlap. Values greater than 0.5 should be used to indicate a positive relationship, values below 0.5 an inverse one. On the null hypothesis that X and Y are identically distributed, $\theta = 0.5$, but the converse does not hold. The θ can be regarded as a measure of separation, or equally, a measure of discriminatory ability (Newcombe 1979). It is equivalent to the area under the receiver operating characteristic curve (AUROC) and the mean ridit (Bross 1958). It has been termed the probability of concordance, common language effect size (McGraw and Wong 1992) and measure of stochastic superiority (Vargha and Delaney 2000). It is considered in the form presented here by Fay and Gennings (1996), and is linearly related to Somers' D which is 2θ-1 (Somers 1962; Edwardes 1995; Newson 2002). Furthermore, in a study such as Ukoli et al. (1993) evaluating height and weight of children against international norms, the mean centile score is equivalent to the U/mn value obtained by comparing the series of interest against the normative series. Hollander and Wolfe (1999) give a historical review.

Hanley and McNeil (1982) presented an asymptotic confidence interval for the AUROC without parametric assumptions, based on an asymptotic variance under the alternative hypothesis derived by Noether (1967, pp. 32–33). A procedure to plot the ROC curve and obtain the AUROC with the Hanley–McNeil interval is available in SPSS from the Graphs menu. Hanley and McNeil also developed a modification based on assumed exponential distributions. Both methods have the deficiencies of producing zero width intervals in extreme cases and limits outwith [0, 1] in near-extreme ones. The Wald variance imputed to U/mn can also be used in hypothesis testing, for a one-sample test of H_0: $\theta = \theta_0$ for some specified value θ_0.

An alternative asymptotic approach was developed by Halperin et al. (1987) and Mee (1990). This method is analogous to the Wilson (1927) score interval for the single proportion: confidence limits are obtained by inversion; that is, solving a quadratic of the form

$$|\theta - \tilde{\theta}| = z\sqrt{\theta(1-\theta)/\tilde{N}_J} \quad (15.3)$$

where $\tilde{N}_J$ denotes a quantity related to the sample sizes which may be termed the pseudo-sample size. This avoids the deficiency of the Wald method in extreme and near-extreme cases, but only because the formula for $\tilde{N}_J$ incorporates an *ad hoc* shift modification to cope with these cases.

Newcombe (2006a) explored some issues relating to the U/mn statistic, and showed the feasibility of developing intervals based on tail areas, albeit of limited practical usefulness. Newcombe (2006b) developed more widely applicable asymptotic methods and evaluated their performance as in Sections 15.3 to 15.6. Subsequent related publications include Ryu and Agresti (2008), Zhou (2008), and Brown et al. (2009).

The measure $\tilde{\theta} = U/mn$ studied here is mathematically identical to the AUROC, nevertheless calculating it and plotting the ROC curve should be regarded as meeting rather different objectives. When the purpose is to characterise the trade-off between sensitivity and specificity of a single test as the cutpoint is altered, as in Section 12.4, the ROC curve as an entity is the obvious summary of the data, rather than a single summary figure.

The quantity θ is used directly as a basis for planning sample size based on the Mann–Whitney test, for example in nQuery Advisor 7.0 (Statistical Solutions 2009), although θ is specified as $\Pr[X < Y]$ for both the "continuous outcome" and "ordered categories" cases.

15.2.1 Two Examples

The generalised Mann–Whitney measure is applicable to categorical and non-categorised data alike. The forensic psychiatry dataset (Table 1.1) considered in Section 1.4.1 serves as an example of a small, non-categorised dataset. With one marked exception, scores for subjects with history of head injury leading to loss of consciousness for over 48 hours were higher than those for the remaining men. The distributional form does not appear close to Gaussian, especially for the head injury cases. (It is conceded that this example does rather stretch the applicability of the *U/mn* measure, in that the difference between the distributions of those with and without head injury is as much a difference in scale as in location—the *U/mn* measure is liable to be rather less useful in such situations.)

TABLE 15.1

Follow-Up of 122 Adults Attending Welsh Language Classes in Cardiff

	Proportion with Successful Outcome
Year 1, term 1	11/37 (30%)
Year 1, term 2	11/35 (31%)
Year 1, term 3	2/5 (40%)
Year 2	9/19 (47%)
Year 3	8/11 (73%)
Beyond year 3	10/15 (67%)

Source: Newcombe, L. P. 2002. The relevance of social context in the education of adult Welsh learners with special reference to Cardiff. Ph.D. thesis, Cardiff University.

Note: Proportions with successful outcome (more than weekly use, not only for media) by level of course.

The median scores of men with and without head injury are 38.5 and 19. Minitab reports the median difference between subjects with and without head injury as the Hodges–Lehmann estimate 17, with 95.2% confidence interval 1 to 27. Because these limits are values extracted from the ordered set of $mn = 78$ signed differences, it is not possible to construct an interval with exactly 95% confidence, the nearest the algorithm can get is 95.2%. And the median and its confidence limits are only interpretable to those familiar with this scale.

Alternatively, the results may be summarised by calculating $\tilde{\theta} = 0.801$. For this, a $100(1 - \alpha)\%$ interval may be calculated for any α; 95% limits are (0.515, 0.933) using method 5 defined in Section 15.3. These figures clearly indicate a very substantial observed effect size, with a wide confidence interval reflecting small sample uncertainty. The upper limit of 0.933 indicates that there might be little overlap between the two distributions in the population from which these observations have been drawn. The lower limit is just above 0.5, corresponding to an obviously trivial difference, yet implying rejection of the null hypothesis at a conventional α level very much in the same way as the corresponding asymptotic p-value of 0.039 or two-tailed exact conditional p-value of 0.036. Familiarity with the scale is not a prerequisite to interpreting these relative measures.

As an example of the application to categorical data, Lynda Pritchard Newcombe (2002) studied adults who were attending Welsh language classes in Cardiff during the academic year 1998–1999. Adult learners of a lesser-used language can make a major contribution to language revitalisation. But one well-recognised major hindrance to this is the fact that relatively few of those who start to learn the language end up *dros y bont* (across the bridge) as fluent speakers. In a cross-sectional study of learners currently attending classes at various levels, a group of 175 students were identified. Approximately 18 months later 122 of them were subsequently followed up, and asked in what ways they were currently using the language, and how frequently. For the purposes of analysis, successful outcome was defined as use at least weekly, excluding media broadcasts in the language. Fifty-one of those followed up were classed as having a successful outcome.

The results obtained are summarised in Table 15.1. There is a clear trend in the proportion with successful outcome, with higher proportions of the more advanced learners still using the language at follow-up. Traditionally, these results would be reported as showing a highly statistically significant association by either a 1 df chi-square test for trend ($X^2 = 10.82$, $p = 0.001$) or a Mann–Whitney two-sided p-value of 0.002. However, neither analysis indicates the strength of the association in a directly interpretable way, nor does the Mann–Whitney U-value itself, which is reported by SPSS as 1234. The SPSS ROC analysis, with "success" defined as the positive value of the state variable, gives AUROC = 0.659, with 95% interval 0.559 to 0.759. The U-value produced by SPSS leads to $\tilde{\theta} = 1234/(51 \times 71) = 0.341$. Because the proportion with successful outcome is higher at more advanced levels of learning, the value that should be reported is the complement of this, 0.659, as produced by the ROC analysis. A better 95% interval is obtained as 0.555 to 0.747 using method 5 defined in Section 15.3. Logistic regression with a linear "exposure" term would not be helpful here as the exposure scale is not metric in a clearly definable way; exposure could be entered as a categorical covariate, but this model would merely produce a perfect fit to the observed proportions.

15.2.2 Relationship between the Generalised Mann–Whitney Measure and the Standardised Difference

For the general Gaussian case with $X \sim N(\mu_1, \sigma_1^2)$ and $Y \sim N(\mu_2, \sigma_2^2)$,

$$\theta = \Phi\left(\frac{\mu_2 - \mu_1}{\sqrt{\sigma_1^2 + \sigma_2^2}}\right). \tag{15.4}$$

where Φ denotes the cumulative distribution function of the standard Gaussian distribution (Simonoff et al. 1986; Noether 1987). In particular, for the homoscedastic Gaussian case, expressed without loss of generality as $X \sim N(0, 1)$ and $Y \sim N(\delta, 1)$, θ reduces to $\Phi(\delta/\sqrt{2})$ (Newcombe 1979). Table 15.2 gives values of δ corresponding to selected values of θ and vice versa.

It is preferable to estimate θ rather than δ, as it is less dependent on distributional assumptions, thus more satisfactory than the standardised difference in extreme cases, as $\hat{\theta} = 1$ suggests δ is large without implying any specific value. The degree of separation implied by a particular value of θ is readily visualised by plotting two Gaussian curves with equal standard

TABLE 15.2

Correspondence between the Generalised Mann–Whitney Measure θ and the Standardised Difference δ in the Homoscedastic Gaussian Case

θ	δ	δ	θ
0.50	0.000	0.00	0.500
0.55	0.178	0.25	0.570
0.60	0.358	0.50	0.638
0.65	0.545	0.75	0.702
0.70	0.742	1.00	0.760
0.75	0.954	1.25	0.812
0.80	1.190	1.50	0.856
0.85	1.466	1.75	0.892
0.90	1.812	2.00	0.921
0.95	2.326	2.50	0.961
0.99	3.290	3.00	0.983
0.999	4.370	3.50	0.993
		4.00	0.998

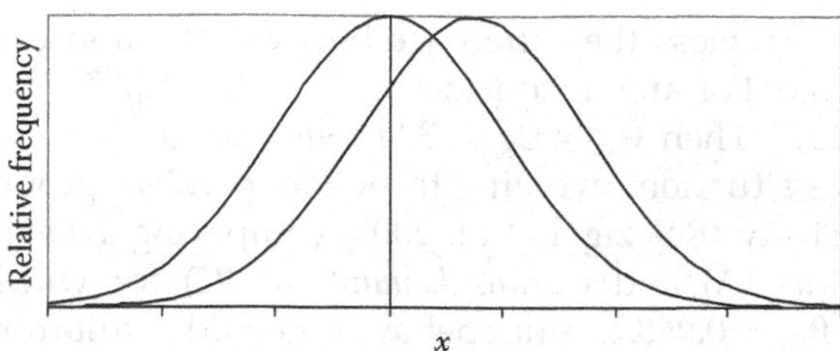

FIGURE 15.1
Relative frequency curves for Gaussian distributions $N(0,1)$ and $N(\delta,1)$ with $\delta = 0.7416$ corresponding to $\theta = 0.7$.

deviations and peaks separated by the corresponding standardised difference (Figure 15.1).

15.2.3 Condorcet and Related Paradoxical Effects

Consider three independent random variables X, Y and Z. In the simplest, homoscedastic Gaussian case with $X \sim N(\mu_1, 1)$, $Y \sim N(\mu_2, 1)$ and $Z \sim N(\mu_3, 1)$, the standardised differences δ_{XY}, δ_{XZ} and δ_{YZ} expressing their separation are simply related by $\delta_{XZ} = \delta_{XY} + \delta_{YZ}$. From the relationship $\theta = \Phi(\delta/\sqrt{2})$, the corresponding θ_{XY}, θ_{XZ} and θ_{YZ} are related by

$$\Phi^{-1}(\theta_{XZ}) = \Phi^{-1}(\theta_{XY}) + \Phi^{-1}(\theta_{YZ}). \tag{15.5}$$

Normally we expect a similar relationship to hold approximately for the $\tilde{\theta}$s, in particular, that when $\tilde{\theta}_{XY}$ and $\tilde{\theta}_{YZ}$ are both greater than 1/2, $\tilde{\theta}_{XZ} \geq \max(\tilde{\theta}_{XY}, \tilde{\theta}_{YZ})$, and also to some degree when the assumptions of Gaussian distributional form and homoscedasticity are relaxed. But very different patterns can sometimes occur.

The most striking of these is when non-transitivity occurs. Since U/mn is simply an alternative way of expressing the information from the Mann–Whitney U statistic, the classical Condorcet (or Escher staircase) non-transitive dominance paradox (Condorcet 1785; Rae 1980; Anonymous 2011a) applies to U/mn just as to the test. Consider the three samples $\boldsymbol{X} = \{3, 5, 7\}$, $\boldsymbol{Y} = \{1, 6, 8\}$ and $\boldsymbol{Z} = \{2, 4, 9\}$. Here a Kruskal–Wallis test comparing the three groups on an equal footing would assign identical rank sums to all. But ranks are altered by omitting uninvolved groups, so here the Mann–Whitney criterion ranks pairs of groups $\boldsymbol{X} < \boldsymbol{Y} < \boldsymbol{Z} < \boldsymbol{X}$. From the fact that $\tilde{\theta}_{XY} = \tilde{\theta}_{YZ} = 5/9$, one would expect $\tilde{\theta}_{XZ}$ to be greater than 5/9, around 0.610 by summating the corresponding δs; in fact it is only 4/9.

Conversely, $\tilde{\theta}_{XZ}$ can be much more extreme than Equation 15.5 would suggest. Because this does not amount to non-transitivity, it is less striking as paradoxical; nevertheless the difference between $\tilde{\theta}_{XZ}$ and either $\tilde{\theta}_{XY}$ or $\tilde{\theta}_{YZ}$ may be very large. For any n, consider $\mathbf{X} = \{1, 2,\ldots, n\}$, $\mathbf{Z} = \{n + 1, n + 2,\ldots, 2n\}$, and $\mathbf{Y} = \mathbf{X}\cup\mathbf{Z}$. Then $\tilde{\theta}_{XY} = \tilde{\theta}_{YZ} = 3/4$ whereas $\tilde{\theta}_{YZ} = 1$. A real example approaching this situation involving three independent groups occurred in a parasitology study (Reiczigel et al. 2003) comparing *Trinoton luridum* (T), *Anatoecus dentatus* (A) and *Carnus hemapterus* (C) for which $\tilde{\theta}_{AT} = 0.759$, $\tilde{\theta}_{CA} = 0.725$ and $\tilde{\theta}_{CT} = 0.968$. Similar behaviour could result from effective triage of a presenting series of subjects Y into subgroups X and Z, for example when a telephone ambulance dispatch prioritisation algorithm for injuries is validated by reference to an injury severity scale. Or equivalently for a non-age-dependent biochemical gene product marker for the autosomal dominant disorder, Huntington's disease, for which penetrance increases from 0 at birth in a sigmoid fashion to approach 100% in elderly heterozygotes (Newcombe 1981). Here, group Z comprises known affected subjects, group Y young clinically unaffected offspring of established heterozygotes, whose genotype is unknown but whose posterior risk of heterozygosity is still at its 50% prior level, and group Z their elderly counterparts whose posterior risk has declined to near zero.

The hypothetical examples above may be extremes for the behaviour of $\tilde{\theta}_{XZ}$ given $\tilde{\theta}_{XY}$ and $\tilde{\theta}_{YZ}$. In most of these examples, one population is simply a mixture of the other two. Nevertheless the parasitology example shows this can occur with genuinely independent groups. The possibility of behaviour so far from additivity on the δ scale should be regarded as a curiosity rather than a serious limitation in practice.

15.2.4 Effect of Discretisation on *U/mn*

Hanley and McNeil (1982) pointed out that $\tilde{\theta}$ values estimated directly as U/mn are not interchangeable with those obtained by fitting continuous distributions to the data. They illustrated this with numerical results and with ROC "curves" for the same data constructed as line segments and smooth convex curves through points. It is recognised (Pearson 1913; Agresti 1976) that discretising continuous data reduces the power available in hypothesis testing with a shift alternative; this occurs predominantly because discretisation tends to reduce θ. Newcombe (2006a) distinguished three cases:

CF continuous distributions, leading (theoretically) to data free of ties

CR continuous distributions, leading (in practice) to rounded or discretised data

DT discrete distributions, leading naturally to tied data.

CF and DT should be regarded as genuinely distinct cases, and the $\tilde{\theta}$s derived from each should be regarded as meaningful estimates, not requiring adjustment. Hanley and McNeil's exemplar dataset (Table 12.3) comprised $m = 58$ ratings of "normal" computed tomography images and $n = 51$ "abnormal" ones, expressed on a five-point Likert scale. Here, two different estimates are available. We may regard the data as arising from case DT and calculate $\tilde{\theta} = 0.893$ directly. Or, following Hanley and McNeil, we may fit a smooth Gaussian-based ROC curve by maximum likelihood, leading to an adjusted estimate, 0.911. Hanley and McNeil do not indicate clearly which figure should be quoted. For two reasons, $\tilde{\theta}$ is the appropriate point estimate. The data are poorly fitted by a homoscedastic Gaussian model. Moreover, the investigators chose to use a five-point scale, they did not consider it meaningful to elicit ratings on a more finely subdivided or continuous scale. Subject to distributional reservations, 0.911 represents the degree of separation that would have been obtained had it been meaningful to express on a continuous scale, but this is counterfactual. In the hybrid case CR it is important to examine the effect of bias that may have been introduced by discretisation. In case DT, the number of categories used is an inherent part of the structure of the data, and ideally confidence intervals for this case should take account of this. Nevertheless the evaluation below shows that asymptotic intervals developed primarily for case CF, with the assumption that the underlying distributions of the two random variables X and Y are absolutely continuous so that ties occur with probability zero, also perform well on data generated from a discrete distribution with five categories.

Newcombe (2006a) examined the degree of bias introduced when data from inherently continuous distributions are discretised or rounded. In every case considered, θ^* calculated from the discretised distributions underestimated θ, to an extent that depends on the degrees of separation and discretisation and distributional form. Hence in interpreting an observed U/mn value in case CR, it is important to bear in mind the resulting uncertainty in the degree of underestimation.

15.3 Definitions of Eight Methods

Let $X_1, \ldots, X_m$ and $Y_1, \ldots, Y_n$ denote independent random samples of sizes m and n from random variables X and Y. The Mann–Whitney test statistic is defined as $U = \sum_{i=1}^{m} \sum_{j=1}^{n} U_{ij}$ where $U_{ij} = 1$ if $Y > X$, 1/2 if $Y = X$, and 0 if $Y < X$.

Then $EU = mn\theta$ where

$$\theta = \Pr[Y > X] + \frac{1}{2}\Pr[Y = X] = \Pr[U_{ij} = 1] + \frac{1}{2}\Pr\left[U_{ij} = \frac{1}{2}\right] \tag{15.6}$$

The empirical estimate of θ is $\tilde{\theta} = U/mn$. The eight asymptotic intervals included in the evaluation are invariant on monotonic increasing transformation of scale.

1. The Hanley–McNeil (1982) interval is based on the general variance under H_1 of the Mann–Whitney statistic derived by Noether (1967, pp. 32–33). Noether expresses an equivalent statistic W in the form $\sum_{i=1}^{m}\sum_{j=1}^{n} s_{ij}$ where $s_{ij} = +1$ if $X_i > Y_j$, -1 if $X_i < Y_j$. This variance depends on θ and on two straddle probabilities. Noether defines p^{+-} as 2 Pr[$X_1 < Y < X_2$] where X_1 and X_2 are sampled from the X distribution and Y is sampled from the Y distribution, all independently. Similarly q^{+-} denotes 2 Pr[$Y_1 < X < Y_2$]. The derivation on page 32 of Noether is very general, encompassing the tied as well as the untied case, but it is subsequently simplified on page 33 for the absolutely continuous case. This expression reduces to the familiar form $mn(m + n + 1)/3$ used in an asymptotic test of the null hypothesis that the two distributions are identical.

   To estimate the relevant quantities from the sample, $\tilde{Q}_1 = \sum_{i=1}^{m} U_{i.}^2/mn^2$ and $\tilde{Q}_2 = \sum_{j=1}^{n} U_{.j}^2/m^2n$ are calculated where $U_{i.} = \sum_{j=1}^{n} U_{ij}$, $U_{.j} = \sum_{i=1}^{m} U_{ij}$. In the absence of ties, $\tilde{Q}_1$ estimates Q_1 = Pr[Y_j and $Y_k > X$] when Y_j and Y_k are sampled *independently*; that is, *with replacement* from $Y_1, Y_2, \ldots, Y_n$ and X is sampled from $X_1, X_2, \ldots, X_m$—a kind of resampling procedure. Similarly $\tilde{Q}_2$ estimates Q_2 = Pr[X_h and $X_i < Y$]. Then Wald limits are calculated as $\tilde{\theta} \pm z\sqrt{V}$ where $V = V_1$, the variance estimate for method 1 which is

$$V_1 = \left\{\tilde{\theta}(1-\tilde{\theta}) + n^*(\tilde{Q}_1 - \tilde{\theta}^2) + m^*(\tilde{Q}_2 - \tilde{\theta}^2)\right\}/mn \qquad (15.7)$$

   where $m^* = m - 1$ and $n^* = n - 1$. As usual, z denotes the appropriate centile of the standard Gaussian distribution.

   This method, and also derived methods 2 and 4 below, share two deficiencies of the Wald interval for proportions. A degenerate, zero width interval occurs when $\tilde{\theta} = 0$ or 1. Furthermore, in near-extreme cases, calculated limits outwith [0, 1] may occur. In practice, overshoot is truncated at 0 or 1. Just as for Wald intervals for proportions, truncation is inadequate because the occurrence of any overlap implies that the value $\theta = 0$ or 1 is impossible, and would be rejected with a zero p-value.

2. Hanley and McNeil also suggested a simplification based on assumed exponential distributions for X and Y. Having demonstrated that the relationship between θ and its standard error does not depend strongly on the distributional form of the data, they proposed to estimate Q_1 and Q_2 by $\tilde{\theta}/(2-\tilde{\theta})$ and $2\tilde{\theta}^2/(1+\tilde{\theta})$. Confidence limits are then calculated as $\tilde{\theta} \pm z\sqrt{V_2(\tilde{\theta})}$ where

$$V_2(\theta) = \theta\,(1-\theta)\,[1 + n^*(1-\theta)/(2-\theta) + m^*\theta/(1+\theta)]/mn \qquad (15.8)$$

Given m and n, the calculated limits now depend on the data only via $\tilde{\theta}$: loosely, $\tilde{\theta}$ may be regarded as "sufficient" for the limits. All methods included in this evaluation except Hanley and McNeil's original method 1 and the Halperin–Mee method 6 have this simplifying property.

3. Just as for proportions the Wilson (1927) score interval is a great improvement on the Wald method; similarly we can improve method 2 by inversion, solving to obtain the two roots of

$$|\theta - \tilde{\theta}| = z\sqrt{V_2(\theta)}. \qquad (15.9)$$

This is no longer closed form but reduces to a quartic for θ, most conveniently solved by iteration.

4. Method 4 modifies method 2 by using $m^* = n^* = (m+n)/2 - 1$ in V_2.
5. Similarly, method 5 symmetrises method 3 by using $m^* = n^* = (m+n)/2 - 1$.
6. The Halperin–Mee method calculates the two quantities

$$\tilde{p}_1 = \sum_{i=1}^{m}\sum_{h\neq i}^{m}\sum_{j=1}^{n} U_{ij}U_{hj}/mn(m-1) \qquad (15.10)$$

$$\tilde{p}_2 = \sum_{i=1}^{m}\sum_{j=1}^{n}\sum_{k\neq j}^{n} U_{ij}U_{ik}/mn(n-1) \qquad (15.11)$$

which in the absence of ties unbiasedly estimate

$$p_1 = \Pr[U_{ij}U_{hj}] = 1\ (h \neq i) \qquad (15.12)$$

$$p_2 = \Pr[U_{ij}U_{ik}] = 1\ (j \neq k). \qquad (15.13)$$

$\tilde{p}_1$ and $\tilde{p}_2$ *superficially* resemble $\tilde{Q}_2$ and $\tilde{Q}_1$ used in method 1, but differ in that Y_j and Y_k are now sampled *without replacement* from $Y_1, Y_2, \ldots, Y_n$; denominators differ correspondingly. They are related by

$$mn^2\tilde{Q}_1 = U + mn(m-1)\tilde{p}_2 \quad (15.14)$$

$$m^2n\tilde{Q}_2 = U + mn(n-1)\tilde{p}_1. \quad (15.15)$$

The interval comprises all θ satisfying

$$|\theta - \tilde{\theta}| \leq z\sqrt{(\theta(1-\theta)/\tilde{N}_J)} \quad (15.16)$$

(Halperin et al. 1987; Equation 2.4 of Mee 1990), where

$$\tilde{\rho}_s = (\tilde{p}_s - \tilde{\theta}^2)/(\tilde{\theta} - \tilde{\theta}^2) \quad (15.17)$$

for $s = 1, 2$ and

$$\tilde{N}_J = mn/\{[(m-1)\tilde{\rho}_1 + 1]/(1-1/n) + [(n-1)\tilde{\rho}_2 + 1]/(1-1/m)\}. \quad (15.18)$$

This interval is identical in form to a Wilson interval for a sample proportion $\tilde{\theta}$ based on $\tilde{N}_J$ observations (although in general neither $\tilde{N}_J$ nor $\tilde{N}_J\tilde{\theta}$ is integer valued), and (unlike methods 3 and 5) is thus symmetrical on a logit scale (Newcombe 1998a).

However, $\tilde{\rho}_1$, $\tilde{\rho}_2$ and $\tilde{N}_J$ are undefined when $\tilde{\theta} = 1$ [or 0], and poorly estimated in near-extreme cases. Mee proposed to then base $\tilde{N}_J$ on a shifted dataset: the smallest possible constant is then subtracted from (or added to) each Y_j to ensure $\min(\tilde{\theta}, 1-\tilde{\theta}) \geq 0.5/\sqrt{mn}$. While this is acceptable for an observed dataset, invariance on monotonic transformation is lost. The present evaluation is designed to use several distributions of origin solely as a means of generating ranked data. Thus to preserve invariance whilst keeping close to Mee's intentions, it was decided after some experimentation to implement this method by ranking the data in the normal manner for a Mann–Whitney test, then shift the Y ranks down (or up) by 1/2, 1,... until $\min(\tilde{\theta}, 1-\tilde{\theta}) \geq 0.5/\sqrt{mn}$.

7. The double beta model is based on a mirror-image pair of beta distributions with cumulative distribution functions

$F(X) = 1 - (1 - x^\alpha)$, $G(y) = y^\alpha$ for $0 \leq x$ and $y \leq 1$ and $\alpha \geq 1$. Then $\theta = 1 - R_1$ and $Q_1 = Q_2 = Q = 1 - 2R_1 + R_2$ where $R_1 = \Gamma^2(\alpha+1)/\Gamma(2\alpha+1)$ and $R_2 = \Gamma(2\alpha+1)\,\Gamma(\alpha+1)/\Gamma(3\alpha+1)$. We solve

$$|\theta - \tilde{\theta}| = z\sqrt{V_7} \tag{15.19}$$

where

$$V_7 = [\theta(1-\theta) + (m+n-2)(Q-\theta^2)]/mn \tag{15.20}$$

essentially by iteration for α, then convert to limits on the θ scale.

To ensure $\alpha \geq 1$, we use $F(x) = x^{\alpha}$, $G(y) = 1 - (1-y)^{\alpha}$ to represent $\theta < 1/2$. Accordingly, interval bisection is based throughout on $\lambda = (1 - 1/\alpha)\,\mathrm{sgn}\,(\theta - 1/2) \in [-1, 1]$.

8. Similarly, in the double Gaussian model, for $X \sim N(0,1)$ and $Y \sim N(\delta,1)$, $\theta = \Phi(\delta/\sqrt{2})$. To obtain Q, when X, Y_1 and Y_2 are chosen independently, let $Z_j = Y_j - X$ ($j = 1, 2$). Then (Z_1, Z_2) is bivariate Gaussian with correlation 1/2, hence $\Pr[Z_1 > a$ and $Z_2 > b]$ is readily programmed. Here V reduces to

$$V_8 = \{(m+n-1)\,\theta\,(1-\theta) - 2(m+n-2)\,T(\delta/\sqrt{2}, 1/\sqrt{3})\}/mn \tag{15.21}$$

where T is the function defined by Owen (1956). Iteration to obtain limits satisfying

$$|\theta - \tilde{\theta}| = z\sqrt{V_8} \tag{15.22}$$

is implemented using interval bisection for δ using "goalposts" at ± 10, then converting back to the θ scale.

15.4 Illustrative Examples

Tables 15.3 and 15.4 present untruncated limits calculated by eight methods for selected datasets. As expected, all methods produce broadly similar intervals for dataset (a) with moderate sample sizes and $\tilde{\theta}$ not extreme. Methods 1, 2 and 4 produce zero width intervals in extreme cases, and overshoot in near-extreme ones. Generally methods 5, 7 and 8 yield similar intervals, whereas those from method 6 can be quite different.

On account of the asymmetry of the underlying exponential model, method 2, and also method 3 derived from it below, do not possess full equivariance (Blyth and Still 1983): the limits are not simply complemented by scale reversal, although they are when group labelling is reversed. This is illustrated by datasets (d) to (g) in Table 15.3. We can conveniently

TABLE 15.3

95% Confidence Limits for the Generalised Mann–Whitney Measure θ Calculated by Eight Methods for Illustrative Datasets

Dataset	m	n	$\tilde{\theta} = U/mn$	Method 1	Method 2	Method 3	Method 4	Method 5	Method 6	Method 7	Method 8
(a) Table 1.1 (Hanley	58	51	0.8932	0.8305	0.8295	0.8102	0.8308	0.8117	0.8171	0.8142	0.8137
and McNeil 1982)				0.9559	0.9568	0.9408	0.9556	0.9400	0.9399	0.9362	0.9366
(b) *XXXXXYYYYY*	5	5	1.0000	1.0000	1.0000	0.6427	1.0000	0.6427	0.6823	0.6438	0.6435
				1.0000	1.0000	1.0000	1.0000	1.0000	1.0000	1.0000	1.0000
(c) *XXXXYXYYYY*	5	5	0.9600	0.8427	0.8245	0.5921	0.8245	0.5921	0.6307	0.5925	0.5924
				1.0773	1.0955	0.9770	1.0955	0.9970	0.9970	0.9952	0.9953
(d) *XYYXY*	2	3	0.6667	0.1009	0.1467	0.2042	0.1366	0.2126	0.1358	0.2148	0.2144
				1.2325	1.1866	0.9286	1.1968	0.9342	0.9622	0.9299	0.9303
(e) *YXXYX*	3	2	0.3333	-0.2325	-0.1866	0.0714	-0.1968	0.0658	0.0378	0.0701	0.0697
				0.8991	0.8533	0.7958	0.8634	0.7874	0.8642	0.7852	0.7856
(f) *YXYYX*	2	3	0.3333	-0.2325	-0.2067	0.0609	-0.1968	0.0658	0.0378	0.0701	0.0697
				0.8991	0.8734	0.7789	0.8634	0.7874	0.8642	0.7852	0.7856
(g) *XYXXY*	3	2	0.6667	0.1009	0.1266	0.2211	0.1366	0.2126	0.1358	0.2148	0.2144
				1.2325	1.2067	0.9391	1.1968	0.9342	0.9622	0.9299	0.9303

Note: Limits calculated by methods 1, 2 and 4 are displayed without truncation to lie within [0,1]. Showing boundary violation for datasets (c)–(g) using these methods and zero width intervals for dataset (b).

TABLE 15.4

95% Confidence Limits for the Generalised Mann–Whitney Measure θ Calculated by Eight Methods for Further Illustrative Datasets

Dataset	m	n	$\tilde{\theta}$	Method 1	Method 2	Method 3	Method 4	Method 5	Method 6	Method 7	Method 8
(h) 18X, 100Y, 2X	20	100	0.9	0.7685	0.8427	0.8149	0.8230	0.7913	0.6925	0.7941	0.7936
				1.0315	0.9573	0.9410	0.9770	0.9529	0.9729	0.9482	0.9486
(i) 10Y, 20X, 90Y	20	100	0.9	0.8412	0.8427	0.8149	0.8230	0.7913	0.8252	0.7941	0.7936
				0.9588	0.9573	0.9410	0.9770	0.9529	0.9449	0.9482	0.9486
(j) 90X, 20Y, 10X	100	20	0.9	0.8412	0.8074	0.7733	0.8230	0.7913	0.8252	0.7941	0.7936
				0.9588	0.9926	0.9605	0.9770	0.9529	0.9449	0.9482	0.9486
(k) 2Y, 100X, 18Y	100	20	0.9	0.7685	0.8074	0.7733	0.8230	0.7913	0.6925	0.7941	0.7936
				1.0315	0.9926	0.9605	0.9770	0.9529	0.9729	0.9482	0.9486
(l) 10X, 100Y, 10X	20	100	0.5	0.2809	0.3608	0.3594	0.3608	0.3669	0.2950	0.3672	0.3671
				0.7191	0.6392	0.6259	0.6392	0.6331	0.7050	0.6328	0.6329
(m) 50Y, 20X, 50Y	20	100	0.5	0.4020	0.3608	0.3594	0.3608	0.3669	0.4034	0.3672	0.3671
				0.5980	0.6392	0.6259	0.6392	0.6331	0.5966	0.6328	0.6329
(n) 50X, 20Y, 50X	100	20	0.5	0.4020	0.3608	0.3741	0.3608	0.3669	0.4034	0.3672	0.3671
				0.5980	0.6392	0.6406	0.6392	0.6331	0.5966	0.6328	0.6329
(o) 10Y, 100X, 10Y	100	20	0.5	0.2809	0.3608	0.3741	0.3608	0.3669	0.2950	0.3672	0.3671
				0.7191	0.6392	0.6406	0.6392	0.6331	0.7050	0.6328	0.6329
(p) 3X, 5Y, 2X	5	5	0.6	0.1706	0.2335	0.2718	0.2335	0.2718	0.2046	0.2734	0.2731
				1.0294	0.9665	0.8537	0.9665	0.8537	0.8974	0.8498	0.8504
(q) 3X, 20Y, 2X	5	20	0.6	0.1706	0.3296	0.3164	0.3184	0.3315	0.2046	0.3323	0.3321
				1.0294	0.8704	0.7942	0.8816	0.8148	0.8974	0.8117	0.8122
(r) 1Y, 12X, 2Y, 2T, 2Y	13	6	0.8012	0.5056	0.5642	0.5140	0.5810	0.5149	0.4051	0.5149	0.5149
(Fearnley 2001)				1.0970	1.0384	0.9408	1.0216	0.9330	0.9598	0.9273	0.9279
(s) Reiczigel (2003) C	9	96	0.9682	0.9366	0.9353	0.8892	0.9063	0.8249	0.9198	0.8318	0.8306
versus T				0.9998	1.0010	0.9870	1.0300	0.9943	0.9878	0.9907	0.9909
(t) Reiczigel (2003) C	18	96	0.5434	0.3574	0.4010	0.3947	0.3981	0.4009	0.3596	0.4011	0.4011
versus B				0.7294	0.6858	0.6680	0.6887	0.6779	0.7161	0.6773	0.6775

Note: Limits calculated by methods 1, 2 and 4 are displayed without truncation to lie within [0,1]. Showing effects of very unequal sample sizes and dispersions

express small datasets by strings of *X*s and *Y*s, thus *XYYXY* denotes dataset (d) in which $m = 2$, $n = 3$ and $X_{(1)} < Y_{(1)} < Y_{(2)} < X_{(2)} < Y_{(3)}$. The calculated limits for dataset (e), which is obtained from (d) by interchanging group labels, are the complements of those for (d). However, dataset (f), obtained from (d) by scale reversal, gives different limits, complementary to those from another related dataset (g) obtained by both interchanging group labels and scale reversal. Datasets (d) and (e) have $V_2(\tilde{\theta}) = 0.0703$, (f) and (g) have $V_2(\tilde{\theta}) = 0.0759$ for method 2. Notwithstanding this asymmetry, a pair of exponential distributions is an appropriate model for two distributions which differ in location and spread simultaneously in a familiar manner. Accordingly, this model is used as one of the distribution pairs in the evaluation.

Observation of non-invariance prompted the idea that these methods might be improved by replacing m^* and n^* by their average, leading to methods 4 and 5. This is a good example of how apparently minor tweaks may be devised which substantially improve the performance of a confidence interval method.

Methods 1 and 6 use information on the interlacing of the two samples whilst $\tilde{\theta}$ is sufficient for the other methods here. This impacts on Q_1 and Q_2 and so forth. Clearly $Q_1 = Q_2$ when the two distributions are mirror images. The greatest disparity occurs when one distribution is degenerate, which may co-exist with very unequal sample sizes m and n. (Mee's evaluation (1990) of coverage was restricted to $m = n$.) Suppose $X \sim U[0, 1]$ and $Y \sim U[\theta - \varepsilon/2, \theta + \varepsilon/2]$ for arbitrarily small $\varepsilon > 0$ to exclude ties. Then $Q_1 = \theta$, $Q_2 = \theta^2$ to within a term of order ε, and V reduces to $\theta(1 - \theta)/m$. Conversely when X is degenerate, V becomes $\theta(1 - \theta)/n$. Table 15.4 shows eight related illustrative datasets (h) to (o) for sample sizes of 20 and 100 which are moderate and substantially but realistically disparate. 18*X*, 100*Y*, 2*X* represents a dataset in which $m = 20$, $n = 100$ and all *Y*s exceed exactly 18 *X*s, and so forth. Pairs of datasets such as (h) and (i) show how, irrespective of $\tilde{\theta}$, the corresponding interval widths for method 1 (before truncation) differ by a factor of $\sqrt{n/m}$ (here $\sqrt{5}$) according to which sample is sandwiched within the other.

Similarly, in method 6, $\tilde{N}_J$ reduces to $m - 1$ when Y is degenerate and $n - 1$ when X is, and thus produces a Wilson interval for $\hat{\theta}$ with denominator $m - 1$ or $n - 1$. Intervals calculated by methods 5, 7 and 8 are intermediate between these. Moreover, methods 1 and 6 give the same intervals for datasets such as 3*X*, *nY*, 2*X* irrespective of n, as exemplified by datasets (*p*) and (*q*). That this is a real issue and not merely a theoretical one is evidenced by the liaison psychiatry study data, example (*r*) (Table 1.1). This dataset may be expressed as 1*Y*, 12*X*, 2*Y*, 2*T*, 2*Y*, the *T*s indicating tied *X* and *Y* scores. The Mee interval included 0.5, in contrast to methods 5, 7 and 8, and despite the significance of the corresponding Mann–Whitney test (two-tailed *p*-value 0.039 SPSS, 0.0436 Minitab) and the fact that no data shift was required. Similarly in the parasitology data, comparisons of *Trinoton luridum* (T) (with low dispersion) and

Brueelia tasniamae (B) (with high dispersion) against *Carnus hemapterus* (C) yielded the results in examples (s) and (t). In example (s), the Mee interval is much narrower than the intervals calculated by methods 5, 7 and 8, and conversely in example (t). The evaluation that follows gives no suggestion that the dependence of methods 6 and 1 on the exact order of observations leads to improved coverage. All pseudo-priors studied primarily involve shift rather than a difference of dispersion or shape. The measure $\tilde{\theta}$ may be of limited relevance when the latter is the main feature of the data, and if deemed appropriate for use at all under those circumstances, method 6 could well be preferable.

These examples also show effects of unequal sample sizes on other methods. For $\tilde{\theta} = 0.9$, both methods 7 and 8 give identical results for datasets (h) and (j) because they are based on mirror-image symmetrical models. So do methods 4 and 5, on account of averaging m and n. Methods 2 and 3 give different results for $m = 20$, $n = 100$ and $m = 100$, $n = 20$ in these cases. For $\tilde{\theta} = 0.5$, the methods that give identical results for datasets (l) and (n) are 7, 8, 4 and 5 as above, and also 2, but not 3. Symmetry occurs for method 2 when $\tilde{\theta} = 0.5$ because the imputed variance is a function of $\tilde{\theta}$ evaluated at the centre of symmetry, whilst method 3 imputes $V(\theta)$, which is not.

15.5 Evaluation

Newcombe (2006b) evaluated the above methods for constructing two-sided intervals for $\theta = \Pr[Y > X] + 1/2\Pr[Y = X]$ for both small and larger sample sizes. All intervals were designed for the absolutely continuous case (CF in Section 15.2.4) in which ties occur with probability 0. Nevertheless, in recognition of the fact that data are invariably recorded in discrete form, tied data are allowed.

It is inherently not possible to evaluate performance as exhaustively as for a single parameter such as a proportion or quantities comparing two proportions; namely, the difference, odds ratio or relative risk, on account of the much greater level of complexity. It is also not possible to reparametrise in terms of the parameter of interest and one or two nuisance parameters—the problem is inherently nonparametric.

The methodology of this evaluation is based largely on the principles set out in earlier chapters. Specifics are based partly on the precedent of Obuchowski and Lieber (2002) who addressed interval estimation for the special case $U/mn = 1$, and determined the lower confidence limit using 10,000 runs for each combination of sample sizes, ROC curve shape (Gaussian only, homoscedastic, or hyper/hyposcedastic by a factor of 2 or 3), and continuous or 5 category ordinal classifications. The present evaluation encompasses both extreme and non-extreme outcomes. As usual, extending to 100,000 runs estimates left and right non-coverage probabilities of the order of 0.025 within ±0.001 with 95% confidence.

We evaluate eight asymptotic interval methods using 6 distributions, 7 combinations of sample sizes m and n and eight values of θ, leading to 336 cells or parameter space points. We obtain the coverage probability and left and right non-coverage probabilities, proportions of PSPs with coverage between 0.93 and 0.97 and below 0.93, mean absolute difference between actual and nominal coverage, summary statistics for interval width, and rates of occurrence of ZWI and overshoot aberrations. In previous evaluations relating to simple proportions and their differences (Newcombe 1998a, b, c) unround sample sizes and parameter values were chosen, since in view of discrete behaviour, round values may not be typical. There is less reason to do so here, because round θ values generally give rise to irrational values for the other parameters arising in this evaluation.

Thus 100,000 datasets were generated for each PSP randomly and independently. For each simulated dataset U/mn is then calculated, with confidence intervals by eight methods as defined above. For each method, summary data are then accrued to estimate coverage, interval width and aberration frequencies for each PSP, then results summarised across all PSPs. Thus the coverage attained for the PSP is calculated as the proportion of the 100,000 datasets for which the interval includes the set value of θ. Left and right non-coverage rates are calculated similarly. The minimum coverage shown in Table 15.6 is thus the minimum of the coverage rates obtained for the 336 PSPs, and so forth.

The following sample size combinations were used

1. $m = n = 5$
2. $m = 5, n = 20$
3. $m = 20, n = 5$
4. $m = n = 20$
5. $m = 20, n = 100$
6. $m = 100, n = 20$
7. $m = n = 100$

The unequal sample size pairs, 2 and 3 and 5 and 6, were chosen to reflect a degree of disparity seldom exceeded in practice. They are interchangeable for some methods, especially when $\theta = 0.5$. Nevertheless the degree of redundancy in the evaluation is quite limited, and all 336 cells are retained in the evaluation.

Eight degrees of shift were considered. Without loss of generality, we choose eight θ values in the range $0.5 \le \theta < 1$; namely, 0.5, 0.6, 0.7, 0.8, 0.9, 0.95, 0.99 and 0.999. (Naturally, $\tilde{\theta}$ and calculated limits can then get below 0.5.) Restricting to $\theta \ge 0.5$ has the effect that left and right non-coverage probabilities are interpreted as measuring mesial and distal non-coverage. For example, when $\theta = 0.8$ and the calculated interval is (0.83, 0.96), this is located too far out from the centre of symmetry, 0.5, hence location is too distal, and non-coverage is mesial. Conversely, if the upper limit is less than 0.8, non-coverage is distal.

Six distribution pairs were considered.

1. Homoscedastic double Gaussian: $X \sim N(0,1)$, $Y \sim N(\mu,1)$.
   $Y - X \sim N(\mu,2)$, so $\theta = \Phi(\mu/\sqrt{2})$. The restriction $\theta \geq 0.5$ above leads to $\mu \geq 0$.
2. Double beta: $X \sim B(1,\alpha)$, $Y \sim B(\alpha,1)$, with $\alpha \geq 1$ *for* $\theta \geq 0.5$
   $F(x) = 1 - (1 - x)^{\alpha}$, $f(x) = \alpha(1 - x)^{\alpha-1}$, $G(y) = y^{\alpha}$, $g(y) = \alpha y^{\alpha-1}$ (x and $y \in [0, 1]$).
   $\theta = 1 - \Gamma^2(\alpha + 1)/\Gamma(2\alpha + 1)$.
3. Hyperscedastic double Gaussian: $X \sim N(4,1)$, $Y \sim N(4c,c^2)$ with $c \geq 1$.
   These are chosen because both distributions have mean – 4SD (i.e., the 0.0032 centile) at zero, so for practical purposes the range for Y extends only to the right of that for X.
   $Y - X \sim N(4(c - 1), c^2 + 1)$ $\theta = \Phi(4(c - 1)/\sqrt{(c^2 + 1)})$
   As $c \to \infty$, $\theta \to \Phi(4) = 0.999968$, not 1, but that is not a practical difficulty here as we are only concerned to represent θ up to 0.999.
   $z = 4(c - 1)/\sqrt{(c^2 + 1)}$ when $c = 1 + \sqrt{[(1 - z^2/16) - 1]}$
4. Hyposcedastic double Gaussian: $X \sim N(-4c,c^2)$, $Y \sim N(-4,1)$ with $c \geq 1$.
   Both distributions have mean + 4 SD = 0.
   $Y - X \sim N(4(c - 1), c^2 + 1)$ and again $\theta = \Phi(4(c - 1)/\sqrt{(c^2 + 1)})$.
5. Double exponential: the general model is $f(x) = \lambda e^{-\lambda x}$, $F(x) = 1 - e^{-\lambda x}$, $g(y) = \mu e^{-\mu x}$, $G(y) = 1 - e^{-y}$ with λ and $\mu > 0$ and x and $y \geq 0$. Then $\theta = \lambda/(\lambda + \mu)$.
   But without important loss of generality we can restrict attention to $\mu = 1$, then $\theta = \lambda/(1 + \lambda)$.
   Here $\lambda \geq 1$ for $\theta \geq 0.5$.
6. 5-category ordinal: $X \sim$ Binomial (4, $1 - p$), $Y \sim$ Binomial (4, p) with $0.5 \leq p \leq 1$.

All six models permit parameter selection to produce θ arbitrarily close to 1. Distribution pairs 1, 2 and 6 are mirror-image symmetrical. Pair 1 is the only pair studied that is a simple shift alternative (Reiczigel et al. 2003). Pair 5 corresponds to the model used by Hanley and McNeil (1982) to set up method 2, and involves positively skew distributions in both groups. Pair 6 is included in order to assess what these methods, designed for discrete data from continuous distributions (case CF), achieve when applied to tied data from discrete distributions (case DT). It would have been desirable to include a pair of discrete, positively skewed distributions on the same support (similar to distribution pair 5 but discrete), a case which frequently occurs in practice, but high values of θ cannot be obtained.

Furthermore, following previous work by others, the proportions of PSPs with coverage in the ranges (0.94, 0.96), (0.93, 0.97) and below 0.93, and the mean absolute difference between actual and nominal coverage were evaluated. All these were reported apart from the first, which in many parts of the evaluation was rarely achieved.

In this evaluation, as occasionally happens in relatively complex contexts, a few of the randomly generated datasets were degenerate and required to be replaced.

15.6 Results of the Evaluation

Table 15.5 presents the results obtained for a typical PSP with data generated from homoscedastic Gaussian distributions, sample sizes of $m = n = 20$, and shift $\theta = 0.8$. Being based on 100,000 randomly generated datasets, the empirical coverage and aberration probabilities expressed to five decimal places are exact. Summaries of this form were produced for each PSP. A qualitatively similar pattern emerged for 90% and 99% intervals.

Table 15.6 summarises the main coverage results for the 336 PSPs. Methods 1, 2 and 4 clearly have very poor coverage. The refinements of methods 3 and especially 5 result in greatly improved coverage. The lowest coverage obtained for method 5 in this evaluation was 0.9010, for the double exponential model at the PSP with $m = 100$, $n = 20$, and $\theta = 0.99$. The mean coverages for methods 6, 7 and 8 were very close to the nominal 0.95 in this evaluation, but the minimum coverage was less favourable than for method 5. Methods 7 and 8 had coverages most often close to 0.95, but method 5 had much the highest proportion of PSPs with coverage of 93% or higher, in line with its relatively favourable minimum coverage. Of the 336 PSPs, 9, 19 and 17 had CP < 0.90 for methods 6, 7 and 8, respectively. These tended to occur with higher values of θ and, surprisingly, m and n. The 20 PSPs for which method 5's CP was between 0.90 and 0.93 were a subset of the 39 for which method 7 and 8's CPs were less than 0.93.

Interval location was most balanced for method 6. For other methods, interval location corresponded to the known properties of Wilson and Wald intervals for proportions. Thus intervals produced by methods 3, 5, 7 and 8 were too mesially located; conversely, intervals produced by methods 1, 2 and 4 were too distally located. Unsurprisingly, all methods produced similar right and left non-coverage when $\theta = 0.5$, becoming increasingly unbalanced as $\theta \to 1$.

As noted in Section 4.4.3, the Box–Cox index is applicable to U/mn as well as to the binomial proportion. Table 15.7 shows this index for confidence intervals calculated by eight methods as above for four datasets, the Hanley–McNeil dataset (a) of Table 15.3 and the Fearnley and Reiczigel datasets (r), (s) and (t) of Table 15.4, following truncation at the boundaries wherever necessary. Methods 5, 7 and 8 which have favourable coverage properties have $\lambda \sim -0.1$, in line with MNCP/NCP ~ 0.2 in Table 15.6. The Halperin–Mee interval

TABLE 15.5

Summary of Coverage, Width and Aberration Results for Eight 95% Intervals for the Generalised Mann–Whitney Measure θ for a Typical Parameter Space Point

Method	Average Width	Minimum Width	Maximum Width	Coverage Probability	Left (Mesial) Non-Coverage	Right (Distal) Non-Coverage	MNCP ÷ NCP	Overshoot Probability	Zero Width Interval Frequency
1	0.270902	0.000000	0.388411	0.91894	0.06937	0.01169	0.856	0.05386	0.00003
2	0.270754	0.000000	0.362284	0.92501	0.06203	0.01296	0.827	0.06750	0.00003
3	0.269100	0.101332	0.337024	0.95590	0.01079	0.03331	0.245	0.00000	0.00000
4	0.270754	0.000000	0.362284	0.92501	0.06203	0.01296	0.827	0.06750	0.00003
5	0.269100	0.101332	0.337024	0.95590	0.01079	0.03331	0.245	0.00000	0.00000
6	0.267192	0.049286	0.367891	0.94997	0.02350	0.02653	0.470	0.00000	0.00000
7	0.264274	0.086994	0.335815	0.95226	0.01230	0.03544	0.258	0.00000	0.00000
8	0.264929	0.088900	0.336128	0.95226	0.01230	0.03544	0.258	0.00000	0.00000

Note: Based on 100,000 randomly generated datasets for one parameter space point: homoscedastic Gaussian distributions, sample sizes $m = n = 20$ and shift $\theta = 0.5$.

TABLE 15.6

Summary of Coverage Characteristics for Eight 95% Intervals for the Generalised Mann–Whitney Measure θ

Method	Mean Coverage	Minimum Coverage	Mean |CP – 0.95|	Proportion of PSPs with CP 93% to 97%	Proportion of PSPs with CP 93% and Above	Mean Left (Mesial) Non-Coverage	Mean Right (Distal) Non-Coverage	Mean MNCP ÷ Mean NCP
1	0.7431	0.0096	0.2071	0.247	0.247	0.2425	0.0145	0.944
2	0.7603	0.0096	0.1927	0.286	0.307	0.2247	0.0150	0.937
3	0.9557	0.7438	0.0181	0.643	0.905	0.0102	0.0340	0.231
4	0.7662	0.0096	0.1859	0.318	0.339	0.2194	0.0143	0.939
5	0.9607	0.9010	0.0160	0.649	0.941	0.0085	0.0308	0.216
6	0.9496	0.8209	0.0173	0.655	0.842	0.0220	0.0283	0.437
7	0.9484	0.8571	0.0138	0.801	0.884	0.0100	0.0415	0.194
8	0.9504	0.8603	0.0139	0.765	0.884	0.0097	0.0398	0.196

Note: Based on 100,000 randomly generated datasets for each of 336 parameter space points representing each combination of eight degrees of shift, six distributions and seven sample size pairs.

TABLE 15.7

Box–Cox Index of Symmetry for Eight 95% Intervals for the Generalised Mann–Whitney Measure θ for Four Illustrative Datasets

Method	Dataset (a) Table 15.3	Dataset (r) Table 15.4	Dataset (s) Table 15.4	Dataset (t) Table 15.4
1	1	0.620	1	1
2	1	0.783	0.959	1
3	−0.070	0.023	−0.334	−1.509
4	1	0.857	0.578	1
5	−0.084	−0.074	−0.036	−0.127
6	0	0	0	0
7	−0.233	−0.153	−0.240	−0.172
8	−0.220	−0.145	−0.226	−0.161

(method 6), being of Wilson form, has $\lambda \equiv 0$ despite MNCP/NCP being 0.44, in line with the results for the Wilson score interval in Table 4.3.

Newcombe (2006b) also showed results restricted to the 56 homoscedastic Gaussian PSPs, the situation in which method 8 might be predicted to perform best. Results are quite similar to those in Table 15.6, with methods 5, 7 and 8 clearly best, and little to choose between them.

Table 15.8 shows selected marginal mean and minimum coverages for methods 5 to 8. Only method 5 produced a mean coverage above 0.95 in all marginals. For method 5, degree of shift was the dominant determinant of coverage, which was conservative at extreme θ. However, for method 6, the dominant determinant of coverage was sample size, which affected the mean coverage much more than for methods 5, 7 and 8. Apparently the discreteness of the shift applied in extreme and adjoining cases to ensure a variance corresponding to θ at most $0.5/\sqrt{mn} = 0.9$ has introduced additional conservatism for $m = n = 5$.

As usual, the dominant determinants of interval width were sample size and degree of shift, so it is only meaningful to compare interval widths of different methods holding these factors constant. Table 15.9 shows average and maximum interval widths for selected combinations of these parameters. Generally, interval width is very similar for methods 5, 7 and 8—slightly higher for method 5, in line with its slightly more conservative coverage. For small n, method 6 produced wider intervals than the other methods for moderate values of θ, but narrower for extreme θ; see also Table 15.4.

Methods 3 and 5 are clearly identical when $m = n$ (e.g., in Table 15.5). When m and n are dissimilar, they can produce quite different intervals. Thus when $\tilde{\theta} = 0.95$, with $m = 20$ and $n = 100$, method 3 produces the interval 0.8885 to 0.9747, much narrower than that for $m = 100$ and $n = 20$, 0.8369 to 0.9862. In both instances method 5 yields 0.8591 to 0.9823. Accordingly, for $\theta = 0.95$, with $m = 20$ and $n = 100$, method 3's mean CP and width were much lower (0.8886 and 0.0843) than for $m = 100$ and $n = 20$ (0.9807 and 0.1464). Corresponding

TABLE 15.8

Mean and Minimum Coverage Probabilities of Methods 5 to 8 for the Generalised Mann–Whitney Measure θ for Selected Subsets of 336 Parameter Space Points

	Method 5		Method 6		Method 7		Method 8	
Subset	**Mean**	**Minimum**	**Mean**	**Minimum**	**Mean**	**Minimum**	**Mean**	**Minimum**
Degree of Shift								
$\theta = 0.5$	0.9517	0.9444	0.9467	0.9289	0.9517	0.9444	0.9517	0.9444
$\theta = 0.7$	0.9568	0.9291	0.9470	0.9099	0.9532	0.9269	0.9535	0.9281
$\theta = 0.9$	0.9599	0.9085	0.9494	0.9012	0.9497	0.8807	0.9511	0.8834
$\theta = 0.999$	0.9778	0.9240	0.9694	0.9240	0.9514	0.8760	0.9593	0.8817
Distributions								
Homoscedastic Gaussian	0.9633	0.9379	0.9513	0.9114	0.9509	0.9352	0.9535	0.9379
Double beta	0.9642	0.9376	0.9521	0.9146	0.9520	0.9346	0.9546	0.9376
Hyperscedastic Gaussian	0.9556	0.9038	0.9456	0.8342	0.9434	0.8645	0.9450	0.8673
Double exponential	0.9547	0.9010	0.9448	0.8209	0.9410	0.8571	0.9428	0.8603
Binomial	0.9707	0.9594	0.9548	0.9202	0.9602	0.9447	0.9620	0.9496
Samples Sizes								
$m = n = 5$	0.9594	0.9366	0.9722	0.9543	0.9547	0.9161	0.9547	0.9161
$m = n = 20$	0.9599	0.9280	0.9538	0.9235	0.9470	0.8799	0.9499	0.8932
$m = n = 100$	0.9589	0.9240	0.9423	0.8313	0.9388	0.8584	0.9420	0.8795

Note: Illustrating dependence on degree of shift, sample size and distribution.

TABLE 15.9

Mean and Maximum Width of Four Intervals for the Generalised Mann–Whitney Measure for Selected Sample Sizes and Degrees of Shift

		Method 5		Method 6		Method 7		Method 8	
Sample Size	**Shift**	**Mean**	**Maximum**	**Mean**	**Maximum**	**Mean**	**Maximum**	**Mean**	**Maximum**
$m = n = 5$	$\theta = 0.6$	0.555	0.589	0.603	0.693	0.549	0.583	0.550	0.584
	$\theta = 0.95$	0.404	0.589	0.388	0.693	0.401	0.583	0.402	0.584
$m = 5, n = 20$	$\theta = 0.6$	0.468	0.490	0.482	0.693	0.463	0.487	0.464	0.488
	$\theta = 0.95$	0.299	0.490	0.217	0.693	0.293	0.487	0.294	0.488
$m = n = 20$	$\theta = 0.6$	0.325	0.337	0.335	0.385	0.323	0.336	0.324	0.336
	$\theta = 0.95$	0.166	0.316	0.140	0.353	0.157	0.313	0.158	0.314
$m = 20, n = 100$	$\theta = 0.6$	0.258	0.266	0.258	0.369	0.257	0.265	0.257	0.266
	$\theta = 0.95$	0.121	0.234	0.093	0.320	0.111	0.231	0.112	0.232
$m = n = 100$	$\theta = 0.6$	0.154	0.158	0.154	0.164	0.153	0.158	0.153	0.158
	$\theta = 0.95$	0.065	0.111	0.059	0.120	0.057	0.108	0.058	0.108

mean CPs for method 5 were 0.9729 and 0.9610; mean interval widths were very similar in both cases, 0.1209 and 0.1203. Evidently the symmetrised version, method 5 performs more satisfactorily than method 3 in these circumstances in particular.

Aberrations only arise from methods 1, 2 and 4, and are then frequent. Method 1 had mean and maximum overshoot rates of 0.304 and 0.947 in this evaluation. These figures increase slightly when exponential-based methods 2 (mean 0.311, max 0.962) and 4 (0.319 and 0.962) are used. Method 1 had mean overshoot rate 0.098 for $\theta = 0.5$, increasing to 0.551 for $\theta = 0.99$, then falling to 0.224 for $\theta = 0.999$; 0.494 when $m = n = 5$, reducing to 0.162 for $m = n = 100$. Methods 2 and 4 produced similar results. Zero width intervals occur together on these three methods, with overall frequency 0.176 in this evaluation, but reaching a maximum of 0.990 for heteroscedastic Gaussian models with $\theta = 0.999$ and $m = n = 5$. The ZWI frequency depends critically on θ (0.776 at $\theta = 0.999$, 0.001 at $\theta = 0.5$) and sample size (0.351 at $m = n = 5$, 0.047 at $m = n = 100$).

In conclusion, method 5, which is based on a method proposed by Hanley and McNeil but with two refinements, is recommended for use as it performs very well, and moreover is readily implemented, as described below.

15.7 Implementation in MS Excel

The spreadsheet VISUALISETHETA.xls converts a value of the generalised Mann–Whitney statistic—which may be either a hypothesised value θ or an empirical value $\tilde{\theta}$—into a diagram similar to Figure 15.1, displaying a pair of identically dispersed Gaussian curves $N(0,1)$ and $N(\delta,1)$ with peaks $\delta = \sqrt{2}\Phi^{-1}(\theta)$ standard deviations apart.

The spreadsheet GENERALISEDMW.xls is displayed in Figure 15.2. It starts with user-input values of the two sample sizes m and n, the generalised Mann–Whitney statistic $\tilde{\theta}$ calculated from the data, and the required confidence level $1 - \alpha$. Interchanging the two sample sizes m and n makes no difference to the results. The $\tilde{\theta}$ would usually be pasted in from other software such as the SPSS ROC procedure. The spreadsheet then gives a confidence interval for θ which is superior to the interval SPSS calculates using method 1, and displays a graphical visualisation of the corresponding degree of separation of two distributions, which extends what is displayed in Figure 15.1 to include the confidence limits. Numerical results displayed are the corresponding standardised difference, $\tilde{\delta} = \sqrt{2}\Phi^{-1}(\tilde{\theta})$, and confidence intervals on both θ and δ scales. The confidence limits (θ_L, θ_U) for θ are calculated by the preferred method 5 as defined above, then transformed into limits (δ_L, δ_U) for δ. The solid curves represent two identically dispersed Gaussian distributions $N(0, 1)$ and $N(\tilde{\delta}, 1)$ where $\tilde{\delta}$ corresponds to $\tilde{\theta}$ as above.

Spreadsheet GENERALISEDMW.

This spreadsheet calculates confidence limits for theta=U/mn by 40 interval bisections.

Uses method 5 of Newcombe RG (2006).
Confidence intervals for an effect size measure based on the Mann-Whitney statistic.
Part 2: Asymptotic methods and evaluation. Statistics in Medicine 25, 559-573.

The spreadsheet displays point and interval estimates on both theta and delta (corresponding standardised difference) scales.
Displays Gaussian curves separated by delta hat, delta lower and delta upper SDs.
Degrees of separation beyond 0.000001 < theta < 0.999999 (corresponding to |delta| > 6.7224) are displayed truncated to this degree of separation.

To use, replace input values for **m, n** and **U/mn** in **bold** as appropriate.
Calculated U/mn from SPSS etc. can be pasted into cell F20.

Two-sided confidence level required **95** %

Data: **m= 58** **n= 51** **U/mn = 0.8932**

Corresponding U= 2642.0

Results:	**Point estimate**	**Confidence limits**	
		lower	**upper**
Theta	0.8932	0.8117	0.9400
Delta	1.7587	1.2504	2.1989

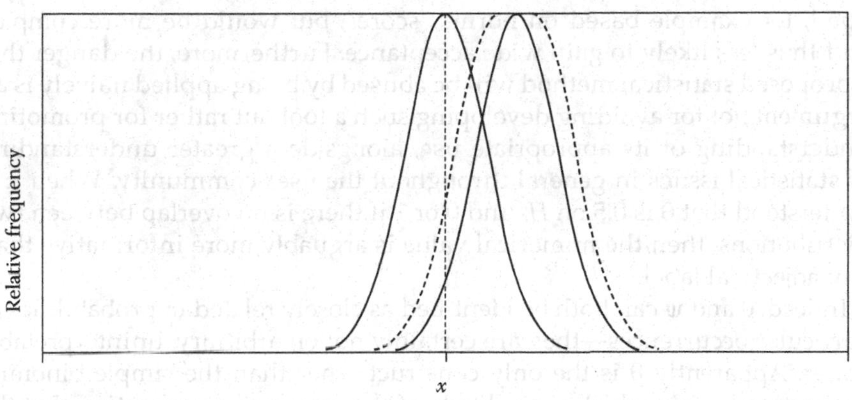

FIGURE 15.2
Spreadsheet GENERALISEDMW.xls to implement calculations for the chosen confidence interval for the unpaired effect size measure $\theta = U/mn$.

The latter is flanked by two broken curves, which display the ascending part of the curve corresponding to the lower limit δ_L and the descending segment corresponding to the upper limit δ_U. Degrees of separation beyond $0.000001 < \theta < 0.999999$ (corresponding to $|\delta| > 6.72$) are displayed truncated to this degree of separation.

15.8 Interpretation

In this chapter $\tilde{\theta} = U/mn$ is developed as a very general and widely applicable, indeed much needed measure of separation of two frequency distributions. The same applies to the measure $\tilde{\psi} = \Delta T/\max$ introduced in the next chapter which generalises the paired Wilcoxon test statistic into an effect size measure. Any novel measure will only achieve wide currency and fulfil its potential if it becomes accepted by both statisticians and the wider research community as appropriate and useful. Scepticism has sometimes been expressed that researchers will be unable to attach a clear meaning to any particular numerical value of non-parametric measures such as θ or ψ. Some measures such as the relative risk are intuitively directly interpretable; the interpretation is less obvious for a measure such as θ or ψ.

The response is, firstly, that a measure of this form is definitely needed, especially when the original scale of measurement is not readily interpreted. Alternative measures to achieve the same objective could be developed, for example based on normal scores, but would be more complex and thus less likely to gain wide acceptance. Furthermore, the danger that a proposed statistical method will be abused by being applied naively is an argument not for avoiding developing such a tool but rather for promoting understanding of its appropriate use, alongside a greater understanding of statistical issues in general throughout the user community. When it is understood that θ is 0.5 on H_0 and 0 or 1 if there is no overlap between two distributions, then the numerical value is arguably more informative than any adjectival label.

Indeed, θ and ψ can both be identified as closely related to probabilities of particular occurrences—they are certainly not on arbitrary, uninterpretable scales. Apparently θ is the only construct other than the simple binomial proportion $p = r/n$ which is so clearly of the nature of a proportion that the BCIS is applicable.

16

Generalised Wilcoxon Measure

16.1 The Rationale for the Generalised Wilcoxon Measure ψ

In previous chapters we have seen how a variety of interpretable effect size measures have been formulated, many of which correspond to simple, widely used statistical tests. Use of one of these, accompanied by a suitable confidence interval, generally leads to more satisfactory inferences than those derived from p-values alone. Some effect size measures are absolute, such as differences between means (in the original units) or between proportions. Others are relative or scaled, including the standardised difference, and a measure θ derived from the Mann–Whitney U statistic as in Chapter 15.

In this chapter we develop a novel relative measure ψ to generalise the Wilcoxon matched-pairs signed-ranks test statistic (Wilcoxon 1945; Hollander and Wolfe 1999, Chapter 3). An estimator $\tilde{\psi}_Z$ based on zero-shifted ranks (ZSRs) is used, with a confidence interval calculated by an inversion method. In earlier versions of this work, confidence intervals were based on a single parameter logistic model, using an estimator based on either Wilcoxon ranks (Newcombe 2007e) or ridit ranks (Newcombe 2010a). The approach described here is simpler and preferable.

In Chapter 15, we considered the case of independent random variables X and Y on the same support. An index $\theta = \Pr[X > Y] + 0.5 \Pr[X = Y]$ was developed to characterise the degree of shift of X relative to Y. A functionally equivalent statistic with potential range from –1 to +1 is Somers' D (Newson 2002) which is defined as $\Pr[X > Y] - \Pr[X < Y] = 2\theta - 1$.

In this chapter we develop an analogous measure for data consisting of paired differences, which converts the Wilcoxon test statistic into a relative effect size measure. Suppose we obtain a sample $\{X_1, X_2,\ldots, X_n\}$ from the distribution. In the Wilcoxon test, normally any zero observations are dropped first, resulting in a possibly reduced sample size $n^* \leq n$. The non-zero observations are then represented by their absolute ranks 1, 2,…, n^*. Sums T_1 and T_2 of signed ranks corresponding to positive and negative observations are formed, and the p-value is determined by the value of T_1. As a corresponding effect size measure scaled to lie within [–1, 1], we may calculate

$$\Delta T/\text{max} = (T_1 - T_2)/\text{max} \tag{16.1}$$

where max represents the highest possible value of T_1 (or equivalently of T_2 or $T_1 + T_2$). A slight modification turns out to be preferable, based on zero shifted ranks as defined below instead of Wilcoxon ranks, in which zeros are not dropped from the sample.

We define the corresponding theoretical parameter value as follows. Suppose that the distribution of the paired differences is continuous and X_1 and X_2 are sampled independently from it. The quantity $\Pr[X_1 + X_2 > 0]$, referred to as eta by Hollander and Wolfe (Equation 3.21, p. 49) and p' by Noether (1987, Section 2.2) is a natural scaled effect size measure corresponding to the paired Wilcoxon test. However, this would be an inappropriate index for the discrete case in which zeros can occur. To cover this case also, we define the population value of ψ as

$$\psi = \Pr[X_1 + X_2 > 0] - \Pr[X_1 + X_2 < 0] \tag{16.2}$$

This is equivalent to Somers' D comparing X_1 and Y_2 where X_1 and X_2 are sampled independently from the same distribution and $Y_2 = -X_2$.

The index ψ takes the value 0 when the distribution of X is symmetrical about 0. Values of +1 and –1 correspond to distributions with strictly positive and negative supports.

We then need to estimate ψ from a sample of n observations, $\{X_1, X_2, \ldots, X_n\}$ using an empirical estimate $\tilde{\psi}$ based on resampling pairs of observations from the sample. It is appropriate here to resample without replacement (i.e., to base $\tilde{\psi}$ only on pairs $\{X_i, X_j\}$ with $i \neq j$), leading to an estimator $\tilde{\psi}_Z$ which is $\Delta T/\text{max}$ based on ZSRs.

While the derivation here is expressed in terms of paired differences, the methodology extends in an obvious way to the equivalent problem of estimation for a single sample that does not arise from paired differences.

The simplest hypothesis test applicable to paired differences is the sign test. Let X denote the random variable from which the paired differences are sampled, with median η, and let $\{X_i, i = 1\ldots n\}$ denote the n values obtained. The numbers of positive, zero and negative X_i are counted. Assuming that no X_i is zero, inferences are based on the binomial distribution, with a p-value based either on an aggregate of tail probabilities or an asymptotic normal approximation. Hollander and Wolfe explain that the simple median of the $\{X_i\}$ is the natural estimator of η corresponding to the sign test. Furthermore, the simplest corresponding sample-size-free measure of effect size is $\zeta = \Pr[X > 0]$, estimated by $\tilde{\zeta}$, the proportion of positive X_is. (As in Chapter 15, the superscript ~ (tilde) will be used to denote an empirical estimate, not ^ (circumflex or hat as this would imply an MLE.) The sample statistic $\tilde{\zeta}$ is a simple binomial proportion, for which a variety of confidence interval methods are available as in Chapter 3. Once again, we need to allow for the possibility of zero differences, so we redefine ζ as $\Pr[X > 0] + 0.5 \Pr[X = 0]$. Then, $\tilde{\zeta}$ is the proportion of positive differences plus half the proportion of zero differences. Confidence intervals may be obtained as in Chapter 9. Compared to the sign test, the Wilcoxon test makes fuller use of the information available by taking into account the relative sizes of positive and negative differences, and is thus generally considerably more powerful than the sign test, which does not. For the same reason, the measure ψ based on the Wilcoxon matched-pairs signed-ranks test statistic, made free of sample size, should be regarded as a more sensitive indicator of shift than ζ.

The index ψ corresponds closely to the Wilcoxon test (albeit with a slight modification) and to the Walsh averages on which the usual Hodges–Lehmann median estimate is based. The Walsh averages (Walsh 1949) are defined as the array $W_{ij} = (X_i + X_j)/2$ for i and $j \in \{1, 2,\ldots, n\}$. Hodges and Lehmann (1963) estimated η by the median of the W_{ij} for $1 \le i \le j \le n$. This is the natural estimate corresponding to the Wilcoxon test (Hollander and Wolfe 1999). Hollander and Wolfe, Conover (1980) and others describe how upper and lower confidence limits for η may be extracted from the ordered set of the $\{W_{ij}\}$, as implemented in Minitab. The sample estimator $\tilde{\psi}$ introduced here is derived from the signs of the Walsh averages.

In extreme cases, $\tilde{\psi} = +1$ or -1 when $\tilde{\zeta} = 0$ or 1. Also, if all the paired differences $X_1,\ldots, X_n$ are equal in magnitude, $\tilde{\psi} = 2\tilde{\zeta} - 1$. Thus $\tilde{\psi}$, like $\tilde{\zeta}$, expresses the preponderance of positive differences over negative ones, but is weighted to take account of their relative magnitudes.

For some parametric models, ψ and ζ may be expressed directly as functions of the parameters. Thus for a Gaussian model with mean μ and variance 1, $\psi = 2\Phi(\sqrt{2}\mu) - 1$ and $\zeta = \Phi(\mu)$. These formulae are analogous to the formula $\theta = \Phi(\delta/\sqrt{2})$ linking the U/mn or AUROC measure θ to the standardised difference δ in the homoscedastic Gaussian case (Section 15.2.2). Table 16.1 shows values of ψ and ζ corresponding to Gaussian models with $\mu = 0.2, 0.4,\ldots, 2.0$. This enables us to identify particular values of ψ with the familiar effect size measure for paired data, the mean of the paired differences divided by their SD, which may aid visualisation of what particular values of ψ represent. It

TABLE 16.1

Relationship of the Generalised Wilcoxon Measure ψ and the Sign Test Measure ζ to the Mean μ for the Gaussian Distribution with Variance 1

μ	ψ	ζ	μ	ψ	ζ
0.0	0.0000	0.5000	1.2	0.9103	0.8849
0.2	0.2227	0.5793	1.4	0.9523	0.9192
0.4	0.4284	0.6554	1.6	0.9763	0.9452
0.6	0.6039	0.7257	1.8	0.9891	0.9641
0.8	0.7421	0.7881	2.0	0.9953	0.9772
1.0	0.8427	0.8413			

also shows the approximate correspondence between ψ and ζ. For $|\psi|$ up to about 0.7, ψ is similar to μ, which may aid interpretation.

Analogous to the spreadsheet VISUALISETHETA.xls developed in Chapter 15, the spreadsheet VISUALISEPSI.xls converts a value of the generalised Wilcoxon statistic—which may be either a hypothesised value ψ or an empirical value $\tilde{\psi}$—into a display of a Gaussian curve $N(\mu, 1)$.

16.2 Paired and Unpaired Effect Size Measures Compared

In the context of hypothesis testing, the necessity to distinguish paired and unpaired data structures is well recognised. The issue is not merely that a paired analysis can lead to a more significant *p*-value or has greater power. The situation is less obvious in relation to effect size measures, as illustrated by several examples involving paired data, in which paired and unpaired measures address different questions, each of which are relevant. Just as for continuous and binary outcomes, the relationship between paired and unpaired analyses depends on the correlation.

Consider the question, "Which is more consistently warmer than London—Madrid or Cape Town?". These three cities were chosen because they lie in similar time zones to make the concept of "the same day" unequivocal in interpretation, but represent both hemispheres, so that one city is in antiphase relative to the others for the annual temperature cycle. Daily average temperatures for 1995 to 2004, expressed in °F and rounded to 1 decimal place, were extracted from the University of Dayton Average daily temperature archive (Kissock 2007). Data were available for all three cities for 3643 of 3653 days. Means were 52.9°F, 58.6°F and 62.0°F for London, Madrid and Cape Town respectively, suggesting Cape Town is more consistently warmer than London than Madrid is. The distributions may be compared on an unpaired basis using $\tilde{\theta}$ as above. The $\tilde{\theta}$ was 0.613 for Madrid versus London and 0.765 for Cape Town versus London. A randomly chosen day in Cape Town has a 76.5% chance of being warmer than an independently randomly chosen day in London, but the corresponding figure for Madrid is only 61.3%.

However, these analyses disregard the paired nature of the data. Madrid was warmer than London on 78.1% of days and cooler than London on 21.4% of days, whereas Cape Town was warmer than London on only 67.0% of days and cooler than London on 32.6% of days. The index $\tilde{\psi}$ takes the value +0.719 for Madrid versus London, but only +0.573 for Cape Town versus London. On the great majority of days, Madrid is warmer than London (the exceptions being mainly in winter due to Madrid's wider temperature range), but on most London summer days the temperature is higher than in Cape Town which is then in winter. Thus Madrid is warmer than London for more of the year than Cape Town is. This rather extreme example demonstrates the need for effect size measures designed specifically for paired data structures. Both measures are meaningful to calculate here, but correspond to different ways of interpreting the question. Of course, the intervals developed in Chapter 15 are not appropriate for the paired case, due to non-independence.

In fact both ζ and ψ are related to θ, in quite different ways. As explained above, ψ is a special case of Somers' D which is $2\theta - 1$. Conversely, Vargha and Delaney (2000) link $\tilde{\zeta}$ to $\tilde{\theta}$ as follows. Consider two samples, $\boldsymbol{X}$ and $\boldsymbol{Y}$, each comprising n observations, and calculate $\tilde{\theta}$ in the usual way. For each of the $n!$ ways of pairing observations across the two samples, calculate $\tilde{\zeta}_j$, $j = 1,\ldots, n!$. Then $\tilde{\theta}$ is the average of the $n!$ $\tilde{\zeta}_j$s. Note that the absence of overlap on an unpaired basis ($\tilde{\theta} = 1$) implies no overlap on a paired basis ($\tilde{\zeta} = 1$ or $\tilde{\psi} = 1$), but not vice versa.

16.3 Estimating the Index ψ

16.3.1 Three Systems of Ranks

In the Wilcoxon test as usually formulated, the absolute values of the paired differences $\{X_i, i = 1, 2,\ldots, n\}$ are represented by their ordinary Wilcoxon ranks $1, 2,\ldots, n$, with corresponding signs attached. Noether (1967, p. 46) observed that many statistics can be formulated of the form $\sum_{i=1}^{n} \omega_i \operatorname{sgn} X_i$, where the ω_i are weights that reflect the absolute distances of the observations from zero; he recommended the simple Wilcoxon ranks, as above. But neither Wilcoxon (1945), Noether nor Hollander and Wolfe (1999) gave any reason why a paired non-parametric test for location shift should necessarily be based on these ordinary ranks, rather than any other possible series of non-negative values monotonically related to the $|X_i|$.

For the purpose of estimating ψ, we show that a formulation based on the ZSRs $0, 1,\ldots, n-1$ is preferable. In the comparison of two independent samples, the Mann–Whitney test, and related measures θ and Somers' D are identical irrespective of whether Wilcoxon or zero-shifted ranks are used. However, the Wilcoxon test statistic, and the corresponding empirical estimate $\tilde{\psi} = \Delta T/\max$, are slightly altered. In the first instance, we assume that neither zeros nor ties as defined below exist in the data. It is well-known that the usual Wilcoxon statistic T_W is then simply derived from the Walsh averages W_{ij}, i and $j \in \{1, 2,\ldots, n\}$ (Tukey 1949). But so is the analogous measure T_Z based on the ZSRs formulation.

We consider three systems of ranks, indexed by $h = 0, 1$ or 2.

The Wilcoxon unsigned ranks $1\ldots n$ are replaced by $1 - h/2, 2 - h/2, \ldots, n - h/2$, for $h = 0, 1$ or 2.

For illustration, we consider the sample $\boldsymbol{X} = \{1, 2, 3, -4, 5\}$, which is already expressed in Wilcoxon signed rank form.

Wilcoxon rank formulation: $h = 0$

The positive and negative rank sums here are $T_1 = 11$ and $T_2 = 4$.

The Wilcoxon test is often formulated in terms of min (T_1, T_2), but it is clearly equivalent to base inferences on $\Delta T = T_1 - T_2$, or (for a two-sided test) its absolute value. An obvious way to convert this into an index ranging on [–1, 1] is to calculate the empirical estimate based on Wilcoxon ranks,

$$\tilde{\psi}_W = \frac{11-4}{11+4} = \frac{7}{15} = 0.4667.$$

As usual, max is defined as the maximum possible value of $T_1 + T_2$, here $n(n + 1)/2$. $T_1 + T_2$ is generally equal to this quantity, but can sometimes be smaller.

Ridit rank formulation (after Bross 1958): $h = 1$

$$\mathbf{X}^{(1)} = \{1/2, 1\ 1/2, 2\ 1/2, -3\ 1/2, 4\ 1/2\}. \qquad \tilde{\psi}_R = \frac{9-3.5}{9+3.5} = \frac{5.5}{12.5} = 0.44.$$

Define the mirror-image sample $\mathbf{Y} = \{Y_1, Y_2, \ldots, Y_n\} = \{-X_1, -X_2, \ldots, -X_n\}$.

Both the above formulations correspond to the same $U_{XY} = 18$, leading to $\frac{U}{mn} = \frac{18}{25} = 0.72$, Somers' D $= \frac{18-7}{25} = 0.44$. The ridit formulation corresponds directly to Somers' D—as far as the sample is concerned.

ZSR formulation: $h = 2$

$$\mathbf{X}^{(2)} = \{0, 1, 2, -3, 4\}. \qquad \tilde{\psi}_Z = \frac{7-3}{7+3} = \frac{4}{10} = 0.4.$$

In all three cases, the numerator and denominator of $\Delta T/\max$ are simple linear functions of h.

$h = 0$ corresponds directly to the Wilcoxon test.

$h = 1$ gives exact alignment with Somers' D comparing samples $\mathbf{X}$ and $\mathbf{Y} = -\mathbf{X}$.

Also, $\tilde{\psi}_R$ is invariant under replication of the sample.

But $h = 2$ is the optimal choice, leading to unbiased estimation of the theoretical ψ. Also, the resulting variance takes a simpler form than in the cases $h = 0$ or 1, because the diagonal terms of the **s** matrix are not included. However, U_{XY} is now 17.5, not 18.

16.3.2 The s Matrix

For the general case, ψ ranging from –1 to +1 is defined as $\Pr[X_1 + X_2 > 0] - \Pr[X_1 + X_2 < 0]$, where X_1 and X_2 are chosen independently at random from the X distribution. The numerator of the empirical estimate $\tilde{\psi} = \Delta T/\max$

may then be obtained by summating entries of the *s* matrix comprising $s_{ij} = \text{sgn}(X_i + X_j)$ for $i, j = 1 \ldots n$. This matrix may be segregated into interpretable zones. Table 16.2 illustrates this process for four examples with $n = 5$, which also appear as examples (l), (m), (n) and (k) in Table 16.3.

Consider first the upper left-hand panel, relating to our small exemplar dataset {1, 2, 3, –4, 5}.

TABLE 16.2

The *s* Matrices for Four Samples with $n = 5$

{1, 2, 3, –4, 5}

X_1 \ X_2	1	2	3	–4	5
1	+1	+1	+1	–1	+1
2	+1	+1	+1	–1	+1
3	+1	+1	+1	–1	+1
–4	–1	–1	–1	–1	–1
5	+1	+1	+1	+1	+1

$\Delta T/\max = (7 - 3)/10 = 0.4$

{–1, 2, 3, –4, 5}

X_1 \ X_2	–1	2	3	–4	5
–1	–1	+1	+1	–1	+1
2	+1	+1	+1	–1	+1
3	+1	+1	+1	–1	+1
–4	–1	–1	–1	–1	+1
5	+1	+1	+1	+1	+1

$\Delta T/\max = (7 - 3)/10 = 0.4$

{0, 1, 2, 3, –4, 5}

X_1 \ X_2	0	2	3	–4	5
0	0	+1	+1	–1	+1
2	+1	+1	+1	–1	+1
3	+1	+1	+1	–1	+1
–4	–1	–1	–1	–1	–1
5	+1	+1	+1	+1	+1

$\Delta T/\max = (7 - 3)/10 = 0.4$

{1, 2, 3, –3, 5}

X_1 \ X_2	1	2	3	–3	5
1	+1	+1	+1	–1	+1
2	+1	+1	+1	–1	+1
3	+1	+1	+1	0	+1
–3	–1	–1	0	1	+1
5	+1	+1	+1	+1	+1

$\Delta T/\max = (7 - 2)/10 = 0.4$

Note: Showing $\Delta T/\max$ for the zero-shifted formulation.

TABLE 16.3

Construction of ΔT/max Based on ZSR for Several Datasets with $n = 5$

Dataset	Data	Signed ZSRs	T_1	T_2	$(T_1-T_2)/10$	95% Limits
(a)	+1, +2, +3, +4, +5	+0, +1, +2, +3, +4	10	0	1.0	−0.0718, +1
(b)	−1, +2, +3, +4, +5	−0, +1, +2, +3, +4	10	0	1.0	
(c)	0, +2, +3, +4, +5	00, +1, +2, +3, +4	10	0	1.0	
(d)	+1, +1, +3, +4, +5	+0.5, +0.5, +2, +3, +4	10	0	1.0	
(e)	0, 0, +3, +4, +5	0 × 0.5, 0 × 0.5, +2, +3, +4	9	0	0.9	−0.1614, +0.9944
(f)	+1, −1, +3, +4, +5	+0.5, −0.5, +2, +3, +4	9.5	0.5	0.9	
(g)	+1, −2, +3, +4, +5	+0, −1, +2, +3, +4	9	1	0.8	−0.2429, +0.9800
(h)	−1, −1, +3, +4, +5	−0.5, −0.5, +2, +3, +4	9	1	0.8	
(i)	+1, +2, −2, +4, +5	+0, +1.5, −1.5, +3, +4	8.5	1.5	0.7	−0.3173, +0.9591
(j)	+1, +2, −3, +4, +5	+0, +1, −2, +3, +4	8	2	0.6	−0.3858, +0.9333
(k)	+1, +2, +3, −3, +5	+0, +1, +2.5, −2.5, +4	7.5	2.5	0.5	−0.4493, +0.9037
(l)	+1, +2, +3, −4, +5	+0, +1, +2, −3, +4	7	3	0.4	−0.5085, +0.8707
(m)	−1, +2, +3, −4, +5	−0, +1, +2, −3, +4	7	3	0.4	
(n)	0, +2, +3, −4, +5	00, +1, +2, −3, +4	7	3	0.4	
(o)	+1, +1, +3, −4, +5	+0.5, +0.5, +2, −3, +4	7	3	0.4	
(p)	+1, +2, +2, −4, +5	+0, +1.5, +1.5, −3, +4	7	3	0.4	
(q)	0, 0, +3, −4, +5	0 × 0.5, 0 × 0.5, +2, −3, +4	6	3	0.3	−0.5640, +0.8348
(r)	+1, −1, +3, −4, +5	+0.5, −0.5, +2, −3, +4	6.5	3.5	0.3	
(s)	+1, +2, +3, +4, −5	+0, +1, +2, +3, −4	6	4	0.2	−0.6161, +0.7963
(t)	+1, −2, +3, −4, +5	+0, −1, +2, −3, +4	6	4	0.2	
(u)	−1, −1, +3, −4, +5	−0.5, −0.5, +2, −3, +4	6	4	0.2	
(v)	0, 0, 0, −4, +5	0 × 1, 0 × 1, 0 × 1, −3, +4	4	3	0.1	−0.6652, +0.7552
(w)	+1, −2, +3, +4, −5	+0, −1, +2, +3, −4	5	5	0.0	−0.7115, +0.7115
(x)	−1, −2, −3, +4, −5	−0, −1, −2, +3, −4	3	7	−0.4	−0.8707, +0.5085

Note: Illustrating effects of zeros and ties. The 95% confidence limits based on the constrained quartic model for q are also shown.

(a) The diagonal cells (shaded) correspond to the sign test. Here, $\tilde{\zeta} = 0.8.$ $\sum_{1 \le i \le 5} s_{ii} = 4 - 1 = 3 \rightarrow \frac{\sum s_{ii}}{\max} = \frac{3}{5} = 0.6 = 2\tilde{\zeta} - 1.$

(b) The off-diagonal cells correspond to $h = 2$. It is immaterial whether we take all $n(n - 1)$ off-diagonal cells or just the $n(n - 1)/2$ cells in the lower triangle. Taking just the lower triangle,

$$\sum_{1 \le i < j \le 5} s_{ij} = 7 - 3 = 4 \rightarrow \frac{\sum s_{ij}}{\max} = \frac{4}{10} = 0.4 = \tilde{\psi}_Z.$$

(c) The lower triangle plus diagonal, with $n(n + 1)/2$ cells, corresponds to $h = 0.$ $\sum_{1 \le i \le j \le 5} s_{ij} = 11 - 4 = 7 \rightarrow \frac{\sum s_{ij}}{\max} = \frac{7}{15} = 0.4667 = \tilde{\psi}_W.$

(d) The whole square, with n^2 cells, corresponds to $h = 1.$ $\sum_{1 \le i, j \le 5} s_{ij} = 18 - 7 = 11 \rightarrow \frac{\sum s_{ij}}{\max} = \frac{11}{25} = 0.44 = \tilde{\psi}_R.$

Thus in the **s** matrix, the diagonal terms estimate the measure ζ which generalises the sign test. The off-diagonal terms correspond to estimating the generalised Wilcoxon measure ψ, and the zero-shifted formulation is the one that corresponds to the off-diagonal terms. From this point on, the zero-shifted formulation is generally assumed.

An equivalent definition, analogous to our empirical definition of $\tilde{\zeta}$ above, is that $\tilde{\psi}_Z$ is the proportion of positive s_{ij} minus the proportion of negative s_{ij}, where both are based on off-diagonal cells only.

16.3.3 A Surprising Consequence of Using ZSRs

Choosing the ZSR formulation has the unexpected consequence that the sign of the numerically smallest difference is usually disregarded. Thus two samples such as {1, 2, 3, –4, 5} and {–1, 2, 3, –4, 5} give rise to identical values of ΔT, in this instance 7 – 3 = 4. Moreover, for the sample {–1, 2, 3, 4, 5}, $\Delta T/\max = 1$. This behaviour occurs only if the lowest observation is not tied. This is the price we pay for unbiased estimation of ψ.

The upper half of Table 16.2 contrasts the **s** matrices for the samples {1, 2, 3, –4, 5} and {–1, 2, 3, –4, 5}. Here, altering the sign of the (unique) lowest observation affects only the first diagonal entry, but none of the off-diagonal ones. The empirical estimate $\tilde{\psi}_Z$ is based on the latter; the former contribute to the statistic $\tilde{\zeta}$ corresponding to the sign test, but not to $\tilde{\psi}_Z$.

This corresponds with $\tilde{\psi}_Z$ estimating unbiasedly ψ defined as $\Pr[X_1 + X_2] > 0 - \Pr[X_1 + X_2] < 0$ when X_1 and X_2 are sampled independently. The estimator $\tilde{\psi}_Z$ is based on all off-diagonal (X_i, X_j) pairs from the sample where

$i \neq j$ (i.e., resampling without replacement). If we use a measure such as $\tilde{\psi}_W$ or $\tilde{\psi}_R$ which includes a contribution from the diagonal terms, the relative contribution of diagonal and off-diagonal terms depends heavily on n, which is obviously unsatisfactory.

The issue of whether resampling the data to obtain our point estimate should be without or with replacement pertains only to ψ, not to the unpaired measure θ, because in that context the Xs and Ys come from different populations.

16.3.4 Tied Data and Zeros

Other major issues relate to tied data and zeros. Throughout the above development we assumed that the distribution of X was absolutely continuous, so that ties occur with probability 0. As all practitioners are aware, in the real world data are invariably recorded with some degree of rounding, consequently the occurrence of ties is the norm, not the exception.

Consider a simple sample comprising five paired differences $\mathbf{X} = \{0, +2, -3, +3, +5\}$. Ties occur when two or more x_i are equal in absolute value, regardless of whether they are of similar or opposite signs. The two observations −3 and +3 constitute a tie. We refer to any observations with $x = 0$ as zeros. A zero corresponds to a particular form of tie in the raw data, the unit in question had equal values for the two variables being compared. The sample may contain one zero, or two or more tied zeros.

When a Wilcoxon test or a sign test is performed, zeros are regarded as uninformative and are dropped; the effective total sample size n is reduced accordingly. Similarly, for the paired 2 × 2 table (Table 8.1) with cell frequencies a, b, c and d totalling to n, the McNemar test is based solely on b and c, irrespective of whether an exact or asymptotic test is used. The odds ratio based on the paired data is most commonly estimated as b/c, once again both this and its confidence interval disregard the information on a and d. Nevertheless, if interest centres on the difference between marginal proportions, $(a + b)/n - (a + c)/n \equiv (b - c)/n$, not $(b - c)/(b + c)$, and the a and d cells must not be disregarded. Indeed, intervals for $(b - c)/n$ that condition on $b + c$ and are derived from an interval for the simple proportion $b/(b + c)$ perform poorly (Newcombe 1998c).

Hollander and Wolfe indicate that while it is usual practice to drop zeros for the Wilcoxon test as above, this is not the only option, and discuss alternatives involving either randomising zeros or conservatising, by counting all zeros as if they were in favour of not rejecting H_0. Furthermore, they recommend that zeros should not be dropped when constructing point and interval estimates for the median difference η.

Correspondingly, in constructing our generalised Wilcoxon index ψ, zeros are not dropped, but are represented by signed ZSRs of zero. Thus, for our sample $\mathbf{X} = \{0, +2, -3, +3, +5\}$ above, we do not drop the zero, but keep n at 5. The unsigned ranks are {1, 2, 3.5, 3.5, 5}, following the usual convention of averaging in the event of ties. The unsigned zero shifted ranks are {0, 1, 2.5, 2.5, 4}. The signed zero shifted ranks become {0, 1, –2.5, 2.5, 4}, leading to $T_1 = 7.5$ and $T_2 = 2.5$. We report $\tilde{\psi}_Z = (7.5 - 2.5)/10 = 0.50$.

If we start with a slightly different sample, **X** = {0, 0, –3, +3, +5}, the unsigned ranks are {1.5, 1.5, 3.5, 3.5, 5}. The unsigned ZSRs are {0.5, 0.5, 2.5, 2.5, 4}. The signed ZSRs become {0, 0, –2.5, 2.5, 4}, leading to $T_1 = 6.5$ and $T_2 = 2.5$. The sum of these is 9, which falls short of $n(n - 1)/2 = 10$ simply because the "signs" associated with the first two signed ZSRs (0.5 and 0.5) are zero. Here, we report $\tilde{\psi}_Z = (6.5 - 2.5)/10 = 0.40$.

If we were to drop all zero ties and reduce n accordingly, this would lead to an exaggerated perception of the strength of effect. Consider the following six datasets:

① {0, 0, 3, 4}	④ {–1, 2, 3, 4}
② {3, 4}	⑤ {1, –2, 3, 4}
③ {1, 2, 3, 4}	⑥ {–1, –2, 3, 4}

For dataset ①, we keep the two zeros and report $\tilde{\psi}_Z = (5 - 0)/6 = 0.833$. Dropping them would result in dataset ② for which $\tilde{\psi}_Z = 1$. Had we actually observed dataset ③, or ②, it would be appropriate to report $\tilde{\psi}_Z = 1$. But, had we been able to record the original variables X and Y to a greater precision, we might equally have observed any of ③, ④, ⑤ or ⑥. These correspond to $\tilde{\psi}_Z = 1$, 1, 0.667 and 0.667, which average to 0.833. This shows why it is appropriate to calculate $\tilde{\psi}$ in this way. All the above calculations work similarly if ordinary or ridit ranks are used to calculate $\tilde{\psi}$.

The effects of tied data and zeros are also evident in the s matrix (Table 16.2). We start with the sample {1, 2, 3, –4, 5} shown in the upper left panel. In the sample {0, 2, 3, –4, 5} (lower left panel), all the off-diagonal entries are unaltered, the only change is that the first diagonal element becomes 0. Once again, $\tilde{\psi}_Z$ remains at $(7 - 3)/10 = 0.4$.

However, for the sample {1, 2, 3, –3, 5} (lower right panel), the off-diagonal entries for (3, –3) and (–3, 3) become 0, so $\tilde{\psi}_Z$ increases to $(7 - 2)/10 = 0.5$.

16.3.5 Examples of Calculation for *n* = 5

Table 16.3 illustrates how the statistic $\tilde{\psi} = \Delta T/\max$ is constructed using ZSRs for a variety of datasets of size 5. These examples illustrate the effects of zeros and tied values in the data. The ΔT can take any integer value from $n(n - 1)/2$ (here, 10) to $-n(n - 1)/2$, and is usually of the same parity as $n(n - 1)/2$.

Also shown are 95% confidence intervals, based on the constrained quartic model introduced in Section 16.4.5. Given n, ΔT is sufficient for the intervals produced by both the constrained quartic model and its precursor, the shifted uniform model developed in Section 16.4.4. Consequently examples are grouped by the value of ΔT.

The logical procedure is

(i) Calculate the signs (+, – or 0).

(ii) Calculate ZSRs of the absolute differences: 0, 1,..., $n-1$ in the absence of ties—or, in the event of ties, split in the usual way by averaging.

(iii) Reunite the signs with the ZSRs, as in the third column of the table. Here, +0, –0 and 00 respresent ZSRs of 0 representing positive, negative and zero observations.

0×0.5 represents a ZSR of 0.5 with a sign of zero.

(iv) Form the sums of positive and negative terms, T_1 and T_2. Terms such as 0×0.5 contribute 0 to T_1 and T_2.

A zero ZSR occurs only when the observation lowest in absolute value (which may be zero) is unique, not when it is tied – compare datasets (l) to (o).

When ties occur away from the smallest observation, the rank averaging process proceeds in exactly the same way as for other non-parametric procedures. Comparing datasets (p) and (k) with (l), we see the effects of ties in the absolute ranks. In dataset (p), the presence of two equal ranks of similar sign does not affect ΔT. Only datasets such as (k), including numerically equal positive and negative observations, can give rise to odd values of $n(n-1)/2 - \Delta T$.

The three datasets (a), (b) and (c) are identical apart from the numerically smallest observation. All three lead to the maximal $\Delta T = +10$, showing how the numerically smallest difference is normally disregarded. Examples (l), (m) and (n) all yield $\Delta T = +4$, for the same reason.

However, this only applies if the numerically smallest difference is unique. If the smallest two unsigned differences are equal, each gets a rank of 0.5. These may have positive or negative signs, as in examples (d) and (f), and hence contribute to T_1 or T_2. When both have the same sign, ΔT is unaffected—compare examples (d) and (a), (o) and (l). Examples (f) and (r) show what happens when they have opposite signs.

If there are multiple zeros, as in examples (e), (q) or (v), the resulting nonzero pooled ranks get zero signs and thus do not contribute to T_1 and T_2. In this case only, $T_1 + T_2$ is smaller than $n(n-1)/2$.

Examples (l) and (x) show how the method is equivariant—reversing the signs of all the data reverses the signs of ΔT/max and the two confidence limits.

16.4 Development of a Confidence Interval for ψ

To develop a confidence interval for ψ, we obtain the Wald variance, analogous to that of the corresponding unpaired effect size measure $\theta = U/mn$ in Section 15.3. For the paired case, let $\{X_1,..., X_n\}$ denote the sample of paired differences. As explained in Section 16.3.4, in contrast to the usual practice when the hypothesis test is applied, tied data points resulting in zero differences are not dropped. Like the corresponding unpaired measure, ΔT may also be expressed as the sum of terms s_{ij} which can take the values +1, −1, or 0 in the event of ties. Here, the s_{ij} take the simple form $\text{sgn}(X_i + X_j)$. The variance of ΔT on H_1 then involves terms such as var s_{ij} and $\text{cov}(s_{ij}, s_{ik})$ and so forth, similar to the expressions on page 32 of Noether (1967).

The resulting variance may be used to formulate a Wald interval. This is not expected to perform well. More refined intervals are based on var ΔT but using a process of inversion, analogous to the Wilson interval for the binomial proportion.

16.4.1 Wald Variance of ΔT for the ZSRs Formulation

Having chosen the ZSR formulation for ΔT, we obtain its variance under H_1 following Noether's derivation for the unpaired case, then verify that it reduces to the correct expression on H_0.

Analogous to the generalised Mann–Whitney measure, the asymptotic variance of ΔT involves ψ and a (single) straddle probability q. This is the probability that $Y_0 = -X_0$ lies between X_1 and X_2, disregarding which of X_1 and X_2 is greater, when X_0, X_1 and X_2 are sampled randomly from the distribution. In the continuous case q is defined as

$$\Pr[X_1 < Y_0 < X_2] + \Pr[X_2 < Y_0 < X_1] = 2\Pr[X_1 < Y_0 < X_2] \qquad (16.3)$$

The parameter ψ is estimated unbiasedly by $\Delta T/\text{max}$, where ΔT is the sum of $n(n-1)/2$ off-diagonal terms

$$\sum_{i,j} s_{ij} = \sum_{1 \le i < j \le n} \text{sgn}(x_i + x_j) \qquad (16.4)$$

For example, when $n = 5$, ΔT reduces to $s_{12} + s_{13} + s_{14} + s_{15} + s_{23} + s_{24} + s_{25} + s_{34} + s_{35} + s_{45}$.

Since the n values constituting the sample, $x_1 \ldots x_n$ are not assumed to be arranged in increasing order of either absolute or algebraic value here, the expectation of each term is ψ, consequently $\mathbf{E}\Delta T = n(n-1)\psi/2$.

We obtain the variance of $\tilde{\psi}$ by evaluating $\mathbf{E}(\Delta T^2)$.

When this is expanded, there are five types of terms, exemplified by

(a) s_{12}^2 (b) $s_{12}s_{13}$ (c) $s_{13}s_{23}$ (d) $s_{12}s_{23}$ (e) $s_{12}s_{34}$

There are $n(n-1)/2$ terms of type (a).

Assuming the absence of ties, terms such as s_{12} can only take the value +1 or −1, whence $s_{ij}^2 \equiv 1$.

But the lower right panel of Table 16.2 show how such terms can be 0 when ties occur.

We derive var $\tilde{\psi}$ assuming the absolutely continuous case here. In Sections 16.6 and 16.7 we demonstrate that boundary-respecting intervals based indirectly on this variance have appropriate coverage properties for both continuous and discrete models.

Furthermore, $\mathbf{E}s_{12}s_{34} = \psi^2$ by independence.

(b) $\mathbf{E}s_{12}s_{13} = \mathbf{E}[\text{sgn}(x_1 + x_2)\ \text{sgn}(x_1 + x_3)]$

(c) $\mathbf{E}s_{13}s_{23} = \mathbf{E}[\text{sgn}(x_1 + x_3)\ \text{sgn}(x_2 + x_3)]$

(d) $\mathbf{E}s_{12}s_{23} = \mathbf{E}[\text{sgn}(x_1 + x_2)\ \text{sgn}(x_2 + x_3)]$

For the paired case, these three quantities are identical. This contrasts with the unpaired case, in which (b) and (c) are $1 - 2q^{+-}$ and $1 - 2p^{+-}$ which are not in general equal.

Here, they are the same, because when Y_i is defined as $-X_i$, $p^{+-} = 2\Pr[X_1 < Y_0 < X_2]$ and $q^{+-} = 2\Pr[Y_1 < X_0 < Y_2]$ are identical.

So (b), (c) and (d) reduce to $2q$ where

$$q = 2\Pr[X_1 < Y_0 < X_2].$$

So the $\{n(n-1)/2\}^2$ terms comprising $\mathbf{E}(\Delta T^2)$ reduce to three components.

There are $n(n-1)/2$ diagonal cells, each with expectation $\mathbf{E}s_{12}^2 = 1$.

There are $n(n-1)(n-2)(n-3)/4$ cells with α, β, γ, δ all different, with expectation $\mathbf{E}s_{12}s_{34} = \psi^2$.

The remaining $n(n-1)(n-2)$ cells have expectation $\mathbf{E}s_{12}s_{13} = 1 - 2q$.

So, in general, $\mathbf{E}(\Delta T^2) = n(n-1)/2 + n(n-1)(n-2)(1-2q) + n(n-1)(n-2)(n-3)\psi^2/4$.

But $\mathbf{E}(\Delta T) = n(n-1)\psi/2$.

$$\text{So var}(\Delta T) = n(n-1)Q/2 \tag{16.5}$$

$$\text{and var}(\tilde{\psi}) = Q/\{n(n-1)/2\} \tag{16.6}$$

$$\text{where } Q = 1 + 2(n-2)(1-2q) - (2n-3)\psi^2 \tag{16.7}$$

On the null hypothesis that the distribution of X is symmetrical about 0, $\psi = 0$ and

$$q = 2\Pr[X_1 < Y_0 < X_2] = 2\Pr[X_1 < X_0 < X_2] = 1/3 \tag{16.8}$$

just as in Noether (1967) for the unpaired case.

So, on H_0, var(ΔT) reduces to

$$n(n-1)[1 + 2(n-2)/3]/2 = n(n-1)(2n-1)/6 \tag{16.9}$$

This is equal to the sum of the squared ZSRs, just as in Noether (1967, p. 47) the variance of ΔT_W based on ordinary Wilcoxon ranks 1, 2,... n under the null hypothesis that the distribution of the differences is symmetrical about 0 is the familiar $\sum_{k=1}^{n} k^2 = \frac{1}{6}n(n+1)(2n+1)$.

16.4.2 Wald Interval for ψ

The Wald interval for ψ then takes the usual form

$$\tilde{\psi} \pm z \times \sqrt{\text{var}(\tilde{\psi})} \tag{16.10}$$

where z denotes the appropriate centile of the standard Gaussian distribution, 1.960 for the usual two-sided 95% interval, and $\text{var}(\tilde{\psi})$ is as defined in Equations 16.6 and 16.7.

First, we consider the dataset {1, 2, 3, −4, 5}, for which $\tilde{\psi}_Z = \frac{T}{\max} = \frac{4}{10} = 0.4$. To obtain the Wald variance estimate, even though $\tilde{\psi}$ is based on resampling without replacement, following Hanley and McNeil (1982) we estimate q by sampling from the $\{X_i\}$ *with* replacement. Sampling *without* replacement would be rather nugatory for the case with $n = 5$.

This leads to var $\tilde{\psi}_Z = 0.2808$, resulting in a 95% interval −0.64 to +1.44. The occurrence of boundary violation is not surprising.

The second example is a comparison of log-transformed T_4 and T_8 cell counts from 20 patients in remission from Hodgkin's disease (Shapiro et al. 1987). These results are used to illustrate the Wilcoxon test by Altman (1991, p. 204). Altman's Table 9.9 includes a non-zero tie, which was introduced by rounding and was useful for his expository purposes, but is eliminated by reverting to the original data.

The differences between natural logs of T_4 and T_8 counts sorted by absolute value are shown in Table 16.4. Here, $\tilde{\psi}_Z = \frac{86}{190} = 0.4526$ and var $\tilde{\psi}_Z = 0.0489$.

The resulting 95% interval, +0.0191 to +0.8862, does not boundary violate for the Hodgkins data. But, in contrast to the better shifted uniform (SU) and constrained quartic (CQ) intervals developed later, the lower limit is above 0, conflicting with $p > 0.1$ as reported by Altman.

TABLE 16.4

Comparison of Log-Transformed T_4 and T_8 Cell Counts from 20 Patients in Remission from Hodgkin's Disease

−0.0250	−0.1901	0.3263	−0.7472
−0.0567	0.1991	−0.5434	0.7836
0.1186	−0.2149	0.5713	0.9478
0.1550	0.2448	0.6925	1.2089
−0.1900	−0.3253	0.6975	1.3398

Source: Data from Shapiro, C. M. et al. 1987. *American Journal of the Medical Sciences* 293, 366–370 and Altman, D. G. 1991. *Practical Statistics for Medical Research*. Chapman & Hall, London, p. 204.

Note: Values shown are differences between natural logs of T_4 and T_8 counts, sorted by absolute value.

16.4.3 The Relationship between q and ψ

For the unpaired measure $\theta = U/mn$, Hanley and McNeil (1982) demonstrated that the relevant straddle probabilities (p^{+-} and q^{+-} in Noether's notation) depend strongly on θ but show very little dependence on the distributional form of X and Y. Similar results hold in the paired case. There is a strong dependence of q on ψ, which is similar for several distributions defined in Sections 16.4.4 and 16.4.5, as illustrated in Table 16.5.

On H_0, $\psi = 0$; the order of X_1, X_2 and Y_0 is purely arbitrary and $q = 1/3$. Conversely, as $\psi \to 1$, clearly $q \to 0$.

The relationship of $Q = \{1 + 2(n-2)(1-2q)\} - \{(2n-3)\psi^2\}$ (Equation 16.7) and hence $\text{var}(\tilde{\psi})$ to ψ follows from this relationship. Figure 16.1 illustrates the relationship of the two bracketed terms to ψ. These two terms both increase

TABLE 16.5

Relationship of q to ψ for the Shifted Uniform, Beta and Gaussian Models

	q Corresponding to ψ for Model				
ψ	Shifted Uniform	Gaussian	B(6,6)	B(3,7)	B(7,3)
0.0	0.333	0.333	0.333	0.332	0.332
0.1	0.331	0.331	0.331	0.331	0.328
0.2	0.323	0.322	0.322	0.323	0.319
0.3	0.310	0.307	0.308	0.309	0.304
0.4	0.290	0.286	0.287	0.290	0.283
0.5	0.264	0.259	0.261	0.264	0.256
0.6	0.231	0.226	0.227	0.230	0.223
0.7	0.190	0.185	0.187	0.190	0.182
0.8	0.140	0.135	0.137	0.139	0.134
0.9	0.079	0.076	0.077	0.078	0.075
1.0	0.0	0.0	0.0	0.0	0.0

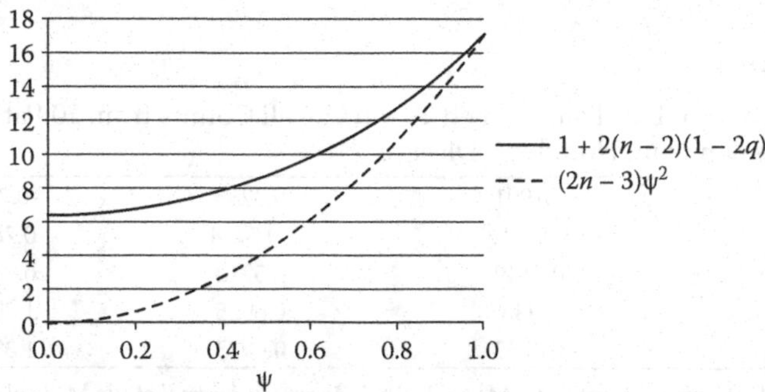

FIGURE 16.1
Relationship of the two terms forming Q, $1 + 2(n-2)(1-2q)$ and $(2n-3)\psi^2$ to ψ. Here, q is based on the constrained quartic model (Section 16.4.5).

as ψ increases from 0, but cancel out when $\psi = 1$. Here, $q(\psi)$ is based on the constrained quartic model introduced in Section 16.4.5, but the relationship takes a similar form for other models.

16.4.4 The Shifted Uniform Model

The quantity q is most conveniently calculated as

$$1 - \Pr[Y_0 < \min(X_1, X_2)] - \Pr[Y_0 > \max(X_1, X_2)] \tag{16.11}$$

But even when just two values are considered, the distribution of the order statistics, $\min(X_1, X_2)$ and $\max(X_1, X_2)$ is generally computationally complex.

Order statistics simplify greatly for two standard distributions, uniform and negative exponential—the latter as exploited by Hanley and McNeil to derive method 2 of Chapter 15. Here, we develop a closed form relationship between ψ and q using shifted uniform model—not in order to act as a realistic model for inferences but purely as an objective method to express q as a closed-form function of ψ.

Consider $X \sim U[\mu - 0.5, \mu + 0.5]$, where $\mu \in [-0.5, 0.5]$ is the mean of the distribution.

$$f(x) = 1 \text{ for } \mu - 0.5 \leq x \leq \mu + 0.5\text{, else } 0.$$

H_0 corresponds to $\mu = 0$.

The distribution of $s = X_1 + X_2$ is symmetrical triangular:

$$f_s(s) = \begin{cases} s - (2\mu - 1) & 2\mu - 1 < s < 2\mu \\ 2\mu + 1 - s & 2\mu < s < 2\mu + 1 \\ \text{else } 0. & \end{cases}$$

Assume without loss of generality that $\mu \geq 0$. Then $\Pr[X_1 + X_2 > 0] = 1 - (2\mu - 1)^2/2$ leading to

$$\psi = 4\mu(1 - \mu) \geq 0 \tag{16.12}$$

This relationship is invertible:

$$\mu = \frac{1}{2}\left(1 - \sqrt{1 - \psi}\right) \tag{16.13}$$

for $\psi \geq 0$.

For $U[\mu - 0.5, \mu + 0.5]$, the distribution of $b = \max(X_1, X_2)$ is wedge-shaped:

$$f_b(b) = 2(b + 0.5 - \mu) = 1 + 2(b - \mu),\ \mu - 0.5 \leq b \leq \mu + 0.5\text{, else } 0.$$

Then, letting $Y_0 = -X_0$, we obtain

$$\Pr\left[Y_0 > \max\left(X_1, X_2\right)\right] = \frac{1}{3}(1-2\mu)^3 .$$

The distribution of $a = \min(X_1, X_2)$ is a mirror image of the distribution of b:

$$f_a(a) = 1 - 2(a - \mu),\ \mu - 0.5 \le a \le \mu + 0.5, \text{ else } 0.$$

This leads to $\Pr\left[Y_0 < \min\left(X_1, X_2\right)\right] = \frac{1}{3} + 2\mu - \frac{8}{3}\mu^3$ and $q = \frac{1}{3} - 4\mu^2 + \frac{16}{3}\mu^3$. Again assuming $\psi \ge 0$, substituting for μ from Equation 16.13 leads to

$$q = \frac{1}{3} - \left(1 - \sqrt{\lambda}\right)^2 + \frac{2}{3}\left(1 - \sqrt{\lambda}\right)^3 \tag{16.14}$$

where $\lambda = 1 - \psi$.

Since for a symmetrical model $q(-\psi) = q(\psi)$, Equation 16.14 applies irrespective of the sign of ψ provided we put

$$\lambda = 1 - |\psi| \tag{16.15}$$

The expression Equation 16.14 for q may be substituted in the variance (Equation 16.6), leading to an inversion interval based on the SU model. Specifically, we solve

$$\tilde{\psi} = \psi \pm z\sqrt{\frac{Q}{n(n-1)/2}} \tag{16.16}$$

for ψ where from Equations 16.7, 16.14 and 16.15 $Q = 1 + 2(n-2)(1-2q) - (2n-3)\psi^2$, $q = \frac{1}{3} - \left(1 - \sqrt{\lambda}\right)^2 + \frac{2}{3}\left(1 - \sqrt{\lambda}\right)^3$ and $\lambda = 1 - |\psi|$.

16.4.5 The Constrained Quartic Model

The relationship of q to ψ may also be estimated numerically for a variety of parametric models.

Hanley and McNeil (1982) showed that for the unpaired case, the relationship of the two relevant straddle probabilities to the parameter of interest, θ does not depend crucially on choice of distribution. We show a similar result for the (single) straddle probability q pertaining to the paired case.

We consider four parametric models:

A Gaussian model $X \sim N(\mu, 1)$.

Three shifted beta models defined by $X + g \sim B(\alpha, \beta)$ for $(\alpha, \beta) = (3,7)$, $(6,6)$ and $(7,3)$, with $0 < g < 1$.

The shifted beta models represent bounded positively skew, symmetrical and negatively skew distributions, and have almost identical variances, 1/52.38, 1/52 and 1/52.38. These three models are also used to generate data in the evaluation for the continuous case (Sections 16.5 and 16.6).

Table 16.5 shows the relationship of q to ψ for the shifted uniform model and these four models. When plotted, the $q(\psi)$ profiles are visually almost indistinguishable, implying that choice of distributional form is not crucial here.

Usually the SU model leads to slightly higher q than the other models, hence lower imputed variance, consequently there is a risk that the inversion interval based on the SU model could be anticonservative. To obviate this, it is preferable to use the minimum value of q across the five models, as a function of ψ. The smooth relationship of this minimised q to ψ may be approximated very closely by a constrained quartic (CQ):

$$q = \lambda^4/3 + \sum_{i=1}^{3} b_i\left(\lambda^i - \lambda^4\right) \tag{16.17}$$

where $\lambda = 1 - |\psi|$.

The form of this relationship ensures that $q = 1/3$ when $\psi = 0$ and $q = 0$ when $\psi = \pm 1$.

The coefficients $b_1 = 0.8202$, $b_2 = -0.8471$ and $b_3 = 0.5618$ are obtained by multiple linear regression of $q - \lambda^4/3$ on $(\lambda^i - \lambda^4)$ for $i = 1, 2, 3$ using data points for $\psi = 0.00\ (0.01)\ 0.99$.

q as a function of ψ is then substituted in the variance (Equation 16.6), leading to an inversion interval based on the CQ model. Once again, we solve

$$\tilde{\psi} = \psi \pm z\sqrt{\frac{Q}{n(n-1)/2}} \tag{16.18}$$

for ψ where $Q = 1 + 2(n-2)(1-2q) - (2n-3)\psi^2$ and $\lambda = 1 - |\psi|$ as before, but now from Equation 16.17 $q = \lambda^4/3 + \sum_{i=1}^{3} b_i\left(\lambda^i - \lambda^4\right)$.

For the Hodgkins dataset, with $\tilde{\psi} = 0.4526$, the 95% SU and CQ inversion intervals are similar, (−0.0588, 0.7462) and (−0.0597, 0.7557), but contrast markedly with the Wald interval, (0.0191, 0.8862).

Both SU and CQ intervals involve single-depth iteration, thus may be implemented in spreadsheet software using a fixed number of interval bisections. The spreadsheet GENWILCOXON.xls, shown in Figure 16.2, is designed to display $\tilde{\psi}_z$ with a CQ interval. It includes a graphical visualisation based on VISUALISEPSI.xls, enhanced to also represent the confidence limits. The input data comprises numbers of negative, positive and zero observations (i.e., paired differences) and Wilcoxon rank sums corresponding to negative

Spreadsheet GENWILCOXON.

This spreadsheet calculates the generalised Wilcoxon measure psi = deltaT/max based on zero shifted ranks.
Confidence limits based on the constrained quartic model are calculated by 40 interval bisections.
To use, replace input values in **bold** as appropriate.
Crude counts and sums of negative and positive Wilcoxon ranks and ties (i.e. zeros)
produced by SPSS's NPAR TESTS routine should be pasted into cells B20:C22.
The zero shifted rank sums correctly distributing any zero ties are shown in cells F20:F21.
The spreadsheet displays point and interval estimates on the scale of psi.
Also on the scale of mu - mean of Gaussian distribution with SD 1 and equivalent psi.
Gaussian curves corresponding to the point estimate and lower and upper limits for mu are displayed.
Designed to be used for |psi| <= 0.99999999 corresponding to |mu| <= 4.0522.
Greater degrees of separation are displayed truncated accordingly.

Two-sided confidence level required **95** %.

Summary statistics:

	N	Wilcoxon rank sum	Wilcoxon mean rank		Zero shifted rank sum
Negative ranks	**8**	**60**	7.5		52
Positive ranks	**12**	**150**	12.5		138
Ties	**0**				
Total	20			Delta T	86
				max	190

Results:	**Point estimate**	**Confidence limits Lower**	**Upper**
Psi	0.4526	−0.0597	0.7557
Mu	0.4255	−0.0529	0.8234

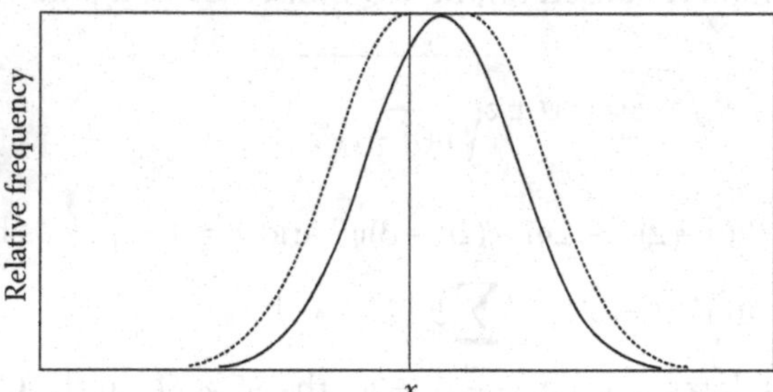

FIGURE 16.2
Spreadsheet GENWILCOXON.xls to implement calculations for the constrained quartic interval for the paired effect size measure $\psi = \Delta T/\max$.

and positive observations, in the format produced by the SPSS NPAR TESTS–WILCOXON routine.

The positive and negative rank sums are first adjusted to reflect the zero shifted formulation and the fact that zeros are not dropped from the analysis. The spreadsheet then calculates $\tilde{\psi}_z$, and also the corresponding Gaussian mean, $\tilde{\mu}_z$ such that the distribution $N(\tilde{\mu}_z, 1)$ has $\psi = \tilde{\psi}_z$, according to the relationship shown in Table 16.1. The confidence limits (ψ_L, ψ_U) for ψ are calculated by the CQ method, then transformed into limits (μ_L, μ_U) for μ.

In the plot, the solid curve represents a Gaussian distribution $N(\tilde{\mu}_z, 1)$ for which $\psi = \tilde{\psi}_z$. It is flanked by two broken curves, which display the ascending part of the curve corresponding to the lower limit μ_L and the descending segment corresponding to the upper limit μ_R. Degrees of shift beyond $|\psi| = 0.99999999$ (corresponding to $|\mu| > 4.0522$) are displayed truncated to this degree of separation.

16.5 Evaluation of Coverage Properties: Continuous Case

The methodology of this evaluation is based largely on the principles set out by Newcombe (1998a). Because neither SU nor CQ intervals can behave anomalously in boundary or near-boundary cases, interest centres on coverage and interval width here.

There were 105 parameter space points used in the evaluation, representing each combination of three factors:

- Five values of ψ, 0.1, 0.3, 0.5, 0.7 and 0.9
- Three distributions B(3,7), B(6,6) and B(7,3), shifted to correspond to the chosen ψ in accordance with Table 16.6
- Seven sample sizes, n = 5, 10, 20, 50, 100, 250 and 500

TABLE 16.6

Shifts Required for 15 Combinations of Distribution and ψ Used in the Evaluation

	Shift Corresponding to Model		
ψ	B(3,7)	B(6,6)	B(7,3)
0.1	0.28144	0.48734	0.69334
0.3	0.25624	0.46125	0.66651
0.5	0.22937	0.43235	0.63587
0.7	0.19769	0.39662	0.59661
0.9	0.14995	0.33834	0.52916

For each of the resulting 105 PSPs independently, 10^7 datasets of the relevant sample size were generated randomly. For each simulated dataset $\tilde{\psi}$ is then calculated, with 95% intervals by both SU and CQ methods as defined above. For each method, summary data are then accumulated to estimate the coverage probability and mean interval width for each PSP. Thus the coverage attained for the PSP is calculated as the proportion of the 10^7 datasets for which the interval includes the preset value of ψ. Left and right non-coverage probabilities are calculated similarly. For all our PSPs, $\psi > 0$, whence these can be interpreted in terms of mesial and distal non-coverage as described in Chapter 4. Use of 10^7 runs estimates a one-sided non-coverage probability of the order of 0.025 to within 0.0001 with 95% confidence.

In this evaluation designed for the continuous case, all randomly chosen datasets are checked carefully for zeros and ties, which are then disaggregated randomly to represent what would happen when we generate data precisely from an absolutely continuous distribution.

16.6 Results of Evaluation for the Continuous Case

Table 16.7 shows the output for a typical PSP, including width and coverage properties for SU and CQ intervals with nominal 95% coverage.

16.6.1 Coverage Probability

Table 16.8 shows the coverage probabilities for 95% SU and CQ intervals, for three beta models, five values of ψ and seven sample sizes. The majority of coverage probabilities are around 0.94 to 0.96, indicating that both the SU and CQ intervals fulfil the objective of a 95% confidence interval.

Coverage probability is generally similar for the three models, with a maximum discrepancy of 0.020 between coverage for B(3,7) and B(7,3) models with $n = 500$ and $\psi = 0.9$. Coverage is usually highest for data

TABLE 16.7

Output for a Specimen Parameter Space Point: B(7,3), $\psi = 0.5$, $n = 500$

```
iopt= 0 nruns= 10000000 n= 500  alpha= 7.0 beta= 3.0 shift= 0.63587 psi00= 0.50
Summary statistics for randomly generated psi values:
Mean  0.500006 SD  0.043653    Min  0.264144     Max 0.728561
Median  0.500569 Quartiles     0.470862  and     0.529780
Method Mean width CP        LNCP     RNCP
   1   0.16448521 0.94156 0.02700 0.03144
   2   0.17025025 0.95030 0.02272 0.02698
```

TABLE 16.8

Coverage Probability for SU (Upper Figure) and CQ (Lower Figure) 95% Intervals for the Generalised Wilcoxon Measure ψ

Model	ψ	$n = 5$	$n = 10$	$n = 20$	$n = 50$	$n = 100$	$n = 250$	$n = 500$
B(3,7)	0.9	0.9410	0.9419	0.9518	0.9555	0.9520	0.9484	0.9472
		0.9410	0.9574	0.9642	0.9689	0.9721	0.9722	0.9721
B(3,7)	0.7	0.9567	0.9539	0.9583	0.9519	0.9502	0.9491	0.9487
		0.9567	0.9539	0.9621	0.9620	0.9611	0.9601	0.9599
B(3,7)	0.5	0.9411	0.9540	0.9554	0.9517	0.9503	0.9499	0.9495
		0.9682	0.9635	0.9602	0.9587	0.9581	0.9577	0.9574
B(3,7)	0.3	0.9499	0.9571	0.9554	0.9510	0.9506	0.9501	0.9500
		0.9499	0.9662	0.9554	0.9558	0.9552	0.9547	0.9547
B(3,7)	0.1	0.9470	0.9558	0.9500	0.9504	0.9500	0.9498	0.9499
		0.9470	0.9558	0.9549	0.9516	0.9512	0.9508	0.9508
B(6,6)	0.9	0.9377	0.9374	0.9481	0.9515	0.9442	0.9395	0.9381
		0.9377	0.9549	0.9614	0.9659	0.9678	0.9667	0.9663
B(6,6)	0.7	0.9583	0.9528	0.9569	0.9470	0.9449	0.9440	0.9435
		0.9583	0.9528	0.9609	0.9581	0.9567	0.9557	0.9554
B(6,6)	0.5	0.9363	0.9546	0.9527	0.9486	0.9473	0.9467	0.9464
		0.9701	0.9641	0.9580	0.9560	0.9555	0.9549	0.9546
B(6,6)	0.3	0.9549	0.9553	0.9537	0.9494	0.9493	0.9486	0.9485
		0.9549	0.9679	0.9537	0.9543	0.9540	0.9534	0.9533
B(6,6)	0.1	0.9555	0.9543	0.9498	0.9503	0.9500	0.9499	0.9499
		0.9555	0.9543	0.9547	0.9514	0.9512	0.9508	0.9509
B(7,3)	0.9	0.9333	0.9303	0.9428	0.9459	0.9344	0.9285	0.9268
		0.9333	0.9505	0.9571	0.9614	0.9619	0.9593	0.9585
B(7,3)	0.7	0.9586	0.9505	0.9541	0.9403	0.9378	0.9367	0.9360
		0.9586	0.9505	0.9583	0.9526	0.9507	0.9495	0.9489
B(7,3)	0.5	0.9320	0.9539	0.9481	0.9437	0.9424	0.9419	0.9416
		0.9716	0.9638	0.9540	0.9517	0.9511	0.9506	0.9503
B(7,3)	0.3	0.9589	0.9522	0.9507	0.9464	0.9462	0.9457	0.9457
		0.9589	0.9688	0.9507	0.9515	0.9510	0.9506	0.9506
B(7,3)	0.1	0.9627	0.9521	0.9484	0.9489	0.9486	0.9484	0.9484
		0.9627	0.9521	0.9534	0.9501	0.9499	0.9494	0.9495

generated from a B(3,7) model, but the opposite can occur when n and ψ are small.

By definition, the CQ interval always contains the SU interval. Consequently the CQ interval usually has a higher coverage probability than the SU interval. Sometimes, especially for small n, the two intervals had identical coverage in this evaluation. This occurs because of the discreteness of the distribution of ΔT. For $n = 5$, the SU and CQ intervals for $\Delta T = 4$ are (−0.4963, 0.8616) and (−0.5085, 0.8707). For other non-negative values of ΔT, the corresponding SU and CQ intervals cover exactly the same subset of the ψ values, 0.9, 0.7, 0.5, 0.3 and 0.1. Consequently the coverage properties for the SU and CQ intervals for $n = 5$ can differ for $\psi = 0.5$, but cannot differ for ψ = 0.9, 0.7, 0.3 or 0.1. In the table as a whole, there are many instances where the coverage probability is below 0.95 for the SU interval, but above 0.95 for the CQ interval. The extra width of the CQ interval is worthwhile as it almost ensures a mean coverage above 0.95 when $n \geq 10$.

16.6.2 Interval Width

Table 16.9 shows the corresponding mean interval widths. Expected width of 95% intervals is almost identical for the symmetrical, positively skew and negatively skew models, with a maximum discrepancy of 0.005. The ratio of widths of CQ and SU intervals ranges from 1.005 for $n = 5$, $\psi = 0.9$ to 1.135 for $n = 500$, $\psi = 0.9$, although much greater discrepancies can occur for discrete distributions, as in Section 16.7.

Mean interval width shows the usual strong dependence on the sample size n. For large n, interval width is about three times as great for $\psi = 0.1$ as for $\psi = 0.9$. For $n = 5$, the expected interval width is only about one-sixth wider for $\psi = 0.1$ than for $\psi = 0.9$. This is because for $n = 5$, the interval width varies relatively little with ΔT, from 1.07 when $\Delta T = 10$ to 1.42 when $\Delta T = 0$ as in Table 16.3—all intervals are too wide to be informative anyway.

16.6.3 Interval Location

Table 16.10 shows the index MNCP/NCP characterising interval location.

From Table 16.3, when $n = 5$ left (mesial) non-coverage cannot occur for any $\psi > 0$, hence MNCP/NCP is always 0. As the sample size increases, there is an improvement in symmetry of coverage. This is most rapid for small values of ψ; even for $n = 500$ the interval is still too mesially located for $\psi = 0.9$. For this and indeed many PSPs, symmetry of coverage is less favourable for the CQ interval than the SU interval.

Interval location as measured by MNCP/NCP differs somewhat between the three models, with excessively mesial location most marked for data generated from the B(3,7) model.

TABLE 16.9

Mean Widths for SU (Upper Figure) and CQ (Lower Figure) 95% Intervals for the Generalised Wilcoxon Measure ψ

Model	ψ	$n = 5$	$n = 10$	$n = 20$	$n = 50$	$n = 100$	$n = 250$	$n = 500$
B(3,7)	0.9	1.125	0.781	0.489	0.253	0.161	0.094	0.065
		1.131	0.796	0.516	0.277	0.179	0.106	0.073
B(3,7)	0.7	1.204	0.914	0.658	0.411	0.287	0.180	0.127
		1.216	0.929	0.676	0.429	0.301	0.189	0.133
B(3,7)	0.5	1.258	1.001	0.762	0.506	0.363	0.232	0.165
		1.273	1.017	0.776	0.518	0.374	0.240	0.170
B(3,7)	0.3	1.292	1.054	0.824	0.562	0.410	0.264	0.188
		1.309	1.073	0.839	0.572	0.416	0.269	0.192
B(3,7)	0.1	1.308	1.079	0.853	0.589	0.432	0.279	0.199
		1.325	1.099	0.869	0.599	0.437	0.282	0.201
B(6,6)	0.9	1.124	0.780	0.488	0.252	0.160	0.094	0.065
		1.130	0.795	0.515	0.277	0.179	0.106	0.073
B(6,6)	0.7	1.202	0.912	0.657	0.411	0.287	0.180	0.127
		1.214	0.927	0.675	0.429	0.301	0.189	0.133
B(6,6)	0.5	1.257	0.999	0.761	0.506	0.363	0.232	0.165
		1.271	1.016	0.775	0.518	0.374	0.240	0.170
B(6,6)	0.3	1.291	1.053	0.823	0.562	0.410	0.264	0.188
		1.307	1.072	0.838	0.572	0.416	0.269	0.192
B(6,6)	0.1	1.308	1.079	0.853	0.589	0.432	0.279	0.199
		1.325	1.099	0.869	0.599	0.437	0.282	0.201
B(7,3)	0.9	1.123	0.779	0.487	0.251	0.160	0.094	0.065
		1.129	0.794	0.514	0.276	0.179	0.106	0.073
B(7,3)	0.7	1.200	0.910	0.655	0.410	0.287	0.180	0.127
		1.211	0.925	0.674	0.428	0.301	0.189	0.133
B(7,3)	0.5	1.254	0.997	0.759	0.505	0.363	0.232	0.164
		1.268	1.014	0.774	0.517	0.374	0.240	0.170
B(7,3)	0.3	1.289	1.052	0.822	0.562	0.410	0.264	0.188
		1.305	1.071	0.838	0.572	0.416	0.269	0.192
B(7,3)	0.1	1.307	1.078	0.852	0.589	0.432	0.279	0.199
		1.324	1.099	0.869	0.599	0.437	0.282	0.201

TABLE 16.10

Interval Location for SU (Upper Figure) and CQ (Lower Figure) 95% Intervals for the Generalised Wilcoxon Measure ψ

Model	ψ	$n = 5$	$n = 10$	$n = 20$	$n = 50$	$n = 100$	$n = 250$	$n = 500$
B(3,7)	0.9	0.00	0.00	0.00	0.00	0.16	0.29	0.36
		0.00	0.00	0.00	0.00	0.05	0.20	0.28
B(3,7)	0.7	0.00	0.00	0.00	0.21	0.30	0.38	0.42
		0.00	0.00	0.00	0.17	0.27	0.36	0.40
B(3,7)	0.5	0.00	0.00	0.17	0.32	0.37	0.42	0.44
		0.00	0.00	0.14	0.29	0.36	0.41	0.44
B(3,7)	0.3	0.00	0.21	0.29	0.38	0.42	0.45	0.47
		0.00	0.00	0.29	0.37	0.42	0.45	0.46
B(3,7)	0.1	0.00	0.33	0.41	0.44	0.46	0.48	0.48
		0.00	0.33	0.41	0.44	0.46	0.48	0.48
B(6,6)	0.9	0.00	0.00	0.00	0.00	0.20	0.32	0.38
		0.00	0.00	0.00	0.00	0.08	0.23	0.31
B(6,6)	0.7	0.00	0.00	0.00	0.24	0.33	0.40	0.43
		0.00	0.00	0.00	0.21	0.30	0.38	0.41
B(6,6)	0.5	0.00	0.00	0.21	0.35	0.39	0.43	0.45
		0.00	0.00	0.19	0.32	0.38	0.42	0.45
B(6,6)	0.3	0.00	0.28	0.34	0.41	0.44	0.46	0.47
		0.00	0.00	0.34	0.40	0.43	0.46	0.47
B(6,6)	0.1	0.00	0.39	0.45	0.47	0.48	0.49	0.49
		0.00	0.39	0.45	0.47	0.48	0.49	0.49
B(7,3)	0.9	0.00	0.00	0.00	0.00	0.23	0.35	0.40
		0.00	0.00	0.00	0.00	0.11	0.27	0.34
B(7,3)	0.7	0.00	0.00	0.00	0.28	0.36	0.41	0.44
		0.00	0.00	0.00	0.25	0.33	0.40	0.43
B(7,3)	0.5	0.00	0.00	0.26	0.38	0.41	0.45	0.46
		0.00	0.00	0.24	0.35	0.40	0.44	0.46
B(7,3)	0.3	0.00	0.35	0.38	0.43	0.46	0.47	0.48
		0.00	0.00	0.38	0.43	0.45	0.47	0.48
B(7,3)	0.1	0.00	0.46	0.49	0.49	0.50	0.50	0.50
		0.00	0.46	0.49	0.49	0.50	0.50	0.50

Note: Location is characterised by MNCP/NCP, the left (mesial) non-coverage probability as a proportion of the total non-coverage probability.

16.7 Coverage Properties for Discrete Distributions

We use the same algorithms as before for the two intervals irrespective of whether zeros or ties occur in the data. When the support scale is discrete, ties or zeros occur with non-zero probability. In Section 16.4.1, the Wald variance of $\tilde{\psi}$ is obtained by evaluating $\mathbf{E}(\Delta T^2)$. This quantity included five types of terms, including squared terms of the form s_{ij}^2 for $i \neq j$. The derivation of the Wald variance assumes that s_{ij} can only take the values +1 or −1, whence $s_{ij}^2 \equiv 1$. However, when ties or zeros occur in the sample, s_{ij} can be zero, as in the lower panels of Table 16.2. In principle, both behaviours affect the Wald variance of $\tilde{\psi}$. In the discrete case, usually the sample size will be greater than the number of scale points—zeros and ties are then inevitable.

It is not feasible to base the evaluation for the discrete case on beta models cut at points in arithmetic progression; namely, 0.2, 0.4, 0.6 and 0.8, because cell probabilities get too low. So we base on an equally spaced five-point scale {−2, −1, 0, 1, 2}, and three base distributions:

Positively skew triangular, with probabilities {0.36, 0.28, 0.2, 0.12, 0.04}
Symmetrical bell-shaped, with probabilities {0.1, 0.2, 0.4, 0.2, 0.1}
Negatively skew triangular, with probabilities {0.04, 0.12, 0.2, 0.28, 0.36}

The scale is then shifted to the right by some quantity g. Shifts of less than 1/4 scale point are immaterial. And a shift of greater than 2 leads to $\psi \equiv 1$ and only strictly positive samples can be formed. So, without loss of generality, for a five-point scale there are only nine possible shifts to consider, namely g = 0, 1/4, 1/2, ..., 2, and −1/4, −1/2, ... −2, of course, but it is redundant to evaluate for these.

Table 16.11 shows precisely calculated ψ and q for 27 combinations of our three distributional forms on a five-point support scale with nine shifts. The q corresponding to ψ based on the constrained quartic model is also shown.

In Table 16.11, q calculated directly is usually (but not invariably) lower than the q corresponding to ψ based on the CQ model. This suggests that when an interval derived from the Wald variance for the absolutely continuous case is applied to discrete data, there is a risk of anticonservatism. Nevertheless the evaluation shows that coverage properties also remain satisfactory in this situation, particularly for the CQ interval.

In the middle block of Table 16.11, corresponding to a symmetrical base distribution, the q corresponding to ψ = 0 is 0.306. This is noteworthy in two ways. As above, it is substantially lower than 1/3, but there is no way that probabilities that are multiples of 0.1 could combine to produce q = 1/3. Furthermore, shifts of 0 and 1/4 both give rise to identical values of q.

The evaluation includes 189 parameter space points, representing the 27 combinations of three distributional forms and nine shifts shown in Table 16.11 with the same seven values of n as before, 5, 10, 20, 50, 100, 250 and

TABLE 16.11

ψ and q for Shifted Positively Skew, Symmetrical and Negatively Skew Discrete Distributions on a Five-Point Support Scale; the q Corresponding to ψ Based on the Constrained Quartic Model Is also Shown

Base Distribution:

x	−2	−1	0	1	2
Pr[x]	0.36	0.28	0.2	0.12	0.04

Shift	ψ	q	CQ q(ψ)
0.00	−0.6400	0.1997	0.2083
0.25	−0.5040	0.2504	0.2548
0.50	−0.3056	0.2842	0.3024
0.75	−0.1072	0.3180	0.3288
1.00	0.1152	0.3064	0.3282
1.25	0.3376	0.2949	0.2961
1.50	0.5392	0.2304	0.2440
1.75	0.7408	0.1659	0.1646
2.00	0.8704	0.0829	0.0932

Base Distribution:

x	−2	−1	0	1	2
Pr[x]	0.1	0.2	0.4	0.2	0.1

Shift	ψ	q	CQ q(ψ)
0.00	0	0.306	0.3333
0.25	0.26	0.306	0.3103
0.50	0.46	0.252	0.2672
0.75	0.66	0.198	0.2003
1.00	0.78	0.138	0.1450
1.25	0.9	0.078	0.0741
1.50	0.94	0.048	0.0463
1.75	0.98	0.018	0.0161
2.00	0.99	0.009	0.0081

Base Distribution:

x	−2	−1	0	1	2
Pr[x]	0.04	0.12	0.2	0.28	0.36

Shift	ψ	q	CQ q(ψ)
0.00	0.6400	0.1997	0.2083
0.25	0.7760	0.1490	0.1470
0.50	0.8464	0.1075	0.1079
0.75	0.9168	0.0660	0.0627
1.00	0.9472	0.0430	0.0410
1.25	0.9776	0.0200	0.0180
1.50	0.9872	0.0115	0.0104
1.75	0.9968	0.0031	0.0026
2.00	0.9984	0.0015	0.0013

500. In contrast to the evaluation for the continuous case, zeros and ties are not disaggregated randomly: zeros are represented by a sign of 0, and when $x_i = x_j$ this contributes 0.5 to the absolute ranks for both observations. For the three discrete models, the values of ψ that can occur are specific to the model. Consequently discrepancies between three models holding ψ fixed generally cannot be evaluated here.

Results were generally in line with those in the evaluation for the continuous case. Coverage probability showed rather greater variation between PSPs than for the continuous case. As before, coverage of the CQ interval was always at least as high as for the SU interval, ranging from about 0.934 for a few PSPs to 0.99993 for $\psi = 0.9984$ with $n = 500$ as above. Table 16.12 shows results for some extreme cases with $\psi = 0.9968$ and 0.9984, in which right non-coverage can become vanishingly small for large n—in almost all instances left non-coverage is zero.

Mean interval width was virtually identical in the continuous and discrete cases for $\psi = 0.9$, the only ψ used in both evaluations. In the discrete evaluation, the CQ/SU width ratio ranged from 1.002 for $n = 5$ with $\psi \geq 0.9872$ to 0.0142/0.0050 = 2.86 for $n = 500$ with $\psi = 0.9984$.

Interval location was usually too mesial, in line with Table 16.10.

Table 16.13 corresponds to Table 16.7 and shows the output for 95% intervals from a pair of PSPs for the discrete case. These examples were chosen to explain the paradoxically conservative behaviour that can occur for the very high values of ψ that can be found with these models, which is displayed in Table 16.12.

Table 16.3 includes 95% CQ limits for ψ with $n = 5$ for all possible values of ΔT/max from 10 to 0. Left non-coverage occurs when the lower limit is greater than the value of the parameter ψ. For $n = 5$, as in Table 16.10 left non-coverage cannot occur for any $\psi > 0$, because the lower limit is below 0 irrespective of ΔT. Consequently the LNCP is always 0 in Table 16.13.

TABLE 16.12

Estimated Right Non-Coverage Probabilities of 95% SU and CQ Intervals for ψ; Discrete Case: Negatively Skew Model, Shift 1.75 and 2, $n = 5$ to 500, for Extreme Values of ψ, 0.9968 and 0.9984

	$\psi = 0.9968$		$\psi = 0.9984$	
Sample Size	**SU Interval**	**CQ Interval**	**SU Interval**	**CQ Interval**
5	0.01477	0.01477	0.01468	0.01468
10	0.05814	0.00619	0.05828	0.00622
20	0.04397	0.00745	0.04387	0.00743
50	0.04890	0.01440	0.04887	0.00358
100	0.04739	0.00680	0.04748	0.00222
250	0.04531	0.00280	0.02441	0.00021
500	0.04890	0.00117	0.03145	0.00007

TABLE 16.13

Output for a Specimen Parameter Space Point for the Discrete Case: Negatively Skew Model, Shift 1.75 and 2, $\psi = 0.9968$ and 0.9984, $n = 5$

```
iopt= 0    nruns= 10000000        n=    5     psi=     0.9968

x            -0.25        0.75         1.75       2.75       3.75
Pr[x]         0.04        0.12         0.20       0.28       0.36

Summary statistics for randomly generated psi values:
Mean    0.996793     SD     0.028141   Min -1.000000     Max     1.000000
Median    1.000000     Quartiles      1.000000   and    1.000000

Method           Mean width               CP                LNCP              RNCP
   1            1.07217779              0.98523            0.00000          0.01477
   2            1.07410340              0.98523            0.00000          0.01477

iopt= 0  nruns= 10000000        n=    5     psi=    0.9984

x             0.00        1.00         2.00       3.00       4.00
Pr[x]         0.04        0.12         0.20       0.28       0.36

Summary statistics for randomly generated psi values:
Mean    0.998409     SD     0.013968   Min. 0.000000     Max     1.000000
Median    1.000000     Quartiles     1.000000   and    1.000000

Method           Mean width               CP                LNCP              RNCP
   1            1.07124497              0.98532            0.00000          0.01468
   2            1.07308397              0.98532            0.00000          0.01468
```

Furthermore, in Table 16.13, the RNCP is approximately 0.015, for both SU and CQ intervals. For these two ψs, the non-coverage rate of the CQ interval becomes very low for large n (Table 16.12). In Table 16.3, the 95% CQ intervals for $\tilde{\psi} = 1.0$ and 0.9 (corresponding to $\Delta T = 10$ and 9) are (–0.0718, 1) and (–0.1614, 0.9944). The two ψs evaluated in Table 16.13, 0.9968 and 0.9984, lie between 1 and 0.9944. So, in the first example with $g = 1\ 3/4$ and $\psi = 0.9968$, $\Delta T = 10$ when the sample contains 0 **or 1** value of –0.25, because of the zero shift. Here, $\text{RNCP} = 1 - \Pr[\Delta T = 10] = 1 - 0.96^5 - 5 \times 0.96^4 \times 0.04 = 0.01476$. Similarly, in the second example with $g = 2$ and $\psi = 0.9984$, $\Delta T = 10$ when the sample contains 0 or 1 zero, leading to the same RNCP of 0.01476. The RNCPs shown in Table 16.13 differ from this figure only by sampling variation based on 10^7 runs.

Exactly the same occurs for the SU interval, because the 95% interval for $\Delta T = 9$ is (–0.1563, 0.9918), once again excluding 0.9968 and 0.9984. In the first example, when the sample contains two values of –0.25, $\Delta T = \mathbf{8}$ and $\tilde{\psi} = 0.8$, so to a first approximation $\psi \approx 0.01476 \times 0.8 + (1 - 0.01476) \times 1 = 0.9970$. In the second example, when the sample contains two zeros, $\Delta T = \mathbf{9}$ and $\tilde{\psi} = 0.9$, leading to $\psi \approx 0.0147 \times 0.9 + (1 - 0.01476) \times 1 = 0.9985$. This explains how two different values of ψ can result in identical coverage properties.

Table 16.12 shows that for $n = 10$ and higher, the RNCP is low for the CQ interval, and gets very low for very large n. The explanation is similar to that for the RNCP of 0.01476 for $n = 5$. Nevertheless, the RNCP is not low for the SU interval—the properties of the CQ and SU intervals diverge for larger n because many more possible values of ΔT are involved than in Table 16.3.

The pattern is similar whether we consider the extreme case with $g = 2$, $\psi = 0.9984$ or the next-to-extreme case with $g = 1\ 3/4$, $\psi = 0.9968$. But unlike for $n = 5$, these two PSPs no longer have identical coverage, for the same reason. For $g = 2$, $\psi = 0.9984$, $n = 500$, the big disparity in RNCP between SU and CQ intervals is reflected in their mean widths of 0.0050 and 0.0142. For these PSPs, left non-coverage occurred vanishingly infrequently, except for $n = 500$ where the LNCP for the SU interval with $g = 1.75$, $\psi = 0.9968$ was 0.00180.

16.8 Discussion

As explained in Chapter 15, a major reason for developing improved confidence intervals for $\theta = U/mn$ was to promote the use of appropriate measures of effect sizes, rather than p-values, as the mainstay of reporting the impact of sampling uncertainty on research findings, following Altman et al. (2000), Grissom and Kim (2011), Cumming (2011) and others. Development of ψ was similarly motivated. But unlike θ, this appears to be a completely novel measure. The development described here demonstrates the practicability of use of a measure such as ψ for the paired situation, complementary to θ for the unpaired one. Remarkably, coverage properties are very similar regardless of the direction of the (admittedly moderate) skewness in our evaluation, although interval location is affected somewhat.

In common with the Wilcoxon statistic T_1, the measure ψ when applied to paired differences is not invariant under monotonic transformation of the original scale of measurement. Thus the Hodgkins' example relates to differences between log-transformed T_4 and T_8 counts. Comparison of untransformed counts would lead to a different $\tilde{\psi}$.

Unlike algebraic transformations such as log, square root, $\sin^{-1}$ and so forth, rank transformations are context-dependent, hence the Condorcet paradox potentially affects $\tilde{\psi}$ as well as $\tilde{\theta}$.

As argued in Section 15.8, the scales for measures such as ψ and θ should be regarded as inherently quantitative, with no need to interpret using adjectival labels such as large, medium and small effect sizes.

There is clearly an important issue whenever two effect size measures are so closely related that the one gets confused with the other—notably, when odds ratios are interpreted as if relative risks, leading to an inflated perception of effect size. In the present context, it will always be important to distinguish clearly between ψ, ζ and θ, notwithstanding the important links between them.

References

Absi, E. G. 2010. Unpublished data.

Absi, E. G., Drage, N. A., Thomas, H. S. et al. 2011. Continuing dental education in radiation protection: Knowledge retention following a postgraduate course. *European Journal of Dental Education* 15: 188–192.

Agresti, A. 1976. The effect of category choice on some ordinal measures of association. *Journal of the American Statistical Association* 71: 49–55.

Agresti, A. and Coull, B. A. 1998. Approximate is better than "exact" for interval estimation of binomial proportions. *American Statistician* 52: 119–126.

Agresti, A. and Coull, B. A. 2000. Approximate is better than "exact" for interval estimation of binomial proportions. *American Statistician* 54: 88.

Agresti, A. and Caffo, B. 2000. Simple and effective confidence intervals for proportions and differences of proportions result from adding 2 successes and 2 failures. *American Statistician* 54: 280–288.

Agresti, A. and Min, Y. 2001. On small-sample confidence intervals for parameters in discrete distributions. *Biometrics* 57: 963–971.

Agresti, A. 2002. *Categorical Data Analysis*. 2nd edition. Wiley, Hoboken, NJ.

Agresti, A. and Hitchcock, D. B. 2005. Bayesian inference for categorical data analysis: A survey. *Statistical Methods & Applications* 14: 297–330.

Agresti, A. and Min, Y. 2005a. Frequentist performance of Bayesian confidence intervals for comparing proportions in 2 × 2 contingency tables. *Biometrics* 61: 515–523.

Agresti, A. and Min, Y. 2005b. Simple improved confidence intervals for comparing matched proportions. *Statistics in Medicine* 24: 729–740.

Algina, J., Keselman, H. J. and Penfield, R. D. 2006. Confidence intervals for an effect size when variances are not equal. *Journal of Modern Applied Statistical Methods* 5: 2–13.

Altman, D. G. 1982. How large a sample? In *Statistics in Practice*. S. M. Gore and D. G. Altman (eds.). British Medical Association, London.

Altman, D. G. 1991. *Practical Statistics for Medical Research*. Chapman & Hall, London.

Altman, D. G. 1998. Confidence intervals for the number needed to treat. *British Medical Journal* 317: 1309–1312.

Altman, D. G., Machin, D., Bryant, T. N. et al. (eds). 2000. *Statistics with Confidence*, 2nd Edition. BMJ Books, London.

Altman, D. G. and Bland, J. M. 2003. Interaction revisited: The difference between two estimates. *British Medical Journal* 326: 219.

American Psychological Association. 2010. *Publication Manual of the American Psychological Association*. Washington, DC.

Anderson, T. W. and Burstein, H. 1967. Approximating the upper binomial confidence limit. *Journal of the American Statistical Association* 62: 857–861.

Angus, J. E. 1987. Confidence coefficient of approximate two-sided confidence intervals for the binomial probability. *Naval Research Logistics* 34: 845–851.

Armitage, P., Berry, G. and Matthews, J. N. S. 2002. *Statistical Methods in Medical Research*, 4th Edition. Blackwell Science, Oxford.

Ayad, F., Ayad, N., Zhang, Y. P. et al. 2009. Comparing the efficacy in reducing dentin hypersensitivity of a new toothpaste containing 8.0% arginine, calcium carbonate, and 1450 ppm fluoride to a commercial sensitive toothpaste containing 2% potassium ion: An eight-week clinical study on Canadian adults. *Journal of Clinical Dentistry* 20: 10–16.

Babor, T. F., de la Fuente, J. R., Saunders, J. et al. 1992. *AUDIT: The Alcohol Use Disorders Identification Test: Guidelines for Use in Primary Health Care*. World Health Organization, Geneva, Switzerland.

Bailey, S. E. R., Bullock, A. D., Cowpe, J. G. et al. 2012. Continuing dental education: Evaluation of the effectiveness of a disinfection and decontamination course. *European Journal of Dental Education* 16: 59–64.

Barnhart, H. X., Haber, M. J. and Lin, L. I. 2007. An overview on assessing agreement with continuous measurements. *Journal of Biopharmaceutical Statistics* 17: 529–569.

Beal, S. L. 1987. Asymptotic confidence intervals for the difference between two binomial parameters for use with small samples. *Biometrics* 43: 941–950.

Bell, T. S., Branston, L. K. and Newcombe, R. G. 1999. *Survey of Non-Attenders for Breast Screening to Investigate Barriers to Mammography*. Unpublished report. Breast Test Wales, Cardiff, Wales.

Bender, R. 2001. Calculating confidence intervals for the number needed to treat. *Controlled Clinical Trials* 22: 102–110.

Bender, R. and Grouven, U. 2008. Interval estimation of the population impact number (PIN). Paper presented at the 29th Annual Conference of the International Society for Clinical Biostatistics, Copenhagen, Denmark.

Berry, G. and Armitage, P. 1995. Mid-p confidence intervals: A brief review. *Statistician* 44: 417–423.

Biggerstaff, B. J. 2000. Comparing diagnostic tests: A simple graphic using likelihood ratios. *Statistics in Medicine* 19: 649–663.

Biggerstaff, B, J. 2008. Confidence intervals for the difference of two proportions estimated from pooled samples. *Journal of Agricultural, Biological, and Environmental Statistics* 13: 478–496.

Biswas, M., Hampton, D., Turkes, A. et al. 2010. Reduced total testosterone concentrations in young healthy South Asian men are partly explained by increased insulin resistance but not by altered adiposity. *Clinical Epidemiology* 73: 457–462.

Blaker, H. 2000. Confidence curves and improved exact confidence intervals for discrete distributions. *Canadian Journal of Statistics* 28: 783–798.

Bland, J. M. and Altman, D. 1986. Statistical methods for assessing agreement between two methods of clinical measurement. *Lancet* 1: 307–310.

Bland, M. 1995. *An Introduction to Medical Statistics*. Oxford University Press, Oxford.

Bloch, D. A. and Kraemer, H. C. 1989. 2 × 2 kappa coefficients: measures of agreement or association. *Biometrics* 45: 269–287.

Blyth, C. R. and Still, H. A. 1983. Binomial confidence intervals. *Journal of the American Statistical Association* 78: 108–116.

Blyth, C. R. 1986. Approximate binomial confidence limits. *Journal of the American Statistical Association* 81: 843–855.

Böhning, D. 1994. Better approximate confidence intervals for a binomial parameter. *Canadian Journal of Statistics* 22: 207–218.

Böhning, D. and Viwatwongkasem, C. 2005. Revisiting proportion estimators. *Statistical Methods in Medical Research* 14: 147–169.

Bonett, D. G. and Price, R. M. 2006. Confidence intervals for a ratio of binomial proportions based on paired data. *Statistics in Medicine* 25: 3039–3047.

Bonett, D. G. and Price, R. M. 2011. Adjusted Wald confidence interval for a difference of binomial proportions based on paired data. *Journal of Educational & Behavioral Statistics*, published online October 17, 2011.

Box, G. E. P. and Cox, D. R. 1964. An analysis of transformations. *Journal of the Royal Statistical Society, Series B* 26: 211–246.

British Association of Surgical Oncology. 2002. Data presented at specialist group meeting, Birmingham.

Bross, I. D. J. 1958. How to use ridit analysis. *Biometrics* 14: 18–38.

Brown, B. M. 1982. Robustness against inequality of variances. *Australian Journal of Statistics* 24: 283–295.

Brown, B. M., Newcombe, R. G. and Zhao, Y. D. 2009. Non-null semiparametric inference for the Mann-Whitney measure. *Journal of Nonparametric Statistics* 21: 743–755.

Brown, L. D., Cai, T. T. and DasGupta, A. 2001. Interval estimation for a proportion. *Statistical Science* 16: 101–133.

Burdick, R. K. and Graybill, F. A. 1992. *Confidence Intervals on Variance Components*. Marcel Dekker, Inc., New York.

Burr, M. L., Fehily, A. M., Gilbert, J. F. et al. 1989. Effects of changes in fat, fish and fibre intakes on death and reinfarction: Diet and reinfarction trial (DART). *Lancet* 2: 757–761.

Burrows, R. F. and Burrows, E. A. 1995. The feasibility of a control population for a randomized controlled trial of seizure prophylaxis in the hypertensive disorders of pregnancy. *American Journal of Obstetrics and Gynecology* 173: 929–935.

Campbell, M. J. and Gardner, M. J. 2000. Medians and their differences. In *Statistics with Confidence*, 2nd Edition. D. G. Altman, D. Machin and T. N. Bryant et al. (eds.). BMJ Books, London.

Carlin, B. P. and Louis, T. A. 1996. *Bayes and Empirical Bayes Methods for Data Analysis*. Chapman & Hall, London.

Carter, R. E, Lin, Y., Lipsitz, S. R. et al. 2010. Relative risk estimated from the ratio of two median unbiased estimates. *Applied Statistics* 59: 657–671.

Cavalli-Sforza, L. L. and Bodmer, W. F. 1971. *The Genetics of Human Populations*. W. H. Freeman & Co., San Francisco.

Charbit, B., Mandelbrot, L., Samain, E. et al. for the PPH Study Group. 2007. The decrease of fibrinogen is an early predictor of the severity of postpartum hemorrhage. *Journal of Thrombosis and Haemostasis* 5: 266–273.

Chatellier, G., Zapetal, E., Lemaitre, D. et al. 1996. The number needed to treat: A clinically useful nomogram in its proper context. *British Medical Journal* 312: 426–429.

Chen, M. and Suryawanshi, S. 2010. Application of methods for assessing risk-benefit in clinical trials. In *JSM Proceedings*, WNAR. Alexandria, VA: American Statistical Association, 3858–3865.

Clarke, D., Newcombe, R. G. and Mansel, R. E. 2004. The learning curve in sentinel node biopsy: The ALMANAC experience. *Annals of Surgical Oncology* 11; 211S–215S.

Clopper, C. J. and Pearson, E. S. 1934. The use of confidence or fiducial limits illustrated in the case of the binomial. *Biometrika* 26: 404–413.

Cochrane, A. L. 1979. 1931–1971: A critical review, with particular reference to the medical profession. In *Medicines for the Year 2000: A Symposium held at the Royal*

College of Physicians. G. Teeling-Smith and N. Wells (eds.), pp. 1–11. Office of Health Economics, London.

Cohen, J. 1960. A coefficient of agreement for nominal scales. *Educational and Psychological Measurement* 20: 37–46.

Cohen, G. R. and Yang, S. Y. 1994. Mid-p confidence intervals for the Poisson expectation. *Statistics in Medicine* 13: 2189–2203.

Cole, T. J. 2000. Sympercents: Symmetric percentage differences on the 100 log(e) scale simplify the presentation of log transformed data. *Statistics in Medicine* 19: 3109–3125.

Condorcet, Marquis de. 1785. *Essai sur l' application de l'analyse d la probability des decisions rendues à la plurality des voix*. L'imprimerie Royale, Paris.

Conover, W. J. 1980. *Practical Non-Parametric Statistics*, 2nd Edition. Wiley, New York.

Copas, J. 1997. Personal communication.

Cox, D. R. 1972. Regression models and life tables. *Journal of the Royal Statistical Society Series B* 34: 187–220.

Cox, D. R. and Hinkley, D. V. 1974. *Theoretical Statistics*. Chapman & Hall, London.

Cox, D. R. and Reid, N. 1992. Parameter orthogonality and approximate conditional inference. *Journal of the Royal Statistical Society Series B* 49: 1–39.

Crawford, N. W., Cincotta, D. R., Lim, A. et al. 2006. A cross-sectional survey of complementary and alternative medicine use by children and adolescents attending the University Hospital of Wales. *BMC Complementary and Alternative Medicine* 6: 16.

Cumming, G. 2009. Inference by eye: Reading the overlap of independent confidence intervals. *Statistics in Medicine* 28: 205–220.

Cumming, G. 2011b. *Understanding the New Statistics. Effect Sizes, Confidence Intervals, and Meta-Analysis*. Routledge, New York.

D'Agostino, R. B., Sullivan, L. and Massaro, J. M. (eds). 2007. *Wiley Encyclopedia of Clinical Trials*. Wiley, Hoboken, NJ.

D'Agostino, R. B., Vasan, R. S., Pencina, M. J. et al. 2008. General cardiovascular risk profile for use in primary care: The Framingham heart study. *Circulation* 117: 743–753.

Daly, L. E. 1998. Confidence intervals made easy: Interval estimation using a substitution method. *American Journal of Epidemiology* 147: 783–790.

Daly, L. E. 2000. Confidence intervals and sample sizes. In *Statistics with Confidence*, 2nd Edition. D. G. Altman, D. Machin and T. N. Bryant et al. (eds). BMJ Books, London.

Davies, C. E., Hill, K. E., Newcombe, R. G. et al. 2007. A prospective study of the microbiology of chronic venous leg ulcers to reevaluate the clinical predictive value of tissue biopsies and swabs. *Wound Repair and Regeneration* 15: 17–22.

Davies, H. T. O., Crombie, I. K. and Tavakoli, M. 1998. When can odds ratios mislead? *British Medical Journal* 316: 989–991.

De Dombal, F. T. and Horrocks, J. C. 1978. Use of receiver operating characteristic curves to evaluate computer confidence threshold and clinical performance in the diagnosis of appendicitis. *Methods of Information in Medicine* 17: 157–161.

Deeks, J. J. 1998. When can odds ratios mislead? *British Medical Journal* 317: 1155–1156.

Deeks, J. J. 2002. Issues in the selection of a summary statistic for meta-analysis of clinical trials with binary outcomes. *Statistics in Medicine* 21: 1575–1600.

Deeks, J. J., Higgins, J. P. T. and Altman, D. G. 2008. Analysing data and undertaking meta-analyses. In *Cochrane Handbook for Systematic Reviews of Interventions*, Version 5.0.1. J. P. T. Higgins and S. Green (eds.). The Cochrane Collaboration.

De Levie, R. 2001. *How to Use Excel in Analytical Chemistry*. Cambridge University Press, Cambridge, United Kingdom.

Desu, M. M. and Raghavarao, D. 2004. *Nonparametric Statistical Methods for Complete and Censored Data*. Chapman & Hall/CRC, Boca Raton, FL, p. 185.

Diem, K. and Lentner, C. (eds.) 1970. *Documenta Geigy Scientific Tables*. Ciba-Geigy, Basel, Switzerland.

Djemal, S., Setchell, D., King, P. et al. 1999. Long-term survival characteristics of 832 resin-retained bridges and splints provided in a post-graduate teaching hospital between 1978 and 1993. *Journal of Oral Rehabilitation* 26: 302–320.

Donner, A. and Zou, G. Y. 2002. Interval estimation for a difference between intraclass kappa statistics. *Biometrics* 58: 209–215.

Donner, A. and Zou, G. Y. 2010. Closed-form confidence intervals for functions of the normal mean and standard deviation. *Statistical Methods in Medical Research*. First published on September 8, 2010 as doi:10.1177/0962280210383082.

Dube, S., Qin, J. and Ramakrishnan, R. 2008. Mathematical analysis of copy number variation in a DNA sample using digital PCR on a nanofluidic device. *PLoS ONE* 3(8): e2876.

Edwardes, M. D. deB. 1995. A confidence interval for Pr(X < Y) – Pr(X > Y) estimated from simple cluster samples. *Biometrics* 51: 571–578.

Edwards, A. G. K., Robling, M. R., Wilkinson, C. et al. 1999. The presentation and management of breast symptoms in general practice in South Wales. *British Journal of General Practice* 49, 447, 811–812.

Edwards, A. W. F. 1992. *Likelihood*. Johns Hopkins University Press, Baltimore, MD.

Edwards, J. H. 1966. Some taxonomic implications of a curious feature of the bivariate normal surface. *British Journal of Preventive and Social Medicine* 20: 42.

Egger, M. 1997. Meta-analysis, principles and procedures. *British Medical Journal* 315: 1533–1537.

Elwyn, G., O'Connor, A. M., Bennett, C. et al. 2009. Assessing the quality of decision support technologies using the International Patient Decision Aid Standards instrument (IPDASi). *PLoS ONE* 4, 3, e4705.

Engels, E. A., Schmid, C. H., Terrin, N. et al. 2000. Heterogeneity and statistical significance in meta-analysis: an empirical study of 125 meta-analyses. *Statistics in Medicine* 19: 1707–1728.

Faergemann, C. 2006. *Interpersonal violence in the Odense municipality,* Denmark 1991–2002. Ph.D. thesis, Odense, University of Southern Denmark.

Fagan, T. 1999. Exact 95% confidence intervals for differences in binomial proportions. *Computers in Biology and Medicine* 29: 83–87.

Fagerland, M. W., Lydersen, S. and Laake, P. 2011. Recommended confidence intervals for two independent binomial proportions. *Statistical Methods in Medical Research* Published online October 13, 2011, DOI: 10.1177/0962280211415469.

Farrier, S. L., Drage, N. A., Newcombe, R. G. et al. 2009. A comparative study of image quality and radiation exposure for dental radiographs produced using a charge-coupled device and a phosphor plate system. *International Endodontic Journal* 42: 900–907.

Fay, M. P. and Gennings, C. 1996. Non-parametric two-sample tests for repeated ordinal responses. *Statistics in Medicine* 15: 429–442.

Fearnley, D. and Williams, T. 2001. A prospective study using the Monroe Dyscontrol Scale as a measure of impulsivity in referrals to a forensic psychiatry service. *Medicine, Science and the Law* 41: 58–62.

Flahault, A., Cadilhac, M. and Thomas, G. 2005. Sample size calculation should be performed for design accuracy in diagnostic test studies. *Journal of Clinical Epidemiology* 58: 859–862.

Fleiss, J. L. 1970. On the asserted invariance of the odds ratio. *British Journal of Preventive and Social Medicine* 24: 45–46.

Fleiss, J. L., Levin, B. and Paik, M. C. 2003. *Statistical Methods for Rates and Proportions*, 3rd Edition. Wiley, Hoboken, NJ.

Fletcher, D., Galloway, R., Chamberlain, D. et al. 2008. Basics in advanced life support: A role for download audit and metronomes. *Resuscitation* 78: 127–134.

Frank, P. I., Morris, J., Hazell, M. L. et al. 2008. Long term prognosis in preschool children with wheeze: longitudinal postal questionnaire study 1993–2004. *British Medical Journal* 336: 1423–1426.

Frenkel, H., Harvey, I. and Newcombe, R. G. 2001. Improving oral health in institutionalised elderly people by educating carers: a randomised controlled trial. *Community Dentistry and Oral Epidemiology* 29: 289–297.

Fujino, Y. 1980. Approximate binomial confidence limits. *Biometrika* 67: 677–681.

Gardner, M. J. and Altman, D. G. 2000. Confidence intervals rather than *P* values. In *Statistics with Confidence*, 2nd Edition. D. G. Altman, D. Machin and T. N. Bryant et al. (eds). BMJ Books, London.

Gart, J. J. and Zweifel, J. R. 1967. On the bias of various estimators of the logit and its variance with application to quantal bioassay. *Biometrika* 54: 181–187.

Ghosh, B. K. 1979. A comparison of some approximate confidence intervals for the binomial parameter. *Journal of the American Statistical Association* 74: 894–900.

Gilks, W. 2004. Bioinformatics: New science—new statistics? *Significance* 1: 7–9.

Goetghebeur, E. J. T. and Pocock, S. J. 1995. Detection and estimation of J-shaped risk-response relationship. *Journal of the Royal Statistical Society Series A* 158: 107–121.

Goldstein, H. and Healy, M. J. R. 1995. The graphical presentation of a collection of means. *Journal of the Royal Statistical Society Series A* 158: 175–177.

Goodfield, M. J. D., Andrew, L. and Evans, E. G. V. 1992. Short-term treatment of dermatophyte onychomycosis with terbinafine. *British Medical Journal* 304: 1151–1154.

Gordon, A. 2002. Personal communication.

Goyal, A., MacNeill, F., Newcombe, R. G. et al. 2009. Results of the UK NEW START sentinel node biopsy training program: A model for future surgical training. 32nd Annual San Antonio Breast Cancer Symposium. *Cancer Research* 69: 538S.

Graham, P. L., Mengersen, K. and Morton, A. P. 2003. Confidence limits for the ratio of two rates based on likelihood scores: non-iterative method. *Statistics in Medicine* 22: 2071–2083.

Graybill, F. A. and Wang, C. M. 1980. Confidence intervals on nonnegative linear combinations of variances. *Journal of the American Statistical Association* 75: 869–873.

Greenland, S. 1988. On sample size and power calculations for studies using confidence intervals. *American Journal of Epidemiology* 128: 231–237.

Griffiths, P. and Hill, I. D. (eds). 1985. *Applied Statistics Algorithms*. Ellis Horwood, Chichester, United Kingdom.

Grissom, R. J. and Kim, J. J. 2011. *Effect Sizes for Research. Univariate and Multivariate Applications*, 2nd Edition. Routledge, New York.
Haldane, J. B. S. 1955. The estimation and significance of the logarithm of a ratio of frequencies. *Annals of Human Genetics* 20: 309–311.
Halperin, M., Gilbert, P. R. and Lachin, J. M. 1987. Distribution-free confidence intervals for $Pr(X_1 < X_2)$. *Biometrics* 43: 71–80.
Hanley, J. A. and McNeil, B. J. 1982. The meaning and use of the area under a receiver operating characteristic (ROC) curve. *Radiology* 143: 29–36.
Hanley, J. A. and Lipmann-Hand, A. 1983. If nothing goes wrong, is everything all right? *Journal of the American Medical Association* 249: 1743–1745.
Hashemi, L., Nandram, B. and Goldberg, R. 1997. Bayesian analysis for a single 2 × 2 table. *Statistics in Medicine* 16: 1311–1328.
Heim, C., Münzer, T. and Listyo, R. 1994. Ondansetron versus Droperidol. Postoperativer therapeutischer Einsatz bei Nausea und Erbrechen. Vergleich von Wirkung, Nebenwirkungen und Akzeptanz bei gynäkologischen, stationären Patientinnen. *Anaesthetist* 43: 504–509.
Heller, R. F. and Dobson, A. A. 2000. Disease impact number and population impact number: Population perspectives to measures of risk and benefit. *British Medical Journal* 321: 950–952.
Higgins, J. P. T. and Green, S. (eds.). 2008. *Cochrane Handbook for Systematic Reviews of Interventions*, Version 5.0.1. The Cochrane Collaboration. www.cochrane-handbook.org.
Hills, M. and Armitage, P. 1979. The two-period cross-over clinical trial. *British Journal of Clinical Pharmacology* 8: 7–20.
Hirji, K. F., Tsiatis, A. A. and Mehta, C. R. 1989. Median unbiased estimation for binary data. *American Statistician* 43: 7–11.
Hodges, J. L. and Lehmann, E. L. 1963. Estimation of location based on ranks. *Annals of Mathematical Statististics* 34: 598–611.
Hodgson, R. J., Alwyn, T., John, B. et al. 2002. The FAST alcohol screening test. *Alcohol and Alcoholism* 37: 61–66.
Hodgson, R. J., John, B., Abbasi, T. et al. 2003. Fast screening for alcohol misuse. *Addictive Behaviors* 28: 1453–1463.
Hollander, M. and Wolfe, D. A. 1999. *Nonparametric Statistical Methods*, 2nd Edition. Wiley, New York.
Hood, K. 1999. Personal communication.
Hope, R. L., Chu, G., Hope, A. H. et al. 1996. Comparison of three faecal occult blood tests in the detection of colorectal neoplasia. *Gut* 39, 722–725.
Howe, W. G. 1974. Approximate confidence limits on the mean of X+Y where X and Y are two tabled independent random variables. *Journal of the American Statistical Association* 69: 789–794.
Hughes, J. A., West, N. X., Parker, D. M. et al. 1999. Development and evaluation of a low erosive blackcurrant drink in vitro and in situ. 1. Comparison with orange juice. *Journal of Dentistry* 27: 285–289.
Iwatsubo, Y., Pairon, J. C., Boutin, C. et al. 1998. Pleural mesothelioma: dose-response relation at low levels of asbestos exposure in a French population-based case-control study. *American Journal of Epidemiology* 148: 133–142.
Jaynes, E. T. 1976. Confidence intervals vs. Bayesian intervals. In *Foundations of Probability Theory, Statistical Inference, and Statistical Theories of Science*. W. L. Harper and C. A. Hooker (eds.). Reidel, Dordrecht, Netherlands.

Jeffreys, H. 1946. An invariant form for the prior probability in estimation problems. *Proceedings of the Royal Society of London. Series A, Mathematical and Physical Sciences* 186: 453–461.

Jewell, D., Sanders, J. and Sharp, D. 2000. The views and anticipated needs of women in early pregnancy. *British Journal of Obstetrics and Gynaecology* 107: 1237–1240.

Jewell, N. P. 1986. On the bias of commonly used measures of association for 2 x 2 tables. *Biometrics* 42: 351–358.

Jia, G. and Song, Y. 2010. Assessing treatment effect in clinical studies with dichotomous endpoints. Paper presented at the Joint Statistical Meeting, Vancouver, BC.

Joseph-Williams, N., Newcombe, R. G., Politi, M. et al. 2012. Toward minimum standards for the certification of patient decision aids: A correlation analysis and modified Delphi consensus process. In submission to *Journal of Clinical Epidemiology*. In preparation.

Julious, S. A. 2005. Two-sided confidence intervals for the single proportion: Comparison of seven methods. *Statistics in Medicine* 24: 3383–3384.

Khan, K. S. and Chien, P. F. W. 1997. Seizure prophylaxis in hypertensive pregnancies: a framework for making clinical decisions. *British Journal of Obstetrics and Gynaecology* 104: 1173–1179.

Khurshid, A. and Ageel, M. I. 2010. Binomial and Poisson confidence intervals and its variants: A bibliography. *Pakistan Journal of Statistics and Operation Research* 6: 76–100.

Kilduff, C. 2009. Validation of Leicester cough monitor in idiopathic pulmonary fibrosis sufferers. Cardiff & Vale University Health Board project 09/CMC/4658.

Kinch, A. P., Warltier, R., Taylor, H. et al. 1988. A clinical trial comparing the failure rates of directly bonded brackets using etch times of 15 or 60 seconds. *American Journal of Orthodontics and Dentofacial Orthopedics* 94: 476–483.

King, G. and Zeng, L. 2002. Estimating risk and rate levels, ratios and differences in case-control studies. *Statistics in Medicine* 21: 1409–1427.

King, P. A., Foster, L.V., Yates, R. J. et al. 2012. Survival characteristics of resin-retained bridgework provided at a UK Teaching Hospital. In preparation for *British Dental Journal*.

Kirk, R. E. 1996. Practical significance: A concept whose time has come. *Educational and Psychological Measurement* 56: 746–759.

Koduah, M., Iles, T. C. and Nix, B. J. 2004. Centile charts I: New method of assessment for univariate reference intervals. *Clinical Chemistry* 50: 901–906.

Koopman, P. A. R. 1984. Confidence intervals for the ratio of two binomial proportions. *Biometrics* 40: 513–517.

Kraemer, H. C. 1992. Evaluating medical tests. Objective and quantitative guidelines. Sage, Newbury Park, CA.

Kraemer, H. C. 2006. Correlation coefficients in medical research: From product moment correlation to the odds ratio. *Statistical Methods in Medical Research* 15: 525–545.

Krishnamoorthy, K. and Peng, J. 2007. Some properties of the exact and score methods for binomial proportion and sample size calculation. *Communications in Statistics—Simulation and Computation* 36: 1171–1186.

Kupper, L. L. and Hafner, K. B. 1989. How appropriate are popular sample size formulas? *American Statistician* 43: 101–105.

Kyle, P. M., Campbell, S., Buckley, D. et al. 1996. A comparison of the inactive urinary kallikrein:creatinine ratio and the angiotensin sensitivity test for the prediction of pre-eclampsia. *British Journal of Obstetrics and Gynaecology* 103: 981–987.

Kyle, P. M., Redman, C. W. G., de Swiet, M. et al. 1997. A comparison of the inactive urinary kallikrein:creatinine ratio and the angiotensin sensitivity test for the prediction of pre-eclampsia. *British Journal of Obstetrics and Gynaecology* 104: 971.

Lancaster, H. O. 1949. The combination of probabilities arising from data in discrete distributions. *Biometrika* 36: 370–382.

Laplace, P. S. 1774. Mémoire sur la probabilité des causes par les événements. *Mémoires de l'Academie Royale de Science* 6: 621–656.

Laplace, P. S. 1812. *Théorie Analytique des Probabilités*. Courcier, Paris.

Laupacis, A., Sackett, D. L. and Roberts, R. S. 1988. An assessment of clinically useful measures of the consequences of treatment. *New England Journal of Medicine* 318: 1728–1733.

Layton, D., Wilton, L. V. and Shakir, S. A. W. 2004. Safety profile of celecoxib as used in general practice in England: Results of a prescription-event monitoring study. *European Journal of Clinical Pharmacology* 60: 489–501.

Lee, Y., Shao, J. and Chow, S. C. 2004. Modified large-sample confidence intervals for linear combinations of variance components: Extension, theory, and application. *Journal of the American Statistical Association* 99: 467–478.

Lewis, G. 2010. Personal communication.

Lewis, M. A. O. 1993. Personal communication.

Li, Y., Koval, J. J., Donner, A. et al. 2010. Interval estimation for the area under the receiver operating characteristic curve when data are subject to error. *Statistics in Medicine* 29: 2521–2531.

Liang, K. Y. and Zeger, S. L. 1988. On the use of concordant pairs in matched case-control designs. *Biometrics* 44: 1145–1156.

Liddell, F. D. K. 1983. Simplified exact analysis of case-referent studies: Matched pairs; dichotomous outcome. *Journal of Epidemiology and Community Health* 37: 82–84.

Lin, H. Y., Newcombe, R. G., Lipsitz, S. R. et al. 2009. Fully-specified bootstrap confidence intervals for the difference of two independent binomial proportions based on the median unbiased estimator. *Statistics in Medicine* 28: 2876–2890.

Lindley, D. V. 1965. *Introduction to Probability and Statistics from a Bayesian Viewpoint. Part 2. Inference*. Cambridge University Press: Cambridge, United Kingdom.

Lloyd, C. J. 1990. Confidence intervals from the difference between two correlated proportions. *Journal of the American Statistical Association* 85: 1154–1158.

Lu, Y. and Bean, J. A. 1995. On the sample size for one-sided equivalence of sensitivities based upon McNemar's test. *Statistics in Medicine* 14: 1831–1839.

Lui, K. J. 2004a. *Statistical Estimation of Epidemiological Risk*. Wiley, Chicester, United Kingdom.

Lui, K. J. 2004b. A simple logical solution to eliminate the limitations of using the number needed to treat. *Evaluation and the Health Professions* 27: 206–214.

Lynd, L. D., O'Brien, B. J. 2004. Advances in risk-benefit evaluation using probabilistic simulation methods: An application to the prophylaxis of deep vein thrombosis. *Journal of Clinical Epidemiology* 57: 795–803.

Machin, D., Campbell, M., Tan, S. B. et al. 2009. *Sample Size Tables for Clinical Studies, 3rd Edition*. Wiley-Blackwell, Chichester, United Kingdom.

Malenka, D. J., Baron, J. A., Johansen, S. et al. 1993. The framing effect of relative and absolute risk. *Journal of General Internal Medicine* 10: 543–548.

Mandel, E. M., Bluestone, C. D. and Rockette, H. E. 1982. Duration of effusion after antibiotic treatment for acute otitis media: Comparison of cefaclor and amoxicillin. *Pediatric Infectious Disease Journal* 1: 310–316.

Mann, H. B. and Whitney, D. R. 1947. On a test of whether one of two random variables is stochastically larger than the other. *Annals of Mathematical Statistics* 18: 50–60.

Mansel, R. E., Fallowfield, L., Kissin, M. et al. 2006. Randomized multicenter trial of sentinel node biopsy in breast cancer: The ALMANAC trial. *Journal of the National Cancer Institute* 98: 1–11.

Mantel, N. and Haenszel, W. 1959. Statistical aspects of the analysis of data from retrospective studies of disease. *Journal of the National Cancer Institute* 22: 719–748.

Mayfield, D., McLeod, G. and Hall, P. 1974. The CAGE questionnaire: Validation of a new alcoholism instrument. *American Journal of Psychiatry* 131: 1121–1123.

McDowell, I. F. W., Clark, Z. E., Bowen, D. J. et al. 1998. Heteroduplex analysis for C677T genotyping of methylene tetrahydrofolate reductase: A case-control study in men with angina. *Netherlands Journal of Medicine* 52 (Suppl.): S26–S27.

McGraw, K. O. and Wong, S. P. 1992. A common language effect size statistic. *Psychological Bulletin* 111: 361–365.

Mckeown, S. 2005. *Does Social Capital Protect Against Adolescent Risk Taking Behaviour? An Exploratory Study*. Masters in Public Health dissertation, Cardiff University, Wales.

McNamee, R. 1999. Personal communication.

McNemar, Q. 1947. Note on the sampling error of the difference between correlated proportions or percentages, *Psychometrika* 17, 153–157.

McQuay, H. J. and Moore, R. A. 1997. Using numerical results from systematic reviews in clinical practice. *Annals of Internal Medicine* 126: 712–720.

Mee, R. W. 1984. Confidence bounds for the difference between two probabilities. *Biometrics* 40: 1175–1176.

Mee, R. W. 1990. Confidence intervals for probabilities and tolerance regions based on a generalisation of the Mann-Whitney statistic. *Journal of the American Statistical Association* 85: 793–800.

Mehta, C. R. and Walsh, S. J. 1992. Comparison of exact, mid-P, and Mantel-Haenszel confidence intervals for the common odds ratio across several 2 × 2 contingency tables. *American Statistician* 46: 146–150.

Miettinen, O. and Nurminen, M. 1985. Comparative analysis of two rates. *Statistics in Medicine* 4: 213–226.

Mir, F. 1990. *The Use of Urine and Saliva in Diagnosing HIV Infection for Epidemiological Surveys*. MPH dissertation, University of Wales College of Medicine.

Miyanaga, Y. 1994. Clinical evaluation of the hydrogen peroxide SCL disinfection system (SCL-D). *Japanese Journal of Soft Contact Lenses* 3: 163–173.

Monroe, R. R. 1970. *Episodic Behavioral Disorder: A Psychodynamic and Neurological Analysis*. Harvard University Press, Cambridge, MA.

Monroe, R. R. 1978. *Brain Dysfunction in Aggressive Criminals*. Lexington Books, Lexington, KY.

Nam, J. M. and Blackwelder, W. C. 2002. Analysis of the ratio of marginal probabilities in a matched-pair setting. *Statistics in Medicine* 21: 689–699.

Nam, J. M. 2009. Efficient interval estimation of a ratio of marginal probabilities in matched-pair data: Non-iterative method. *Statistics in Medicine* 28: 2929–2935.

National Radiological Protection Board. 1994. *Guidelines on Radiology Standards for Primary Dental Care*. Documents of the NRPB 5(3).

Newcombe, L. P. 2002. *The Relevance of Social Context in the Education of Adult Welsh Learners with Special Reference to Cardiff*. Ph.D. thesis, Cardiff University, Wales.

Newcombe, R. G. 1979. *A Critical Review of Risk Prediction, with Special Reference to the Perinatal Period*. Ph.D. thesis, University of Wales.

Newcombe, R. G. 1981. A life table for onset of Huntington's Chorea. *Annals of Human Genetics* 45: 375–385.

Newcombe, R. G. 1987. Towards a reduction in publication bias. *British Medical Journal* 295: 656–659.

Newcombe, R. G. and Duff, G. R. 1987. Eyes or patients? Traps for the unwary in the statistical analysis of ophthalmological studies. *British Journal of Ophthalmology* 71: 645–646.

Newcombe, R. G. 1992a. Confidence intervals: enlightening or mystifying? *British Medical Journal* 304: 381–382.

Newcombe, R. G. 1992b. Latin square designs for crossover studies balanced for carryover effects. *Statistics in Medicine* 11: 560.

Newcombe, R. G. 1996. The relationship between chi-square statistics from matched and unmatched analyses. *Journal of Clinical Epidemiology* 49: 1325.

Newcombe, R. G. 1998a. Two-sided confidence intervals for the single proportion: Comparison of seven methods. *Statistics in Medicine* 17: 857–872.

Newcombe, R. G. 1998b. Interval estimation for the difference between independent proportions: Comparison of eleven methods. *Statistics in Medicine* 17: 873–890.

Newcombe, R. G. 1998c. Improved confidence intervals for the difference between binomial proportions based on paired data. *Statistics in Medicine* 17: 2635–2650.

Newcombe, R. G. 1999. Confidence intervals for the number needed to treat—Absolute risk reduction is less likely to be misunderstood. *British Medical Journal* 318: 1765.

Newcombe, R. G. and Altman, D. G. 2000. Proportions and their differences. In *Statistics with Confidence*, 2nd Edition. D. G. Altman, D. Machin and T. N. Bryant et al. (eds.) BMJ Books, London.

Newcombe, R. G. 2001a. Logit confidence intervals and the inverse sinh transformation. *American Statistician* 55: 200–202.

Newcombe, R. G. 2001b. Simultaneous comparison of sensitivity and specificity of two tests in the paired design: Straightforward graphical approach. *Statistics in Medicine* 20: 907–915.

Newcombe, R. G. 2001c. Estimating the difference between differences: Measurement of additive scale interaction for proportions. *Statistics in Medicine* 20: 2885–2893.

Newcombe, R. G. 2003a. Confidence intervals for the mean of a variable taking the values 0, 1 and 2. *Statistics in Medicine* 22: 2737–2750.

Newcombe, R. G. 2003b. Confidence limits for the ratio of two rates based on likelihood scores: Non-iterative method. *Statistics in Medicine* 22: 2085–2086.

Newcombe, R. G. 2006a. Confidence intervals for an effect size measure based on the Mann-Whitney statistic. Part 1: General issues and tail area based methods. *Statistics in Medicine* 25: 543–557.

Newcombe, R. G. 2006b. Confidence intervals for an effect size measure based on the Mann-Whitney statistic. Part 2: Asymptotic methods and evaluation. *Statistics in Medicine* 25: 559–573.

Newcombe, R. G. 2006c. A deficiency of the odds ratio as a measure of effect size. *Statistics in Medicine* 25: 4235–4240.

Newcombe, R. G. 2007a. Bayesian estimation of false negative rate in a clinical trial of sentinel node biopsy. *Statistics in Medicine* 26: 3429–3442.

Newcombe, R. G. 2007b. Comments on "Confidence intervals for a ratio of binomial proportions based on paired data". *Statistics in Medicine* 26: 4684–4685.

Newcombe, R. G. 2007c. An evaluation of 14 confidence interval methods for the single proportion. Results published in part as Newcombe (2011b), the remainder is unpublished.

Newcombe, R. G. 2007d. Absolute risk reduction. In *Wiley Encyclopedia of Clinical Trials*. R. B. D'Agostino, L. Sullivan and J. M. Massaro (eds.). Wiley, Hoboken, NJ.

Newcombe, R. G. 2007e. A relative measure of effect size for paired data generalising the Wilcoxon matched-pairs signed-ranks test statistic. International Society for Clinical Biostatistics conference, Alexandroupolis, Greece.

Newcombe, R. G., Stroud, A. E. and Wiles, C. M. 2008a. How much does prior information sway the diagnostic process? Test Evaluation Symposium, University of Birmingham.

Newcombe, R. G. 2008b. A second odds ratio paradox. Unpublished.

Newcombe, R. G. 2008c. Nine out of ten cat owners … Reporting proportions and related quantities. In *Public Opinion Research Focus*. L. O. Petrieff and R. V. Miller (eds.). Nova, Hauppauge, NY, pp. 123–134.

Newcombe, R. G. and Farrier, S. L. 2008d. A generalisation of the tail-based p-value to characterise the conformity of trinomial proportions to prescribed norms. *Statistical Methods in Medical Research* 17: 609–619.

Newcombe, R G. 2010a. A relative measure of effect size for paired data generalising the Wilcoxon matched-pairs signed-ranks test statistic. In *JSM Proceedings*, WNAR. Alexandria, VA: American Statistical Association, 1254–1268.

Newcombe, R. G. and Nurminen, M. M. 2011a. In defence of score intervals for proportions and their differences. *Communications in Statistics—Theory & Methods* 40: 1271–1282.

Newcombe, R. G. 2011b. Measures of location for confidence intervals for proportions. *Communications in Statistics—Theory & Methods* 40: 1743–1767.

Newcombe, R. G. 2011c. Propagating imprecision: Combining confidence intervals from independent sources. *Communications in Statistics—Theory & Methods* 40: 3154–3180.

Newson, R. 2002. Parameters behind "nonparametric" statistics: Kendall's tau, Somers' D and median differences. *Stata Journal* 2: 45–64.

NHS Breast Screening Programme. 1996. *Quality Assurance Guidelines for Surgeons in Breast Cancer Screening*. NHSBSP Publications, 20.

Noether, G. E. 1967. *Elements of Nonparametric Statistics*. Wiley, New York.

Noether, G. E. 1987. Sample size determination for some common nonparametric tests. *Journal of the American Statistical Association* 82: 645–647.

Norton, P. G. and Dunn, E. V. 1985. Snoring as a risk factor for disease: an epidemiological survey. *British Medical Journal* 201: 630–632.

Obuchowski, N. A. and Lieber, M. L. 2002. Confidence bounds when the estimated ROC area is 1.0. *Acta Radiologica* 9: 526–530.

O'Connor, A. M., Bennett, C. L., Stacey, D. et al. 2009. *Decision Aids for People Facing Health Treatment or Screening Decisions (Review)*. The Cochrane Library, Issue 3.

Office for Official Publications of the European Communities. European Communities. 2004. *Radiation Protection. European Guidelines on Radiation Protection in Dental Radiology. The Safe use of Radiographs in Dental Practice*. Issue 136.

Oldroyd, J., White, M., Unwin, N. C. et al. 1999. Changes in insulin sensitivity in people with impaired glucose tolerance: 6 months follow-up of a randomised controlled trial. *Diabetes Medicine* 16 (Suppl. 1): 217.

Olejnik, S. and Algina, J. 2000. Measures of effect size for comparative studies: Applications, interpretations and limitations. *Contemporary Educational Psychology* 25: 241–286.

Owen, D. B. 1956. Tables for computing bivariate normal probabilities. *Annals of Mathematical Statistics* 27: 1075–1090.

Pal, N., Axisa, B., Yusof, S. et al. 2008. Volume and outcome for major upper GI surgery in England. *Journal of Gastrointestinal Surgery* 12: 353–357.

Parzen, M., Lipsitz, S., Ibrahim, J. et al. 2002. An estimate of the odds ratio that always exists. *Journal of Computational and Graphical Statistics* 11: 420–436.

Paul, S. R. and Zaihra, T. 2008. Interval estimation of risk difference for data sampled from clusters. *Statistics in Medicine* 27: 4207–4220.

Pearson, K. 1913. On the measurement of the influence of "broad categories" on correlation. *Biometrika* 9: 116–139.

Peattie, S., Peattie, K., Newcombe, R. G. et al. 2011. Social marketing to extinguish fire-setting behaviour. In *Proceedings of the 10th International Congress of the International Association on Public and Nonprofit Marketing*. I. V. Pereira (ed). Oporto, Portugal.

Pei, Y., Tang, M. L. Wong, W. K. et al. 2012. Confidence intervals for correlated proportion differences from paired data in a two-arm randomised clinical trial. *Statistical Methods in Medical Research* 21: 167–187.

Pepe, M. S. 1992. Inference using surrogate outcome data and a validation sample. *Biometrika* 79: 355–365.

Pereira, I. V. (ed). 2011. *Proceedings of the 10th International Congress of the International Association on Public and Nonprofit Marketing*. Oporto, Portugal, June 16–17, 2011.

Peskun, P. 1993. A new confidence interval method based on the normal approximation for the difference of two binomial probabilities. *Journal of the American Statistical Association* 88: 656–661.

Petrieff, L. O. and Miller, R. V. (eds.). 2008. *Public Opinion Research Focus*. Nova, Hauppauge, NY.

Pires, A. M. and Amado, C. 2008. Interval estimators for a binomial proportion: Comparison of twenty methods. *REVSTAT Statistical Journal* 6: 165–197.

Plugge, E., Yudkin, P. and Douglas, N. 2009. Changes in women's use of illicit drugs following imprisonment. *Addiction* 104: 215–222.

Pratt, J. W. 1968. A normal approximation for binomial, F, beta, and other common, related tail probabilities. II. *Journal of the American Statistical Association* 63: 1457–1483.

Price, R. M. and Bonett, D. G. 2008. Confidence intervals for a ratio of two independent binomial proportions. *Statistics in Medicine* 27: 5497–5508.

Rae, D. 1980. An altimeter for Mr. Escher's Stairway: A comment on William H. Riker's "Implications from the Disequilibrium of Majority Rule for the Study of Institutions". *American Political Science Review* 74: 451–455.

Ramasundarahettige, C. F., Donner, A. and Zou, G. Y. 2009. Confidence interval construction for a difference between two dependent intraclass correlation coefficients. *Statistics in Medicine* 28: 1041–1053.

Rascol, O., Brooks, D., Korczyn, A. D. et al. 2000. A five year study of the incidence of dyskinesia in patients with early Parkinson's disease who were treated with ropinirole or levodopa. *New England Journal of Medicine* 342: 1484–1491.

Reiczigel, J. 2003. Confidence intervals for the binomial parameter: Some new considerations. *Statistics in Medicine* 22: 611–621.

Reiczigel, J., Rozsa, L. and Zakarias, I. 2003. Bootstrap Wilcoxon-Mann-Whitney test for use in relation with non-shift alternatives. *Controlled Clinical Trials* 24: (Suppl. 3): p. 291.

Reiczigel, J. and Abonyi-Toth, Z. 2008. Confidence sets for two binomial proportions and confidence intervals for the difference and/or ratio of proportions. *Computational Statistics Data & Analysis* 52: 5046–5053.

Reilly, M., Daly, L. and Hutchinson, M. 1993. An epidemiological study of Wilson's disease in the Republic of Ireland. *Journal of Neurology, Neurosurgery & Psychiatry* 56: 298–300.

Richardson, G., Van Woerden, H., Edwards, R. et al. 2011. Community based cardiovascular risk reduction—Age and the Framingham risk score. *British Journal of Cardiology* 18: 180–184.

Rindskopf, D. 2000. Approximate is better than "exact" for interval estimation of binomial proportions. *American Statistician* 54: 88.

Rindskopf, D. 2010. Logistic regression with floor and ceiling effects. In *JSM Proceedings*, WNAR. Alexandria, VA: American Statistical Association, 806–815.

Rittgen, W. and Becker, N. 2000. SMR analysis of historical follow-up studies with missing death certificates. *Biometrics* 56: 1164–1169.

Roberts, A. G., Elder, G. H., Newcombe, R. G. et al. 1988. Heterogeneity of familial porphyria cutanea tarda. *Journal of Medical Genetics* 25: 669–676.

Rothman, K. J. and Greenland, S. 1998. *Modern Epidemiology*, 2nd Edition. Lippincott-Raven, Philadelphia.

Rümke, C. L. 1974. Implications of the statement: No side effects were observed. *New England Journal of Medicine* 292: 372–373.

Ruxton, G. D. and Neuhäuser, M. 2010. When should we use one-tailed hypothesis testing? *Methods in Ecology & Evolution* 1: 114–117.

Ryu, E. J. and Agresti, A. 2008. Modeling and inference for an ordinal effect size measure. *Statistics in Medicine* 27: 1703–1717.

Sackett, D. L., Deeks, J. J. and Altman, D. G. 1996. Down with odds ratios! *Evidence-Based Medicine* 1: 164–166.

Santner, T. J. and Duffy, D. E. 1989. *The Statistical Analysis of Discrete Data*. Springer, New York.

Santner, T. J., Pradhan, V., Senchaudhuri, P. et al. 2007. Small-sample comparisons of confidence intervals for the difference of two independent binomial proportions. *Computational Statistics & Data Analysis* 51: 5791–5799.

Schaeffer, J., Burch, N., Björnsson, Y. et al. 2007. Checkers is solved. *Science* 317: 1518–1522.

Schechtman, E. 2002. Odds ratio, relative risk, absolute risk reduction, and the number needed to treat—Which of these should we use? *Value in Health* 5: 431–436.

Schiff, T., Dotson, M., Cohen, S. et al. 1994. Efficacy of a dentifrice containing potassium nitrate, soluble pyrophosphate, PVM/MA copolymer, and sodium fluoride on dentinal hypersensitivity: A twelve-week clinical study. *Journal of Clinical Dentistry* 5: 87–92.

Schiff, T., Delgado, E., Zhang, Y. P. et al. 2009. The clinical effect of a single direct topical application of a dentifrice containing 8.0% arginine, calcium carbonate, and 1450 ppm fluoride on dentin hypersensitivity: The use of a cotton swab applicator versus the use of a fingertip. *Journal of Clinical Dentistry* 20: 131–136.

Schlesselman, J. J. 1982. *Case-Control Studies: Design, Conduct, Analysis*. Oxford University Press, New York.

Schräder, P., Grouven, U. and Bender, R. 2007. Können Mindestmengen für Knieprothesen anhand von Routinedaten berechnet werden? Ergebnisse einer Schwellenwertanalyse mit Daten der externen stationären Qualitätssicherung. *Orthopäde* 36: 570–576.

Schulz, K. F., Altman, D. G. and Moher, D. for the CONSORT Group. 2010. CONSORT 2010 Statement: Updated guidelines for reporting parallel group randomised trials. *British Medical Journal* 340: c332.

Schwartz, L. M., Woloshin, S. and Welch, H. G. 1999. Misunderstandings about the effects of race and sex on physicians' referrals for cardiac catheterization. *New England Journal of Medicine* 341: 279–283.

Scott, W. A. 1955. Reliability of content analysis: The case of nominal scale coding. *Public Opinion Quarterly* 19: 321–325.

Sen, P. K. 1963. On the estimation of relative potency in dilution (-direct) assays by distribution-free methods. *Biometrics* 19: 532–552.

Shapiro, C. M., Beckmann, E., Christiansen, N. et al. 1987. Immunologic status of patients in remission from Hodgkin's disease and disseminated malignancies. *American Journal of the Medical Sciences* 293, 366–370.

Shi, L. and Bai, P. 2008. Bayesian confidence interval for the difference of two proportions in the matched-paired design. *Communications in Statistics—Theory and Methods* 37: 2034–2051.

Shi, L., Sun, H. Y. and Bai, P. 2009. Bayesian confidence interval for difference of the proportions in a 2×2 table with structural zero. *Journal of Applied Statistics* 36: 483–494.

Simonoff, J. S., Hochberg, Y. and Reiser, B. 1986. Alternative estimation procedures for Pr(X<Y) in categorized data. *Biometrics* 42: 895–907.

Sinclair, J. C. and Bracken, M. B. 1994. Clinically useful measures of effect in binary analyses of randomized trials. *Journal of Clinical Epidemiology* 47: 881–889.

Smith, S. G., Touquet, R., Wright, S. et al. 1996. Detection of alcohol misusing patients in accident and emergency departments: The Paddington Alcohol Test. *Journal of Accident and Emergency Medicine* 5: 308–312.

Smyth, E. T. M., McIlvenny, G., Enstone, J. et al. 2008. Four Country Healthcare Associated Infection Prevalence Survey 2006: Overview of the Results. *Journal of Hospital Infection* 69: 230–248.

Somers, R. H. 1962. A new asymmetric measure of association for ordinal variables. *American Sociological Review* 27: 799–811.

Stang, A., Poole, C. and Bender, R. 2010. Common problems related to the use of number needed to treat *Journal of Clinical Epidemiology* 63: 820–825.

Statistical Solutions. 2009. *nQuery Advisor 7.0*. Boston, http://www.statistical-solutions-software.com/products-page/nquery-advisor-sample-size-software/.

Stone, M. 1969. The role of significance testing. Some data with a message. *Biometrika* 56: 485–493.

Stroud, A. E., Lawrie, B. W. and Wiles, C. M. 2002. Inter and intra rater reliability of cervical auscultation to detect aspiration in patients with dysphagia. *Clinical Rehabilitation* 16: 640–645.

Tang, M. L., Tang, N. S. and Chan, I. S. F. 2005. Confidence interval construction for proportion difference in small sample paired studies. *Statistics in Medicine* 24: 3565–3579.

Tang, M. L., Ling, M. H., Ling, L. et al. 2010a. Confidence intervals for a difference between proportions based on paired data. *Statistics in Medicine* 29: 86–98.

Tang, M. L., Li, H. Q. and Tang, N. S. 2010b. Confidence interval construction for proportion ratio in paired studies based on hybrid method. *Statistical Methods in Medical Research*, first published on September 27, 2010 as doi:10.1177/0962280210384714.

Tang, M. L., Qiu, S. F. and Poon, W. Y. 2012. Confidence interval construction for disease prevalence based on partial validation series. *Computational Statistics and Data Analysis* 56: 1200–1220.

Tang, M. L., He, X. and Tian, G. L. 2012. A confidence interval approach for equivalence/noninferiority trials: An application in otorhinolaryngologic study. *Communications in Statistics—Simulation and Computation*, in press.

Tang, N. S. and Tang, M. L. 2003. Statistical inference for risk difference in an incomplete correlated 2×2 table. *Biometrical Journal* 45: 34–46.

Tang, N. S., Tang, M. L. and Carey, V. J. 2004. Confidence interval for rate ratio in a 2×2 table with structural zero: An application in assessing false-negative rate ratio when combining two diagnostic tests. *Biometrics* 60: 550–555.

Tang, N. S, Qiu, S. F., Tang, M. L. et al. 2011. Asymptotic confidence interval construction for proportion difference in medical studies with bilateral data. *Statistical Methods in Medical Research* 20: 233–259.

Tango, T. 1998. Equivalence test and confidence interval for the difference in proportions for the paired-sample design. *Statistics in Medicine* 17: 891–908.

Tango, T. 1999. Improved confidence intervals for the difference between binomial proportions based on paired data. *Statistics in Medicine* 18: 3511–3513.

Teeling-Smith, G. and Wells, N. (eds.). 1979. *Medicines for the Year 2000: A Symposium Held at the Royal College of Physicians*. London: Office of Health Economics.

Tenenbein, A. A. 1970. A double sampling scheme for estimating from binomial data with misclassifications. *Journal of the American Statistical Association* 65: 1350–1361.

Tramèr, M. R., Moore, R. A., Reynolds, D. J. M. et al. 1997. A quantitative systematic review of ondansetron in treatment of established postoperative nausea and vomiting. *British Medical Journal* 314: 1088–1092.

Tryon, W. W. 2001. Evaluating statistical difference, equivalence, and indeterminacy using inferential confidence intervals: An integrated alternative method of conducting null hypothesis statistical tests. *Psychological Methods* 6: 371–386.

Tryon, W. W. and Lewis, C. 2009. Evaluating independent proportions for statistical difference, equivalence, indeterminacy, and trivial difference using inferential confidence intervals. *Journal of Educational and Behavioral Statistics* 34: 171–189.

Tukey, J. W. 1949. The simplest signed-rank tests. Memo Report 17, Statistical Research Group, Princeton University.

Turnbull, L. W., Brown, S. R., Olivier, C. et al. 2010. Multicentre randomised controlled trial examining the cost-effectiveness of contrast-enhanced high field magnetic resonance imaging in women with primary breast cancer scheduled for wide local excision (COMICE). *Health Technology Assessment* 14: 1.

Turnbull, P. J., Stimson, G. V. and Dolan, K A. 1992. Prevalence of HIV infection among ex-prisoners in England. *British Medical Journal* 304: 90–91.

Tuyl, F. A. W. M. 2007. *Estimation of the Binomial Parameter: In Defence of Bayes*. Ph.D. thesis, University of Newcastle, New South Wales, Australia.

Ukoli, F. A, Adams-Campbell, L. L., Ononu, J. et al. 1993. Nutritional status of urban Nigerian school children relative to the NCHS reference population. *East African Medical Journal* 70: 409–413.

Vach, W., Gerke, O. and Høilund-Carlsen, P. F. 2012. Three principles to define the success of a diagnostic study could be identified. *Journal of Clinical Oncology* 65: 293–300.

Vaeth, M. and Poulsen, S. 1998. Comments on a commentary: Statistical evaluation of split mouth caries trials. *Community Dentistry and Oral Epidemiology* 26: 80–83.

Van Rooyen, S., Godlee, F. et al. 1999. Effect of open peer review on quality of reviews and on reviewers' recommendations: a randomised trial. *British Medical Journal* 318: 23–27.

Vargha, A. and Delaney, H. D. 2000. A critique and improvement of the CL common language effect size statistics of McGraw and Wong. *Journal of Educational and Behavioral Statistics* 25: 101–132.

Vollset, S. E. 1993. Confidence intervals for a binomial proportion. *Statistics in Medicine* 12: 809–824.

Wald, A. 1943. Tests of statistical hypotheses concerning several parameters when the number of observations is large. *Transactions of the American Mathematical Society* 54: 426–482.

Walsh, J. E. 1949. Some significance tests for the median which are valid under very general conditions. *Annals of Mathematical Statistics* 20: 64–81.

Walters, D. E. 1985. An examination of the conservative nature of "classical" confidence limits for a proportion. *Biometrical Journal* 27: 851–861.

Watson, G. S. and Nguyen, H. 1985. A confidence region in a ternary diagram from point counts. *Journal of Mathematical Geology* 17: 209–213.

Watson, G. S. 1987. Confidence regions in ternary diagrams 2. *Journal of Mathematical Geology* 19: 347–348.

Wellcome Trust Case Control Consortium. 2007. Genome-wide association study of 14,000 cases of seven common diseases and 3,000 shared controls. *Nature* 447: 661–678.

Weltje, G. J. 2002. Quantitative analysis of detrital modes: Statistically rigorous confidence regions in ternary diagrams and their use in sedimentary petrology. *Earth Science Reviews* 57: 211–253.

Wichmann, B. A. and Hill, I. D. 1985. An efficient and portable pseudo-random number generator. In *Applied Statistics Algorithms*. P. Griffiths and I. D. Hill (eds.). Ellis Horwood, Chichester, United Kingdom.

Wilcoxon, F. 1945. Individual comparisons by ranking methods. *Biometrics Bulletin* 1: 80–83.

Wiles, C. M., Newcombe, R. G., Fuller, K. J. et al. 2001. A controlled randomised crossover trial of the effects of physiotherapy on mobility in chronic multiple sclerosis. *Journal of Neurology, Neurosurgery and Psychiatry* 70:174–179.

Wilson, E. B. 1927. Probable inference, the law of succession, and statistical inference. *Journal of the American Statistical Association* 22: 209–212.

Woolf, B. 1955. On estimating the relationship between blood group and disease. *Annals of Human Genetics* 19: 251–253.

Yates, R., Moran, J., Addy, M. et al. 1997. The comparative effect of acidified sodium chlorite and chlorhexidine mouthrinses on plaque removal and salivary bacterial counts. *Journal of Clinical Periodontology* 24: 603–609.

Yule, G. U. 1900. On the association of attributes in statistics. *Philosophical Transactions of the Royal Society of London, Series A* 194: 257–319.

Zhang, J. and Yu, K. F. 1998. What's the relative risk. *Journal of the American Medical Association* 280: 1690–1691.

Zhou, W. 2008. Statistical inference for $P(X<Y)$. *Statistics in Medicine* 27: 257–279.

Zhou, X. H., Li, C. M. and Yang, Z. 2008. Improving interval estimation of binomial proportions. *Philosophical Transactions of the Royal Society, Series A* 366: 2405–2418.

Zhu, Y., Xu, Y., Wei, Y. et al. 2008. Association of IL-1B gene polymorphisms with nasopharyngeal carcinoma in a Chinese population. *Clinical Oncology* 20: 207–211.

Zou, G. Y. and Donner, A. 2004. A simple alternative confidence interval for the difference between two proportions. *Controlled Clinical Trials* 25: 3–12.

Zou, G. Y. 2007. Toward using confidence intervals to compare correlations. *Psychological Methods* 12: 399–413.

Zou, G. Y. and Donner, A. 2008a. Construction of confidence limits about effect measures: A general approach. *Statistics in Medicine* 27: 1693–1702.

Zou, G. Y. 2008b. On the estimation of additive interaction by use of the four-by-two table and beyond. *American Journal of Epidemiology* 168: 212–224.

Zou, G. Y., Huang, W. and Zhang X. 2009a. A note on confidence interval estimation for a linear function of binomial proportions. *Computational Statistics & Data Analysis* 53: 1080–1085.

Zou, G. Y., Taleban, J. and Huo, C. Y. 2009b. Confidence interval estimation for lognormal data with application to health economics. *Computational Statistics and Data Analysis* 53: 3755–3764.

Zou, G. Y., Huo, C. Y. and Taleban, J. 2009c. Simple confidence intervals for lognormal means and their differences with environmental applications. *Environmetrics* 20: 172–180.

Zou, G. Y. 2009d. Assessment of risks by predicting counterfactuals. *Statistics in Medicine* 28: 3761–3781.

Zou, G. Y. and Donner, A. 2010a. A generalization of Fieller's theorem for ratios of non-normal variables and some practical applications. In *Biometrics: Methods, Applications and Analyses*. H. Schuster and W. Metzger (eds.). Nova, Hauppauge, NY, pp. 197–216.

Zou, G. Y., Donner, A. and Taleban, J. 2010b. A dirty dozen: Confidence interval estimation for 12 parameters in the one-way random effects model. Paper presented at the Joint Statistical Meeting, Vancouver.

Zou, G. Y. 2010c. Confidence interval estimation under inverse sampling. *Computational Statistics and Data Analysis* 54: 55–64.

Zou, G. Y. 2011. Confidence interval estimation for the Bland-Altman limits of agreement with multiple observations per individual. *Statistical Methods in Medical Research*. Published online June 24, 2011, DOI:10.1177/0962280211402548.

Zou, G. Y. 2012. Sample size formulas for estimating intraclass correlation coefficients with precision and assurance. *Statistics in Medicine*, in submission.

Online Documents (All Accessed December 2, 2011)

Anonymous. 2011a. Voting paradox. http://en.wikipedia.org/wiki/Voting_paradox

Anonymous. 2011b. First-move advantage in chess. http://en.wikipedia.org/wiki/First-move_advantage_in_chess#cite_note-173

Cumming, G. 2011a. ABC Radio broadcast October 9, 2011. http://www.abc.net.au/rn/ockhamsrazor/stories/2011/3333636.htm

Dewey, ME. 2001. Who was Bonferroni? http://www.aghmed.fsnet.co.uk/bonf/feb01s.pdf

Forte, B. 2002. A4 vs US letter. http://betweenborders.com/wordsmithing/a4-vs-us-letter/

Goertzel, T. G. 2010. Guide to computing margins of error for percentages and means. http://crab.rutgers.edu/~goertzel/marginsoferror.htm

Kissock, K. 2007. Average daily temperature archive. http://academic.udayton.edu/kissock/http/Weather/default.htm

Macdonald, P. D. M. 1997. Confidence intervals for the mean of a Poisson distribution. http://www.math.mcmaster.ca/peter/s743/poissonalpha.html

Newcombe, R. G. 2010b. A simple demonstration of Pythagoras' theorem using P-pentominos. https://sites.google.com/site/ppentominoesandpythagoras

Sullivan, K. 2007. SMR analysis version 4.11.19. http://www.sph.emory.edu/~cdckms/exact-midP-SMR.html

UK Biobank. 2006. UK Biobank: Protocol for a large-scale prospective epidemiological resource. http://www.ukbiobank.ac.uk/docs/UKBProtocol.pdf

Vlachos, P. 2005. StatLib—Applied statistics algorithms. http://lib.stat.cmu.edu/apstat/

Appendix 1: Glossary of Some Statistical Terms

Binary variable: A variable which can take only two values, typically 1 (denoting a positive response) and 0 (denoting a negative response).

Binomial distribution: The distribution normally used to model a *binary variable*. Let n denote the sample size, and let π denote the true probability of a positive response. Let R denote the number of positive responses in the sample.

Then $\Pr[R = r] = \binom{n}{r} \pi^r (1-\pi)^{n-r}$ for $r = 0, 1, \ldots, n$.

Here, $\binom{n}{r}$ denotes $\frac{n!}{r!(n-r)!}$ where $n! = n \times (n-1)!$ and $1! = 0! = 1$.

Bonferroni correction: The mathematically and conceptually simplest procedure to rescale significance levels to adjust for *multiple comparisons*. In its simplest form, if k hypothesis tests are carried out, the p-value is required to be below α/k, not α, to be regarded as statistically significant. A slight refinement uses $1 - (1 - \alpha)^{1/k}$ instead of α/k.

Boundary-respecting: For some parameters, the support space has a lower bound or an upper bound, or both. For example, *proportions* must lie between 0 and 1, *odds ratios* and *relative risks* must be non-negative. Boundary-respecting confidence interval methods ensure that calculated lower and upper limits both lie within the range of validity.

Chi-square tests: A class of elementary hypothesis tests commonly used for binary or categorical data, in which the test statistic has an asymptotic chi-square distribution with ν degrees of freedom on the null hypothesis. The ν does not depend on the sample size, but is derived from the numbers of rows and columns of the contingency table summarising the data. For the simplest case of a 2 × 2 table, $\nu = 1$.

Coefficient of variation: The coefficient of variation is often used to quantify the degree of variation in relative terms. It is defined as the SD divided by the mean, and is usually expressed as a percentage.

Confidence interval: A confidence interval for a parameter θ is a range of values, derived from a sample of data, which is designed to have a prespecified probability $1-\alpha$ of including the true value of θ. Confidence intervals are usually two-sided, extending both below and above the point estimate. The two values L and U delimiting the interval are called confidence limits. The 95% confidence intervals are most commonly used, with $1\text{-}\alpha = 0.95$.

Conservative: A confidence interval method is conservative if the attained coverage is usually greater than the nominal coverage level.

Coverage probability: The probability that a confidence interval for a parameter θ attains its objective of including the true value. For proportions and related quantities, the coverage probability is an erratic function of θ; consequently the nominal coverage probability $1\text{-}\alpha$ is generally not achieved. Some statisticians maintain that the nominal coverage level should represent a minimum over all possible parameter values. Others seek to align $1\text{-}\alpha$ with the mean coverage over an appropriate range of parameter values.

Credible interval: Credible intervals are the Bayesian analogue of confidence intervals. A 95% credible interval is delimited by the 2.5 and 97.5 centiles of the posterior distribution. Bayesian intervals constructed in this way often have good properties when viewed as confidence intervals.

Delta method: A method to calculate an asymptotic standard error, confidence limits or hypothesis test based on an algebraic function of a quantity whose standard error is known. See Section 3.4.2 for an algebraic definition.

Deviance: A deviance criterion based on minus twice the log of the maximised likelihood is used to compare goodness of fit of nested models such as logistic models.

Distal and mesial location: Some parameters have a symmetrical support scale. For example, the support for the binomial parameter is the interval [0, 1], with a natural symmetry about the midpoint of the scale at 0.5. Suppose the true proportion π is 0.3. If the confidence interval for π based on a sample is 0.1 to 0.2, this is too far away from the scale midpoint to include π. This interval is said to be too distal: mesial non-coverage has occurred. Conversely, an interval from 0.4 to 0.7 is too mesially located to include π: this is distal non-coverage.

Exact: The Clopper–Pearson interval for the simple binomial proportion π is often described as "exact", in two senses. Each limit may be derived by equating a sum of exactly calculated probabilities in the tail of the distribution with $\alpha/2$. Furthermore, the resulting interval is strictly *conservative*. Neither of these properties implies that the coverage of the interval is precisely at the nominal level α for all values of π.

Homoscedastic: A homoscedastic model postulates an equal degree of variation in two or more groups; typically $\sigma_1 = \sigma_2 = \ldots = \sigma_k$.

Hypothesis test: A hypothesis test sets up two competing hypotheses, the null hypothesis H_0 and the alternative hypothesis H_1 to account for the data observed in a sample. By default, H_0 states that there is no difference (between two means, between two proportions, etc.) or no association (between two variables), in the population from which the sample is drawn, and H_1 states that H_0 is false. For illustration, suppose the test aims to demonstrate that two means or two proportions differ. A p-value is calculated, which is the probability of observing a difference as extreme as that calculated from the sample, or more so, if H_0 is true. If the p-value is below some chosen value α (often 0.05), H_0 is rejected and the difference is reported to be statistically significant.

Kruskal–Wallis test: The Kruskal–Wallis test is the non-parametric analogue of one-way analysis of variance. It is a generalisation of the Mann–Whitney test, and is used to compare location between three or more independent samples.

Mann–Whitney test: A commonly used non-parametric alternative to the t-test comparing two independent samples, $\{X_i, i = 1, \ldots, m\}$ and $\{Y_j, j = 1, \ldots, n\}$. In the usual test formulation, the test statistic is the quantity U which counts the number of pairs (X_i, Y_j) with $X_i > Y_j$ plus half the number of pairs with $X_i = Y_j$. An equivalent quantity U/mn may be used as a corresponding effect size measure; it is equivalent to the area under the *ROC curve*.

Markov Chain Monte Carlo (MCMC): A computationally intensive Bayesian method of analysis, applicable to a wide range of models. MCMC is used to construct the posterior distribution and derived quantities including low and high centiles delimiting a credible interval.

Mesial: See distal.

Mantel–Haenszel method: A procedure to calculate a pooled odds ratio in a stratified analysis, combining information across several fourfold tables.

Mixed model: A mixed model is a statistical model containing both fixed effects and random effects. Mixed models relate particularly to settings where repeated measurements are made on the same units, or where measurements are made on clusters of related units.

Multiple comparisons: In many studies, more than one hypothesis test is carried out. Consequently the probability on H_0 that at least one of these tests will yield a statistically significant difference is much greater than α. This phenomenon is a major factor impacting on the use of hypothesis tests for drawing inferences. Several approaches to mitigate this are available, including defining a limited set of comparisons as of prior importance, and procedures to rescale significance levels such as the *Bonferroni correction*.

Nuisance parameter: A nuisance parameter is a parameter in a model which is not of primary concern to estimate, yet which plays an important role in the model. For example, in a comparison of two binomial proportions π_1 and π_2, interest may centre on the difference, $\Delta = \pi_1 - \pi_2$. The sum of the two proportions $\pi_1 + \pi_2$ (or equivalently their average) then has the status of a nuisance parameter. It imposes constraints on Δ, for example $|\Delta| \leq \pi_1 + \pi_2$. Several approaches are available to make inferences about Δ in the presence of such a nuisance parameter.

Number needed to treat: Let p_1 and p_2 denote the proportions of cases with an adverse outcome in two treatment groups 1 and 2, typically in a clinical trial. Then the number needed to treat (NNT) is defined as $\frac{1}{p_1 - p_2}$. For example, if $p_1 = 0.4$ and $p_2 = 0.2$, NNT = 5, meaning that five patients need to be treated by method 2 instead of method 1 to prevent one adverse outcome.

Odds ratio: A ratio measure comparing two proportions π_1 and π_2, defined as $\frac{\pi_1/(1-\pi_1)}{\pi_2/(1-\pi_2)}$.

The odds ratio has several advantages over the simple ratio of the two proportions, π_1/π_2. Unlike the relative risk, it can be estimated in a retrospective case-control study, and it plays an important role in logistic regression. However, it also has several disadvantages.

One-sided test: A test in which the alternative hypothesis H_1 is directional, usually of the form $\Delta > 0$ or $\Delta < 0$. When an asymptotic test is used the resulting one-sided *p*-value is half the usual two-sided *p*-value.

***p*-value:** See hypothesis test.

Power: The power of a *hypothesis test* is the probability of correctly rejecting the null hypothesis when it is false.

Proportion: The term "proportion" is commonly used in two quite different senses.

Consider the following examples.

1. In a series of 200 premenopausal women with breast cancer, 10 have the BRCA1 gene, and the remaining 190 do not. The proportion in this series is 5%.
2. Water forms a 60% proportion of the weight of the adult human body.

In this book, the word "proportion" is always used with a meaning as in example 1, to summarise a *binary variable.*

Receiver operating characteristic (ROC) curve: A graphical display of the degree of separation between two samples for a continuous or discrete variable. When applied to variables used in medical tests, the ROC plots *sensitivity* against 1-*specificity* for different cutoff values of the variable.

Relative risk or risk ratio: The ratio of the probabilities of an unfavourable outcome in two groups.

Robustness: Robust statistical methods do not depend heavily on the data conforming closely to the assumptions embodied in the model. For example, inferences concerning means based on the t-distribution are quite robust to departures from Gaussian distributional form, whilst inferences concerning variances based on the chi-square distribution are not robust. Non-parametric or distribution-free methods are designed to be more robust than their parametric counterparts.

Sensitivity and specificity: Two proportions used to characterise the performance of a diagnostic or screening test. The sensitivity measures what proportion of diseased individuals are correctly classified as such by the test. The specificity measures what proportion of unaffected individuals are correctly classified as such by the test. It is desirable for both sensitivity and specificity to be high. But there is a trade-off between them; moving a test threshold to increase sensitivity reduces the specificity, and vice versa.

Skewness: Many frequency distributions are poorly approximated by symmetrical models such as the Gaussian distribution. A positively skew distribution has a long tail to the right. A negatively skew distribution has a long tail to the left. Generally, if a physiological variable tends to be lowered in disease (such as haemoglobin or birthweight), the resulting distribution is negatively skewed. If it tends to be increased (such as bacterial colony counts), a positively skew distribution results. The lognormal distribution is positively skewed.

Standard error: The standard error of an estimate expresses its imprecision due to being based on a finite sample size. For example, the standard error of a sample mean is $s/\sqrt{n}$ where s denotes the standard deviation and n denotes the sample size. The standard error is used to calculate confidence intervals and hypothesis tests.

Standardised difference: This measure, sometimes referred to as Cohen's d, is the difference between two means divided by the within-groups standard deviation.

Sufficient statistics: In many models, the likelihood function reduces to a function of a small number of sufficient statistics which contain all the information in the sample that is relevant for inferences. For the binomial distribution, given the sample size n, the number of positive responses r, or equivalently the sample proportion $p = r/n$, is sufficient for inferences concerning the binomial parameter π. For the Gaussian distribution, the sample mean and SD are jointly sufficient.

***t*-test:** A class of elementary hypothesis tests commonly used to compare means of two samples, for the usual situation where the standard deviation is unknown and needs to be estimated from the data. Tests are available for both independent and individually paired samples. The test statistic, calculated by dividing the observed difference by its standard error, is referred to the central t-distribution with ν degrees of freedom. ν is derived from the sample size(s) and is usually large.

Two-sided test: In most contexts, *hypothesis tests* are assumed to be two-sided by default, evaluating a null hypothesis of the form $H_0 : \Delta = 0$ against an alternative H_1: $\Delta \neq 0$ (i.e., Δ could be either positive or negative). Two-sided hypothesis tests have a natural duality with the usual two-sided *confidence intervals*.

Walsh averages: For a sample of data, $\{X_i, i = 1, \ldots, n\}$, the Walsh averages are defined as the array $W_{ij} = (X_i + X_j)/2$ for i and $j \in \{1, 2, \ldots, n\}$. They are used in non-parametric procedures.

Appendix 2: Introduction to Logarithms and Exponentials

Logarithms are a very important class of mathematical functions which may be unfamiliar to some readers. Several important statistical calculations involve logarithms. In Section 2.4 we consider the need to log-transform some positively skewed continuous variables, which approximate to lognormal (log-Gaussian) distributional form. In Chapters 10 and 11 we introduce confidence intervals for ratio measures such as relative risks and odds ratios. It is the logarithms of these quantities that have good sampling properties. An important transformation for a proportion p known as the logit transformation is defined as the natural log of $p/(1-p)$. Some confidence intervals for proportions are symmetrical on the logit scale. Use of logit transformation is basic to a regression model known as logistic regression, which is appropriate to model the dependence of a binary variable on one or several explanatory variables.

Logarithms turn multiplication into addition and division into subtraction. Logarithms may be calculated to various bases, the most important being $e = 2.718\ldots$ and 10. Log to the base 10 is written $\log_{10}$. Log to the base e is written $\log_e$, or more simply ln (natural log). Logarithms calculated to one base are simply multiples of those calculated to another base, so for the purposes of scale transformation they are equivalent in use. It is easiest to see the properties of logarithms to base 10, so we will examine their main properties using this base. It is equally feasible to use either base 10 or e for scale transformation for continuous variables, and sometimes logs to base 10 lead to a clearer interpretation, as in Section 11.5.2. Other statistical calculations involving logarithms are most simply performed using logs to base e.

Exponentiation is a mathematical operation which generalises the idea of raising a number to a power. Logarithms also turn exponentiation into multiplication. Exponentiation is dual to taking logarithms: if $y = e^x$, then $x = \ln(y)$.

Log functions can be applied to any number greater than zero; as $y \to 0$, $\ln(y) \to -\infty$. Figure A2.1 shows log functions to the two most commonly used bases, e and 10.

Table A2.1 shows logs to base 10 (rounded to 1 decimal place) for selected numbers between 0.001 and 1000. Note that $100 = 10^2$, $1000 = 10^3$, $0.1 = 10^{-1}$, and so forth. The following basic properties of logs apply irrespective of which base is used.

$$\log(1) = 0 \qquad \log(x^k) = k\log(x)$$
$$\log(xy) = \log(x) + \log(y) \qquad \log(\sqrt{x}) = \log(x)/2$$
$$\log(x/y) = \log(x) - \log(y) \qquad \log(1/x) = -\log(x).$$

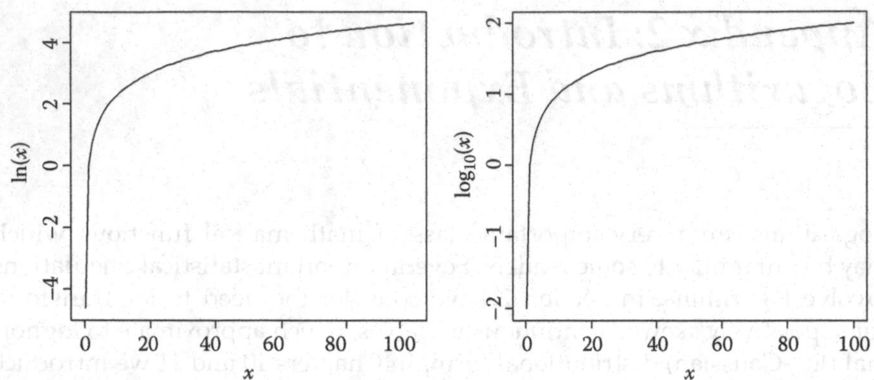

FIGURE A2.1
Logarithm functions to base e = 2.71828 (natural logs) and to base 10.

The reader can easily verify that the figures in the table satisfy these relationships—usually exactly. For example:

$$\log(4 \times 5) = \log(4) + \log(5)$$
$$\log(8 \div 2) = \log(8) - \log(2)$$
$$\log(2^3) = 3 \times \log(2)$$

TABLE A2.1
Logs to Base 10 (Rounded to 1 Decimal Place) for Selected Numbers between 0.001 and 1000

y	$\log_{10}(y)$	y	$\log_{10}(y)$
0.001	−3.0	1	0.0
0.002	−2.7	2	0.3
0.004	−2.4	4	0.6
0.005	−2.3	5	0.7
0.008	−2.1	8	0.9
0.01	−2.0	10	1.0
0.02	−1.7	20	1.3
0.04	−1.4	40	1.6
0.05	−1.3	50	1.7
0.08	−1.1	80	1.9
0.1	−1.0	100	2.0
0.2	−0.7	200	2.3
0.4	−0.4	400	2.6
0.5	−0.3	500	2.7
0.8	−0.1	800	2.9
1	0.0	1000	3.0

log(√4) = 1/2 log(4)
log(1/5) = –log(5).

The table goes as far as logs for 1000 and 0.001, to show that in fact the logs shown here have been rounded a little too drastically, in order to provide a very simple demonstration of how logs work. All these logs have been rounded to just one decimal place. In fact the log of 2 is not exactly 0.3, but is 0.30103.... If we calculate 10 × log(2) using log(2) = 0.3 from this table, this gives 3.0 which is the log of 1000, but in fact 2^{10} is 1024, not 1000.

The choice of 10 as the base for calculating logarithms is often a very convenient one, and is the basis for expressing acidity as a pH and sound intensity in decibels, as described later. But it is not the only choice, and for many purposes, it is not the most convenient choice. Logs to the base 2 are a natural choice in a dilution experiment, although generally software does not include such a function. The most important system of logarithms consists of logs to the base *e*, which are often called natural logs, and denoted by the function ln. This base, 2.71828... is known as ***e*** in honour of the Swiss mathematician Leonhard Euler (1707–1783). The function exp(*x*) = e^x is known as the exponential function. It is defined as the infinite sum

$$e^x = \sum_{n=0}^{\infty} \frac{x^n}{n!} \tag{A2.1}$$

where as usual *n*! means *n* × (*n* – 1) × ... × 1, and both 1! and 0! are 1. The exponential and natural log functions are inverse to each other: if *y* = exp(*x*), then *x* = ln (*y*). These functions are widely available in statistical and spreadsheet software and on electronic calculators. Figure A2.2 shows the exponential function, alongside the antilog function $x \to 10^x$.

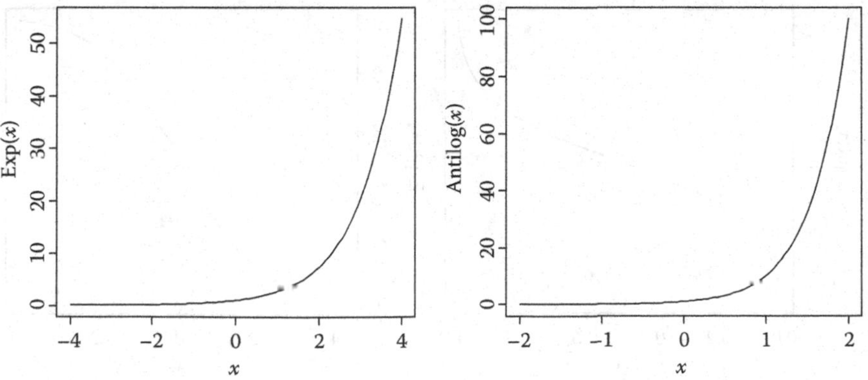

FIGURE A2.2
The exponential function e^x and the antilog function 10^x.

Calculations performed using logs to different bases are equivalent—it does not matter what base is used, provided all calculations consistently use the same base. For any x, $\log_{10}(x) = 0.4343 \times \ln(x)$ and $\ln(x) = 2.3026 \times \log_{10}(x)$. This is why the graphs for these two functions look so similar (Figure A2.1).

The "unit" on the log scale is a very large unit—representing a factor of either 10 or nearly 3. For quantities that can vary by several orders of magnitude, such as the acidity or alkalinity of a solution expressed as a pH, this is reasonable. In other contexts, a smaller unit can be more manageable, accordingly log-transformed sound intensity is not normally expressed directly in $\log_{10}$ units but is multiplied by 10 to give a value in decibels. For statistical analyses of a continuous variable using a log transformation as described in Section 2.4, a suitable transformation is $x \rightarrow 100 \times \ln(x)$. This transformation has the advantage that small differences on the log-transformed scale can be interpreted directly as percentage differences, without any need to back-transform. Cole (2000) suggested a slight modification of such differences into sympercents, which leads to differences, standard deviations and regression coefficients of y that are equivalent to symmetric percentage differences, standard deviations and regression coefficients of x.

It would not make sense to log-transform proportions as such. In general, statistical methods that are applicable to p, the proportion of "yes" responses should be equally applicable to q, the proportion of "no" responses. But there is no useful relationship between $\log(q)$ and $\log(p)$. Instead of a simple log transformation, the logit transformation holds a special place in the analysis of proportions. The function $\text{logit}(p)$ is defined as $\ln(p/(1-p))$, for $0 < p < 1$. The logit function approaches $-\infty$ or $+\infty$ as $p \rightarrow 0$ or 1. It has the desirable property that $\text{logit}(q) = -\text{logit}(p)$. The inverse of the logit function, sometimes termed expit as in R software, is defined by $\text{expit}(x) = e^x/(1 + e^x) = 1/(1 + e^{-x})$. Figure A2.3 shows the logit and expit functions.

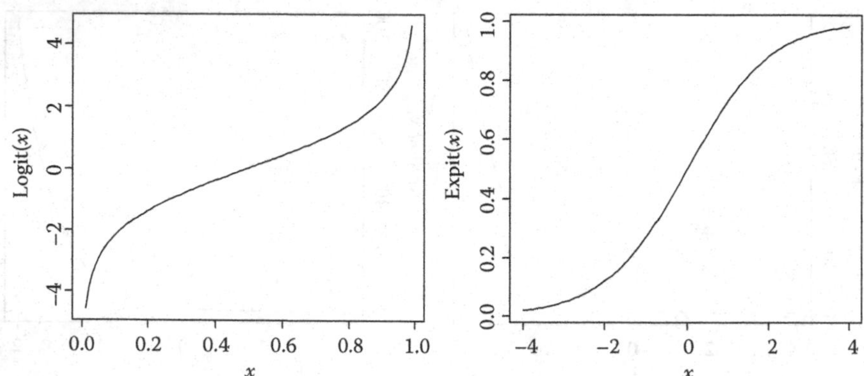

FIGURE A2.3
The logit function $\ln(x/(1-x))$ and the expit function $e^x/(1 + e^x) = 1/(1 + e^{-x})$.

The expressions "log scale" and "ratio scale" are used synonymously. Many familiar scales in everyday use in a variety of contexts are essentially of this kind, including the following.

- Acidity or alkalinity, measured by pH. This is essentially minus the logarithm (to base 10) of the hydrogen (H+) ion concentration.
- Sound intensity measured in decibels. This is the log (to base 10) of the ratio of the power intensity of the sound to that of a predetermined standard, multiplied by 10.
- Paper size. A4 is part of the ISO 216 series of related paper sizes known more commonly as the A-series. In Europe A4 is the most familiar paper size, and has the dimensions 297 × 210 mm. These seem highly arbitrary, unround figures. But in fact they are perfectly logical. A4 is really $2^{-1.75} \times 2^{-2.25}$ metres. The base size of the system is A0, which is $2^{0.25} \times 2^{-0.25}$ metres. This has two properties: the area is 1 square metre, and the aspect ratio is $\sqrt{2}$. A1 is obtained by halving A0 to create a sheet with dimensions $2^{-0.25} \times 2^{-0.75}$ metres, and so forth. A4 has area 2^{-4} square metres, which makes it easy to dispense paper by weighing, thus the weight of an A4 sheet of standard 80 gsm printer paper is simply 5g. The aspect ratio of $\sqrt{2}$ was a design feature ensuring that an A3 sheet may be photocopied, reducing in size to A4, without any change in the ratio of vertical to horizontal measurements, so that circles remain circles and do not become ellipses. Thus the figures following the "A", 0, 1, 2, 3, 4 and so forth, essentially relate to log area (to the base 1/2). However, the paper size marketed as letter size in the United States is an approximation to A4 in Imperial units, 8.5 × 11 inches (i.e., 215.9 × 279.4 mm), which does not have this favourable aspect ratio of $\sqrt{2}$ (Forte 2002).
- Musical pitch. A violinist plays two notes using the same string. The first note is played with the string unstopped, the second with the string stopped halfway along its length. This halves the wavelength, and doubles the frequency of the note produced. The second note is an octave above the first. If the string is shortened to one-quarter of its length, the wavelength is halved again, and the frequency is doubled again—the note produced is two octaves above the base note for the string. The interval known as a perfect fifth—for example, from C to G—corresponds to a frequency ratio of 1.5. A major third, from C to E, corresponds to a frequency ratio of 1.25. However, keyboard instruments use the equitempered scale introduced by J. S. Bach, in which the octave is divided into 12 semitones, an interval of a semitone corresponds to a frequency ratio of $2^{1/12}$ which is approximately 1.0595. On this scale, the "perfect fifth" is in fact slightly flattened, with a frequency ratio of $2^{7/12}$ = 1.4983. The distinction between the true perfect fifth and its flattened equitempered counterpart is detectable only by highly trained ears.

Consequently, if we were to measure the highest frequency of audible sound in a sample of individuals, expressed in Hz, this measurement would be a natural candidate for log transformation. Conversely, if we were to measure the intensity in decibels of the quietest 1000 Hz sound that each subject could hear, we would not normally consider log-transforming this variable, as it is already in log-transformed form. Measurements of acidity or alkalinity expressed in pH would not usually be candidates for log transformation, for the same reason.

Index

D

Z

Printed in the United States
by Baker & Taylor Publisher Services